Recent Advances in Redox Active Plant and Microbial Products

Claus Jacob · Gilbert Kirsch
Alan Slusarenko · Paul G. Winyard
Torsten Burkholz
Editors

Recent Advances in Redox Active Plant and Microbial Products

From Basic Chemistry to Widespread Applications in Medicine and Agriculture

Springer

Editors
Claus Jacob
Torsten Burkholz
Division of Bioorganic Chemistry
School of Pharmacy
University of Saarland
Saarbruecken, Saarland
Germany

Gilbert Kirsch
SRSMC UMR 7565
Université de Lorraine
Metz, Moselle
France

Alan Slusarenko
Plant Physiology (Bio III)
RWTH Aachen University
Aachen
Germany

Paul G. Winyard
Medical School
University of Exeter
Exeter
UK

ISBN 978-94-017-8952-3 ISBN 978-94-017-8953-0 (eBook)
DOI 10.1007/978-94-017-8953-0

Library of Congress Control Number: 2014948063

Springer Dordrecht Heidelberg New York London

Printed on acid-free paper

Springer is part of Springer Science+Business Media (www.springer.com)

This book is dedicated to the memory of Bert Lester Vallee (1919–2010), a fine biological chemist, lord of the zinc proteins and skilled teacher, who was an inspiration to many. Bert's life has been a true reflection of twentieth century science. His personality, wit and, above all, humor were unrivalled. Sadly, Bert died in 2010 whilst working on his final publication, at a time when the "RedCat" network was active. Bert outlived his very first funding agency, the League of Nations, by more than 60 years. His spirit lives on and has also inspired this book. As Bert would put it: "This is not about the shrubs of a country club, but about a damn serious piece of research which is hot like a two-barrel shotgun"

Bert at his famous Christmas Party 1997. Foto courtesy of Claus Jacob

Preface

Reactive oxygen species and reactive nitrogen species have received widespread attention, ranging from the research field of bioinorganic chemistry through biochemistry, pharmacology, toxicology, cell biology, all the way to medicine with issues of health and disease, leading to aspects of nutrition and lifestyle. This interest also extends to the plant world and to viruses and bacteria and other organisms. There is an important crosstalk between this latter field and the health aspects, namely the properties and use of redox active plant and microbial products, a rapidly developing field in redox science.

With this background, the editors of this monograph, Claus Jacob, Torsten Burkholz, Gilbert Kirsch, Alan Slusarenko, and Paul G. Winyard, have to be congratulated for assembling current knowledge in these areas. The wide range of topics reflects the impact of redox active products, from sources in biology, molecular identification of individual compounds, to classes of secondary plant products such as flavonoids, various sulfur compounds, and coumarins. Importantly, the basic knowledge on the sometimes overstressed terms of redox active and oxidative stress is provided. Biological processes are comprehensively considered, such as chemoprevention, inflammation, and other medical uses.

The book covers an impressive range, presented by experts in the field, so that we can be very thankful to the authors and editors for providing this timely information suitable for a widespread range of readers.

Duesseldorf, December 2013 Helmut Sies

The 'Explanatory Box'

Dear Reader,

As you may have noticed already, this book is not an ordinary one. It is neither written by one or two authors, as one would expect from an authoritative piece in the field, nor is it the kind of compilation of more or less loose chapters one usually finds in the aftermath of conferences as proceedings or special issues. Indeed, this book itself is rather special. When first confronted with the topic at hand, the editors were excited about the prospect of a book covering various aspects—and angles—of redox active secondary metabolites. There was a general enthusiasm that one should address emerging areas in the field, ranging from the search for new products in remote areas of the world to hands-on product development and commercialization. Back in the real world, however, the editors soon noticed that they may well be able to provide some expertise in some areas of redox active secondary metabolites, yet were probably ignorant in many others.

At this point, a cunning plan was devised, inspired by the kind of life class drawing one of the editors is particularly fond of. Like a painter, the editors decided to sketch the overall picture first, which would clearly resemble the ultimate work, yet despite its many lines and curves, was still lacking the details and firework of bright colors to turn it into a true masterpiece. The individual highlights were therefore added later on, not by the editors themselves, but by genuine experts in their respective fields. As a result, a rather special kind of book has emerged which tells a comprehensive, well-structured narrative, yet at the same time is able to rely on the expertise of many and not just a few experts.

One notable 'side effect' of this particular tactic is, of course, that the book now contains various chapters that excel on cutting-edge research and development(s). The chapters necessarily live up to the considerable complexity presented by modern research into secondary metabolites (e.g., analytical techniques, intracellular staining methods, and aspects of commercial development). Occasionally, these chapters present issues that may be more difficult to understand for a reader not already familiar with the matter at hand. Such difficulties are unavoidable, of course, as a range of different, often just emerging aspects of redox active secondary metabolites are discussed, and this discussion is delivered by experts in the field. In order to smoothen some of the more difficult passages in the book, we have decided to introduce a number of so-called 'Explanatory Boxes' at specific positions in the individual chapters. These boxes are designed to assist the reader with relevant background information and also to explain more complex issues in a simple, easy-to-understand language. While the boxes are not designed as substitutes for the chapters they are placed in, they should enable the reader to understand better the context and also the content of these chapters.

Contents

Chapter 1
Introduction

Claus Jacob, Torsten Burkholz, Gilbert Kirsch, Alan Slusarenko and Paul G. Winyard

Nature provides a treasure chest which is filled with a huge range of diverse and exciting natural products. Among them, we can find a wide variety of primary and secondary metabolites which originate from plants, fungi, bacteria, and animals and for us possess interesting practical uses in the context of nutrition, medicine, and agriculture. In between the many and extraordinarily diverse natural compounds, redox active secondary metabolites stand out for a number of reasons and, not surprisingly, have recently attracted considerable interest in research and development. These compounds have also been at the center of the European Marie Curie Initial Training Network "RedCat" whose activities are reflected throughout this book (see Explanatory Box 1).

Explanatory Box 1: The "RedCat" Initial Training Network

As part of Framework 7, the European Commission has provided substantial funding to promote the training of Early Stage and Experienced Researchers (ESR, ER, respectively) under the Marie Curie Action. The training network

C. Jacob (✉) · T. Burkholz
Division of Bioorganic Chemistry, School of Pharmacy,
University of Saarland, 66123 Saarbruecken, Germany
e-mail: c.jacob@mx.uni-saarland.de

G. Kirsch
SRSMC UMR 7565 (former Laboratoire d'Ingénierie Moléculaire
et Biochimie Pharmacologique), Université de Lorraine, 57070 Metz, France

A. Slusarenko
Department of Plant Physiology, Molecular Phytopathology,
Institut fuer Biologie III, RWTH Aachen, 52056 Aachen, Germany

P. G. Winyard
University of Exeter Medical School, St. Luke's Campus, Exeter EX1 2LU, UK
e-mail: p.g.winyard@exeter.ac.uk

C. Jacob et al. (eds.), *Recent Advances in Redox Active Plant and Microbial Products*,
DOI: 10.1007/978-94-017-8953-0_1, © Springer Science+Business Media Dordrecht 2014

"RedCat" has been one of those international, multisite networks, which in 2008–2012 has provided training for ten ESR and four ER at ten partner institutions (and eight associated partners) across Europe. "RedCat" has been coordinated by Claus Jacob, who is also one of the editors of this book. The acronym actually stands for 'redox catalysts', and in many ways finds its reflection in the topics covered by this book. Indeed, all editors of the book and most of the contributors were involved heavily in "RedCat" training and research activities, either as partners, collaborators, hosts for ESR and ER placements, teachers, or external advisors of the network.[1]

Overall, the network has been quite a success. It has delivered training and research in a serious, yet thoroughly enjoyable manner and has covered many aspects of science and science-related subjects, including highlights such as communication, Intellectual Property and commercialization, epistemology, Bioethics, Science and Society, and Science and Arts. One of the highlights of the network was the mobility of ESR and ER across Europe. Research placements in host laboratories abroad were daily network business, turning many investigators not only into excellent scientists, but into cunning linguists at the same time. These activities are reflected in the considerable output of the network. "RedCat" has already generated an impressive number of—often jointly authored—publications and forged many long-lasting scientific cooperations and personal friendships. It has also lived up to its promise to train a good number of emerging scientists who are now actively pursuing their independent research careers.

In any case, all ESR and ER have passed their rites of passage to independent scientists. They now know how to "lead the dog so it does not crap on its leash", as Claus Jacob metaphorically says after successful PhD—or Diploma—vivas.

While the "RedCat" spirit—and the "RedCat" website—are still alive, and "RedCat" publications are presently extending the literature on natural products and biological aspects of redox active substances, some of the "Cats" have moved on and diversified in new networks, such as the Interreg IVa network "Corena" covering the Saarland, Lorraine, and Luxembourg or a new graduate training network "NutriOx" which now exists between the Universities of Saarland, Burgundy, and Lorraine.

This book is a fine summary of some of the most successful "RedCat" activities. Indeed, many of the chapters have been co-authored by the

[1] "RedCat" stands for *red*ox *cat*alysis. It describes an Initial Training Network for early stage and experienced researchers funded under the Framework 7 'People' program of the European Union (Grant Agreement No 215009, Natural Products and related Redox Catalysts: Basic Research and Applications in Medicine and Agriculture). The web page is available using the following url: http://www.redcat-itn.eu/.

network's own ESR and ER. To them, the book will provide a lasting memory of the fantastic times they had while being turned into the scientists of the future. For the coordinator and some of the supervisors, the book will raise fond memories of a very special time when massive science and mental sickness were closer together than ever before (and after).

The basis for this special interest may be found in human cell physiology, which is rich in redox related processes. Research during the last couple of decades has demonstrated that cellular redox processes are of paramount importance, not only in the context of energy metabolism but also regarding the overall 'well-being' of the cell and organism as a whole. Disturbances in the intracellular redox balance, for instance in form of oxidative stress (OS), may disrupt normal cellular function and ultimately cause severe damage to the cell (the latter may undergo apoptosis), tissue, and organism affected. Indeed, for several decades, OS has been considered as a major cause of many human diseases, including auto-inflammatory and neurodegenerative diseases, diabetes, glaucoma, and cancer to mention just a few. While such a simple causal relationship between OS and human diseases no longer holds true, it is safe to state that many human disorders and ailments are directly or indirectly associated with a disturbed intracellular redox balance. At the same time, it has become apparent that redox processes are not only at the root or a consequence of disease, but also play a vital and thoroughly beneficial role in healthy cells, tissue, and organisms. In a more subtle manner, cellular redox events are indeed involved in widespread cellular signaling, regulation, and control processes. In fact, some of the early studies in the field of cellular redox regulation have even linked major cellular 'life and death decisions', such as proliferation, differentiation, and apoptosis to the cytosolic electrochemical potential regulated by the glutathione/glutathione disulfide (GSH/GSSG) redox pair.

More recent studies have refined this model by considering cellular redox events at a molecular level, for instance in form of oxidative protein modifications. Within this context, the oxidation of cysteine residues in proteins appears to play a major role. Various studies have highlighted the special impact that oxidative cysteine modifications, such as *S*-thiolation, disulfide, or sulfenic formation have on protein function and enzyme activity—and subsequently also on cellular processes, such as cell signaling. Not surprisingly, a number of new and often still speculative concepts and hypotheses have been postulated during the last couple of years to address this unique intracellular sulfur redox network, and some of them, such as the 'cellular redoxome', the 'cellular sulfenome', and the 'cellular thiolstat', will be discussed briefly as part of Chap. 9.

At the same time, there have been various attempts to control such cellular redox processes by endogenous compounds. Most obvious, perhaps, is the widespread and sometimes almost excessive use of so-called antioxidants, i.e., compounds that apparently counteract OS and hence somehow may protect cells, tissues, and even the whole organism against oxidative damage. Such antioxidants feature prominently, for instance, in fruit juices, lemonades, tea bags, mueslis, and anti-aging, anti-wrinkling skin crèmes where they are said to provide an extra benefit to the consumer. Such antioxidants are almost exclusively derived from edible (parts of) plants, which is hardly surprising, as such compounds are usually readily available and in many respects safe for human consumption or at least for topical applications. They include some of the usual suspects, such as vitamins A, C, and E, a couple of coenzymes, flavonoids, and organic sulfur compounds, which will be discussed later on in this book as part of individual chapters. While the current hype surrounding the practical uses of such antioxidants as chemopreventive, anti-aging, anti-inflammatory, or even anticancer agents is often not fully justified from a purely scientific perspective, antioxidants in food and food supplements are a hot topic in research and represent a rapidly expanding market for a wide range of new products. Bearing in mind that the aging organism is marred by OS and a loss of natural antioxidant defenses, the full potential of antioxidants in an aging society becomes apparent.

The various issues associated with such antioxidants are manifold, complex, and diverse, and range from the more or less problematic labeling of certain food products with ORAC (Oxygen Radical Absorbance Capacity) values and the marketing of antioxidant food supplements for the elderly to the use of antioxidants as free radical scavengers in anti-aging skin crèmes and cosmetics. Within this context, one must bear in mind that the size of the antioxidant market is enormous, and a range of entirely new products (and associated manufacturing processes) have been emerging in this field, which go well beyond the traditional antioxidant vitamin preparations or ACE juices (see also Chap. 2). These developments may be exemplified by the sudden use of grape seeds for the manufacture of grape seed flour: The latter is rich in polyphenols and has turned a former waste product (i.e., grape seeds) into a hitherto rarely considered product and—economically speaking—into a gold mine, which has already made its way into most households in the form of quite common products, such as grape seed flour-enriched pasta.

This simple example of polyphenol-rich grape seed flour as 'green gold' highlights the huge impact redox active secondary metabolites currently exert on nutrition, research, economics, and society as a whole. Not surprisingly, there are other secondary metabolites besides the polyphenols which already play a similar role in society or are likely to acquire a certain importance soon. As part of this book, we will consider a range of such rather promising natural products, for instance as part of Chap. 6. In doing so, we will focus on the chemical properties, biochemical actions, and biological activities which may (or may not) justify such a practical use. Importantly, our discussion will not be driven by or limited to the usual antioxidants, such as certain vitamins or flavonoids, but will endeavor to explore a range of different, often less known antioxidants, forming a scientific perspective on equally interesting, emerging redox active secondary metabolites. Some of these compounds, such as the organic sulfur compounds (OSCs), exhibit extraordinary biological activities and open up potential applications in the field of medicine and agriculture, which go well beyond a traditional antioxidant or chemopreventive activity. Other compounds, such as xanthohumol or certain natural isothiocyanates, even seem to influence epigenetic events and hence contribute to the emerging field of nutri-epigenetics. In contrast, various oligomeric and polymeric metabolites, such as the proanthocynanidins and tannins, are not only redox active but also associate with proteins and enzymes, an interaction that may ultimately result in antimicrobial activity and may even contribute to a reduced calorie uptake in the gut.

These examples should be sufficient to demonstrate that this book on secondary metabolites is not just yet another compilation of traditional plant products or antioxidants. We will rather consider redox active secondary metabolites from various scientific perspectives, including their identification, chemical analysis, and isolation, their biosynthetic pathways and synthesis in the laboratory as well as their biological activity, biochemical mode(s) of action and practical applications. Here, we will also pay tribute to modern nano-technological methods able to improve bioavailability and witness how a natural product, such as a garlic extract, can be turned into a commercial success story. This facet-rich agglomerate of newly emerging themes requires specialist knowledge from diverse disciplines, including plant biology, bioanalytics, synthetic chemistry, cell biology, medicine, agriculture, and economics. We have therefore recruited a number of internationally renowned experts who have kindly contributed their in-depth knowledge and experiences to individual chapters of this book. As a result, the contributors are able to discuss a wide range of selected topics related to redox active secondary metabolites from vastly different, and often unusual and exciting angles—and with the kind of in-depth knowledge which is clearly required as part of a serious scientific discourse. While this book cannot serve as a complete or ultimate encyclopedia listing all known redox active natural products, it aims to highlight some of the most recent and exciting developments in the field in order to stimulate future research.

First, we will therefore take a more historic perspective on the 'antioxidant story', its basis, main notions, and lessons for the present and the future. Here, Agnieszka Bartoszek from the Gdansk University of Technology will argue that the idea of antioxidants as agents beneficial to human health has developed from an initial, perhaps premature hypothesis to a more sophisticated biochemical regulatory concept which is changing human nutritional behavior and lifestyle to the better. These conceptual changes over the last decades also provide a fertile ground for more speculative and innovative approaches to redox active secondary metabolites which will be discussed in many of the subsequent chapters.

We will continue with a closer look at the identification and characterization of natural substances using cutting-edge analytical techniques. One chapter contributed by Patrick Chaimbault from the University of Lorraine in France will consider recent advances in employing techniques such as mass spectrometry for the identification of natural substances. During the last decade, such techniques have developed at a breathtaking pace, and it is now possible to identify certain plant metabolites even in intact plant materials.

After a closer look at the bioanalytical methods currently available for the identification of secondary metabolites, we will turn our attention to the various biological activities—and possible practical applications—which have been associated with these products during the last decade or two. Here, we will first explore the true meaning of 'redox activity', especially in the context of antioxidant activity and redox modulation. This chapter, contributed by Torsten Burkholz and Claus Jacob from the University of Saarland in Saarbruecken, Germany, will infuse concepts of (redox) chemistry into the biochemical debate in order to resolve some of the myths of biological redox chemistry. We will, for instance, differentiate between simple chemical oxidizing or reducing agents and subsequent cellular pro- or antioxidant responses. We will also provide a brief discussion on the hot issue of the ORAC value.

Equipped with these more fundamental insights we will then turn our attention to the vast and often misty field of OS, antioxidants, chemoprevention, and some of the surprisingly beneficial roles ascribed to pro-oxidants. Contributed by Lars-Oliver Klotz from the Friedrich-Schiller-University, Jena, this chapter will provide a critical view of some of the notations commonly associated with antioxidants. It will reject the rather outdated view of antioxidants as 'good' and oxidants as 'bad' species and promote a more differentiated view, which is based on subtle cellular redox signaling and regulatory events affected and fine-tuned by pro- as well as antioxidants. Ultimately, both pro- and antioxidants may be equally beneficial or detrimental, depending on the specific situation and circumstances and, of course, the amounts consumed.

This more general chapter is followed by three contributions which discuss individual redox modulating natural compounds. Each of these chapters, and the compounds discussed therein, represents a particular aspect of redox active secondary metabolites which is of considerable interest to human nutrition, yet is based on entirely different underlying biochemical processes. This parade of selected natural compounds will begin with flavonoids and stilbenes, compounds that occur in many fruits and berries and seem to exert protective effects on the human body, such as the ones commonly implicated in the "French Paradox". Contributed by Artur Silva and colleagues from the University of Aveiro in Portugal, this particular chapter will build the bridge between the natural compounds found in plants and their potential uses on the one hand, and the ability of synthetic chemists to produce such natural substances as well as their tailor-made derivatives in high amounts and excellent purity in the organic synthesis laboratory.

While this chapter focuses on the chemistry and biological activity of natural compounds involved in the "French Paradox", the subsequent chapter by Norbert Latruffe and his colleagues from the University of Burgundy at Dijon, France, takes a closer look at the various health benefits associated with the stilbene compound resveratrol, which is found in grapes and grape products, especially in red wine. This chapter reviews the latest literature in order to assess if the consumption of red wine may indeed protect from cardiovascular or inflammatory diseases.

As inflammation is commonly associated with a sharp increase in reactive oxygen and reactive nitrogen species (ROS and RNS), it is hardly surprising that many redox active natural products have been associated with anti-inflammatory effects in the past. Here, polyphenolic compounds, such as colorful proanthocyanidins found in many fruits and berries, are of particular interest. Chapter 8 contributed by Hadi Ebrahimanijad from the Shahid Bahonar University, Kerman, Iran together with Claus Jacob and his team (Saarbruecken) will therefore take a closer look at oligomeric and polymeric substances, such as the polyphenolic proanthocyanidins. The latter feature a range of emergent properties which are the direct result of the oligo- and polymeric character of the compounds and cannot be found in the monomeric counterparts. Besides exhibiting a powerful redox behavior due to the presence of a multitude of redox centers in these polymers, these substances are also able to non-covalently interact with proteins and enzymes. The resulting inhibitory effects on proteins and enzymes are highly interesting, for instance, from the perspective of antimicrobial activity and in the context of the control of digestive enzymes.

After highlighting the importance of some of the nutri-chemicals for human health, and discussing the (bio-)chemical basis of such activities, the reader's attention will then be turned to more practical applications of redox active secondary metabolites and the plant products in which they are present. Here, our focus will reside with organic sulfur compounds. Despite often being grouped together by natural product scientists, OSCs are chemically extraordinarily diverse and associated with a wide range of activities and possible applications in medicine and agriculture. In order to understand why many OSCs are highly, yet often also selectively active in biological systems, we will first need to consider their specific, sulfur-based reactivity and how this reactivity relates to specific cellular targets, such as the network of cysteine proteins which together form the 'cellular thiolstat'. After this brief and more general overview of sulfur-based biological redox systems and some of their modulators, which is presented by Martin Gruhlke and Alan Slusarenko, the team from the RWTH Aachen will take us on a journey to the intricate and often rather surprising chemistry of thiosulfinates and polysulfanes—and how this chemistry translates into biological activity and possible applications in medicine and agriculture. Subsequently, Chris Hamilton and his colleagues from the University of East Anglia will introduce us to bacillothiol and mycothiol, two bacterial thiols that have emerged as potential targets of novel, thiol-modulating antibacterial agents. Besides proposing new targets for therapeutic redox modulation, this chapter underlines the importance of thiol-based cellular redox systems in various organisms and showcases promising avenues for a more or less selective attack at such thiol-containing molecules.

The subsequent chapter is written by Awais Anwar and his colleagues from the British company ECOSpray Ltd., which during the last decade has developed a range of interesting garlic-derived products for applications in the field of eco-friendly agriculture. Awais and his colleagues discuss the value and potential uses of such natural materials, and how a simple agricultural product such as garlic can be turned into a commercially viable and ecologically sound product.

The section on OSCs and their potential uses is followed by a fine selection of chapters that try to push the boundaries of redox active natural product research to new limits and to open up new applications for such compounds.

As part of this section, we will consider several cutting-edge approaches to generate new compounds or to turn existing products into new 'forms' which enable practical applications. The first chapter of this section, contributed by Peter Olofson from Redoxis AB in Sweden and his colleagues from Sweden and Denmark, considers a rather simple yet effective means of 'processing' an existing natural product, in this case chlorophyll, into a new compound with potential practical application. Indeed, the production of phytol from chlorophyll by digestion in herbivorous animals represents an elegant and, at the same time, traditional means of unlocking the potential of natural products and developing new ones. While phytol in itself is not redox active, it nonetheless seems to influence various cellular processes that ultimately may also regulate redox events involved in inflammation. This chapter is followed by several other contributions which consider more eloquent biotechnological, clinical, and nanotechnological approaches. The chapter contributed by Frédéric Bourgaud from the University of Lorraine illustrates the more biological avenue, which considers the biosynthetic pathway of a natural product, in this case a novel furanocoumarin. Once the enzymes involved in the biosynthesis of such a molecule are known, it is, in principle, possible to fine-tune the structure of the natural product my mutation of the biosynthetic machinery. The subsequent chapter by Gilbert Kirsch and colleagues (Metz), in contrast, chooses a more chemical approach which starts with the synthesis of coumarin derivatives and ultimately generates tailor-made derivatives in the form of natural product hybrid molecules.

The final chapter of this section provides a third angle to custom-made natural products especially with regard to solubility and bioavailability. Rather than modifying the chemical structure of the molecule, however, the approach described by Cornelia Keck and Karl-Herbert Schaefer from the University of Applied Sciences in Kaiserslautern, Germany, considers the size of the natural product particles. Here, nanotechnology opens the door to a whole new field of natural product research as it endows good solubility and bioavailability to a range of otherwise difficult to use natural products. To date, the practical applications of the 'non-nano' forms of these products have been limited by the poor solubility and hence limited efficacy in complex organisms, including in humans.

Importantly, the notion of redox active natural products is not limited to small molecules. Redox active enzymes also play an increasing—and increasingly diverse—role in chemistry, biotechnology, and medicine. The final two contributions therefore consider practical uses of redox active enzymes. The chapter by Jennifer Littlechild and colleagues from the University of Exeter in the UK reviews the properties and practical uses of certain haloperoxidases in industrial-scale chemical synthesis. The chapter by Peter Meiser and colleagues from Ursapharm Arzneimittel GmbH in Guedingen, Germany looks at the Bromelain enzymes derived from pineapple as a powerful enzyme cocktail used against injuries and inflammation.

After this exciting encounter with some of the still more speculative, yet clearly cutting-edge approaches and technologies in natural product research, it will be time to summarize the current state of this field of research and to provide a brief outlook on imminent potential developments. Here, the scientific perspective is clearly important, yet it cannot be considered in isolation and without paying suitable attention to the ongoing changes in society, medicine, and nutritional habits, as well as to agricultural demands and emerging economic niches. So there is our summary of the journey provided by this book—enjoy!

Authors Biography

Claus Jacob (born 1969) has been trained as a synthetic (in)organic and biological chemist at the Universities of Kaiserslautern, Leicester, Oxford, and Harvard. He graduated with a 1st class B.Sc. (Hons.) degree from the University of Leicester in 1993, and with a D.Phil. from the University of Oxford in 1997 ("Genetic engineering of redox active enzymes", supervisor Prof. Allen Hill FRS). He subsequently joined the institute of Prof. Bert Vallee at Harvard Medical School as a Feodor Lynen Fellow (Alexander von Humboldt-Foundation) to study processes controlling intracellular zinc homeostasis. During this time, he also obtained a Magister Artium degree in Philosophy, History, and Psychology from the University of Hagen in Germany (M.A. dissertation on Protochemistry as constructivist foundation of chemistry). He left the US in 1999 to spend some time with Prof. Helmut Sies at the Heinrich-Heine-University in Duesseldorf, Germany, as part of a BASF Research Fellowship from the German Merit Foundation.

Claus started his independent scientific career as lecturer at the University of Exeter in the UK in 1999 and in 2005 moved to the University of Saarland where he currently holds the position of Professor of Bioorganic Chemistry. Claus is an expert in redox active compounds and their impact on biological systems and to date has published over 100 publications in this field. Over the years, his research has focused on Reactive Sulfur Species (RSS) and the cellular thiolstat, terms his team has introduced in 2001 and 2010, respectively. Besides his strong interest in redox active sulfur, Claus has also developed an active research program on synthetic 'sensor/effector' redox modulators based on selenium and tellurium, on redox active plant metabolites and on nanoscopic redox particles. His research includes synthetic and analytical chemistry, biological activity studies and 'intracellular diagnostics' to decipher and map out intracellular events and mode(s) of actions. Claus has coordinated the EU Marie Curie Initial Training Network "RedCat" (2008–2012), has been a partner in the technology transfer project "Corena" (2009–2012) and is currently partially in charge of the natural products project "NutriOx".

Throughout the years, Claus has undertaken many projects to become a highly skilled undertaker, but never a true philosopher, yet his more philosophical and cunning linguistic outpourings are famous and he still maintains a keen interest in various aspects related to the philosophy of chemistry.

Torsten Burkholz (born 1979) has been trained as an inorganic and medicinal chemist at the University of Saarland in Saarbruecken, Germany, graduating with a German "Diplom" in Chemistry. After completing his PhD studies at the University of Saarland in the fields of Chemistry and Pharmacy under the supervision of Prof. Claus Jacob in 2010, he joined the European Marie Curie Initial Training Network "RedCat" as postdoctoral Experienced Researcher, conducting research in the field of Cell Biology in the group of Prof. Paul G. Winyard at the Peninsula College of Medicine and Dentistry, Exeter, UK. In 2012 Torsten moved back to the University of Saarland where he currently holds the position of an "Akademischer Rat" in Bioorganic Chemistry.

As part of this senior position, Torsten is managing the research laboratory of Bioorganic Chemistry, as well as the relevant teaching and the examinations of undergraduate students. Together with Prof. Claus Jacob, he has to date published over 30 publications in the field of oxidative stress, chalcogen containing natural compounds, and their biological activity and, more recently, on the cellular thiolstat. Torsten's ongoing research includes synthetic and analytical chemistry, biological activity studies, and 'intracellular' diagnostics.

In 2013, Torsten was appointed as Visiting Professor at the University of Applied Sciences Kaiserslautern, where he is lecturing Pharmacology. In the same year, he established his own small company named "Dr. Burkholz Life Science Consulting UG" which provides scientific consultations for small and medium-sized companies in the field of nutrition, food supplements, and natural compounds. His company also offers training and consulting in Inorganic and Analytical Chemistry, as well as in Physics and Pharmacology.

Gilbert Kirsch (born 1947) has been trained as an organic chemist at the Universities of Strasbourg and Metz. He started his academic career in 1973 at the University of Metz (now University of Lorraine) where he currently holds the position of Professor of Organic Chemistry. He has been a postdoc at Oak Ridge National Laboratory (TN) in the Nuclear Medicine Group and was also an invited scientist at Kodak (Rochester, NY) as well as invited professor at the University of Minho (Portugal) and Emory University (Atlanta, GA).

Gilbert's interests reside in heterocyclic chemistry, especially in the field of five-membered aromatic systems (thiophenes, selenophenes, tellurophenes, thiazoles, selenazoles) and their benzo-condensed derivatives. Lately, he developed synthetic work in the field of coumarins, looking at biological activities (CDC25 phosphatase inhibition). From his research, he published over 200 papers, wrote different chapters in books, like Patai's Functional Group Series, in Houben-Weyl, in Chemistry of Heterocyclic Compounds (J. Wiley Interscience), and in Springer's Selenium and Tellurium Chemistry. He holds also few patents in the field of heart imaging and sulfur-containing tire additives. Gilbert has coordinated an Interreg program (acronym "Corena") on natural compounds for Medicine and Agriculture, and has participated in the EU ITN Marie Curie program "RedCat". He is participating in the regional programs Bioprolor and Biocaptech and in a national French ANR program on *para*-hydrogen.

Alan Slusarenko is Head of the Plant Physiology Department at RWTH Aachen University. His research has centered on resistance mechanisms of *Arabidopsis* to infection and more recently on Natural Products in Plant Protection. Alan obtained a PhD in Plant Pathology from Imperial College in 1981 and was a lecturer in the Department of Plant Biology at Hull University in the UK from 1983 until moving in 1988 to an Assistant Professorship in Molecular Plant Pathology at the University of Zuerich in Switzerland and subsequently in 1995 to the Chair of Plant Physiology at RWTH Aachen in Germany.

Paul Winyard is Professor of Experimental Medicine at the University of Exeter Medical School (formerly Peninsula College of Medicine and Dentistry), Exeter, UK, where he has been based since 2002. Previously, he held the same title at St Bartholomew's and the Royal London School of Medicine and Dentistry, London, and was a Visiting Professor at the University of California, San Francisco (2000–2001). Paul originally trained as a biochemist, and his current research interests center on the role of oxidative/nitrative stress and redox signaling in chronic inflammatory diseases such as rheumatoid arthritis. Paul leads an internationally recognized research group, having published over 200 research papers in the field of oxidative stress in inflammation. In particular, his research has focused on the development of novel free radical-related assays and therapeutic strategies, and the translation of these developments into clinical diagnostic assays and pre-clinical and early-phase clinical studies.

Paul is a co-inventor in relation to seven patents, and is a Senior Editor of the *Journal of Inflammation*. He is also a member of the editorial boards of *Redox Biology*, *Frontiers in Oxidant Physiology*, *Frontiers in Inflammation*, *Current Pharmaceutical Design* and the *Open Inflammation Journal*. He is a committee member of the *British Inflammation Research Association* (*BIRAs*), and serves on a number of UK and European research grant awarding committees.

Part I
Connecting Section Between Chapters 1 and 2

The previous chapter has provided a brief introduction to the rather wide, diverse, and often complex field of redox active secondary metabolites and the research and practical applications associated with it. It is now time to turn our attention to the various sources and implications of such products especially in the context of human health. Indeed, as the next chapter will reveal, there are numerous natural sources of secondary metabolites, which include many plants (and parts thereof), but also fungi and bacteria. While fruits and berries are perhaps the best known examples of agricultural products rich in redox active substances, such as flavonoids and related 'antioxidants', other plants are also rich in such bioreactive molecules, for instance in organic sulfur compounds, which are found in garlic, onions, and many other edible plants and mushrooms. Many plants also contain electrophilic substances, such as xanthohumol found in hop, which are redox active in their own respect. These compounds do not act as electron acceptors in a strict sense, i.e., via electron transfer, but engage in addition and substitution reactions. The latter result in oxidative modifications of electron-rich, nucleophilic sites, such as thiols and amines in proteins, enzymes, membranes, DNA, and RNA.

Importantly, many of these redox active metabolites are present in reasonable amounts in rather common, widely cultivatable and often edible plants or parts thereof (as is the case for many flavonoids, resveratrol, organic sulfur compounds, xanthohumol, proanthocyanidins, etc.). This particular aspect distinguishes such more common metabolites from 'rare' natural products, which are found only in exotic plants and there in tiny amounts only (e.g., alkaloids). Actual and potential practical uses of the substances considered here are therefore not only interesting because of the pronounced biological activities associated with many of them (especially in the context of 'antioxidant' activity and suspected health benefits). They are also interesting because of their widespread availability and hence the economic feasibility of potential applications. Indeed, some of the following chapters will demonstrate how such natural products can be converted successfully into commercial products and considerable economic success stories.

Chapter 2
Antioxidants: A Premature Scientific Hypothesis that Reshuffled the Traditional Food Pyramid

Agnieszka Bartoszek

Keywords Antioxidants • Epigenomics • Functional food • Nutrichemicals • Thio-redox status

The curious story of dietary antioxidants, i.e., reducing agents found in considerable amounts in edible plants and often not necessarily understood fully in the context of redox chemistry, can be traced back to the famous statistical analysis of carcinogenic risk factors by Doll and Peto (1981). Their work revealed among others the positive correlation between cancer incidence and meat consumption. In particular, the statistical strength of meat's impact on human health created a kind of a shock that such an obvious component of the human diet may be as carcinogenic as cigarette smoke; unfortunately subsequent epidemiological studies were repeatedly confirming this worrisome association (WCRF/AICR 1997). The emerging results of consecutive statistical analyses derived from ever larger human studies, however, uncovered another, more optimistic association, i.e., a decreased risk of cancer (and also of other chronic diseases) in populations whose diets were rich in plant-based foods, especially in all types of colorful fruits and vegetables, and thus rich in (redox active) secondary metabolites (for sources see Explanatory Box 1).

Explanatory Box 1: Sources of Secondary Metabolites

Secondary metabolites are organic compounds that are not directly involved in the normal growth, development, or reproduction of the organism they originate from. They are found extensively in nature, especially, but not exclusively in plants, where they serve numerous and often diverse purposes,

A. Bartoszek (✉)
Department of Food Chemistry, Technology and Biotechnology, Gdańsk University of Technology, Narutowicza Street 11/12, Gdańsk, Poland
e-mail: agnieszka.bartoszek@pg.gda.pl

C. Jacob et al. (eds.), *Recent Advances in Redox Active Plant and Microbial Products*,
DOI: 10.1007/978-94-017-8953-0_2, © Springer Science+Business Media Dordrecht 2014

ranging from host defense and clearing competitors from respective pastures to attracting mates, pollinators, and distributors of their seeds. Numerous examples of redox active secondary metabolites will be discussed as part of this and subsequent chapters of the book. Here, we will briefly mention a few of the most obvious and perhaps exciting ones. Well-known examples of common redox active secondary metabolites include certain vitamins, such as ascorbic acid (vitamin C), which is contained in high amount in citrus fruits (such as oranges, lemons, mandarins), as well as α-tocopherol (vitamin E) and its isomers, which are found in many vegetable oils and in grape seeds. In many fruits, nuts, and vegetables, these vitamins are joined by the large and diverse family of polyphenols, which are found, for instance, in grapes, tea and coffee, cacao, apples, tomatoes, soy beans, peanuts, almonds, and in pomegranates. Colorful berries, such as blueberries, black currants, chokeberries, raspberries, goji berries, and jostaberries, are particularly rich in polyphenolic flavonoids—and indeed, their color is often due to the presence of such compounds.

Such colorful vegetables, fruits, and berries also contain a range of additional redox active groups of compounds, including anthocyanidins (for instance found in cranberries) and proanthocyanidins. As many of these compounds are poorly water soluble and located in the skin or seeds, caution should be taken as juices derived from such fruits, etc., may not be particularly rich in these substances. This also holds true for resveratrol and other stilbenes, which are found in the skin of red grapes and in red (but not in white) wine. Other redox active secondary metabolites are more omnipresent and easier to access for consumption, including many flavonoids. Catechins, for instance, are present as catechin gallate in green or black tea, and as (+)-catechin and (−)-epicatechin in cacao, while quercetin is a predominant component in apples and in many berries.

Many of these more readily available metabolites have been considered as beneficial ingredients in ‘functional foods’. In common perception, quercetin is the ‘healthy ingredient’ in apples, apple juice, and apple wine, (epi) catechins are the beneficial ingredients of cacao and (hot) chocolate, and resveratrol turns wine and xanthohumol (a prenylated chalconoid contained in hops) turns beer healthy. Such ‘health claims’ ascribed to a particular ingredient of a natural product are often problematic, of course, especially once processed foods rich in sugar or (often alcoholic) beverages are concerned. Nonetheless, many redox active secondary metabolites can indeed be found in food stores. Prominent examples include the various sulfur-containing secondary metabolites, i.e., organic sulfur compounds, such as allicin and polysulfanes in garlic and onions, cabbage and dithiins, and asparagusic acid in asparagus, allyl isothiocyanate (mustard oil) in mustard seeds, sulforaphane in broccoli and lenthionine, a rather unusual organic compound, in Shiitake mushrooms (chemically, lenthionine is a relative of varacin, a cyclic polysulfane found in marine ascidiacea from the family *Polycitor*).

Often, spices and flavorings represent a valuable source of redox active secondary metabolites with suspected health benefits, which include curcumin from the turmeric root, capsaicin from chili peppers, and hydroxytyrosol (and derived compounds) found in olive oil. From a nutritional and culinary perspective, the use of such spices—instead of salt—is therefore advisable for a number of good reasons. Such considerations also include the extended use of vegetables as part of our daily diet. Indeed, rather common vegetables sometimes are surprisingly rich in redox active secondary metabolites. Worth mentioning are, for instance, beetroots, a very tasty and healthy vegetable used extensively in Eastern European cuisine, which should not be confused with 'beefroots' or 'meatroots'. These roots contain a fair amount of the 'antioxidant' betalain betanin, which gives beetroots their characteristic color (and not an anthocyanidine). Common tomatoes are also rich in 'antioxidant' substances, including lycopenes and tomatines. Indeed, α-tomatine is a fine example of a hybrid molecule, where elements of a steroid merge with aspects of polyphenolic redox activity. Similar redox active steroidal structures are found in ginseng (ginsenosides) and other saponin-containing plants (such as the soapwort plant) and in lower organisms.

As the hunt for new and perhaps more efficient 'antioxidants' is currently on, ever more exotic sources and molecules are being investigated, including propolis and many colorful flowers (such as the purple coneflower). Indeed, the flowers of plants, together with the petals and leaves, should not be underestimated as they often contain more biologically active ingredients than the corresponding fruits or seeds (phytol will be discussed later on). Some of these flowers—or products thereof—are even edible, for instance dandelion flowers and elderflower (extracts). Redox active metabolites found in flowers often differ from the ones found in fruits and berries, and some of these substances, such as aurones, therefore will be discussed in more detail as part of subsequent chapters.

In parallel, research on the molecular foundations of chronic diseases, such as cardiovascular or neurological disorders, diabetes, and cancer, lead to the observation that all of these illnesses may have one particular risk factor in common, namely a disturbance in redox homeostasis, commonly referred to as 'oxidative stress' (OS, see Chap. 1). OS describes a condition in which the production of damaging reactive oxygen species (ROS, present in the body as an inevitable consequence of respiration, exacerbated by the inflammatory response and by exogenous factors such as environmental pollutants) exceeds the capacity of the body's antioxidant defenses to neutralize them (see Chap. 5).

The two sets of scientific observations became combined into one hypothesis which elegantly explained how different diseases may be prevented simultaneously by plant-borne secondary metabolites with divergent chemical structures. The presumption was that if OS brought about by an organism's weakened antioxidative barrier accelerates processes leading to the diseased state, the factors neutralizing oxidants should prevent or at least slow down disease progression. Numerous plant secondary metabolites are characterized by a low reduction potential, so their health promoting capability emerging from epidemiological investigations was immediately associated with antioxidant activity, a property reflected by the variety of chemical structures. There were also two additional circumstances which warranted the antioxidant hypothesis instantaneous public attention and scientific consensus. First, the promotion of low caloric plant foods was in line with the strategy of reducing obesity that began to dramatically increase at the end of the twentieth century. Second, the measurement of—the total of—redox active substances (or their 'redox capacity'), be it in mixtures like those in foods or after purification, is very straightforward. Soon, a number of easy and convenient methods to assess antioxidant activity was developed which could be adopted by any research or industrial laboratory.

Once this antioxidant hypothesis was formulated and its implications deduced, scientists worldwide began to verify it in the course of specifically designed research projects involving a plethora of *in vitro* and *in vivo* models, as well as controlled human intervention studies and clinical approaches. This part of the antioxidant story does not differ from scientific deductive business as usual—take a hypothesis, examine experimentally its assumptions, and either confirm initial assumptions or reject or at least modify them. This time, however, the story leaked out of research laboratories virtually with a massive tsunami effect. The developed societies, in particular, despite their access to good nutrition and most of all advanced and extensive medical care, and despite experiencing a reasonably comfortable life with corresponding eagerness to enjoy attractive leisure activities, turned out not only not to be devoid of life-threatening illnesses, on the contrary, the richer the country, the more chronic diseases seemed to take their toll, with serious forms of cancer included. In response, the wealthy societies demanded from the scientific community that a miraculous though scientifically supported cure would be elaborated to ensure a long and healthy lifetime. Antioxidants were just what could satisfy these expectations. First media, then consumers, and soon after industry embraced the combat against OS as the straight route to trouble-free longevity. Antioxidants and free radicals began to be pronounced even by those who had no knowledge of the true meaning of these terms—let alone redox chemistry. The chemical definition of a reducing agent had undergone multitude of reinterpretations depending on to whom it was addressed (see also Chap. 4). In food chemistry, *antioxidants are compounds that are used to increase the 'oxidative stability' of foods*. In the case of medical community, according to the Webster New World Medical Dictionary: *antioxidant is a substance that reduces damage due to oxygen, such as that caused by free radicals; well-known antioxidants include enzymes and other substances, such as vitamin C, vitamin E, and beta-carotene, which are capable of counteracting the damaging effects of oxidation*. The Internet portal 'About.com Chemistry' in the Chemistry Glossary Index also gives the definition stressing antiradical action of an 'antioxidant' as: *an enzyme or other organic molecule that can counteract the damaging effects of oxygen in tissues; although the term technically applies to molecules reacting with oxygen, it is often applied to molecules that protect from any free radical* (*molecules with unpaired electron*).

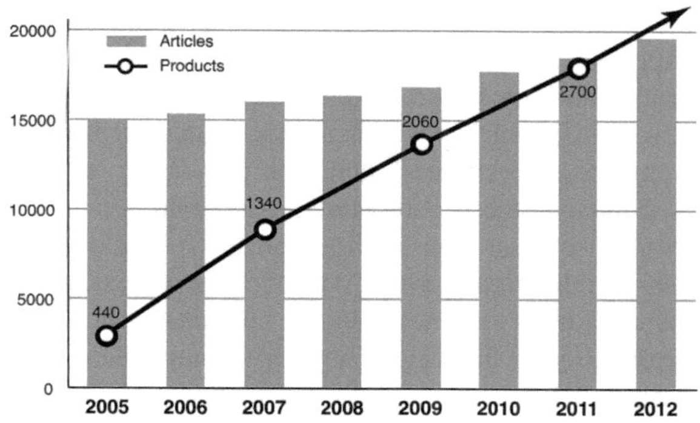

Fig. 1 The number of scientific articles listed in PubMed containing the term 'antioxidant' among keywords and products marketed with the information on antioxidant content according to Global New Product Database

These blurred definitions no longer were related strictly to the reduction potential of compounds. Actually, the term antioxidant has no longer retained a proper chemical connotation at all. Its popular meaning has become equivalent to 'bioactive phytochemical' or 'bioactive food component' regardless of redox activity.

Perhaps the most unexpected development was, that the very word 'antioxidant' occurred to be an extremely well selling marketing slogan. Indeed, the number and variety of products launched bearing a distinct antioxidant claim has been on the constant increase, as has the number of research papers found in the PubMed database (http://www.ncbi.nlm.nih.gov/pubmed/) containing the term 'antioxidant' among the keywords (Fig. 1).

Furthermore, the American Chemical Association noticed a shift in researchers' interest from detrimental to various beneficial food components with papers specifically related to antioxidants appearing in the ACS supported food journals in growing quantity: 45 (5 %) in 1997, 362 (20 %) in 2008 and 391 (18 %) in 2010 (Seiber and Kleinschmidt 2012). The ability to scavenge free radicals (used interchangeably with ROS) by a given food component or phytochemical became synonymous with the term 'functional food', which implies that a food product offers to consumers some benefits beyond simple nutrition. In 2005, in the food and beverage industry, another slogan was coined—'superfruit' or less frequently 'superveg'—to annotate the edible plants with particularly high antioxidant activity (determined usually by the ORAC or FRAP method).[1]

At a rather breathtaking pace, the antioxidant hypothesis was translated into a flourishing business whose annual market value according to Euromonitor (December 2013) in Europe alone is estimated at about 34 billion Euro.

[1] ORAC stands for 'Oxygen Radical Absorbance Capacity' and describes a method of measuring antioxidant capacities in biological samples *in vitro*. ORAC values of fruits and vegetables were published by the United States Department of Agriculture (USDA), but were withdrawn in 2012 as biologically invalid (USDA, Oxygen Radical Absorbance Capacity (ORAC) of Selected Foods, 2010), stating that no physiological proof *in vivo* existed in support of the free-radical theory. The Ferric Reducing Ability of Plasma (FRAP, also Ferric ion Reducing Antioxidant Power)-assay is an standard antioxidant capacity assay that uses Trolox (6-hydroxy-2,5,7,8-tetramethylchroman-2-carboxylic acid, water soluble analog of vitamin E) as a standard benchmark.

Research in the field of food and nutrition, as well as some branches of experimental medicine focussing on bioactive phytochemicals, have been developing at the same speed (see Explanatory Box 2). Numerous mechanisms have been discovered by which phytochemicals could potentially influence the human organism, from altering genome function to preventing or fighting microbial threat. The antioxidant hypothesis, however, has resisted scientific verification in human studies. While it is clear from *in vitro* and *in vivo* experiments that antioxidative nutraceuticals (another term born as a result of the antioxidant hypothesis) can reduce oxidative damage, trials completed in the mid-1990s showed no protection against cancer or cardiovascular disease by high-dose supplements of beta-carotene; to make matters worse, such high doses of this particular 'antioxidant' actually seemed to increase the risk of lung cancer in smokers. Two subsequent meta-analyses of human cohort and case–control studies with vitamin E (Miller et al. 2005) and with a range of micronutrient supplements (Bjelakovic et al. 2007) also concluded that low doses of antioxidants have no significant effect, and high doses might actually increase mortality and incidence of disease. Further large, long-term studies conducted to test whether antioxidant supplements (vitamin C, A, E, beta-carotene) taken alone or in combination, and for periods of at least a few years, may help prevent chronic diseases in people, neither gave any promising results.

Explanatory Box 2: The Origin of Modern Pharmaceuticals

This chapter opens the discussion on potential health benefits associated with redox active secondary metabolites. Indeed, such compounds often exhibit 'antioxidant' properties *in vitro* and hence are suspected of chemopreventive or otherwise protective properties. Despite the rather complex issues surrounding the 'antioxidant hypothesis' (several of which will be discussed critically as part of this chapter), the more general idea that natural products could act as protective or even therapeutic agents is not unreasonable. Indeed, a brief look at the origins of currently licensed and used pharmaceutical agents demonstrates that about a quarter to a third of our drugs are of natural origin. Such agents are either genuine natural compounds, such as the alkaloid chinin, or semi-synthetic compounds based on natural products and modified to increase activity or bioavailability or to reduce side effects, such as acetyl-salicylic acid (Aspirin). These traditional, small-molecule natural products, which also include many anticancer drugs and antibiotics, have recently been joined by other 'biologicals', i.e., larger molecules with a pronounced and often selective action, including peptides (here a traditional example is insulin) and (native or modified) antibodies. These natural products in a narrower sense of the word are joined by various mimics, i.e., synthetic compounds intentionally resembling (aspects of) a natural substance.

The success of natural products in drug development can be explained by a number of factors. Natural products have often developed during evolution to endow their producers with a particular advantage. These compounds possess high and often selective activity (for instance against predators, bacteria, fungi), which can be exploited for human uses as well. At the same time, Folk Medicine has in a way already tested such compounds for centuries, and these unintentional 'clinical trials' in humans provide extremely valuable information about activity, application, dosage, and possible side effects.

Nonetheless, a continued importance of natural products as leads in drug development is by no means self-evident. At the end of the last century, powerful new methods were introduced to drug development which in this field could well have pushed aside natural products research altogether. Here, computational methods, in silico drug design, combinatorial chemistry with consecutive compound libraries, and fast screening methods relying heavily on robots have provided powerful and fast new tools to generate and screen immense numbers of new, synthetic compounds. In reality, neither the more logical, rational design approaches nor the more random gunshot methods relying on massive numbers of compounds and screens, however, have resulted in the kind of successes that have been anticipated at the beginning of the 1990s. Indeed, among the over 200 recently produced new therapeutic agents, about one-third can still be classified as 'natural', and two-thirds as synthetic. In fact, natural product (re-)search for biologically active ingredients is currently experiencing a significant renaissance at several ends. Such natural products not only provide leads for new therapeutics, but also for eco-friendly pesticides gentle to the environment and food-chain and as part of functional foods and food supplements in the context of a healthy and healthily aging society. Here, and also increasingly in the field of cosmetics, 'naturals' are clearly superior to 'synthetics' for a number of reasons, which also address issues other than pure efficiency, such as safety, protection of the environment, social acceptance, sustainability, and often (local) economic wealth generation cycles. Some of these issues will be discussed as part of the chapters dealing with product development and commercialization. It should be noted that redox active secondary metabolites are of particular interest here, as they are able to modulate numerous biological processes, some of which are central to the prevention of or fight against diseases and also play a major role in the aging organism. Last but not least, the sheer richness of such metabolites, especially in our food, evidently provides considerable scope for their extensive practical uses and widespread applications.

The question emerges what has gone wrong. Why have strategies used successfully during the development of medical treatments failed in the case of antioxidant food components? Are the reducing phytochemicals indeed without any impact on human health or have the essential points been overlooked while designing the human studies? Some researchers are inclined to admit that beneficial biological properties observed for certain antioxidant food components are associated not that much with their redox activity, but are determined by other, more specific pharmacophoric features of their chemical structure. Others, me in that number, attempt to recognize shortages in the current human studies and to propose ways of overcoming them in future investigations. Two recent papers from the Bast group have explained why at least in the case of flavonoids a pleiotropic research approach is required (Weseler and Bast 2012). These authors have formulated ten major misconceptions concerning antioxidants (Bast and Haenen 2013). Here, I will try to add some thoughts that have not been raised in the mentioned papers, concentrating rather on some basic needs of living organisms than on antioxidants themselves.

Let us recall the famous essay by Dobzhansky (1973)—"Nothing in Biology makes sense except in the light of evolution". I would paraphrase this title for the sake of the current chapter—"Nothing makes sense in Biology but in the context of chemistry". Should antioxidants be placed on such chemistry backgrounds and why may this be important for interpreting former and planning subsequent experiments? The frequently raised issue concerning antioxidants is their source; a number of sources have been listed with particular emphasis on colorful plants. Nonetheless, the simplest answer is that reducing compounds are actually omnipresent in aerobic organisms. They are found in prokaryotic microorganisms, fungi, plants, and animals including humans, because substances controlling the reduced state in a cell are prerequisites of survival. As soon as oxygen appeared in the Earth's atmosphere, the single-celled organisms had to protect themselves against its toxicity. And this has not changed until now. Availability of free oxygen, however, created also an opportunity for fundamental improvement in efficiency of life's energy-generating systems.

The route from protection against oxygen to its successful utilization may be traced along the evolutionary timeline of proteins containing the metallo-porphyrin component. For example, hemoglobins are found in organisms representing all kingdoms and are involved in different aspects of oxygen interactions with the organism: scavenging to prevent toxicity, transport, storage, redox reactions, and energy production (Hardison 1999). These proteins are not alone in their functions. They are accompanied by low molecular weight reducing compounds, whose reduction potential may decide with which type of ROS they ultimately interact with most efficiently. What could not be taken into account when proposing the antioxidant hypothesis was that oxygen-derived species are not only unwanted by-products of oxidative metabolism. Ample evidence obtained in recent years documents that ROS can have not only deleterious effects, but also constitute necessary elements in chains of signal transduction pathways (Bartosz 2009). Their main and widely recognized effects concern the inhibition of protein tyrosine phosphatases, thus enhancing the activity of protein tyrosine kinases involved, i.e., in mediation of the effects of growth factors and cytokines (Forman et al. 2010). There is evidence that ROS are produced by specialized enzymes (NADPH oxidases, Nox) not only in the context of defense (in phagocytes), but also in many other cell types, and here for signaling purposes (Ago et al. 2011). In their recent review, Hernandez-Garcia et al. (2010) describe the most important functions of ROS for life, which include regulation of cell division, cell differentiation, transport, and apoptosis. It thus becomes evident that the ingested antioxidants play not necessarily only beneficial roles associated with protection against oxidative damage, but their overabundance may actually become disruptive once the redox homeostasis of the cell is deregulated.

Furthermore, the most recent discoveries have pointed to the epigenetic importance of 5′-methylcytosine (5-mC). Cytosine methylation has long been known to modulate gene expression and consequently cellular identity. This DNA modification, often embracing promotor regions, is a stable chemical alteration that represses transcription in eukaryotic organisms (Klose and Bird 2006). The reverse process, the removal of such methyl groups, allows cells to regain the ability to express silenced genes, or to recover their totipotent potential in the case of germ cells. While some demethylation may occur via a passive process deleting 5-mC from DNA upon repeated replications, active demethylation remained elusive up to 2011. Here, current evidence supports the existence of iterative oxidation of 5-methylcytosine to 5-carboxycytosine (Nabel et al. 2012). The first step of this process generating 5-OH-methylcytosine still seems to be involved in suppression of gene expression (Tahiliani et al. 2009). The subsequent products of oxidation—5-formylcytosine and 5-carboxycytosine—were demonstrated to be substrates for base excision repair (BER) and have been immediately accepted as intermediates of *active demethylation* (He et al. 2011; Ito et al. 2011). A similar role has been also proposed for 8-oxoguanine, which up till now was regarded solely as a marker of genotoxic exposure to ROS. Currently, however, it is considered as an element of redox sensitive epigenetic regulation (Guz et al. 2012). The dual role of genotoxins—including agents known to induce DNA oxidation—as possible methylome disruptors and remediators has been suggested as well (Lewandowska and Bartoszek 2011). An altered pattern of DNA methylation is observed in many diseases, especially in cancer and neurological disorders, which are prevented by dietary antioxidants. This association may point to the role of redox homeostasis in the maintenance of the correct shapes of human epigenomes over lifetime involving pro- and antioxidant mechanisms not predicted before (Brewer 2010).

It follows that oxygen based redox processes are not only important for the harvest of energy, but are behind the most vital mechanisms ensuring proper function of cells, tissues and the organism as a whole. It can be speculated that discrete values or levels of the cellular redox potential controlled by different ROS trigger certain sets of mechanisms, e.g., via redox sensitive transcription factors influencing gene expression. Fine-tuning of these levels may be offered by sulfur containing compounds creating a larger thio-redox system that would be able to respond to even tiny changes in the organism's homeostasis (such a system will be introduced in Chap. 7 in form of the 'cellular thiolstat'). So should not the reduction potential of dietary antioxidants be considered while discussing their impact on human wellbeing? If the major endogenous reducing agent GSH is taken as a reference point, could any exogenous compound with the higher reduction potential (and usually lower bioavailability) really make a noticeable difference in the neutralization/sequestration of excessive ROS as long as the physiological content of GSH is not significantly impaired? So far, such rather fundamental considerations are not taken into account at all. The antioxidant activity of dietary redox active components is simply based on their ability to reduce either free radicals or oxidants that are not even found in nature. In contrast, the relation between TEAC[2] or ORAC values or physiologically relevant thio-redox potentials, in the sense sketched out above, has neither been established nor even considered in earnest.

As indicated above, in living organisms the inevitable formation of ROS during oxidative respiration is used to control other biological processes, which can be dependent for instance on energy supply, like cell growth and division. In this particular case, the increased ROS concentration might signal the energetic readiness of the cell for proliferation. Such a multidirectional exploitation is also executed in the case of cellular antioxidants. GSH, apart from being a thiol-based redox buffering molecule, is also a major player in the detoxification of xenobiotics, as well as of potentially undesirable products of the metabolism. Here, it reacts with electrophilic substances which otherwise could bind to nucleophilic centers of proteins and nucleic acids hence endangering their proper structure. Moreover, hydrophobic compounds conjugated to GSH are substrates for efflux pumps present in membranes regulating the cholesterol level and preventing intracellular accumulation of toxic substances. Similarly, flavonoids abundantly produced in plants not only protect these plants against UV exposure but also help to capture additional energy from sun radiation, attract pollinators, and discourage pathogens. In view of the apparently failing antioxidant hypothesis, it has been postulated that health promoting properties of plant secondary metabolites in the human organism are connected with other aspects than protection against ROS. The food synergy concept, for instance, additionally stresses the evolutionary importance of non-random mixtures of numerous molecules that occur naturally in foods and affect human health in a concerted fashion (Jacobs et al. 2012). But can the antioxidant hypothesis be regarded as properly tested at the current stage and therefore should it be abandoned altogether due to the lack of immediate confirmation?

The above-mentioned human studies, which provided a negative message on health benefits of supplementation with antioxidants, did not consider the relation between the reduction potential of compounds tested and for instance that of GSH. But this is not the only problem with the design of these investigations. All antioxidants used were vitamins (A, C, E, β-carotene), nutrients which are necessary for the organism's survival and have to be absorbed from dietary sources in which their concentration naturally is not very high. It can be presumed that in the case of such indispensable compounds, specialist mechanisms have evolved to ensure their efficient uptake from the alimentary tract. In the case of supplementation, such mechanisms will enable relatively efficacious absorption of the vitamins ingested, hence resulting in their enhanced impact on the thio-redox homeostasis. Yet, it is not unequivocally proven that a low reduction potential is indeed relevant for their function or whether it may actually lead to undesirable

[2] TEAC stands for 'Trolox Equivalent Antioxidant Capacity', most commonly measured using the ABTS (2,2′-azino-bis(3-ethylbenzothiazoline-6-sulphonic acid) decolorization assay.

side effects, as discussed recently for vitamin C (Kuo 2013). The negative results observed in the case of supplementation with vitamins may be thus explained in at least two ways. First, one related to their antioxidant nature, when a substantial bioavailability may lead to an imbalance in cellular thio-redox status impairing various redox dependent processes. Second, the excessive supply of vitamins may deregulate some of the organism's functions related directly to their physiological role. In contrast, non-nutrient antioxidants, such as phenolic plant components, are extremely poorly absorbed and it can be hypothesized that their bioavailability might be more flexibly adjusted to the actual requirement of maintaining a proper thio-redox status. The latter is in a way confirmed by observations during research aimed at the alleviation of side effects of anticancer therapy. The experiments carried out with the aid of *in vivo* models demonstrated efficacy of synthetic antioxidants, as well as of various non-nutrient antioxidant phytochemicals, in situations of OS. For instance, a diminished cardiotoxicity of doxorubicin has been observed. This drug causes the formation of ROS and hence significant side effects which may be counteracted by the use of certain antioxidants. Clearly, such a cardioprotective effect was seen for reducing compounds with divergent chemical structures (flavonoids, lycopene, betalains, *N*-acetylcysteine, etc., Piasek et al. 2009). In clinical practice, antioxidants (but not vitamins) are routinely administered to patients along with the respective anticancer drug to curb ROS and thereby prevent cardiac damage.

Another hint concerning the importance of thiol-specific redox modulatory effects for health benefits of phytochemicals comes from epidemiological studies on the relationship between cancer incidence and dietary pattern. The most consistent inverse correlation between risk of cancer and the amount of food consumed was observed for *brassica* vegetables. It is generally accepted that anticarcinogenic action of these plants is brought about by organic sulfur compounds—*glucosinolates*—that are degraded by the enzyme myrosinase to a number of products among which isothiocyanates exhibit the strongest chemopreventive potential (Dinkova-Kostova and Kostov 2012). The mechanism of action of these thiol-specific modifiers is strictly related to thio-redox homeostasis. Isothiocyanates influence the thiolstat by their ability to react with SH groups of cysteine residues in glutathione, as well as 'antioxidant' proteins including thioredoxin. Most importantly, these phytochemicals, by reacting with specific SH groups, are able to restructure protein complexes releasing nuclear factors such as Nrf2 and NF-κB. As a result, the former is translocated to the nucleus and triggers the expression of cytoprotective genes mediated by the antioxidant response element (ARE). The inhibited translocation of NF-κB prevents the expression of pro-inflammatory genes. Both mechanisms are of key relevance in the prophylaxis of chronic diseases which explains the above-mentioned beneficial role of diet rich in *brassica* vegetables revealed by epidemiological observations. To conclude, the best established chemopreventive dietary ingredients are those that display the ability to influence the cellular (or more generally the organism's) thio-redox status. Can this observation be translated into a more general statement, i.e., are only the substances that are capable of impacting on the thio-redox balance also able to bring about a significant health benefit? The answer to this question remains to be determined. It is noteworthy, however, that the chemopreventive efficacy, that is the limitation of cancer incidence in humans, was seen for whole vegetables and not for isolated phytochemicals.

The two promising lines of evidence described above suggest that we may indeed be missing some important pieces of the redox conundrum. The discrepancy between experimental research and epidemiological studies vividly demonstrates that there are major gaps in our current understanding of the role reducing agents play in human wellness. Consequently, the proper exploitation of this issue needs some verified tools to predict the biological activity of antioxidants in the human organism, both when still in the digestive tract and after absorption. It will also be necessary to predict and further investigate the ability of such antioxidants to protect foodstuffs against oxidative spoilage which so far has represented a major interest in food science.

Maybe it would be justified to forget for a while about the antioxidant hypothesis in its current shape and form. Then, taking advantage of the vast amount of data collected over the past two decades, it may be advisable to carry out a shift from the deductive to a more inductive approach and scrutinise without wishful thinking what story all these results are really telling. Here, one should take into account the expanding knowledge on possible evolutionary roles of ROS and mechanisms controlling the thio-redox homeostasis—this time also not forgetting about chemistry.

Finally, let us admit that the perhaps premature antioxidant hypothesis has an unprecedented positive impact on human nutrition, lifestyle and, most of all, the awareness of responsibility for one's own health. Millions of people have revolutionized their personal food pyramids and adopted diets richer in fruits and vegetables. Even more astonishing is the fact that the food producers, instead of sticking to the cheapest solutions, began the search for ingredients and technologies leading to the production of health-related quality, or as this industry prefers to call it, functional foods. Furthermore, a growing number of nonprofit organizations have been founded in cooperation with research institutions and governmental agencies to advocate a healthier lifestyle. Education on proper nutrition, rich in plant antioxidants, is now beginning as early as in the kindergarten. Would it be possible to create such a compelling impact without the catchy term ANTIOXIDANT? That is indeed very doubtful. Therefore, it is the great responsibility of the academic community not to waste these achievements for the sake of a not fully recognized and, in any case debatable, scientific accuracy.

References

Ago T, Kuroda J, Kamouchi M, Sadoshima J, Kitazono T (2011) Pathophysiological roles of NADPH oxidase/Nox family proteins in the vascular system—review and perspective. Circ J 75(8):1791–1800

Bartosz G (2009) Reactive oxygen species: destroyers or messengers? Biochem Pharmacol 77(8):1303–1315

Bast A, Haenen GRMM (2013) Ten misconceptions about antioxidants. Trends Pharmacol Sci 34(8):430–436

Bjelakovic G, Nikolova D, Gluud LL, Simonetti RG, Gluud C (2007) Mortality in randomized trials of antioxidant supplements for primary and secondary prevention: systematic review and meta-analysis. Jama-J Am Med Assoc 297(8):842–857

Brewer GJ (2010) Epigenetic oxidative redox shift (EORS) theory of aging unifies the free radical and insulin signaling theories. Exp Gerontol 45(3):173–179

Dinkova-Kostova AT, Kostov RV (2012) Glucosinolates and isothiocyanates in health and disease. Trends Mol Med 18(6):337–347

Dobzhansky T (1973) Nothing in biology makes sense except in the light of evolution. Am Biol Teach 35:125–129

Doll R, Peto R (1981) The causes of cancer: quantitative estimates of avoidable risks of cancer in the United-States today. J Natl Cancer Inst 66(6):1192–1308

Forman HJ, Maiorino M, Ursini F (2010) Signaling functions of reactive oxygen species. Biochemistry 49(5):835–842

Guz J, Jurgowiak M, Olinski R (2012) Oxidation and deamination of nucleobases as an epigenetic tool. Postep Hig Med Dosw 66:275–286

Hardison R (1999) The evolution of hemoglobin. Am Sci 87(2):126–137

He YF, Li BZ, Li Z, Liu P, Wang Y, Tang QY, Ding JP, Jia YY, Chen ZC, Li L, Sun Y, Li XX, Dai Q, Song CX, Zhang KL, He C, Xu GL (2011) Tet-mediated formation of 5-carboxylcytosine and its excision by TDG in mammalian DNA. Science 333(6047):1303–1307

Hernandez-Garcia D, Wood CD, Castro-Obregon S, Covarrubias L (2010) Reactive oxygen species: a radical role in development? Free Radical Bio Med 49(2):130–143

Ito S, Shen L, Dai Q, Wu SC, Collins LB, Swenberg JA, He C, Zhang Y (2011) Tet proteins can convert 5-methylcytosine to 5-formylcytosine and 5-carboxylcytosine. Science 333(6047):1300–1303

Jacobs DR, Tapsell LC, Temple NJ (2012) Food synergy: the key to balancing the nutrition research effort. Public Health Rev 33(2):507–529

Klose RJ, Bird AP (2006) Genomic DNA methylation: the mark and its mediators. Trends Biochem Sci 31(2):89–97

Kuo SM (2013) The multifaceted biological roles of vitamin C. J Nutr Food Sci 3(5)

Lewandowska J, Bartoszek A (2011) DNA methylation in cancer development, diagnosis and therapy-multiple opportunities for genotoxic agents to act as methylome disruptors or remediators. Mutagenesis 26(4):475–487

Miller ER, Pastor-Barriuso R, Dalal D, Riemersma RA, Appel LJ, Guallar E (2005) Meta-analysis: high-dosage vitamin E supplementation may increase all-cause mortality. Ann Int Med 142(1):37–46

Nabel CS, Manning SA, Kohli RM (2012) The curious chemical biology of cytosine: deamination, methylation, and oxidation as modulators of genomic potential. ACS Chem Biol 7(1):20–30

Piasek A, Bartoszek A, Namiesnik J (2009) Phytochemicals that counteract the cardiotoxic side effects of cancer chemotherapy. Postep Hig Med Dosw 63:142–158

Seiber JN, Kleinschmidt L (2012) From detrimental to beneficial constituents in foods: tracking the publication trends in JAFC. J Agr Food Chem 60(27):6644–6647

Tahiliani M, Koh KP, Shen YH, Pastor WA, Bandukwala H, Brudno Y, Agarwal S, Iyer LM, Liu DR, Aravind L, Rao A (2009) Conversion of 5-methylcytosine to 5-hydroxymethylcytosine in mammalian DNA by MLL partner TET1. Science 324(5929):930–935

WCRF/AICR (1997) Food, nutrition and the prevention of cancer: a global perspective

Weseler AR, Bast A (2012) Pleiotropic-acting nutrients require integrative investigational approaches: the example of flavonoids. J Agr Food Chem 60(36):8941–8946

Author Biography

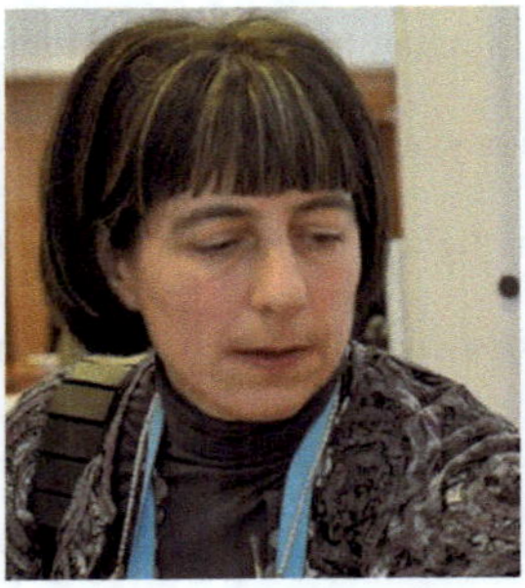

Agnieszka Bartoszek graduated from the Chemical Faculty, Gdańsk University of Technology (GUT) in 1981, receiving the degree of MSc in Organic Chemistry and the title of Engineer. In the same year, she enrolled for doctoral studies at the Department of Pharmaceutical Technology and Biochemistry, GUT, where she joined the research team headed by Prof. Jerzy Konopa and was involved in research on the mechanism of action of DNA-reactive antitumor compounds. In 1990, she received her PhD degree based on a thesis concerning DNA adduct formation by a group of antitumor derivatives of 1-nitro-9-aminoacridine, which was analyzed by the then just developed ^{32}P-post-labelling assay. During 1988–1989, she worked for the Imperial Cancer Research Fund in Edinburgh, at the Laboratory of Molecular Pharmacology and Drug Metabolism, then Medical Oncology Unit. Her major research concerned the biological consequences of activation of antitumor drugs doxorubicin and mitomycin C by NADPH cytochrome P450. In 1990, she obtained an academic teacher position at the Chemical Faculty, GUT, where she works until now and is currently involved in several research projects and extensive teaching.

Agnieszka's main research interests initially concentrated on the mechanism of action of antitumor drugs and compounds, especially their capability of binding to DNA. Another line of scientific investigation, adopted after moving to the Department of Food Chemistry, Technology and Biotechnology, GUT, concerns chemopreventive properties of selected food components and dietary intervention during cancer chemotherapy. The projects on the chemopreventive potential of traditional foods resulted in a close cooperation with the food industry. They are aimed at the design and elaboration of technologies for health-oriented functional food products enriched in plant-based anticarcinogenic phytochemicals.

Part II
Connecting Section Between Chapters 2 and 3

The previous chapter has highlighted some of the more common natural sources of secondary metabolites and also discussed biological activities associated with them. Therefore, the question that arises is how such active plant metabolites can be identified and possibly purified for further research or even for practical applications. Indeed, this particular question is not easy to answer, and belittles the massive efforts often required to move from a leaf or bark of an exotic tree all the way to a pharmaceutical or cosmetic product.

Traditionally, such compounds have been found because of some existing prior knowledge of activities associated with plants or fungi. This knowledge may well have been embedded within Folk Medicine or certain spiritual rituals dating back many centuries. While it has allowed to identify interesting plants (and parts thereof), it has not provided any scientific basis with regard to the active ingredient(s), the exact biological activities associated with them, and the biochemical mechanisms responsible for such activity. Indeed, in order to investigate such crucial aspects related to activity (and its future refinement), the active ingredients have to be isolated in a *pure* form and characterized *chemically*.

In the past, such isolation, purification, and characterization procedures have been extremely cumbersome and often have taken many years to complete. Today, a barrage of modern technology is at our disposal to identify specific compounds (or patterns thereof) in plants and plant products, and to separate these individual compounds by sophisticated methods. These procedures often function in combination with activity screens on the one hand and structure elucidation (e.g., via mass spectrometry) on the other. Indeed, by employing such techniques in combination, it is now possible to identify and purify an active ingredient from a plant within a few days, rather than within months or years.

The following chapter will therefore introduce us to some of the cutting-edge technologies, which are currently being used to hunt for biologically active ingredients in biological materials. As it is virtually impossible to cover all analytical techniques available in this field in just one chapter, this presentation is necessarily limited to a few key techniques, which have been selected because they are widely and readily applicable, fast, and comparably straightforward to apply.

These methods also include 'hyphenated' combinations of analytical techniques, which often involve sophisticated and innovative analysis by mass spectroscopy.

Chapter 3
The Modern Art of Identification of Natural Substances in Whole Plants

Patrick Chaimbault

Keywords (Bio-)analytical methods • Mass spectrometry • Separation methods • Structural • Elucidation • Supercritical liquid extraction

1 Introduction

Although the oxidative reaction is essential for life, this process is responsible for important stresses causing serious cellular damage to DNA, lipids, and proteins. These damages can result in cancers, cardiovascular, or neurodegenerative diseases. Among the armory available against Reactive Oxygen Species (ROS), antioxidant secondary metabolites of plants can be involved in cell defense. One of the main sources of natural antioxidants is the dietary intake. It is well known nowadays that antioxidants are important constituents of vegetables and fruits. For example, vitamin C (ascorbic acid) contained in high amounts in citrus fruits (oranges, lemons, mandarins, etc.,) is able to neutralize ROS and so participate in cell defense, but also protects against rancidity which affects the color or/and aroma of foodstuffs. Furthermore, quite a lot of natural antioxidants are also of interest for anti-aging cosmetics.

Thus, the identification of natural antioxidants is an increasingly important subject. Their isolation from plants and their structural elucidation require the use of several techniques and the development of different analytical methods including extraction procedures. These procedures have to be adapted at each stage (extraction, separation, detection, etc.,) based on the physicochemical properties (solvent solubility, polarity, hydrophobicity, etc.,) of the antioxidants considered.

P. Chaimbault (✉)
Laboratoire de Chimie et Physique Approche Multiéchelle des Milieux Complexes, LCP-A2MC (EA 4632) ICPM, Université de Lorraine, 57078 Metz, France
e-mail: patrick.chaimbault@univ-lorraine.fr

C. Jacob et al. (eds.), *Recent Advances in Redox Active Plant and Microbial Products*, DOI: 10.1007/978-94-017-8953-0_3, © Springer Science+Business Media Dordrecht 2014

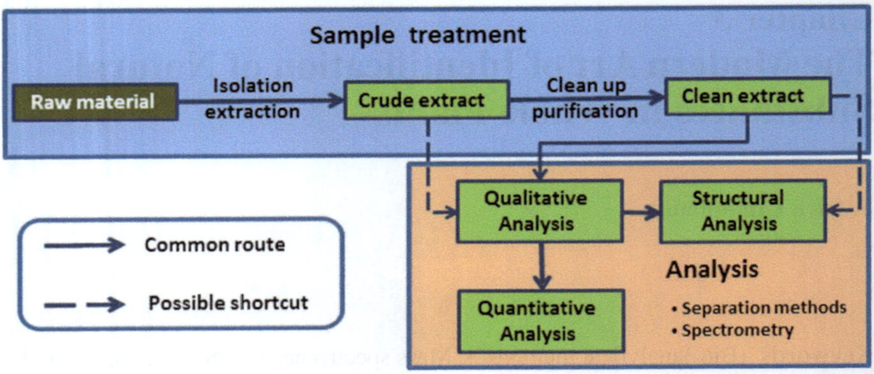

Fig. 1 General scheme of plant analysis. Shortcuts are possible depending on the quality or the purity obtained after sample treatment

This chapter provides an overview of the current techniques used to identify interesting natural products (see also Explanatory Box 1). Whenever it is possible, examples of applications are given for some antioxidants. Following the scheme presented in Fig. 1, the chapter focuses on their isolation (extraction and purification), separation, and detection for qualitative or quantitative analysis. The last part is dedicated to hyphenation between separative methods and mass spectrometry for the identification of secondary metabolites in plant extracts.

Explanatory Box 1: (Bio-)Analytical Techniques

Natural product research relies heavily on a barrage of powerful analytical techniques. The latter enable, for instance, the identification and characterization of a given compound, including structural elucidation. The latter is often far from trivial, as many natural products contain one or several chiral centers. Analytical techniques commonly employed in this context include elemental analysis to determine the elemental composition of a given compound (e.g., content of carbon, nitrogen, hydrogen, sulfur), (high resolution) mass spectrometry (HR-MS) to determine its molecular weight, aspects of purity, and, if applicable, specific isotope splittings, UV/Vis and IR spectroscopy to identify the presence of specific functional groups and various types of one-dimensional or two-dimensional nuclear magnetic resonance (NMR) to investigate the structure of the compound.

Further methods may be used depending on the particular physicochemical properties of a compound, for instance polarimetry or circular dichroism in case of chiral compounds, refractometry in the case of liquids, electron spin resonance (ESR) in the case of radicals, and many transition metal ions and electrochemical methods in the case of redox active compounds. Indeed, electrochemical methods such as Cyclic Voltammetry and Differential Pulse Polarography have recently been at the forefront of investigations into the redox behavior of many secondary metabolites, their (*in vitro*) interactions with biomolecules, and the redox sensitivity of biological targets (proteins, enzymes).

At the same time, analysis is paramount when it comes to the purity of a given natural product. Again, methods such as HR-MS and NMR may be used, together with various types of chromatography. Here, high-performance liquid chromatography (HPLC) and gas chromatography (GC) are commonly used methods also mentioned in the text.

This 'analysis' of the natural compound itself is matched by the analysis of biological activity and biochemical mode(s) of action. Here, rather different methods are employed. Biological activity is usually determined in rapid screens based on isolated proteins and enzymes, cells, microorganisms, viruses, or even small animals (such as nematodes). In the last couple of years, in silico screening has become important, as computational methods become more and more predictive and hence competitive. Such virtual screens enable researchers to reduce the number of real biological screens and to focus on a dramatically reduced number of carefully selected compounds. At the same time, the methods available to track down intracellular processes triggered by various natural and unnatural compounds, such as specific cellular signaling pathways, have become more sophisticated. It is now possible to use a whole barrage of staining and labeling techniques to investigate individual cellular responses, from proliferation, differentiation, and apoptosis to the up- and down-regulation of individual proteins using fluorescently labeled antibodies. As we will see in some of the following chapters, it is also possible to use fluorescent staining to detect cellular components, such as various Reactive Oxygen Species, posttranslational cysteine modifications, Ca^{2+} influx, and content of reduced glutathione to name just a few. These methods are based on fluorescence and microscopy. Other methods used to track down compounds in cells include Energy Dispersive X-ray Analysis (EDX), which is indicative of the presence of certain elements and can now also be applied to biological samples without the need of any labeling. These methods are complemented by other proteomic and genomic analytical techniques, such as isotope labeling in combination with mass spectrometry, Western Blots to quantify (time resolved) changes in protein levels, and stains indicative of the mitochondrial membrane potential $\Delta\Psi_M$.

The development of analytical techniques to track down and characterize natural products is currently attracting considerable attention. At the same time, screening techniques to determine the biological activity of compounds are becoming ever more sophisticated. They are joined by highly effective methods for 'intracellular diagnostics' and target evaluation. Here, areas such as proteomics, genomics, chemogenomics, and life cell imaging are growing rapidly.

2 Extraction of Antioxidants from Plants

Extraction is one of the most imperative steps for the analysis of natural products involving chromatography. The procedures involved are so that the molecules of interest initially contained in the plant tissues are solubilized in a liquid phase. The resulting solution can be more or less directly injected into the chromatographic system or submitted to other analytical pretreatments (e.g., liquid–liquid extraction, solid phase extraction, etc.,) depending on the quality required of the purification. The ideal extraction procedure must be simple, rapid, and efficient. Nevertheless, the active molecules must be carefully preserved from possible degradation.

2.1 Conventional Extraction Procedures

Since the earliest times, man has always performed extraction procedures of active molecules from plants for medicinal purposes. For many drugs, he had to prepare brews by maceration in cold water or decoction, infusion, or percolation of plants using warm or boiling water. Increasing the temperature generally improves the efficiency of the extraction procedure because heat increases the diffusion and solubility of the molecules in the solvent. During the process, the solvent penetrates the tissues and dissolves the metabolites. A pre-grinding of the plant (especially for very hard and dense samples such as nuts or bark) is usually carried out to facilitate the disruption of the cell walls and to increase the surface exchange between the solid matrix and the solvent (Heldman and Hartel 1997). The solution moves through the solid matrix to its surface and becomes dispersed in the solvent (Fellows 2000). The solid matrix can be eliminated afterwards by filtration, decantation, etc. The solvent plays a key role in this process and consequently it must be carefully chosen regarding the hydrophilic–lipophilic balance of the metabolite to be extracted. The Folch extraction is undoubtedly one of the most popular solid liquid extraction procedures (Folch et al. 1957). The tissue is mixed with chloroform/methanol (2/1, v/v) in the following proportion: 1 g in 20 ml of solvent mixture. The whole mixture is shaken for 15–20 min at room temperature allowing the extraction of apolar antioxidants such as carotenes and carotenoids. By contrast, alcohols (methanol, ethanol) or hydro-alcoholic mixtures give better results for more polar compounds such as polyphenols (Jokic et al. 2010).

The traditional extraction procedures also include some slightly more complex methods such as hydrodistillation or Sohxlet extraction. The latter is probably the most regarded among all the extraction methods and it often constitutes the reference for comparison between other available procedures (accelerated solvent extraction, microwave-assisted extraction, etc.,). It is a solid–liquid extraction procedure presenting the benefits of a constant source of renewed solvent by distillation. The extracting solvent is thus never saturated and the extraction yield is improved as the sample is progressively depleted of its molecules of interest.

In spite of its efficiency, this method requires time (several hours even in its automated version) and high amounts of solvent (hundreds of milliliters). Moreover, the constant heat (use of boiling solvent for hours) may also induce the thermal degradation of active molecules reducing the activity of the final extract (Wang and Weller 2006). However, the classic Sohxlet remains an efficient low-cost method because it only requires laboratory glassware and an electric heater. Furthermore, the extract composition can be easily modulated according to the nature of the solvent used (Zarnowski and Suzuki 2004). Finally, in some cases, the highest yields are even obtained with a Soxhlet apparatus. This was observed, for example, for the methanolic extraction of protocatechuic, *p*-hydroxy-benzoic, vanillic, and ferulic acids in *Sambucus nigra L.* inflorescence (Waksmundzka-Hainos et al. 2007).

2.2 Improved Extraction Methods

Alternative extraction methods were introduced to overcome the inconveniences of the traditional methods. They were developed in order to prevent active products from thermal degradation, to improve the extraction yields, but also to reduce solvent consummation, time, and costs.

2.2.1 Supercritical Fluid Extraction

In certain ways, the supercritical fluid extraction (SFE) is a form of solid–liquid extraction where the extracting solvent is not a liquid at atmospheric pressure but a supercritical fluid. Such a fluid is at an intermediate physical state between liquid and gas (see Explanatory Box 2). It adopts the density of a liquid, whereas its viscosity and diffusion rate correspond to a gas. Consequently, the diffusion and solubility of the molecules in the solvent (i.e., the supercritical fluid itself) are more efficient. Reaching a supercritical state supposes that the temperature and pressure of the fluid are raised over their critical value (Fig. 2a). For carbon dioxide, which is probably the most used supercritical fluid, both parameters are relatively low: 31 °C and 74 bars, respectively. These conditions are obtained and maintained, thanks to a complex equipment (Fig. 2b) comprising a pump delivering the high-pressurized fluid (e.g., a liquid chromatography pump), temperature controllers, pressure regulators, and special SFE cells. Thus, this extraction method is more expensive than traditional extraction methods.

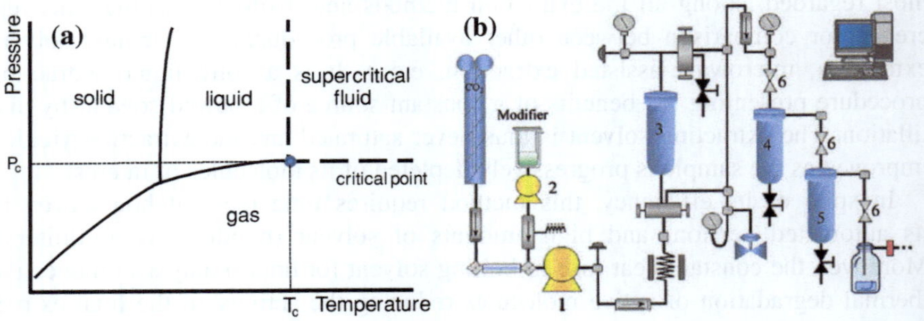

Fig. 2 **a** Physical state of a fluid as a function of pressure and temperature (P_c = critical pressure, T_c = critical temperature). **b** Scheme of a SFE equipment. Diagram of a supercritical fluid extraction pilot plant equipped with two fractionation cells. *1* CO_2 pump, *2* modifier pump, *3* solid samples extraction cell, *4* fractionation cell 1, *5* fractionation cell 2, *6* valve, reprinted with permission (Herrero et al. 2006)

Explanatory Box 2: Supercritical Fluids

Supercritical fluids in general, and supercritical carbon dioxide (CO_2) in particular, have recently attracted considerable attention in natural products research. These materials occur once a liquid reaches its 'critical point' which is defined by a specific temperature and pressure and can be illustrated in a so-called 'phase diagram'. Such diagrams are widely known and contain areas where a given compound is either solid, liquid, or gaseous (i.e., occurs in one of the traditional states of matter). The critical point, however, marks a specific temperature–pressure point beyond which a rather different phase occurs. Instead of evaporating to form a gas, the liquid turns 'supercritical'. Most common liquids do not turn supercritical under normal everyday conditions. Water, for instance, may well turn supercritical, but only at high temperature and under substantial pressure. CO_2, in contrast, turns supercritical at just over 300 K (at around 30 °C) and 72 atmosphere pressure. It is therefore an ideal supercritical fluid for practical applications, as it is easily produced and also chemically fairly inert. Curiously, CO_2 is unusual as it does not occur in a liquid phase under normal pressure at all, but sublimes at −78 °C.

Such supercritical materials show properties of liquids as well as gases, and are very 'temperamental' once either their temperature or pressure are changed. Indeed, various properties of these supercritical fluids reside between the ones of classical fluids and gases, such as density, viscosity, and diffusivity. Hence it is possible to exploit the liquid-like and gas-like properties of such supercritical materials for a range of practical applications.

It is possible, for instance, to use supercritical CO_2 for the extraction of caffeine from coffee beans or of essential oils from natural sources. Here, the ability to 'extract' caffeine or oils is clearly associated with the liquid-like behavior of supercritical CO_2, while the subsequent removal of the 'solvent' (as a gas and without any residues) takes advantage of the gaseous behavior of the CO_2. It is even possible to use supercritical CO_2 for the separation of compounds, as it is easily possible to modify its density by 'fine-tuning' temperature and pressure.

SFE processes as follows. The sample is introduced into the extracting vessel. The supercritical fluid, regulated at the desired pressure and temperature, passes through the sample and dissolves the molecules of interest. The dissolving power of the fluid is lowered by decrementing the applied pressure and the extracted molecules in the collector. In the case of sophisticated equipments, several collectors may be associated in series to improve the efficiency of the sample fractionation and when SFE is carried out at industrial scale, the CO_2 is possibly recycled. Finally, CO_2 is a gas at room temperature, so its complete elimination from the extract is easily achieved. Finally, a dry plant extract is obtained without any organic solvent residues.

As mentioned above, carbon dioxide is the most popular fluid. Supercritical CO_2 preferentially extracts apolar compounds, but by changing its density by modulating the applied pressure and temperature, its extracting power can be optimized for other metabolites as well. This technique is widely used to extract phytonutrients such as caffeine, (Park et al. 2007) aroma, (Blanch et al. 1995; Polesello et al. 1993; Augusto et al. 2003) spices, (Illes et al. 1999), and lipids (Eggers 1996) but also antioxidants such as astaxanthine (Lopez et al. 2004), carotenes (Illes et al. 1999; Longo et al. 2012), tocotrienols, tocopherols (Illes et al. 1999), and resveratrol (Benova et al. 2010). Furthermore, because SFE is performed in the absence of light and oxygen, it prevents possible degradation or isomerization of these compounds during the extraction step. Active metabolites poorly soluble in CO_2, such as phenolic and glycosidic compounds, however, cannot be extracted from the plant limiting the interest of SFE for many applications. A small percentage of a polar organic modifier (up to 20 %) is miscible to the supercritical CO_2. The addition of a co-solvent (e.g., methanol, ethanol, acetonitrile) increases the solubility of polar compounds (Valcarcel and Tena 1997) but sometimes reduces the selectivity of the extraction. In addition, the use of an organic modifier requires a slightly higher temperature to reach the supercritical state and this could be disadvantageous for thermally labile compounds. In addition, the dried extract, finally obtained may still contain traces of organic solvent residues.

The use of subcritical water is a possible alternative to CO_2 for the extraction of polar metabolites. Subcritical water extraction (SWE) is generally carried out

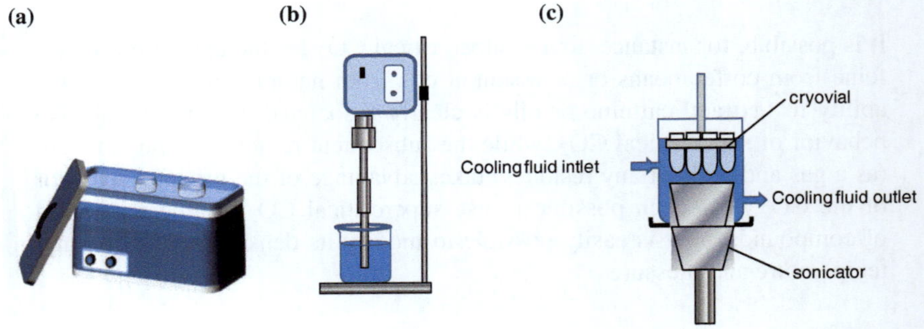

Fig. 3 Devices for ultrasonic assisted extraction. **a** Ultrasonicbath. **b** Ultrasonicprobe. **c** Cup-horn

at high temperature (100 to 374 °C). At the same time, the high pressure applied (10–60 bars) maintains its liquid state. Increasing the water temperature lowers its dielectric constant allowing the solubilization of less polar molecules. Consequently, the polarity of water can be easily adjusted from a polar to an apolar-like solvent just by varying both parameters. Thus SWE has been used for the extraction of catechins and proanthocyanidins (Garcia-Marino et al. 2006). For this application, SWE is more efficient than a solid–liquid extraction performed in a water–methanol mixture (25/75, v/v). The extraction of anthraquinone of *Morinda citrifolia* by SWE is also more efficient than ethanol or ultrasound-assisted extraction for the extraction (Pongnaravane et al. 2006). Although SWE improves the extraction ability for polar compounds, it remains unsuitable for thermolabile molecules (Lang and Wai 2001). Furthermore, although SWE only uses a 'green' solvent, maintaining the high temperature and and the pressure is energy consuming.

2.2.2 Ultrasonic Assisted Extraction

The ultrasonic assisted extraction (UAE) is a fairly simple and easy-to-use method. Sample and solvents are mixed together in a vessel and the mixture is then either placed in an ultrasonic bath or directly submitted to ultrasound using a diving probe (Fig. 3). The cup-horn sonicators are ideal for small volume multisampling extractions. In a study concerning the ultrasound assisted extraction of ginseng saponins from ginseng roots, Wu et al. (2001) compared the performance between an ultrasonic bath and a horn probe. They obtained similar results in terms of extraction yield and rate with both devices.

The extraction mechanism has already been described in detail (Luque-Garcia and Luque de Castro 2003; Soni et al. 2010). Sound waves at frequencies higher than 20 kHz are applied. The sound waves travel into the matter in which the mechanical vibrations generated induce expansion and compression cycles. The expansion creates bubbles in the solvent. They grow and finally collapse when they reach a critical volume. Rapid gas and vapor compression inside the bubbles produce a very high temperature and pressure (Suslick 1994). When coming close to a solid matrix, the bubbles become asymmetric and when they collapse, their implosion produces high-speed jets of liquid impacting the solid surface. These high-speed jets lead to the detachment and expulsion of small particles of matrix facilitating the transfer of the molecules in the extracting solvent.

In order to optimize an UAE, several parameters have to be considered. In a study on extraction of oil in flaxseeds by ultrasound, Zhang et al. (2008) showed that the higher the ultrasonic power, the higher the yields. This parameter, however, has also to be regulated as the yields may decrease above the optimal power. The extraction time is shortened and an excess of sonication can even decrease the quality of extracts. The same goes for temperature (Wang and Weller 2006). Very often, all these parameters are easily optimized using an experimental design. For example, the highest antioxidant activity of polyphenols [(−)-epicatechin, procyanidin B2, chlorogenic acid, and procyanidin B1] extracted from unripe apples are obtained for an ultrasonic power of 519.39 W, an extraction time of 30 min, and an extraction temperature of 50 °C in ethanol 50 % (Yue et al. 2012).

The main benefits of UAE over conventional Soxhlet extraction are higher efficiency (higher yields in less time) (Zhang et al. 2008) and the leaching of thermolabile compounds. Compared to other advanced techniques such as SFE or accelerated solvent extraction (ASE), the pressure is generally lower. Hence, the UAE equipment is cheaper than that for SFE and ASE. Because UAE can be carried out with polar or apolar solvents, it allows extracting compounds whatever their polarity, whereas supercritical CO_2 is rather limited to apolar molecules. Thus UAE is used increasingly for the extraction of natural compounds in plants. The mechanical activity of the ultrasound improves the diffusion of solvents into the tissue and facilitates the disruption of the cell wall. The solubilization of the metabolites is thus favored. The extraction of tirucallane-like triterpenes from *Juliania adstringens* was achieved with ethanol as solvent (Makino et al. 2004). Hromadkova et al. performed the extraction of the hemicellulose components of buckwheat hulls in alkali solutions (Hromadkova and Ebringerova 2003). The literature also mentions applications concerning antioxidants. For example, two ultrasonic extraction methods were optimized by Adam et al. (2009) for the isolation of esculetin, scopoletin, 7-hydroxycoumarin, rutin, xanthotoxin, 5-methoxypsoralen, and quercetin from plants such as *Mentha* species, *Ruta graveolens L.*, *Achyllea millefolium L.*, *Plantago lanceolata L.*, and *Coriandrum sativum L.* Glycosylated phenolic antioxidants such as myricitrin-3-rhamnoside, quercitrin-3-rhamnoside, europetin-3-rhamnoside, or kaempferol-3-rhamnoside were extracted from *Acacia confusa* flowers and buds (Tung et al. 2011). In this case, the best results were obtained applying 12 sonication cycles of 10 min each.

2.2.3 Accelerated Solvent Extraction

Accelerated solvent extraction (ASE) is a solid–liquid extraction process that bears similarities to SFE. Indeed, this patented technique (Dionex) is performed

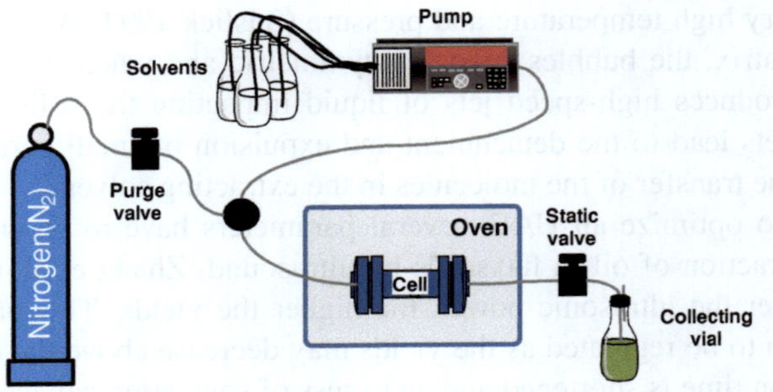

Fig. 4 Equipment for accelerated solvent extraction (*ASE*). The sample is weighed and placed into the extraction cell. The cell is filled with solvent, pressurized, and heated in the oven. The hot solvent extracts the metabolites by direct contact with the sample by both static and dynamic action. After the extraction is complete, the solvent is removed from the cell to the collecting vial with the assurance of compressed nitrogen. The sample is ready for further analysis

at elevated temperatures, usually between 50 and 200 °C and at pressures between 10 and 15 MPa to maintain the solvent in a liquid state below its critical conditions. Subcritical water can also be used in an ASE apparatus (Fig. 4) (Eskilsson et al. 2004) but generally ASE is carried out with common organic solvents (pure or mixtures). The use of adapted stainless cells (Dionium^{TM}) also allows the sample pretreatments at relatively extreme pH (0.1 M of H_2SO_4, HCl or KOH) (Dorich et al. 2008).

Thus, ASE allows extracting apolar as well as polar compounds depending on the choice of the solvent. For example, Mulbry et al. (2009) optimized an oil extraction process for algae studying several compositions of solvent starting from hexane to methanol–chloroform mixtures. They showed that ASE yields higher values for total oil content than the traditional Folch method. Other studies showed that the same goes for oxysterols in food (Boselli et al. 2001) or carotenoids in plants (Denery et al. 2004). Compared to techniques performed at atmospheric pressure (Soxhlet or UAE), extracts are obtained in a very short time and with a little solvent consumption. The extraction time is reduced from several hours (for Sohxlet) to a few tens of minutes and ASE requires around ten times less solvent. In addition, as for all pressurized solvent extraction methods, the extraction efficiency is improved. The temperature, however, must be carefully controlled, otherwise there is still a risk of decomposition for thermolabile metabolites. Solvent composition, temperature, applied pressure, and the number of cycles are parameters associated to the optimization of the extraction yields (Cicchetti and Chaintreau 2009). These parameters can easily be optimized, thanks to experimental design. Using response surface methodology (RSM), Hossain et al. optimized the composition of rosemary, marjoram, and oregano extracts. The optimum temperature (allowing the highest antioxidant activity for the extract) was 129 °C and the optimal percentage of methanol in the hydro-alcoholic solvent was around 56 % for marjoram and rosemary, whereas it was only 33 % for oregano (Hossain et al. 2011). This study concerns several phenolic antioxidants including rosamarinic, caffeic, and gallic acids.

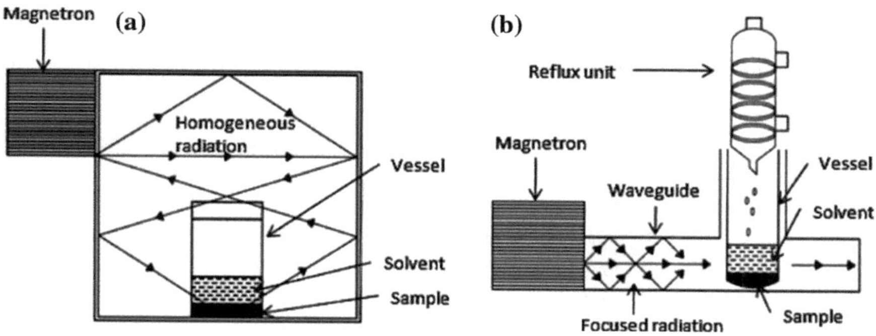

Fig. 5 Apparatus for microwave assisted solid extraction **a** Closed-type microwave system and **b** open-type microwave system. Reprinted with permission (Chan et al. 2011)

2.2.4 Microwave Assisted Solid Extraction

In the landscape of the extraction of organic compounds, microwaves appeared in the middle of the 1980s (Ganzler et al. 1986). Extraction is carried out in a microwave oven whose power (600–1000 W) and frequencies (~2,450 MHz) are the same as for domestic ovens. Commercially available MASE systems (Fig. 5) use either closed extraction vessels allowing a pressure and temperature control or open vessels topped with a condenser (Chan et al. 2011; Mandal et al. 2007). Both systems have their own advantages and drawbacks (Table 1).

Whatever the apparatus, a part of the electromagnetic energy absorbed by the extracting material is converted into heat energy via two distinct and simultaneously occurring mechanisms: (1) the friction between molecules due to the electrophoretic migration of ions under the changing magnetic field and (2) the constant realignment (more than 10^5 times per s) of the dipolar moment of molecules (solvent and/or metabolites) with the magnetic field. Both clearly indicate that only dielectric material or solvents are subjected to microwaves. The main advantage of the MASE over the methods does not only lie in its rapidity but also on its heating process. Compared to classic heating, microwaves are able to directly reach the heart of the sample. Thus, this technique is not only a surface method and through heating the method guarantees better extraction efficiency.

Several parameters strongly affect the final quality and the yield of the extract. The solvent must be carefully chosen taking into account its absorbing properties and the solubility of the extracted metabolites. Generally, organic solvents with high dielectric constants (i.e., methanol, ethanol) exhibit the highest performances. Very often, small amounts of water in the extracting solvent improve cell penetration and the heating of the plant matrix. For example, the optimum solvent composition for the extraction of chlorogenic and geniposidic acid from *Eucommia ulmodie* is aqueous methanol 80/20 (v/v) (Li et al. 2004). By contrast, a solvent with a low dielectric constant (i.e., hexane) is often less efficient in MASE compared to the corresponding Soxhlet procedure (Alfaro et al. 2003). Such a solvent,

Table 1 Advantages and drawbacks of microwave assisted solid extraction (Mandal et al. 2007)

MASE system	Advantages	Drawbacks
Open vessel	• Safety (operating at atmospheric pressure) • Possible addition of reagents during the extraction process • Solvent excess easily removed • Large samples can be processed • Easy cooling down and no depressurization • Low cost • Possible fully automation • Suitable for thermolabile species	• Less precision than closed-vessel systems • Throughput is lower in most open systems • Extraction times usually longer than for closed vessels
Closed vessel	• High temperatures reached allowing the decrease of the treatment time • Loss of volatile substances during irradiation is avoided • Less solvent is required (no evaporation occurs) • Less hazardous during acid extractions (fumes are confined in the closed vessel) • High throughput (a few ten samples simultaneously)	• Risks of explosion at high pressures • Treatment of limited amount of material • Some material (PTFE) does not withstand high temperatures • No addition of reagents or solvents during the procedure • Cooling down after the treatment and before the opening of the vessel (to avoid loss of volatile compounds)

however, can be mixed to a good microwave absorbing solvent to improve the solubilization of apolar compounds. For example, the mixture of ethanol (the absorbing solvent) and hexane 1/3 (v/v) was the optimal composition for the extraction of solanesol from tobacco leaves (Zhou and Liu 2006). In this case, hexane helps the solubilizing of this apolar metabolite. Zhao et al. (2012) recently used graphite powder sealed in glass tubes as microwave absorbing agent. The samples, the solvent, and the absorption microwave tubes are transferred into closed vessels and then submitted to microwave heating. Such an approach allows the use of nonpolar solvents (i.e., hexane, cyclohexane, isooctane, petroleum ether, etc.,) in microwave assisted extraction. The solvent volume is also an important parameter. Even if the solvent to sample mass ratio can strongly vary from an application to another, ratios in the range of 10:1 (ml/mg) to 20:1 (ml/mg) are usually the optimal proportions. Yet whatever the case, the solvent volume has to be sufficient to completely immerse the solid sample. Finally, the constitutive water of the plant can also be used as extracting solvent itself. This kind of procedure is named solvent-free microwave-assisted extraction (SFMAE) and is generally carried out at atmospheric pressure. Here, nothing else (i.e., no additional solvent) is required besides the samples' natural moisture. This moisture serves as heating source leading to the disruption of the cell walls and to the release of analytes. Many applications have already been carried out using this approach, e.g., for essential oils (Lucchesi et al. 2004a, b). SFMAE, however, can also be performed under pressure. Michel et al. (2011a) developed and optimized (extraction time, irradiation power, number of cycles) the pressurized solvent-free microwave assisted extraction (PSFME) technique for antioxidative compounds contained in sea buckthorn (*Hippophaë rhamnoides L.*) berries. Compared to other extraction techniques (including ASE), PSFME leads to the most active and richest extract regarding its phenolic content. Molecules such as quercetine and its glycosylated derivatives were successfully extracted, including isorhamnetin, which could not be extracted reasonably with any other techniques.

As for the other described methods above, the extraction time must also be carefully optimized. A too short or too long extraction time strongly affects the yields. The extraction time is generally shorter than for other methods. Usually, a few tens of minutes are sufficient but the extraction time can sometimes be reduced to a few tens of seconds. Thereafter, overexposure to heat often leads to thermal degradation of the active molecules. The same goes for the microwave power. In fact, the best conditions of extraction are a compromise between time and power. High power requires low extraction times and vice versa. The consequences, however, obviously are not the same. Indeed, rapid exposure at high microwave power induces a rapid cell wall disruption leaching not only the analytes but also unwanted impurities, whereas long time exposure at low power is more selective in the extracted compounds (Mandal et al. 2007). In some cases, the microwave power was found less decisive. For example, Gao et al. (2006) showed that the extraction yields of flavonoids from *S. medusa* were not significantly affected by the microwave power above 400 W. They used a 1,200 W-microwave power allowing a shorter extraction time and higher efficiency than a dynamic solvent extraction without microwave assistance.

The particle size of the matrix is also a crucial parameter. As for other procedures, reducing the particle size provides a higher surface of contact with the solvent improving the extraction efficiency. Furthermore, in MAE, having small particles increases the efficiency of penetration of the microwave into the plant tissues. Usually, fine grinding of the solids is ranged between a few hundred micrometers to a few millimeters. The smallest particles pose the most difficulties regarding their removal from the final extract.

Generally, because of the efficient heart heating and the short extraction time avoiding thermal degradation, MASE leads to very interesting results in terms of yields and sample activity. For example, Uquiche et al. improved the recovery (up to 45.3 %), quality (higher content in unsaturated fatty acids), and stability to oxidation of Chilean hazelnut oils after a microwave pretreatment prior to the pressing step. They assigned the higher stability to the oxidative deterioration to a possible inactivation of the oxidative enzymes (Uquiche et al. 2008). Prior to their quantitative analysis by liquid chromatography coupled to mass spectrometry, Sánchez-Ávila et al. achieved the extraction of some pentacyclic triterpenes of interest from olive leaves (erythrodiol, uvaol, oleanolic, ursolic and maslinic acids) by microwave assisted extraction in only 5 min (Sanchez-Avila et al. 2009), whereas it took more than 5 h by maceration and around 20 min by UAE. MASE also showed excellent results for the extraction of antioxidants such as polyphenols. It has been successfully applied to the extraction of hydroxycinnamic acids and flavanols from green tea leaves leading to better results than the corresponding conventional hot water extraction (Nkhili et al. 2009).

Microwaves can also be combined with other techniques. For example, coupling microwave with ultrasonic extraction is complementary and may present many advantages. Performing a simultaneous ultrasonic/microwave assisted extraction (UMAE), Lou et al. showed that the extract of burdock leaves exhibited higher antioxidant and antibacterial activities than the corresponding maceration extract. Moreover, UMAE only requires 30 s. In these extracts, they also evidenced a higher total content in phenolic compounds (chlorogenic, *o*-hydrobenzoic, caffeic, ferulic, *p*-coumaric acids and rutin) than in the extract achieved by maceration (Lou et al. 2011).

2.3 Concluding Remarks on Extraction

Sample extraction is a critical step of any qualitative or quantitative analysis by separative methods. Besides, the same applies to the activity of the final extract. It is hard to foresee, however, which would be the best extraction methods. In spite of hard heating and time consumption, Soxhlet remains a good basic approach. Among the improved extraction methods, none of them appears to be superior to another. They all have their own advantages and drawbacks and whatever the methods, the parameters governing the final quality of the extract have to be carefully optimized. Finally, the chosen methods really depend on the application, the molecules, and many other parameters. Indeed, for example, the highest extraction yields of protocatechuic, *p*-hydroxybenzoic, and gallic acids from *Polygonum aviculare* foliage were obtained by MASE in a closed system, whereas the highest yield for ferulic acid was obtained by UAE (Wang and Weller 2006). Many other similar examples are found in the literature. As already mentioned, several extraction approaches can be combined together. In addition, they can even be coupled online with purification techniques. For example, Yang et al. (2007) coupled continuous ultrasound assisted extraction, solid-phase extraction, and liquid chromatography for the determination of danshensu sodium salt and four tanshinones in the roots of *Salvia miltiorrhiza bunge*.

3 Clean-Up and Purification Techniques

The solid–liquid extraction methods are more or less selective regarding several parameters, such as the nature of the matrix (leaves, roots, bark, etc.,) and the conditions applied (temperature, pressure, etc.,). The same goes for the qualitative and/or quantitative analysis of the produced extracts by a separation with gas or liquid chromatography. The specificity of detection and the selectivity of these methods strongly depend on the systems chosen, including the nature of the stationary and mobile phases, and obviously the detector. The extracts generally still contain many unwanted molecules that may interfere with the analytes of interest during the analytical step and there is a real need for their removal.

3.1 Clean-Up Techniques

Very often, a clean-up procedure needs to be carried out in order to eliminate the interfering contaminants from the extracts before their analysis. Moreover, this step may also be a good opportunity for pre-concentration to improve their limit of detection. Many techniques can be implemented for this purpose. Among these techniques, the liquid–liquid extraction (LLE), solid-phase extraction (SPE), and microextraction (SPME) are widely used.

3.1.1 Liquid–Liquid Extraction

Liquid–liquid extraction is one of the traditional ways of sample purification used to reduce matrix interferences. The sample in solution is put in contact with a second nonmiscible solvent, vigorously mixed, and then decanted until the return to two independent liquid phases. According to their affinity for the two solvents, the compounds split between the two liquid phases. The clean-up procedure consists in the elimination of the contaminants which preferentially stand aside from the analytes of interest in one of the phases. In practice, one of the phases is usually an aqueous plant extract, while the other phase is composed of a less polar organic solvent. The parameters that strongly affect the final results are the nature of the organic solvent, the pH (which changes the charge state of weak acidic or basic analytes and thus their water solubility), the temperature, and the presence of complexing compounds in the extract. The compound solubility is governed by their chemical nature and the polarity of the solvent. The polarity however, can be very different from one compound to another, therefore it is difficult to develop an LLE procedure suitable to recover all plant actives. Nevertheless, this fact may even be used as a benefit. For example, phenolic compounds have been extracted with petroleum ether, ethyl acetate, or diethyl ether allowing the removal of lipids and unwanted polyphenols (Fuleki and Francis 1968). Many other examples of a clean-up procedure can be found in the literature (Garcia-Salas et al. 2010). For example, Muñoz-González et al. (2012) carried out a liquid–liquid extraction of phenolic acids using ethylacetate in acidic condition prior to their analysis by gas chromatography coupled to mass spectrometry (GC-MS). A similar approach was conducted for the extraction of resveratrol and other stilbenoids from beer (Jerkovic et al. 2008) before their analysis by liquid chromatography coupled to mass spectrometry (LC-MS)

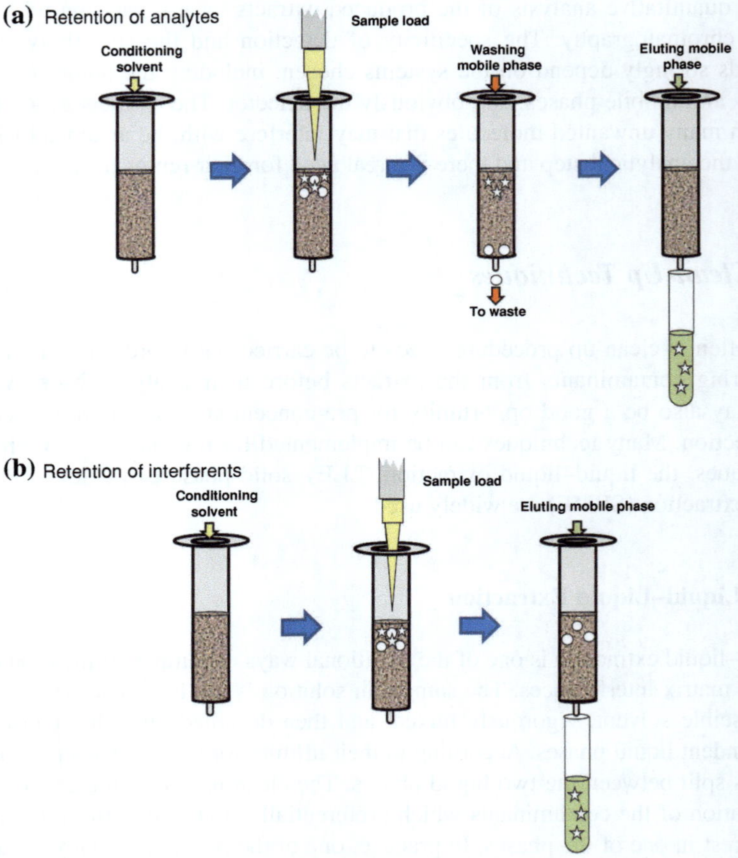

Fig. 6 Principle of the SPE: **a** for analytes more retained than the interfering compounds and **b** for analytes less retained than the interfering compounds

The main drawback of LLE is the requirement for expensive and hazardous organic solvents. High volatile solvents such as dichloromethane or hexane are particularly interesting to carry out evaporation to dryness, but are unfavorable regarding cost and the environment.

3.1.2 Solid-Phase Extraction and Solid-Phase Microextraction

Solid-phase extraction is strongly inspired from the open column system used for years in the laboratories of organic chemistry. The main difference is its size reduction (small cartridges) and the quality of the packing material. The two principles of extraction are explained in Fig. 6.

The SPE cartridge is first conditioned with solvents for good wetting of the adsorbent material. The solvent is adapted to allow the anchorage of the molecule

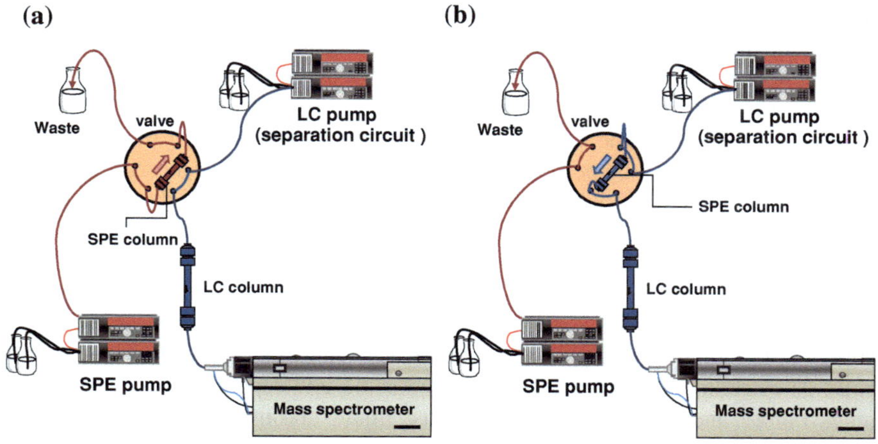

Fig. 7 On-line SPE: **a** The sample is loaded on the SPE column for preconcentration with the assistance of the SPE circuit (*brown*), while the analytical circuit is equilibrating (*blue*). **b** After a 1/6th turn of the valve, the SPE column is back flushed and the sample is injected onto the column for analysis

of interest in the stationary phase. The solvent flows through the adsorbent either by gravity, overpressure applied at the top of the cartridge, or depression at its bottom regarding the equipment. The sample is then loaded onto the top of the cartridge. The cartridge is sometimes dried after penetration of the mixture. The next step then depends on how the analyte of interest and the contaminants behave toward the adsorbent. If the analytes have a higher affinity for the stationary phase than the interfering molecules, the cartridge is washed with a solvent allowing only the elution of the impurities. In a second step, the molecules of interest are recovered with a small volume of a solvent possessing strong elution power. In the second scheme, the analytes have weaker affinity for the stationary phase than the interfering molecules. The compounds of interest are eluted with a small solvent volume without releasing the unwanted molecules.

Currently, SPE is undoubtedly the most popular sample clean-up procedure because of its simplicity and versatility. Indeed, as in liquid chromatography, there is a large set of commercially available stationary phases exhibiting many different adsorbing properties. The clean-up can be performed on silica ensuring the retention of polar solutes or reversed phases (e.g., octyl- or octadecylsilyl supports) for the retention of apolar compounds. Strong as well as weak cation or anion exchangers are also available for the retention of ionic and ionizable compounds. Some providers even propose mixed beds composed of a reversed phase and ion exchangers or porous graphitized carbons as adsorbent. There are also many different cartridge models available for the treatment of a few microliters to a few ten milliliter sample volumes including 96-deep well plates for high throughput extraction. SPE can be performed either manually, by automated systems (sample preparation robots) or even online with liquid chromatography (Fig. 7). Generally, online SPE is mostly performed for their determination of natural compounds in biological fluids such as plasma or urine. For example, an online SPE-LC-MS/MS method was developed for the determination of salidroside (one of the major phenolic glycosides of *Rhodiola sp.* displaying antioxidant activity) in rat plasma (Chang et al. 2007). The cost of SPE is strongly connected to the equipment. Nevertheless, SPE can be performed manually at low cost using a cartridge (prices ranging from 2 to 10 €) with a syringe adapter or with an SPE chamber (around 1,000 €) and a vacuum pump. The cartridges can also be recycled after use.

SPE has been used widely in the field of antioxidants. For example, tocopherols were extracted from vegetable oils on porous polymer SPE cartridges by Beldean-Galea et al. (2010) before their separation with capillary gas chromatography. Highly polar compounds may also be concerned. Thus, vitamin C, (−) epicatechin, and (+) catechin were extracted from juice using end-capped reversed-phase columns before analysis by high-performance liquid chromatography coupled to a diode array detector and a mass spectrometer (Shui and Leong 2004). Elimination of sugars from coffee pulp extracts was carried out on C18 cartridges in order to purify polyphenols before their analysis by normal-phase HPLC and matrix assisted laser desorption ionization time-of-flight mass spectrometry (MALDI-TOFMS) (Ramirez-Coronel et al. 2004). SPE has also been employed for resveratrol (Malovana et al. 2001). SPE cartridges can even be filled with molecularly imprinted polymers (MIP). These polymers exhibit the highest selectivity toward a targeted compound or a family of compounds compared to traditional adsorbents. For example, Claude et al. (2008) synthesized a very selective MIP for solid-phase extraction of the 18-β-glycyrrhetinic acid, a triterpene acid, from the liquorice roots. The described SPE protocol allows the elimination of 93 % of the matrix (including the fatty acid fraction) during the washing step and 98 % of the 18-β-glycyrrhetinic acid is recovered during the elution step. A second example is the fabrication of highly selective MIPs for the specific preconcentration of the flavonoids (quercetine and rutin) from white and red wine achieved by Theodoridis et al. (2006).

When the compounds of interest are particularly diluted in a complex matrix and rather volatile, the preconcentration by solid phase microextraction (SPME) becomes more suitable (Hinshaw 2003). Furthermore, SPME eliminates most of the drawbacks associated with extracting organic solvents. This technique is based on the partition of the analytes between a liquid or a gas phase and an adsorbent polymer laid on a silica fiber. This procedure is applicable to solid, liquid, and gaseous samples. SPME is involved in various domains such as the extraction of volatile organic compounds (Larroque et al. 2006) or pesticides (Sagratini et al. 2007; Cortes-Aguado et al. 2008; Schurek et al. 2008) in environmental chemistry. SPME is performed either online or offline prior to analysis by GC or LC. The principle of the offline SPME is depicted in Fig. 8.

The extraction efficiency mainly relates to several physicochemical parameters. Temperature, adsorbing time, and choice of the adsorbent (Table 2) need to be carefully optimized. Low molecular weight or volatile compounds usually require a 100 μm polydimethylsiloxane (PDMS)-coated fiber, whereas an 85 μm polyacrylate-coated fiber is more suitable for very polar analytes. The user should be aware that the fiber may be rapidly saturated with concentrated samples and

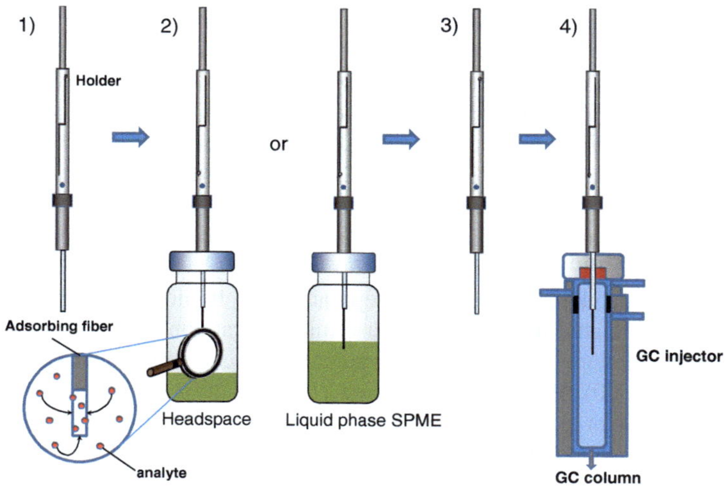

Fig. 8 Principle of manual offline solid phase microextraction (*SPME*): *1* The adsorbent fiber is retracted inside of the needle of the SMPE holder. *2* The needle of the holder is introduced through the rubber *top* of the vial and the fiber is pulled out until the molecules of interest are adsorbed. Two modes are possible: headspace (adsorption from the gas phase) and liquid phase (direct adsorption from the liquid phase). *3* The adsorbent fiber is retracted. *4* The needle of the holder is put through the rubber septum of the GC injector, the fiber is pulled out of the needle. Heat of the GC injector causes adsorbed molecules to vaporize. They are then transferred into the column where they are separated

Table 2 Fibers of common use in SPME and their application domains (PDMS: Polydimethylsiloxane; DVB: Divinylbenzene)

Nature of the adsorbent	Coating (μm)	Applications
PDMS	100	LC and GC (volatile and semi-volatile compounds)
Carbowax-DVB	65	GC (polar volatile and semi-volatile compound)
PDMS-DVB	60	LC and GC (volatile and semi-volatile alcohols or amines)
Carboxen-PDMS	75	GC (trace of volatile compounds)
DVB-Carboxen-PDMS	30–50	GC (volatile and semi-volatile compounds)
Polyacrylate	85	LC (polar compounds)

that compounds may compete with each other for the adsorption. This is a particularly sensitive issue for quantitative analysis as the analytes exhibiting the highest affinity for the sorbent and may affect the accurate quantification of the other compounds (Gorecki et al. 1999).

Stirring and pH are also associated parameters. SPME can be performed by soaking the fiber in the liquid extract. This technique, however, is usually applied to the extraction of volatile compounds from nonvolatile samples isolated in the headspace or gas portion of a sample vial.

Many applications are found in the literature concerning natural compounds. These applications mainly concern volatile natural compounds and their concentration by headspace solid-phase micro-extraction (HS-SPME) performed prior to a GC analysis. Numerous SPME methods have been developed in food chemistry for the analysis of aroma (Sides et al. 2000; Kataoka et al. 2000; Coleman and Dube 2005) and spices (Perez et al. 2007). HS-SPME is also suitable for the analysis of essential oils (Kovacevic and Kac 2001; Paolini et al. 2008). It has also been applied to volatile organic sulfur aroma compounds in black-and-white truffle (Pelusio et al. 1995), and in beer (Xiao et al. 2006). Some applications have been also developed in SMPE for a separation by liquid chromatography. For example, an automated in-tube solid phase microextraction coupled capillary LC-MS has been implemented for the determination of catechins and caffeine in tea by Wu et al. (2000). Polyphenols in wine (Flamini 2003) or phytohormones in plant extracts (Liu et al. 2007b) can also be analyzed by LC after an SPME step.

Several other related techniques may also be mentioned such as stir bar sorptive extraction (SBSE), liquid–liquid microextraction (LLME) or liquid-phase microextraction (LPME). Some applications were also developed in the field of natural compounds with many examples in the field of food flavor chemistry (Jelen et al. 2012; Xiao et al. 2006).

3.2 Purification

For structural elucidation by NMR or biological tests, a scale-up isolation of the natural active molecules needs to be carried out in order to obtain them in larger amounts and in higher purity. In the laboratory, several techniques may be involved (Sticher 2008). Flash chromatography remains a basic approach. Although flash chromatography is less resolute than most instrumental techniques, it exhibits the advantage of easy implementation at low cost.

3.2.1 Planar Chromatography

Preparative planar chromatography or thin-layer chromatography (TLC) can be involved in the purification process by increasing the thickness of the stationary phase layer (0.5 to 2 mm) compared to analytical plates. One of the main advantages of preparative TLC lies in the possibility of recovering the separated compounds by scratching the TLC plate. Then the compounds are dissolved and the silica stationary support is eliminated by filtration. This approach however remains limited to a few milligrams of material.

Overpressure layer chromatography (OPLC) is one of the forced flow TLC techniques. It requires specific equipment (Fig. 9) but the separation is faster and the resolution is highly improved (Nyiredy 2001). OPLC can be used either for analytical or semi-preparative purposes (Bryson and Papillard 2004; Tyihak et al. 2012).

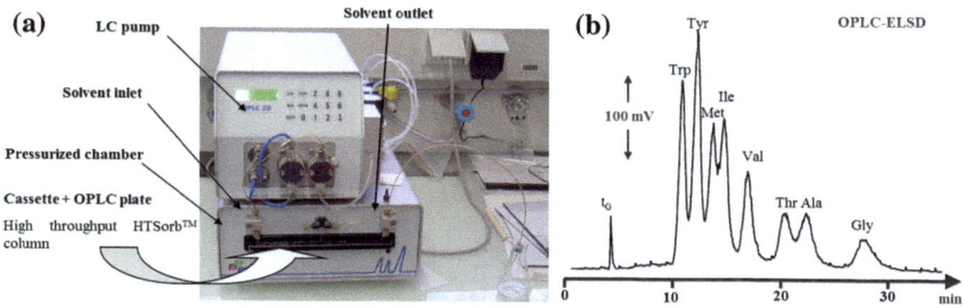

Fig. 9 Overpressure layer chromatography (*OPLC*). **a** Apparatus: The OPLC plate is spotted with the liquid sample. This step can be performed manually or with an automatic TLC Sampler. The OPLC plate is put in a cassette. The assembly is inserted in the pressurized chamber and a high counter pressure is applied onto the cassette (up to 50 bars). The LC pump flows the mobile phase through the plate to carry out the separation. Detection is ensured either by an offline procedure (UV lamp, coloration with reagents) or with online detection. **b** Fractionation of 8 amino acids by OPLC coupled to an ELSD: mobile phase: acetonitrile/water/acetic acid (75/25/0.1, v/v/v). On-line Injection (Rheodyne 7125 valve): 20 µL of a solution at 300 mg/L of each amino acid. OPLC plate: 5 × 20 cm backed aluminum with a silica gel (thickness particle size 11 µm. Online Detection: Evaporative Light Scattering Detector (*ELSD*) Sedex 75 (SEDERE) (pressure = 2.2 bars, Temperature = 45 °C, Photomutiplicator = 8) (Chaimbault, unpublished results)

The OPLC preparation unit allows a transposition from the analytical to a few ten mg semi-preparative scale. OPLC has been used for the purification of active compounds from a plant extracts as for example the main components (thymol, carvacrol, (−)-linalool, diethyl-phthalate, and α-terpineol) of *Thymus vulgaris L.* essential oil (Moricz et al. 2012a). Another similar application is isolation by OPLC of some antibacterial components from chamomile (*Matricaria recutica L.*) flower extract before their offline identification by GC–MS (Moricz et al. 2012b).

3.2.2 Preparative Liquid Chromatography

Generally, instrumental methods of purification are superior to the others in terms of resolution and separation speed. Among instrumental methods, preparative LC (prep-LC) is one of the most commonly used. This is probably due to the fact that purification can be achieved by transposing a chromatographic analytical profile into a semi-prep or prep LC method. The overload of the column however often leads to critical peak deformations. Consequently, the chromatographic resolution decreases with a risk of remixing of the separated compounds. Nevertheless, this technique allows the purification of a few milligrams to a few ten milligrams (semi-prep LC) to hundred milligrams of the desired compounds (prep-LC). The variety of commercially available supports (silica, reverse phase, ion exchange, etc.,) makes the prep-LC a versatile method for the isolation of apolar, polar, and charged compounds. The same LC equipment can sometimes be used for analytical or preparative applications depending on the column size. The amount of stationary phase used is much higher so that the column diameters are usually in

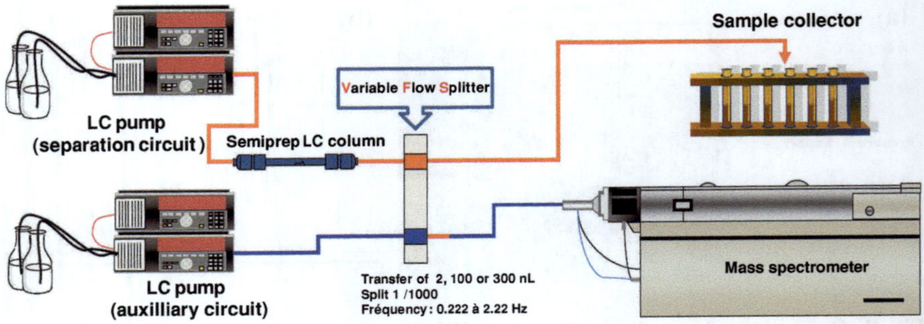

Fig. 10 Principle of prep or semiprep LC with online detection by mass spectrometry using a variable flow splitter (*VFS*). The MS detector can be replaced by an ELSD (Himbert et al. 2003). The frequency of the VFS is adjusted to such frequency that only a small amount of the extract (e.g., 1/1,000) is sent to MS for identification. The main part of the purified fraction is sent to the collector

range from 10 to 100 mm. The average particle size of semi-prep LC stationary phases is typically between 5 and 10 μm and may reach 100 μm for prep-LC. The main drawback of the prep-LC is the high solvent consumption. Indeed, the flow rate in semi-prep LC is usually in the range of 2 mL/min to a few tens mL/min all along the separation process. Semi-prep and prep LC can be coupled with the common detection methods (UV-Visible absorbance, fluorescence) but also with evaporative light scattering detection (ELSD) (Liu et al. 2007a) and mass spectrometry using a passive or active splitting of the mobile phase (Variable Flow Splitter, Fig. 10) (Nichols and Foster 2009; Himbert et al. 2003). MS is particularly interesting as it provides online control of the purity. Semi-prep and prep LC have been employed in many applications concerning plants. For example, Liu et al. (2007a) isolated ginsenosides from crude extracts of *Panax notoginseng* by semi-preparative HPLC coupled with an ELSD. The sample load was 300–400 mg and the final purity of each isolated compound was more than 97 %. Jiao et al. (2007) purified tricin from an antioxidant product derived from bamboo leaves. Around 3 g of tricin was prepared from 174 g of a crude fraction.

3.2.3 Countercurrent Chromatography

Countercurrent Chromatography (CCC) is an interesting alternative to prep-LC for the purification of natural molecules. CCC is now a generic term including several types of chromatography based on the same principle. The two main techniques are the centrifugal partition chromatography (CPC) developed by Nunogaki (Murayama et al. 1982) and the high-speed countercurrent chromatography (HSCCC) set up by Hito (Ito et al. 1982). The main difference between CPC and HSCCC is the instrumental design (which will not be discussed in this chapter). CCC is an instrumental LLE process allowing hundreds of automatic successive extractions, thanks to a special column possessing one or more cartridges interlinked by channels.

The fractionation consists of a continuous-flow partition of compounds between two immiscible liquid phases. The first liquid phase is the stationary phase, whereas the second behaves as the mobile phase. As for LLE, the solutes are separated according to their partition coefficients between the two solvent phases based on their hydrophilic–lipophilic balance. Evidence is increasing that the nature and volumes of both liquid phases strongly influence the final chromatographic resolution. In CPC, the liquid stationary phase is immobilized, thanks to the centrifugal force applied by the spinning motion of the coil column. The mobile phase is passed through the stationary phase by an LC pump. The main advantage of CCC over prep LC is that no solid stationary support is required to perform the separation. Thus, there is no irreversible sample adsorption and the recovery is close to 100 %. Moreover, the role of the two liquid phases can be switched during a run. The liquid stationary phase becomes the mobile phase and vice versa allowing the highly retained compound to be eluted from the column.

CCC is particularly recommended when samples contain solid particles or high-molecular constituents and when the solutions contain high concentrations of acids, bases, or salts in order to prevent the LC column from plugging and to avoid any interferences (Berthod et al. 2009).

To check the purity of the collected fractions, offline controls can be carried out by TLC (Sannomiya et al. 2004) or LC. Yet, online detections are possible using the same detectors as for LC. Thus, the detection can be achieved by UV (including a diode array detector), (Peng et al. 2005; Goncalves et al. 2011) evaporative light scattering detection (Cao et al. 2003; Ha et al. 2011) and obviously mass spectrometry with a passive T split (Gutzeit et al. 2007) or a variable flow splitter (Toribio et al. 2009).

CPC provides successful analytical chromatographic separations but because of its high charge capacity, CCC is above all an excellent purification technique. The preparative scale strongly depends on the rotor volume (Srivastava 2008). Purifications between a few milligrams to several hundred grams can be considered under certain conditions, the purification of a few kilograms can also be achieved (Sutherland and Fisher 2009). Insufficient resolution can be improved using multidimensional CCC (2DCCC) (Lu et al. 2007). Furthermore, the hyphenation of CCC with LC allows fast online checking of fraction purity (Liang et al. 2011; Michel et al. 2011b). This kind of coupling is not so obvious and requires specific experimental setups to overcome the differences in pressure and flow rate in the CCC and LC circuits.

To the present day, many applications on analysis and purification of natural molecules in plants have been published (Marston and Hostettmann 2006). Water soluble as well as insoluble metabolites may be addressed by this technique (Wanasundara et al. 2005). For example, Schræder et al. developed an HSCCC method for the isolation of phytosterols and they obtained sitostanol and β-sitosterol fractions with purity higher than 99 % (Schroder and Vetter 2011). Yu et al. (2011) proposed a CPC method for the purification of coniferyl ferulate from the extracts of *Angelica sinensis* oil (Fig. 11). The purity of the fraction was checked by LC UV and the identification was carried out by MS, 1H, and 13C NMR (Yu et al. 2011).

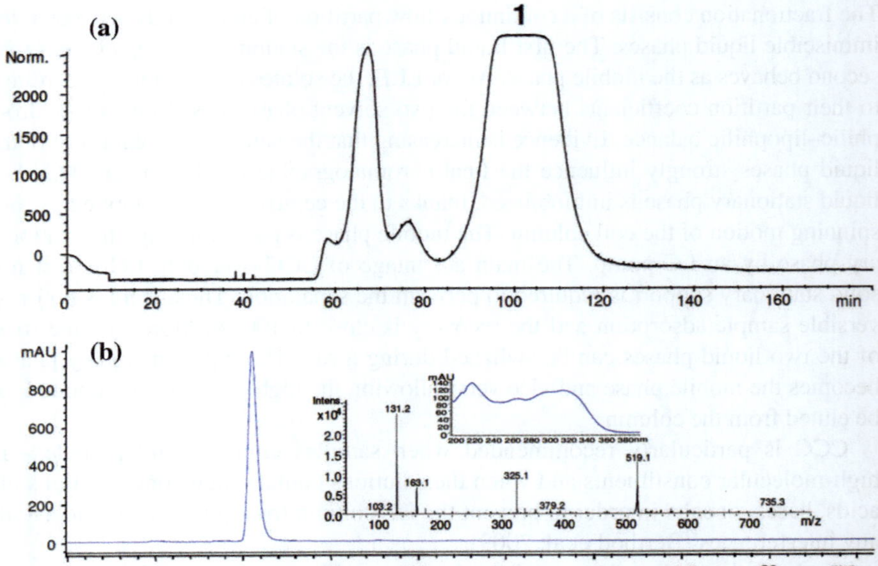

Fig. 11 High-performance CPC chromatogram of a coniferyl ferulate rich fraction (**a**), HPLC chromatogram of peak fraction 1 (**b**) with UV spectra and MS data. HPCPC solvent system: n-hexane-ethyl acetate-ethanol-water (5:5:5:5, v/v) in elution; station phase:upper organic phase; mobile phase:lower aqueous phase; descending mode; flow rate:2.0 ml/min; rotor speed 1,500 rpm, detector wavelength: 318 nm. The HPLC condition was adapted from an HPLC analysis performed on a Zorbax ODS C18 column (250 × 4.6 mm I.D., 5 mm) with a Zorbax ODS C18 guard column (12.5 × 4.6 mm I.D., 5 mm); mobile phase: 1 % aqueous acetic acid (**a**) and acetonitrile (**b**) using a gradient program of 50 % in 0 to 15 min, 50 to 100 % B in 15 to 18 min, and 100 % B in 18 to 23 min. The flow rate was 1 ml/min and the injection volume was 10 ml. The column was operated at 25 °C. ESI-MS conditions are as follows: drying gas N_2, 10 l/min; temperature, 350 °C; pressure of nebulizer, 40 psi; source voltage, 4.0 kV; Scan range, 50–800 m/z, positive mode (Yu et al. 2011)

Many applications are related to flavonoids and other polyphenols (Gutzeit et al. 2007; Lu et al. 2007; Michel et al. 2011b). Among the highest polar metabolites, glycosylated compounds such as glucosinolates (Toribio et al. 2007) or saponins (Shirota et al. 2008) have also been purified by CPC.

4 Separative Methods

After extraction, the sample or fraction content needs to be qualitatively and quantitatively characterized. At the very minimum, the purity must be checked but very often a plant extract remains a complex mixture that needs to be analyzed further in more detail. Among the analytical arsenal available, chromatographic techniques (including capillary electrophoresis) are essential because they allow the separation of the different constituents of complex samples before their identification (currently, structural identification is regularly carried out online by mass spectrometry and even by NMR) and quantification by a variable set of more or less specific detectors.

Chromatography techniques are known for long. Gas chromatography (GC) was first developed by Martin and Synge (Nobel prize laureates in 1952) in 1941 (Martin and Synge 1941) and high-performance liquid chromatography (HPLC) got its real kick-start in the 1970s. Since they were created, the relentless technological progress continuously improves chromatography in many aspects. Whatever the physicochemical properties of the molecules (i.e., apolar vs. polar, neutral vs. ionic, small molecules vs. polymers, etc.,), nowadays there is always a system suitable for their separation. In the same way, the set of detectors available for chromatography has considerably advanced and with time, these detectors have detectors have become more and more sensitive. Both aspects contribute to an increasing interest in chromatography.

This part gives an overview of the different techniques available for the analysis of natural products and describes current trends in this area. One of these trends is the systematic coupling of separative methods with mass spectrometry. The hyphenation of separative methods with MS is the subject of the final part of this chapter.

4.1 Choice of the Separative Technique

Separations of analytes are always achieved according to their differences in physicochemical properties. Two approaches are possible:

- Chromatography is a forced flow separation technique relying on the partition of the analytes between the stationary and the mobile phase. The compounds exhibiting high affinity for the stationary phase are more retained than the others. By choosing the nature of the stationary phase or changing the physicochemical properties of the mobile phase, chromatographists modulate the retention so that separation is achieved with the required resolution.
- In capillary electrophoresis (CE), separation is obtained by the differential migration of the analytes under the influence of an electric field. One could think that this technique is restricted to ions but several variants have been developed to achieve the separation of neutral compounds and enantiomers as well (Altria 1996b).

The choice of the right separative technique (including the detector) is key to a successful analysis. The technique is chosen in accordance with the physicochemical properties of the analytes. Their chemical modification by a specific reagent is possible to turn them compatible with the analytical method (derivatization). Such derivatization, however, generally induces more drawbacks than benefits.

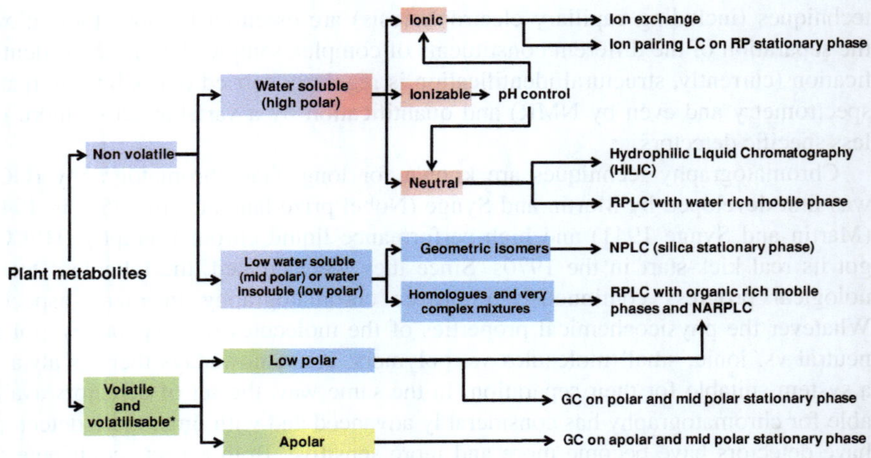

Fig. 12 Selection of the chromatographic system according to the physicochemical properties of the analytes. Principal combinations are indicated and the possible use of supercritical fluid chromatography (*SFC*) or TLC is not indicated. *RP* reversed phase. *NARP* nonaqueous reversed phase. *NP* Normal phase. *Asterisk* for thermo resistant compounds

In chromatography, small molecules (MW < 3,000 Da) are separated depending on their physicochemical properties. Many different types of chromatography are applicable to their separation (Fig. 12). Obviously, there are often alternative solutions to any given problem but this kind of decision tree remains a good starting point to the development of a future separation method.

4.2 Gas Chromatography

4.2.1 Generalities

Gas Chromatography (GC) is one of the most commonly used analytic techniques in the laboratories. The mobile phase (carrier gas) consists in an inert gas (e.g., helium or nitrogen) and the stationary phase consists in a column either packed with a solid support or coated with a high boiling polymer. The latter behave as an immobilized liquid. Nowadays, analyses are performed using coated capillary columns because they are much more efficient than packed columns.

Retention is a function of the partition of compounds between the mobile and stationary phases. The higher the temperature, the more the compound is in the gas phase. Consequently, the retention time decreases as the temperature increases because the analyte interacts less with the stationary phase. It is easily understandable that this technique is rather suitable for volatile thermostable compounds. As the carrier gas is inert, the selectivity toward compounds is principally based on the nature of the stationary phase although parameters such as the temperature, film thickness of stationary phase, and column length may also be considered for the development of separation. In chromatography, retention relies on the concept that "birds of a feather flock together." Separations on apolar supports (e.g., 100 % methylsiloxane polymer and methylsiloxane polymers with 5 % of phenyl groups) are governed by the temperature profile and the apolar character of the molecules. These supports are employed for all general-purpose applications. They are appreciable for biological samples because they are resistant to contamination and for analysis of low volatile compounds because they are stable at high temperature (up to 325 °C in isotherm and 350 °C in the temperature program). These columns, however, are less performing for distinguishing analytes according to their polar groups. Typically, the analysis of free fatty acids (or fatty acid methyl esters) varying in their unsaturation number and position requires polar stationary phases (e.g., polyethylene glycol) for separation. By contrast, polyethylene stationary phases do not resist high temperature (T_{max}: 250 °C). Such columns can be replaced by the newly developed ionic liquid stationary phases that withstand higher temperatures (Sun and Armstrong 2010). Depending on the application, the column polarity can be progressively increased with the percentage of phenyl or cyanopropyl groups in the methylsiloxane polymer. Chiral stationary phases are also commercially available and they can be implemented for separation of volatile enantiomers in plants (Konig and Hochmuth 2004).

Apart from the separation considerations, the instrumentation itself also plays a key role. Generally, plant analysis involves a preanalytical extraction procedure so that the final extract is dissolved in a liquid medium. The solvent must be volatile enough to ensure the total spontaneous transformation of the sample into a gas phase. Then, the gas phase is transferred into the column. The sample is introduced under pressure at the inlet of the column via the injector.

Several different GC injectors can be implemented in plant analysis (del Rio et al. 2007) but the split/splitless injector remains probably the most used because of its versatility. The split mode can be used for injection of highly concentrated samples, whereas the splitless mode is selected for trace analysis. The sample introduction into the injector is achieved with a syringe (1 to 2 μL) for liquid extract or as previously mentioned, by desorption of an SPME fiber (Kim et al. 2003; Nunes et al. 2006). The injector temperature must be high enough to ensure the instantaneous volatilization of the sample and its correct transfer into the column. Inappropriate temperatures lead to excessive analyte discrimination during injection and to peak broadening during analysis.

The molecules are detected at the outlet of the column. Even if mass spectrometry is currently probably the most widely application used detection method, the flame ionization detector (FID) remains of interest because it is universal for hydrocarbon compounds (Buchmabauer et al. 1997; Hasanloo et al. 2008; Rather et al. 2012). Other detectors may present some interest in plant analysis such as the thermo-ionic detector also called NPD because it specifically detects nitrogen- (i.e., alkaloids) and phosphorus-containing compounds. This detector is most often used for the specific quantification of nitrogen- and phosphorus-containing pesticides in plants (Tekel et al. 2001; Hirahara et al. 2005). The same applies to the

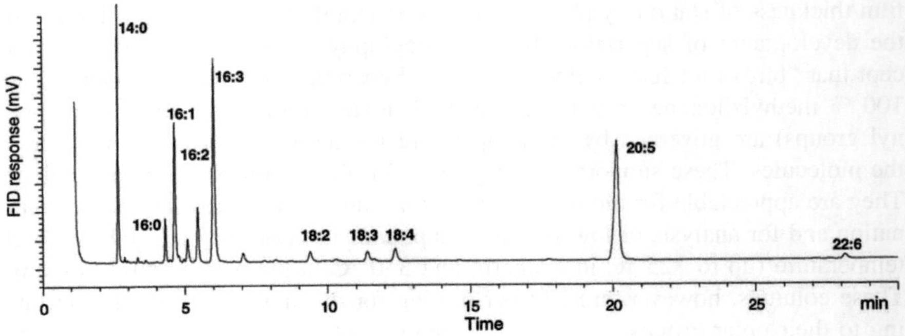

Fig. 13 Gas chromatogram of a mixture of FAMEs in a microalgae-oil extract. Column: DBwax (60 m × 0.32 mm ID), Carrier gas: helium; flow rate: 3 ml/min. Injection split: 1/20. Column temperature: 180 °C during 10 min, then increased to 230 °C by 20 °C/min. The final temperature was kept for 20 min. Detection: flame ionization detector (*FID*). Reprinted with permission from Viron et al. (2000)

electron capture detector and its use for the quantification of chlorinated pesticides (Ismail et al. 1993; Cao et al. 2008).

4.2.2 Applications to Plant Analysis

Gas chromatography has been involved in plant analysis for a long time. Obviously, flavors and fragrances are commonly analyzed by this technique (De Medici et al. 2006; Verhoeven et al. 2012). Apolar compounds such as phytosterols (Grunwald 1970; Du and Ahn 2002; Dhara et al. 2010), tocopherols, and tocotrienols (Du and Ahn 2002; Ruperez et al. 2001), and fatty acids (free or as fatty acid methyl esters) (Fig. 13) (Hasanloo et al. 2008; Seppanen-Laakso et al. 2002; Ng 2002; Viron et al. 2000) are also analyzed by GC-MS. Carotenoids are less volatile and thermodegradable. Therefore, gas chromatography is not really suitable for their analysis (Rodriguez-Amaya 2001). They are rather analyzed by liquid and supercritical fluid chromatography.

Some small polar metabolites can be analyzed by GC. Zou et al., who developed a simple and rapid procedure for determination of caffeine in beverages by GC/MS. The limit of detection of this method was 0.001 mg/L (Zou and Li 2006). GC coupled to mass spectrometry is also used for identification and quantification of alkaloids such as cocaine and its metabolites (Jenkins et al. 1996).

In the past, gas chromatography was also used for analysis of water-soluble metabolites such as sugars and amino acids. Although they are nonvolatile, it is possible to analyze these thermodegradable molecules after a derivatization step (e.g., methylation, silylation, acetylation, etc.,). Some more or less specific reagents are employed to substitute the labile hydrogen atoms in the molecules. Deprived of intermolecular hydrogen bonds, these metabolites become volatile and more resistant toward heat. This approach is still used in laboratories but tends to be abandoned with the progress in analytical chemistry. New methods are now available in liquid chromatography. For instance, with ELSD or electrospray MS, amino acids can be analyzed in their native form (i.e., underivatized) by LC (Chaimbault et al. 1999; Petritis et al. 1999).

4.2.3 Current Trends

GC technology is still undergoing further development. As previously mentioned, one of the current tendencies is its systematic coupling with MS detection. In the recent years, research has also been focused on fast GC and comprehensive multidimensional GC (GC × GC) allowing GC runs to go faster (high throughput analysis) and ensuring that the exploration of particularly complex samples is improved, respectively.

- *Fast-GC*: the world shortage of helium and the relentless miniaturization of the apparatus facilitated the development of fast GC. Both points are more or less linked together. On one hand, hydrogen replaces helium as carrier gas. Hydrogen is particularly interesting because of its high fluidity. Thus, H_2 flows more rapidly through the column. Additionally, the chromatographic efficiency is improved. On the other hand, the size reduction of columns in all dimensions (i.e., short and/or wide bore columns) also enhances the chromatographic speed and efficiency with decreased mobile phase consumption. In fast GC, the internal diameter is decreased from 250 μm (commonly used with He) to 100 μm. Consequently, the flow rate of the carrier gas is decreased by a factor of 2.5 to 3.5 but maintaining it in such condition requires higher inlet pressure. This inconvenience can be easily overcome with H_2. Finally, the reduction of column inner diameter and the use of the hydrogen as carrier gas reduces analysis time at constant chromatographic resolution (Korytar et al. 2002). An increase of speed by factor of 3–5 can easily be achieved with fast GC. Regarding the sensitivity, the low load ability of the narrow-bore column involves a reduction of the injected volume to avoid peak broadening. Nevertheless, in fast GC, the analytes elute as narrower and, hence, higher peaks resulting in improvement in the limits of detection. Fast GC has advantageously been applied to plant analysis. Mondello et al. (2003) developed a fast- GC analytical method for fats and oils. They compared the fast GC chromatographic profile transposed from a GC method performed on a polar column (Supelcowax-10) of classic dimension (30 m × 0.25 mm ID × 0.25 μm film thickness) using helium as carrier gas. They obtained a similar chromatogram using the same column in a narrow-bore version (10 m × 0.1 mm ID × 0.1 μm film thickness) and H_2 as carrier gas except that the analysis time is reduced by a factor of 5 (15 min instead of 75 min). In another study carried out by the same team, a fast GC analysis of citrus essential oils was carried out with a speed gain of almost 14-fold in comparison with traditional GC procedures (Mondello et al. 2004). Around 60 volatile metabolites including terpenes were analyzed in only 3.3 min.

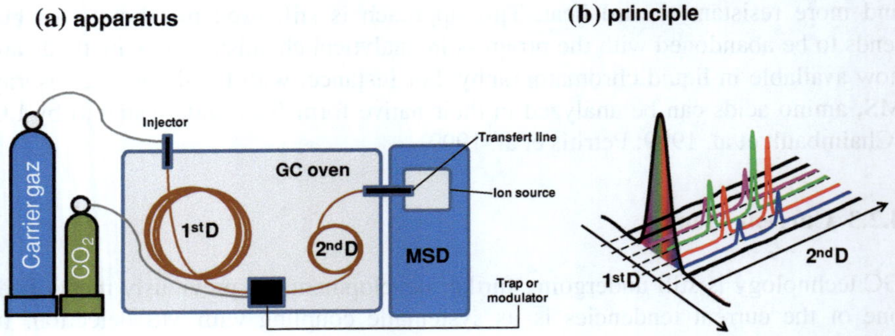

Fig. 14 Multidimensional GC (GC × GC with MS detection). **a** Scheme of a GC × GC apparatus. **b** Principle of separation using two dimensions: The sample is injected onto the first column (first D) and the compounds eluted are trapped at its outlet thanks to a time-programmed modulator (frozen by CO_2 expansion). When the modulator is warmed, the trapped molecules are injected onto the second dimension where co-eluted analytes are resolved

- *GC × GC*: One dimension is very often insufficient to separate all the compounds of interest in particularly complex samples. Thus, chromatographists sought to improve the resolution. One possibility is to increase the column length, yet this approach is rapidly limited. The other approach consists in coupling chromatographic columns so that the analysis is performed in two dimensions. This approach requires specific equipment (Fig. 14) involving a key body, which is mostly a cryogenic moving trap (Marriott and Kinghorn 1997) or a dual-jet modulator (Beens et al. 2001) depending on the apparatus. Basically, this device captures the eluted molecules at the outlet of the first column (first dimension) during a programmed time. Meanwhile, the analysis is performed on the second column (second dimension). When the analysis in the second dimension is finished, the trap releases the trapped molecules onto the second dimension. Additionally, the cryo-modulation provides an increase in the S/N ratios by a refocusing effect within each analysis carried out on the second dimension. The functioning of the system supposes that (1) the stationary phases of both dimensions are different (first dimension apolar and second dimension polar or vice versa) and (2) that the column size of the second dimension is much smaller than the first dimension so that the second dimension analysis is performed in a few seconds (e.g., 5–10 s). The larger part of the separation is ensured by the first dimension and the separation of co-eluted compounds is achieved by the second dimension. The operation of such a system requires a careful optimization (column lengths, trapping time, temperature, etc.,) and severe parameter controls. The resulting chromatogram appears as a 2D contour plot (Fig. 15) with a color scale corresponding to the respective peak heights of analytes. GC × GC is very often coupled to mass spectrometry detection but analysis can be carried out with flame ionization and electron capture detection as well (Augusto et al. 2010; Corporation 2004).

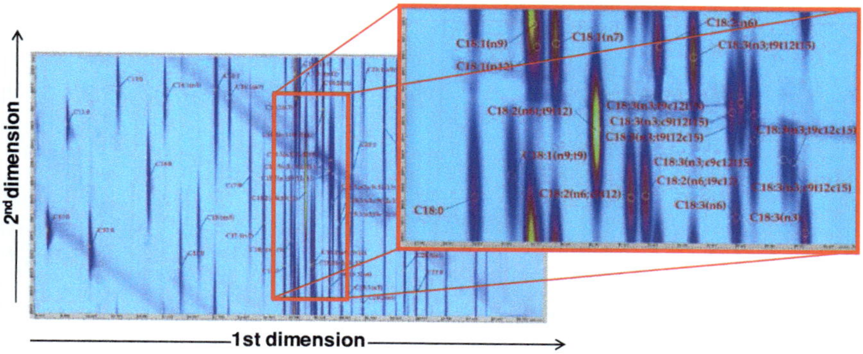

Fig. 15 GC × GC-FID 2D chromatogram of the standard mixture of 45 FAMEs. The first dimension is polar: BPX-70 column (30 m × 0.25 mm × 0.25 μm). The second dimension is apolar: ZB5-MS column (10, 5 and 2 m × 0.25 mm × 0.25 μm). Temperature program: 120–230 °C at 20 °C/min 230–260 °C (10 min) at 20 °C/min. Carrier gas flow rate in the first dimension = 0.5 and 12 mL/min in the second dimension. Modulation time = 1.40 s. Adapted from de Geus et al. (2001)

Historically, multidimensional GC has been developed for the petrochemical industry (Phillips and Beens 1999; Bertsch 2000; Bertoncini et al. 2005) and environmental chemistry, including the analysis of pesticide residues (Banerjee et al. 2008). In the field of natural compounds extracted from plants, GC × GC above all has been used for the characterization of volatiles in essential oils in many species such as *Rosa damascena* Mill. (Ozel et al. 2006), *Origanum micranthum* (Gogus et al. 2005), *Eucalyptus camaldulensis* (Ozel et al. 2008), fresh and aged tea plant (Tranchida et al. 2010) or roasted and unroasted fruit from *Pistacia terebinthus* (Gogus et al. 2011). The other main family of natural compounds concerns fatty acids (de Geus et al. 2001; Manzano et al. 2011) and fatty alcohols as methyl derivatives (Manzano et al. 2011). At last, polar metabolites (e.g., shikimic acid, citric acid, many sugars, etc.,) have been also analyzed as trimethylsilylated derivatives by GC × GC in a metabolomic study concerning *Populous tremula* (Jover et al. 2005).

4.3 Chromatographic Methods Using a Liquid Mobile Phase

4.3.1 Generalities

Whenever the metabolites of interest are nonvolatile or thermolabile, the separation methods involving a liquid mobile phase are undoubtedly the most appropriate. These methods include thin-layer chromatography (TLC), liquid chromatography (LC), and supercritical fluid chromatography (SFC), which in many respects shows similarities with LC.

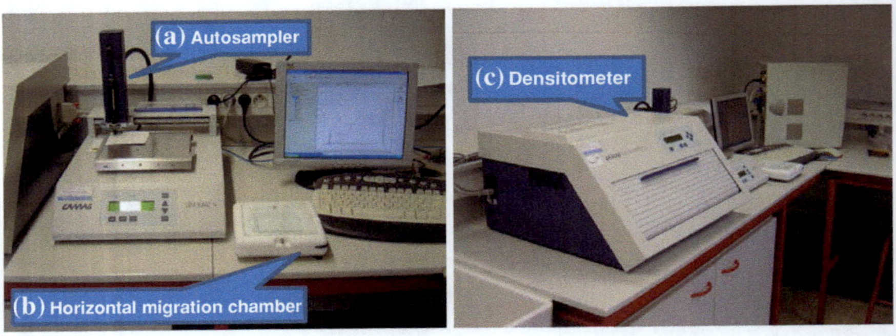

Fig. 16 HPTLC basic instruments from CAMAG. **a** Autosampler: The sample is deposited onto the HPTLC plate using a syringe. The sample is band wide sprayed (suitable for automatic detection). The spraying gas is generally nitrogen. The number of bands and their width as well as the sprayed volume and deposition speed are programmed depending on the number of deposited samples, nature of solvent (volatility), and the concentration of analytes, respectively. **b** Horizontal migration chamber: separations can be carried out from two sides in unsaturated, saturated, and sandwich configuration but single-sided developments are also possible. **c** UV–Visible scanner for densitometric evaluation of TLC/HPTLC chromatograms. The scanner can work either in absorbance, reflectance, or fluorescence mode

4.3.2 Thin-Layer Chromatography and High-Performance Thin-Layer Chromatography

In thin-layer chromatography, also called planar chromatography, separations are carried out on a flat stationary phase supported by a glass plate or aluminum foil. The most used stationary phase is likely silica gel but applications can be also developed on reversed phase, (Poukensrenwart et al. 1992) ion exchange (Luo et al. 1998; Karuna and Sashidhar 1999), and even chiral materials for enantiomer separation (Srivastava et al. 2008). High-performance TLC is the instrumental version of TLC. Their main differences lie in the stationary phase particle sizes, plate dimensions, and migration mode. The TLC plate contains 5–20 μm particles constituting a 250 μm film thickness and the migration is vertical. HPTLC is performed on 100 μm thick films constituted of 5 μm particles and separation takes place in a horizontal development chamber (Fig. 14). HPTLC exhibits a two times higher efficiency than TLC over a short distance of migration (up to about 10 cm) (Nyiredy 2001). Except for OPLC (described in Sect. 3.2.1), the mobile phase migrates through the stationary phase by capillary action. Sample application determines the quality of the analysis. Although HPLC provides higher resolution power (i.e., higher numbers of theoretical plates) than TLC or HPTLC, many aspects of planar chromatography remain of interest in the area of plant analysis. Furthermore, when the resolution power is not the limiting factor, HPTLC and HPLC lead to very comparable results (Farina et al. 1995; Sharma et al. 2007; Urakova et al. 2008). Due to its offline principle, the (HP)TLC system offers versatility and a low cost analysis. The sample deposition is either manual or automated (Fig. 16). Here, the use of an automatic sampler leads to more reliable results in quantitative analysis. The sample preparation is simple and there is no risk of clogging by small solid particles (no previous filtration is required). Samples and standards are processed together facilitating the identification of compounds. Furthermore, several analyses may be run simultaneously lowering analysis time and the consumption of the mobile phase. Last but not least, HPTLC also allows multidimensional separations simply by rotating the plate by one quarter turn and performing further separation with another mobile phase (Waksmundzka-Hajnos et al. 2008).

TLC plates constitute open systems. Thus, the basic detection is cheaper than for LC: a visual detection is often sufficient (direct or indirect for non-UV-absorbing compounds after a post-run derivatization with appropriate reagents). Sophisticated detectors, however, are also available commercially, such as UV-visible and fluorescence densitometers (Fig. 16). These scanners allow accurate quantification of metabolites with the help of calibration curves. At last, it is now even possible to perform offline MS detection on TLC plates either by MALDI-TOFMS (Wilson 1999; Crecelius et al. 2003) or electrospray-MS (Morlock and Jautz 2008).

TLC and HPTLC are perfect approaches for roughing out and formulating basic knowledge on a given complex sample. Many applications can be found concerning natural compounds in pharmacy, cosmetics, food chemistry, etc. Several applications concern antioxidants such as anthocyanidins (Farina et al. 1995), phenolic compounds (Sharma et al. 2007), and flavonoids (Cvek et al. 2007), caffeoyl derivatives such as chlorogenic acid (Urakova et al. 2008) or curcumin (Ashraf et al. 2012). HPTLC analysis allows the separation of high polar compounds such as glycosylated metabolites (e.g., phenylpropanoid glycosides from *Ballota nigra L.*) (Janicsak et al. 2007) or apolar natural molecules such as pentacyclic triterpens (e.g., taraxerol in *Clitoria ternatea*) (Kumar et al. 2008) and sterols (Shanker et al. 2008).

4.3.3 Liquid Chromatography

High-Performance Liquid Chromatography (HPLC or more simply LC) is undoubtedly the most widespread separation technique. This popularity is likely due to its unique versatility in terms of application and its high resolutive power. Indeed, almost every class of molecules can be analyzed by LC. Furthermore, the recent progress in LC results in a relentless improvement in efficiency (i.e., increase in the theoretical plate number) with a simultaneous increase in analytical speed. Thus, for example, Ultra Performance Liquid Chromatography (UPLC) is a high throughput LC technique exhibiting separative power comparable to capillary GC. As for GC, comprehensive multidimensional LC (LC $\times$ LC) allows the analysis of particularly complex samples increasing the peak capacity of a single chromatographic run.

(a) General aspects

Whatever the LC technology (classic LC or fast LC), the stationary phase is composed of more or less porous spherical particles packed in a column. The retention mechanism depends on the nature of the support. In the field of small molecules, two primary physicochemical characteristics are involved in the LC separations: polarity and electrical charge. Based on the concept that "birds of a feather flock together," water-soluble molecules are highly retained by polar supports [normal phase (NPLC) and hydrophilic interaction liquid chromatography (HILIC)]. Inversely, low-polar compounds are highly retained on apolar stationary phases [reversed-phase liquid chromatography (RPLC)]. Ionic molecules can also be separated using ion-exchange chromatography but ion-pairing chromatography on RP supports or mixed bed supports (e.g., RP mixed with ion exchange supports) may also be implemented to perform separation of organic ions. The separation is based on the differential interaction between the analytes and the stationary phase. Thus, the choice of the stationary phase is key to success in the separation but contrary to GC, the composition of the mobile phase (eluent) is also of importance. In RPLC and HILIC for instance, the retention and selectivity is modulated by the hydrophilic–lipophilic balance of the mobile phase and additional parameters such as pH, ionic strength, and even the temperature often have to be adjusted to achieve successful separation.

All the separation modes above evoked may be used for natural compound analysis. Low-polar compounds can be analyzed either by NPLC or RPLC. NPLC is particularly interesting for the separation of geometrical isomers and several reports have described for a long time the use of silica as adsorbent for the analysis of tocopherol and tocotrienol isomers (Weber 1984; Balz et al. 1993; Tangney et al. 1979; Chase et al. 1994). Their retention is ensured by the use of a mobile phase containing a high proportion of apolar solvent (e.g., 98.8 % hexane) and a small proportion of polar modifier (e.g., 1.2 % isopropanol) (Weber 1984). Separation of apolar compounds can also be achieved on porous graphitized carbon (PGC). This support additionally enables the separation of geometric isomers, but contrary to silica PGC, is above all an apolar adsorbent. The retention on PGC, however, is very sensitive to the position of electron-rich groups. Therefore, PGC is often used for geometric isomers. For instance, Rhourri-Frih et al. (2012) investigated the potential of PGC for separation of triterpenoids in natural resins. Here, PGC exhibited a unique selectivity toward ursolic, betulinic, and oleanolic acids. RPLC is the most widespread separation mode because of its versatility and robustness toward complex mixtures. Indeed, the full range of metabolite polarity can be analyzed with RP columns. For example, Abidi (1999) proposed the resolution of *cis/trans* isomers of tocotrienols on RP support. The elution of apolar metabolites requires a high percentage of organic modifiers up to 100 % (nonaqueous reversed-phase liquid chromatography) in the case of carotenes (Lesellier et al. 1989) or fat soluble vitamins (Gentili et al. 2012). The retention of metabolites of intermediary polarity is achieved by increasing the aqueous fraction in the mobile phase. Many applications concerning antioxidants and other classes of plant compounds may be cited. In this field, Wichitnithad et al. (2009) developed a simple and sensitive isocratic RPLC-UV method for the simultaneous quantification of curcuminoids (curcumin, desmethoxycurcumin and *bis*-desmethoxycurcumin) in commercial turmeric extracts with sub-µg/mL limits of detection. In order to investigate whether flavonoids and stilbenes of red wine benefit health, Stecher et al. (2001) developed a RPLC method for the determination of rutin, quercetin, myricetin kaempferol, *cis*- and *trans*-resveratrol. Rutin is a quercetine glycoside (α-L-rhamnopyranosyl-(1–6)-β-D-glucopyranose). This polar metabolite is less favorably retained in RPLC than the other polyphenols analyzed. As a consequence, the authors used an aqueous methanol gradient elution starting from a water-rich mixture allowing a single-run analysis of compounds exhibiting different polarities. Hydrophilic compounds can be retained in RPLC with a water rich (up to 100 %) mobile phase, however, polar supports are very often more appropriate for their separation. A few ten percents of water are required in the mobile phase for their elution turning the NPLC into HILIC. Although the NPLC and HILIC stationary phase may be the same (e.g., silica, diol or amino propyl supports), their separation mechanisms are somewhat different. In HILIC, water (coming from the mobile phase) forms a dynamic layer adsorbed onto the stationary phase. The polar analytes are separated by their differential partitions between this adsorbed layer and the mobile phase (Wang and He 2011). Depending on the nature of the stationary phases, an anion (e.g., amino bonded support) or cation (e.g., silica) exchange mechanism can complete the resolution. Many polar plant metabolites such as sugars and oligosaccharides (Remoroza et al. 2012; Leijdekkers et al. 2011), amino acids, and small peptides (Tolstikov and Fiehn 2002), water-soluble vitamins (thiamine, riboflavin, nicotinic acid, nicotinamide, pyridoxine, folic acid, cyanocobalamin, and ascorbic acid) (Karatapanis et al. 2009) have been separated in this way. Ionic compounds can be separated using ion-exchange or ion-pairing chromatography on RPLC support (IP-RPLC). For example, Antonopoulos et al. proposed a separation of enzymatically digested κ-carrageenans using an anion exchange mechanism (Antonopoulos et al. 2004a) and Arnault et al. (2003) developed an IP-RPLC method for the simultaneous analysis of alliin, deoxyalliin, allicin, and dipeptide precursors in garlic products.

The success of LC may also be attributed to a very extensive range of available detectors. The refractive index detector (RID) is universal but limited to simple concentrated samples analyzed under isocratic conditions. In fact, the basic detection of LC is UV–visible (UV–Vis) absorbance. UV–Vis detection [including diode array detectors (DAD)] is widely used in the field of natural compounds as many of them possess a chromophore. This is the case for antioxidants such a resveratrol (Lamuelaraventos et al. 1995), flavonoids (Colombo et al. 2006), curcumin (Wichitnithad et al. 2009), tocopherols (Seker et al. 2012), etc. Whenever a molecule has a flat conjugated polycyclic chromophore, fluorescence may be used for their sensitive and specific detection. For example, fluorescence detection has been used for a long time for the analysis of tocopherols and tocotrienol (Abidi 1999). Alternative detections can be implemented for compounds lacking a chromophore.

Electrochemical detection (ECD) is an extremely selective and sensitive detection technique for metabolites possessing reducable or oxidizable functional groups (e.g., *o*- or *p*-quinones, ketones, aldehydes, etc.,). Their detection is ensured by the electrical current resulting from oxidation or reduction reactions in the detector cell. Such a detection method allows the quantification of sugars, (Corradini et al. 2012) and amino acids (Marioli and Sereno 1996; Agrafiotou et al. 2009). The use of ECD is obviously not reserved to nonabsorbing compounds and it may possibly also be applied to the detection of catechins (Maoela et al. 2009) resveratrol (Kolouchova-Hanzlikova et al. 2004) and many other antioxidants in medicinal plants. Furthermore, as for many specific detectors, ECD provides a huge sensitivity. In their comparative study, Subagio et al. (2001) found detection limits for ECD about 75 times lower than for UV in the case of catechins. Detection limits of 3 and 15 µg/L were obtained by ECD for *trans*-resveratrol and *cis*-resveratrol, respectively (Kolouchova-Hanzlikova et al. 2004). Another alternative is the evaporative light scattering detector (ELSD) or the corona charged aerosol detector (Corona CAD). Both detection principles are based on the nebulization of the mobile phase. This starting point requires the use of mobile phases composed of highly volatile solvents (including acids or bases added for pH and ionic strength adjustment). The principle of detection is different in both cases. With ELSD, the aerosol is composed of neutral micro droplets. The solvent is dried in a drift tube and the solid particles of analytes scatter light according to their size (mass sensitive). The scattered light is detected by a photomultiplier at fixed angle from incident light (Dreux et al. 1996; Megoulas and Koupparis 2005). Consequently, ELSD may be considered as a universal detector. It is more sensitive than the RID and almost not affected by gradient elution (as long as the volatility of the mobile phase is respected). ELSD has been used widely for the detection of natural compounds lacking a chromophore, such as lipids (e.g., triglycerides, fatty acids, steryl glucosides, etc.,) (Stolyhwo et al. 1985; Marcato and Cecchin 1996; Moreau et al. 2008), carbohydrates (Herbreteau et al. 1992; Antonopoulos et al. 2004b; Nogueira et al. 2005) and amino acids (Chaimbault et al. 2000; Petritis et al. 1999). Nevertheless, ELSD was also used for UV-absorbing compounds, such as phenolic terpens (Bicchi et al. 2000) or resveratrol-like antioxidants (e.g., astringin, piceid, and isorhapontin) (Co et al. 2012). With Corona CAD, the aerosol is composed of charged micro droplets. In some respects, the Corona CAD may be considered as an atmospheric pressure chemical ionization-mass spectrometer (APCI-MS) without any mass analyzer. The Corona CAD detector detects any nonvolatile or semi-volatile analytes under the condition that these compounds are ionizable. The Corona CAD is generally found to be three to 6 times more sensitive than the ELSD (Vervoort et al. 2008), even 12 times for complex for lipid samples (Hazotte et al. 2007), but the ELSD is less sensitive toward an increase in the salt concentration in the mobile phase (Vervoort et al. 2008). The area of application for Corona CAD is the same as for ELSD. Many applications have been carried out for analysis of lipids (Moreau 2006; Lisa et al. 2007) and oligosaccharides (Inagaki et al. 2007; Asa 2006).

(b) Current trends

As for GC, the key words in LC development are: faster, more sensitive, and still more resolute. Hence the consequences are the same as in GC. More and more applications have been developed with mass spectrometry; considerable efforts have been put into the development of new stationary phases and instruments for fast LC and multidimensional LC (LC × LC) are now operational for applications to very complex matrices.

- *Fast LC*: High throughput LC started in 1999 with the development of monolithic columns (Guiochon 2007). Due to their wide pores, monolithic columns can work at high flow rates (up to 10 mL/min) without generating high backpressures. So, they were introduced as an alternative to conventional packed columns (4.6 mm I.D. columns packed with spherical particles within a diameter range of 5 to 10 μm). In addition to the analytical speed, the length of column is not pressure limited meaning that several columns may be coupled together to improve peak resolution (Novakova et al. 2004). Monolithic columns are constituted of a single porous rod of silica (functionalized or not). The largest pores (2 μm-sized macropores) allow the rapid circulation of the mobile phase at low pressure while the smallest pores (130 Å-sized mesopores) provide a sufficient exchange area to ensure analyte retention. Thus, monolithic columns exhibit similar chromatographic resolution as conventional columns but in a shorter analysis time. A substantial interest of monolithic columns is the use of conventional LC equipment. Thus many applications have been transposed easily in the past on this support. Flavonoids and isoflavones from tomato were analyzed in less than 10 min on a reversed-phase monolithic column (Biesaga et al. 2009) and soy extracts (Rostagno et al. 2007). Monolithic columns, however, are slowly being abandoned in favor of (1) ultra performance liquid chromatography (UPLC) and (2) core-shell particles.
  The size reduction of particles (sub 2 μm diameters) highly improves column efficiency (i.e., by an order of a magnitude order with hundred thousand theoretical plates per meter) and thus leads to better resolution. Consequently, the first means to speed-up separation is the shortening of the column length (50 mm instead of 150 mm for conventional support). Furthermore, the use of smaller particle size leads to a possible increase in the flow rate almost without any peak broadening. The downsize is a dramatical increase in backpressure. This problem can be overcome in two ways.
  The first solution consists in developing appropriate analytical equipment (pumps, injectors, columns) with strong resistance to overpressure reaching 1,000 bars (or even higher) (Gritti and Guiochon 2012) and high-speed detectors. This approach called ultra performance liquid chromatography (UPLC) has been followed by many instrument manufacturers and HILIC, ion-exchange, reversed and normal stationary phases are now available for applications under ultrahigh pressure conditions (Nunez et al. 2012). Thus, all previously developed methods on conventional material may be transposed to UPLC. An example of phenolic metabolite analysis via UPLC is presented in Fig. 17. Nováková et al. (2010)

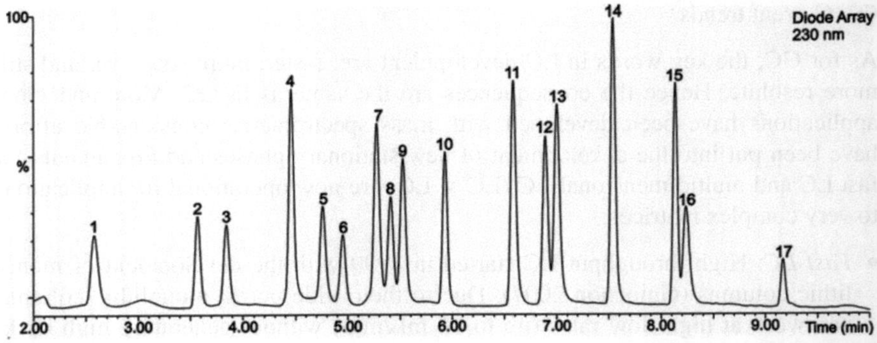

Fig. 17 UPLC–UV chromatogram of a mixture of the following 17 phenolic acids detected at 230 nm: *1* gallic acid, *2* 3,5-dihydroxybenzoic acid, *3* protocatechuic acid, *4* chlorogenic acid, *5* gentisic acid, *6* 4-hydroxybenzoic acid, *7* caffeic acid, *8* vanillic acid, *9* syringic acid, *10* 3-hydroxybenzoic acid, *11* 4-coumaric acid, *12* sinapic acid, *13* ferulic acid, *14* 3-coumaric acid, *15* 2-coumaric acid, *16* salicylic acid and *17* trans-cinnamic acid. Reprinted with permission (Gruz et al. 2008)

simultaneously determined caffeic and chlorogenic acids, umbelliferone (coumarin), and nine other flavonoid derivatives at a few nmol/L level in Chamomile (*Matricaria recutita L.*) flowers and tea extracts by UPLC–MS/MS in around 20 min. Lerma-García et al. (2010) quantified vegetable sterols (ergosterol, cholesterol, campesterol, stigmasterol, β-sitosterol, and lanosterol) in oil with limits of detection below 0.07 μg/mL by LC-MS in less than 5 min.

The second solution to overcome high pressure is the use of fused-core particle packed columns. Their most significant advantage over sub 2 μm packed material is their relatively larger permeability. Thus, this kind of support allows at least twice faster and more efficient separation of analytes than conventional columns at half the backpressure compared to the UPLC column. The use of fused-core particles is relatively new in chromatographic separation of natural compounds. In this area, 20 phenolic compounds (gallic and caffeic acid derivatives, flavonoids including their glycosylated metabolites) were analyzed in tea, mate, and coffee in less than 5 min by UPLC-UV-fluorescence (Rostagno et al. 2011). Using an RP fused-core support, Olszewska (2012) quantified ten flavonoid aglycones with high sensitivity (LODs from 0.115 to 0.525 ng injected) in *Ginkgo biloba*, *Betula pendula* and a variety of *Sorbus* species.

- LC × LC: Comprehensive multidimensional LC has been developed exactly for the same reasons as GC. Even if LC has become more and more efficient in the last couple of years, one dimension remains often insufficient to explore the composition of complex samples. Consequently, there is a need to increase the peak capacity of the analysis. This is especially the case with the appearance of 'omic' disciplines such as proteomics or metabolomics. The general principle of 2D-LC is the same as in GC. The basic equipment, however, is obviously different. The use of a cryo-modulator is not applicable and this device is

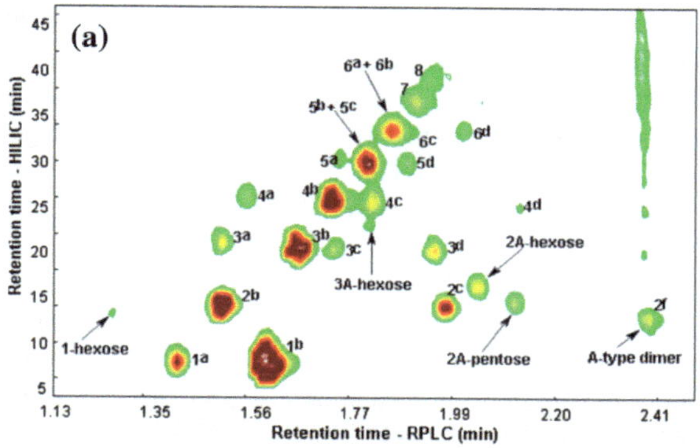

Fig. 18 Fluorescence contour plot obtained for the HILIC × RP-LC analysis of cocoa procyanidins. Numbers correspond to the degree of polymerization of procyanidine isomers. 'a' indicates A-type procyanidin. Reprinted with permission (Kalili and de Villiers 2009)

replaced by a more or less complex set of valves fitted with loops (Groskreutz et al. 2012). The monolithic columns (Lubda et al. 2001) and more recently the fused-core columns (Dugo et al. 2008) are advantageously used for rapid analysis in the second dimension. One of the main problems is related to the choice of the orthogonal separation modes because of possible solvent incompatibility (the first dimension mobile phase becomes the second dimension injection solvent). Possible solutions have been provided in the literature (Stoll et al. 2007; Cesla et al. 2009). For example, LC × LC was applied to the determination of polyphenols in wine (Donato et al. 2011), cocoa (Kalili and de Villiers 2009) (Fig. 18) and flavonols of green tea using a combination of the HILIC mode in the first dimension and the RPLC mode in the second dimension (de Villiers et al. 2010). In the case of the 2D-LC analysis of cocoa procyanidins, the peak capacity is around 3,500 (Kalili and de Villiers 2009).

4.3.4 Supercritical Fluid Chromatography

Whatever the domain of application, supercritical fluid chromatography (SFC) is less commonly used than LC. Yet, this technique exhibits some advantages in separative sciences. SFC is a hybrid technique using a supercritical fluid as mobile phase. Physicochemical properties of such a fluid have already been evoked in Sect. 2.2.1 for SFE. The principal supercritical fluid for chromatography is carbon dioxide. In SFC, CO_2 exhibits smart solubilizing properties, low viscosity and high diffusivity. SFC is usually 10 to 20 times faster than conventional LC (Rosset et al. 1991) and also provides higher efficiency. New stationary LC phases such as core-shell can be implemented to perform separation enhancing its chromatographic performances even more (Lesellier 2012). Thus, SFC can be considered as the first fast chromatographic technique for nonvolatile compounds. Supercritical CO_2, however, possess a low dielectric constant and consequently rather acts as normal-phase LC. Furthermore, SFC is applicable to compounds of low to moderate polarity. Temperature and pressure affect the solubility of the analytes in supercritical CO_2 and thus modulate the retention and the selectivity in SFC. Additionally, a small percentage of polar modifiers (e.g., methanol) can be added in the mobile phase to elute more polar compounds. SFC is compatible with GC detectors such as FID, NPD, electron capture (Richter et al. 1989), Fourier transform infrared (FTIR) (Calvey et al. 1995) but also with LC detectors like UV–Vis, ELSD (Dreux and Lafosse 1997) and even mass spectrometry (Herbreteau et al. 1999).

SFC replaces advantageously GC for nonvolatile and thermolabile compounds. Some applications in the area of plant natural product research can be found in the literature. Not surprisingly, most of the separations are concerning apolar and low-polar metabolites such as *cis*- and *trans*-carotene (Lesellier et al. 1991), tocopherols (Jiang et al. 2003), fat-soluble vitamins (Turne et al. 2001), allicin in garlic extracts (Calvey et al. 1994) or carnosic acid from rosemary (Ramirez et al. 2004). Most of the time however, natural compounds are commonly extracted by SFE and analyzed by LC.

4.4 Capillary Electrophoresis

Capillary electrophoresis is a typical method for ion analysis. Contrary to the previously described methods requiring forced flow elution, electrophoresis uses a completely different mechanism of separation based on the migration of charged species subjected to an intense electric field. The separation takes place in a fused silica capillary which is generally a few ten cm to one meter long and has an internal diameter varying from 50 up to 100 μm. Except for capillary electrochromatography, the separation does not involve partition of molecules between a stationary phase and a mobile phase. The difference of mobility enabling the separation (electromigration) of analytes is solely due to their difference in charge over size ratio (Altria 1996a). Thus, the highest charge, results in the highest mobility, and, inversely the highest hydrodynamic radius (size) results in the lowest mobility. The capillary is filled with ion strength and pH controlled buffers ensuring conductivity during the analysis and a constant charge state for weak acidic or basic compounds. The inner wall of the capillary is lined with more or less deprotonated silanols (depending on the pH) and to maintain electroneutrality, cations of the buffer build up near the surface. When the separating voltage is applied across the tube, the cations naturally migrate toward the cathode. This phenomenon called electroosmotic flow (Altria 1996a) is superimposed over the electromigration so that cation analysis is speeded up. For anion analysis, it is better is to invert the flow using cationic surfactants.

Capillary electrophoresis has been applied to the analysis of natural compounds either in food chemistry or in phytochemistry. Many applications concern the analysis of polyphenols in propolis (Volpi 2004; Hilhorst et al. 1998) or in plant extracts (Kristo et al. 2002; Crego et al. 2004; Vaher and Koel 2003) with UV detection or mass spectrometry (Gomez-Romero et al. 2007; Hurtado-Fernandez et al. 2010).

A current trend in CE is the miniaturization and in this area. Several authors have already proposed CE-microchip methods to identify phenolic compounds in green tea (Hompesch et al. 2005), in wine (Scampicchio et al. 2004), pears and apples (Blasco et al. 2005) using an amperometric detection.

Micellar electrokinetic chromatography (MEKC) is an interesting variant extending capillary electrophoresis to the separation of neutral compounds. In this technique, ionic surfactants are added to the separation buffer at a concentration so that corresponding micelles are formed. Under the influence of the electric field, the micelles migrate toward the counter electrode. Neutral compounds are partitioned between the buffer and the micelles. When they enter the micelle they migrate at its velocity and when they are out, they move at the electroosmotic flow speed. Depending on this equilibrium, neutral compounds are separated (Otsuka and Terabe 1996). This approach has also been implemented for natural compounds in plants. Zhu et al. (2008) achieved the separation of six flavonoids (tangeretin, nobiletin, hesperetin, naringenin, hesperidin, and naringin) in less than 15 min with sub-ng/mL LOD using a conductivity detector. They applied this for the detection of these substances in extracts of *Fructus aurantii Immaturus* (Zhu et al. 2008). Jiang et al. developed a MEKC-UV method for the simultaneous determination of ten bioactive flavonoids in food containing propolis and *Ginkgo biloba*. They added 16 mM β-cyclodextrin to their buffer to complete the separation. The LOD were ranging from 0.15 to 1.36 μg/mL (Jiang et al. 2008). Lee et al. quantified glucoraphanin in MEKC-UV DAD at 230 nm (Lee et al. 2010). The glucoraphanin migrates around 5 min with a theoretical plate number of 380,000/m allowing its quantification in complex vegetal matrices (broccoli seeds, florets, and Brussels sprouts). Many other applications involving MECK can be found in the literature (El Deeb et al. 2011) and its possible coupling with atmospheric pressure photoionization-mass spectrometry (Marchi et al. 2009) offers new opportunities in the field of plant analysis.

Capillary electrophoresis is a rapid analytical method exhibiting high separation efficiency, low solvent consumption, versatility, and simplicity (Cifuentes 2006) but some specific features have hindered its development in the past. One of its main drawbacks is the difficulty to couple capillary electrophoresis with mass spectrometry. The first reason is due to electric compatibility between apparatus but it can easily be overcome by plugging both apparatus to common ground. The second difficulty was the scan speed of mass spectrometers which was too slow for the high peak efficiency of CE. This is now resolved with the new generation of mass spectrometers developed for fast LC. The last point concerns the (low) flow rate incompatibility between the MS ion sources (i.e., electrospray) and CE. Several solutions have been provided to overcome this problem among which are the sheath flow interface, the sheathless interface and sheathless interface fitted with a gold wire electrode (Cai and Henion 1995). The fine adjustment of these interfaces, however, was generally hard to solve. Recently, a universal plug-and-spray adapter derived from Moini's work (Moini 2007) was developed offering new opportunities for further user-friendly developments in CE-MS.

New technological advances, as well as novel instrument configurations could reinforce CE-MS robustness in the future and would stimulate its development especially for plant matrix analysis.

5 Analysis by Hyphenated Techniques

The most significant advances in analytical chemistry and particularly in separative sciences are probably due to the relentless progresses in their hyphenation with spectrometry. Obviously, gas chromatography was associated for a long time to infrared (Pichard et al. 1990) and mass spectrometry but in the last 15 years, the huge technological change lies in the coupling with liquid phase separation techniques. Mass spectrometry is the most widespread technique in this field and its expansion is undoubtedly not over yet. Nuclear Magnetic Resonance (NMR) can also be coupled with liquid chromatography. Very often, LC-NMR is associated with LC-MS for structural elucidation, particularly for isomers which exhibit similar mass spectra. For example, an LC-DAD-MS/SPE-NMR method was developed for the structural elucidation of iridoid glycoside isomers from *Harpagophytum procumbens* (Seger et al. 2005). LC-NMR-MS was also used for the separation and characterization of secoisolariciresinol diglucoside isomers in flaxseed (Fritsche et al. 2002) or antioxidants (carnosol, carnosaldehyde, epiisorosmanol, carnosic, and 12-methoxycarnosic acid) in rosemary extracts (Pukalskas et al. 2005). The use of NMR however remains limited probably due to its high costs (including deuterated solvents) and rather low sensitivity compared to MS. Moreover, for many applications, mass spectrometry is sufficient to resolve the questions raised. This last part of this chapter therefore focuses on hyphenation with mass spectrometry.

5.1 Why Hyphenation with Mass Spectrometry?

The hyphenation between separative methods and structural spectrometry techniques (i.e., FTIR, NMR and MS) always increases analytical peak capacities in a single run and in several respects; mass spectrometry appears to be the perfect detection method for chromatography. Indeed, the ideal detector should exhibit the following qualities:

- **It must be universal**. In mass spectrometry, the critical parameter is ion production so that a mass-over-charge ratio is detected. Currently, the commercially available ion sources are covering all areas of molecular size (from small molecules to polymers) and polarity (from apolar compounds to ions) (Fig. 19). The only requirement is the compatibility between the separative methods and those ion sources. Electron impact (EI) and chemical ionization

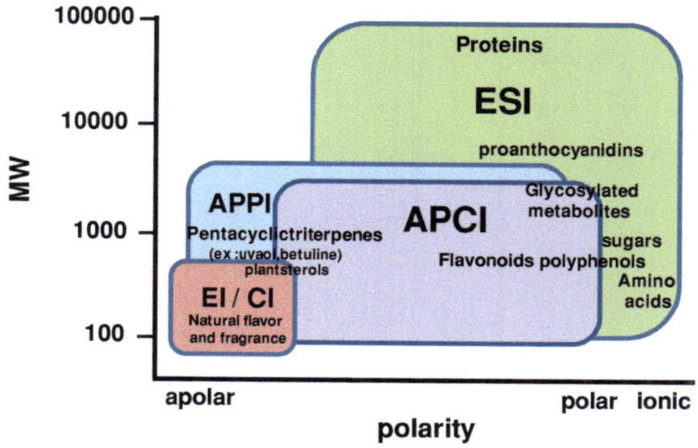

Fig. 19 Range of use of the MS ion sources as a function of molecular weight and polarity. Readers are cautioned that the areas are only given as tendencies. EI: electron impact, CI: chemical ionization, ESI: electrospray ionization, APCI: atmospheric pressure chemical ionization, APPI: atmospheric pressure photoionization. Adapted from Syage et al. (2008), De Hoffman et al. (1994)

(CI) is compatible with GC because of the vacuum required in both sources. Atmospheric pressure ionization sources (API) are compatible with separation techniques requiring a liquid mobile phase.

- **The ideal detector must be** sensitive and stable over time. Moreover, it must provide signals proportional to the analyte concentration (amount) in the sample. Mass spectrometry is commonly used for the quantification of analytes at sub-ppb levels thanks to selected ion monitoring (SIM) or multiple reaction monitoring (MRM) in tandem MS (Cao et al. 2006). The signal stability, however, clearly relates to the nature of the sample. Biological samples tend to clog the mass spectrometer optoelectronics resulting in a more or less progressive decrease in the signal. This drawback can be reduced substantially by a clean-up procedure of the sample before its analysis. The decrease in signal during quantitative analysis is also compensated by the use of internal standards (ideally stable isotope labeled products).
- **The ideal detector must not** induce artifacts. In MS, physical artifacts generally originate from bad settings or a compressed gas supply. Chemical artifacts are often due to the well-known ion suppression phenomenon which may have several sources. The presence of less volatile compounds can decrease droplet evaporation. The matrix can compete with the analytes during the ionization process and affects the amount of the analyte ions in the gas phase reaching the detector (Annesley 2003). The importance of the ion suppression is also depending on the ion source. Thus, APCI and APPI are less subjected to ion suppression due to the matrix effect than ESI (Marchi et al. 2009; Jessome and Volmer 2006). Nevertheless, the problem is often overcome using a sample clean-up such as SPE or equivalent.

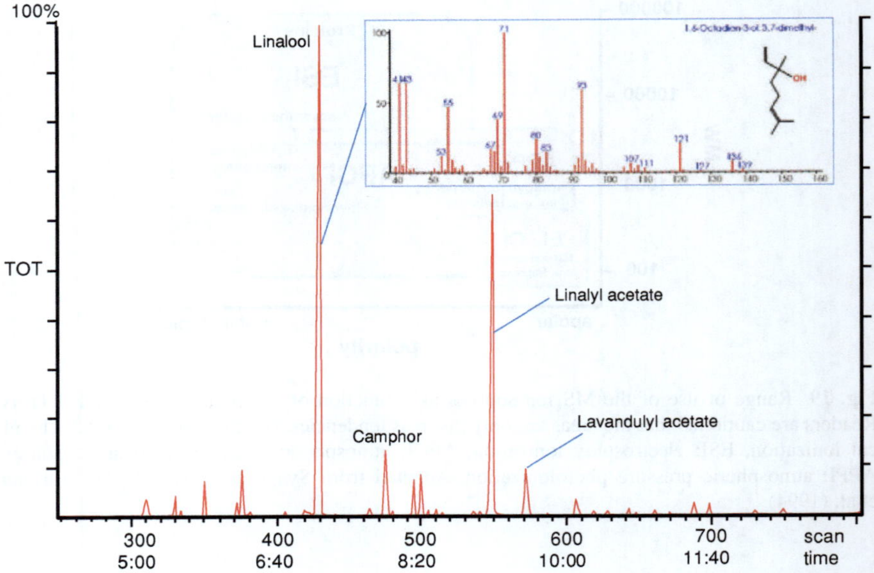

Fig. 20 Analysis of lavender oil by GC-EI-MS. Linalool (and other indexed compounds in this figure) was identified after request to spectrum library (e.g., NIST library). The accuracy of the identification is assessed by statistic scores in terms of purity, fit and reversed fit between the experimental and the library spectra

- **The ideal detector should** provide structural information. Interestingly, structural information is precisely the rationale of mass spectrometry (Fig. 20). In GC-EI-MS, the structural elucidation can also be assisted by spectrum libraries. Readers are cautioned, however, that the final attribution is under the responsibility of the chemist. Indeed, libraries tend to provide identifications on request. The chemist, however, must check whether the identification ultimately also makes sense. For all other ionization techniques, usually no libraries are available because of the lack in reproducibility of spectra within the apparatus. Efforts are currently underway to provide a solution to this problem. Here, intern libraries may be constructed from standards in order to identify compounds in complex mixtures. With very mild ionization methods such as ESI, structural information is either obtained from the in-source fragmentation of the pseudo molecular ions (i.e., the protonated or deprotonated molecules) (Cole 1997) or by tandem mass spectrometry. With the new generation of MS devices, obtaining of structural information is also facilitated by software which is able on request to automatically trigger MS/MS experiments above a predefined threshold.

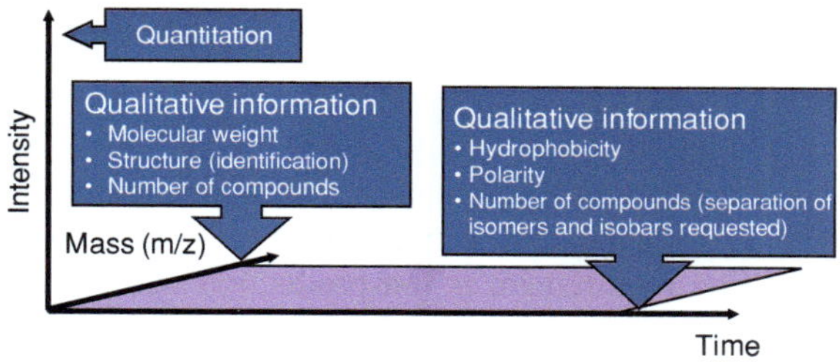

Fig. 21 Principle of separative methods hyphenated to mass spectrometry. The peak capacity of a single run is increased thanks to multidimensional analysis (at least two dimensions, even more with tandem MS). Both dimensions are complementary and each one counterbalances the lack of information of the other

- **The ideal detector should** be able to distinguish co-eluted compounds and to deconvolute their signals. During a run, the specific ion currents corresponding to each compound can be extracted from the total ion current. The problem of co-elution is therefore resolved unless co-eluted analytes are isomers (even in high resolution MS) or isobaric (a problem with low resolution MS). If the mass spectrometer is not fitted with an ion mobility device, in case of isomers, a separate separative method may be included (Fig. 21).
- **The ideal detector should** be specific and only detect the desired compounds. Here, the main question is how to be both universal and specific. The mass spectrometer succeeds in such a challenge. It is universal in full scan and becomes specific in Selected Ion Monitoring (SIM) and even more in tandem mass spectrometry thanks to the Multiple Reaction Monitoring (MRM) mode. In that particular case, all compounds that do not show the required daughter ion(s) resulting from the targeted mother ion are not detected. As previously mentioned however, the compounds transparent to the detection are still eluted from the column and may induce ion suppression (i.e., decrease in signal intensity due to ionization competition between analytes and the unwanted compounds in the matrix).
- **The ideal detector should** be inexpensive, easy to use, and not destructive. These points are more or less in agreement with MS detection. The MS detector is destructive. It is so sensitive, however, that only a small part of the sample is required for detection in semi-prep LC. Except for GC-MS, the use of a splitting system allows collection of the main fraction of the sample at the column outlet. The price and the ease of use strongly depend on the apparatus. GC-MS instruments are obviously cheaper and more user-friendly than Fourier Transform Ion Cyclotron Resonance Mass Spectrometry (LC-FTICRMS).

5.2 *Interfacing Mass Spectrometry with Separation Techniques*

One sensitive issue concerning hyphenation is how compatible mass spectrometry is with regard to the separative techniques. Two problems may be taken into consideration. The first requirement is ion production without which MS cannot be performed (MS carries out mass-over-charge ratio measurements). The second problem is high vacuum required to analyze the ions produced. The different separative methods do not exhibit the same degree of compatibility in this field and the same applies for ion production techniques.

Electron impact and chemical ionization also require high vacuum to be implemented. Consequently, these ion sources are fully compatible only with capillary GC. The mobile phase does not bring too much matter into the source and small excess gas is easily purged from the source by the efficient pumping system (e.g., turbo molecular pumps). The GC column is directly plumbed to the ion source and the interface just consists of a heated transfer line between the GC oven and the ion source. The role of the transfer line is to keep the compounds in the gas phase when they leave the oven.

The coupling with liquid separative methods historically was more problematic until the emergence of the API sources (ESI: electrospray ionization, APCI: atmospheric pressure chemical ionization and APPI: atmospheric pressure photoionization). Atmospheric pressure ionization has compelled MS manufacturers to develop dedicated interfaces allowing the evacuation of a large gas volume so that a sufficient vacuum is maintained for MS analyzers. In a first stage, the only real requirement in MS coupling is a volatile mobile phase. It is also well known, however, that the composition of the mobile phase strongly influences ionization efficiency in ESI, APCI, and APPI (Kostiainen and Kauppila 2009). Therefore, all the LC modes (RPLC, NPLC, etc.,) do not exhibit the same degree of compatibility with API-MS (Table 3). For example, apolar solvents do not suit ESI because of their low conductivity. Thus NPLC, contrary to APPI, leads to poor results when it is directly coupled to ESI. By contrast, although salts and ion-pairing agents may cause ion suppression, ion-exchange chromatography (Chen et al. 2008, 2009) and ion-pairing reversed-phase liquid chromatography (Chaimbault et al. 1999; de Person et al. 2008) remain sufficiently compatible with ESI-MS. RPLC and HILIC are the most universal LC modes for hyphenation. The only circumspection is to avoid too concentrated buffers for pH or ionic strength control. HILIC is preferable to RPLC for highly polar compound analysis with ESI because the retention of such compounds is favored in an organic-rich aqueous mobile phase, which in turn improves ion desorption in the source (Wang and He 2011).

Compatibility of the mobile phase (including the flow rate) with the ion source alone is not sufficient to achieve the highest sensitivity of detection. The result also depends on the physicochemical properties of the analytes themselves. In fact, the sensitivity is strongly related to the difference of proton affinity between analytes and the mobile phase solvents and additives in the gas phase (in APPI, the

Table 3 Compatibility of electrospray (ESI), atmospheric pressure chemical ionization (APCI) and atmospheric pressure photoionization (APPI) in LC-MS

	Ion sources			
Chromatography	ESI	APCI	APPI	Comments
NPLC	±	++	+++	Low ionization rate in ESI with mobile phases containing high percentages of apolar solvents (e.g., hexane)
RPLC	+++	+++	+++	Volatile additives such as acetic or formic acid, ammonia, ammonium acetate, and format can be used for pH control at a few ten mM concentrations APPI: methanol is preferred to acetonitrile and acidic or basic buffers tend to decrease the MS signal (Kostiainen and Kauppila 2009)
HILIC	+++	+++	+++	Same remarks as for RPLC
IP-RPLC	++	−	−	IP-RPLC chromatography means little for APCI and APPI as both ionization processes are rather applicable to neutral and low-polar compounds. Additionally, strong acidic as well as basic additives decrease ionization efficiency. Volatile IP agents are causing ion suppression but can be used with ESI-MS: • IP agent for cation separations: perfluorocarboxylic acids from trifluoroacetic (TFA) to pentadecafluorooctanoic acid. TFA and heptafluorobutyric acid (HFBA) usually cause severe ion suppression whereas the longest homologues may improve the MS signal (Chaimbault et al. 1999; de Person et al. 2008) • IP agents for anion separations: *N,N*-dimethyl hexylamine, trialkyl amines (e.g., triethylamine, tributylamine) (Pruvost et al. 2001; Storm et al. 1999)
IEC	+	−	−	Ion evaporation is less efficient in aqueous mobile phases than in organic-rich aqueous eluents. Mobile phases require highly concentrated saline buffers that induce ion suppression

NPLC normal-phase liquid chromatography, *RPLC* reversed-phase liquid chromatography, *HILIC* hydrophilic interaction liquid chromatography, *IP-RPLC* ion-paring reversed-phase liquid chromatography, *IEC* ion-exchange chromatography

nature of the dopant also plays a key role). Thus, in a comparative study between APCI and APPI conducted in order to determine the amount of pentacyclic triterpenes (e.g., α and β-amyrin, friedelin, erythrodiol, etc.,) in bark samples by LC-MS, Rhourri-Frih et al. found that APPI was overall more sensitive than APCI in positive ion mode (LODs in the range of 0.005 to 0.015 mg/L for APPI and

0.002 to 0.84 mg/L for APCI in SIM). APCI showed the greatest sensitivity for acidic triterpenes (ursolic, betulinic, and oleanolic acids) in negative ion mode. As expected, ESI failed to ionize low-polar compounds such as betulin or uvaol (see also Fig. 19) even though the methanol mobile phase was fully compatible with this ion source (Rhourri-Frih et al. 2009). Furthermore, Rauha et al. (2001) demonstrated the importance of optimizing the composition of the mobile phase according to the ion source. The authors optimized the composition of the mobile phase for ESI, APCI, and APPI-MS detection of flavonoids [(+)-catechin, isorhamnetin, vitexin, isoquercitrin, luteolin-3′-7-diglucoside]. Under optimized conditions, the limits of detection were of the same order of magnitude for ESI, APCI, and APPI in positive and negative ion mode even if overall, ESI in negative mode resulted in better LODs.

The analysis of antioxidants by LC-MS does not show any particular difficulties and many applications can be found in the literature involving the different API sources. The ion source is generally used in good agreement with Fig. 19. Carotenoids (Hao et al. 2005) and fat-soluble vitamins (Stoggl et al. 2001) are analyzed in LC-MS with APCI or APPI. For example, Stoggl et al. (2001) achieved quantification in the femtomole range of vitamin E and some other tocopherol derivatives in food and phytopharmaceutical preparations by HPLC-APCI-MS/MS. Recently, six fat-soluble vitamins (A, E, D2, D3, K1 and K2) and six water-soluble vitamins (B1, B2, B3, B6, B7 and B12) were analyzed simultaneously in LC-APPI-MS/MS. This study demonstrates that APPI is clearly a versatile ion source allowing ionization of both polar and apolar analytes. Obviously, electrospray is ideally suited for detection at low levels (ng/g of fruits or vegetable) of water-soluble vitamins (Gentili et al. 2008). Polyphenols are usually detected as deprotonated molecules (due to the phenol groups) either by APCI (Justesen 2000) or ESI (Cho et al. 2004; Calderon et al. 2009). Calderon et al. (2009) developed a screening method for antioxidants in vegetal material by RPLC-ESI-MS. In this study, the authors identified catechins, epicatechins, their dimers and trimers (proanthocyanidins), some gallic acid derivatives, and a few other glycosylated flavonoids (derived from quercetin) in cocoa.

The other separative methods described in this chapter can also be coupled to MS. Supercritical fluid chromatography is very often interfaced with APCI (Dost and Davidson 2000; Manninen and Laakso 1997; Taylor 2009) but it can also be coupled with ESI-MS (Salvador et al. 1999). For example, Matsubara et al. (2012) developed a highly sensitive and rapid profiling method for carotenoids and their epoxidized products by SFC-ESI-MS/MS in 20 min with detection limits in the femto-molar range order. Capillary electrophoresis is most often coupled to MS with ESI as ion source but it can be also coupled with atmospheric pressure photoionization-mass spectrometry (Marchi et al. 2009). APPI shows even a higher robustness than ESI even toward low and nonvolatile buffers (Mol et al. 2005; Hommerson et al. 2009). The best sensitivities, however, are obviously still obtained with volatile buffers at low salt concentrations (Kostiainen and Kauppila 2009; Hakala et al. 2003). At last, as previously mentioned (Sect. 4.4), HPTLC can be interfaced with MALDI-TOFMS and even electrospray-MS.

6 Conclusions

This chapter has tried to provide an overview of the current techniques used for extraction, purification, and analysis of plant secondary metabolites. Many techniques may be implemented at any level but it is impossible to provide general rules on their choice for new applications. Indeed, there are very often several solutions to achieve a satisfying analysis and the right choice of a technique strongly depends on the starting material and the metabolites of interest. Nevertheless, it must be admitted that the current state of art in analytical chemistry probably provides an adequate solution to perform the development of the individual methods required. Furthermore, continuous research in this field will undoubtedly offer new and precise exploration tools for complex samples, such as plant extract. Besides, the 'peak capacity' race has already started with the following *leitmotiv*: always faster, more informative, and more sensitive. The democratization of hyphenation between separative methods and structural elucidation techniques—and in this context notably mass spectrometry—is only the first step. The expansion of rapid separation and comprehensive multidimensional techniques is ongoing and the relentless technological progress will undoubtedly improve matters further.

References

Abidi SL (1999) Reversed-phase retention characteristics of tocotrienol antioxidants. J Chromatogr A 844(1–2):67–75

Adam M, Dobias P, Eisner A, Ventura K (2009) Extraction of antioxidants from plants using ultrasonic methods and their antioxidant capacity. J Sep Sci 32(2):288–294

Agrafiotou P, Sotiropoulos S, Pappa-Louisi A (2009) Direct RP-HPLC determination of underivatized amino acids with online dual UV absorbance, fluorescence, and multiple electrochemical detection. J Sep Sci 32(7):949–954

Alfaro MJ, Belanger JMR, Padilla FC, Pare JRJ (2003) Influence of solvent, matrix dielectric properties, and applied power on the liquid-phase microwave-assisted processes (MAP (TM)) extraction of ginger (*Zingiber officinale*). Food Res Int 36(5):499–504

Altria KD (1996a) Fundamentals of capillary electrophoresis theory. In: Altria KD (ed) Capillary electrophoresis guidebook, methods in molecular biology, vol 52. Humana Press, Totowa, pp 3–13

Altria KD (1996b) Methods in molecular biology, capillary electrophoresis guidebook. Humanan Press, Totowa

Annesley TM (2003) Ion suppression in mass spectrometry. Clin Chem 49(7):1041–1044

Antonopoulos A, Herbreteau B, Lafosse M, Helbert W (2004) Comparative analysis of enzymatically digested kappa-carrageenans, using liquid chromatography on ion-exchange and porous graphitic carbon columns coupled to an evaporative light scattering detector. J Chromatogr A 1023(2):231–238

Arnault I, Christides JP, Mandon N, Haffner T, Kahane R, Auger J (2003) High-performance ion-pair chromatography method for simultaneous analysis of alliin, deoxyalliin, allicin and dipeptide precursors in garlic products using multiple mass spectrometry and UV detection. J Chromatogr A 991(1):69–75

Asa D (2006) Carbohydrate and oligosaccharide analysis with a universal HPLC detector. Am Lab 38(7):16

Ashraf K, Mujeeb M, Ahmad A, Amir M, Mallick MN, Sharma D (2012) Validated HPTLC analysis method for quantification of variability in content of curcumin in *Curcuma longa L* (turmeric) collected from different geographical region of India. Asian Pac J Trop Biomed 2(2):S584–S588

Augusto F, Lopes ALE, Zini CA (2003) Sampling and sample preparation for analysis of aromas and fragrances. Trac-Trend Anal Chem 22(3):160–169

Augusto F, Poppi RJ, Pedroso MP, de Godoy LAF, Hantao LW (2010) GC × GC-FID for qualitative and quantitative analysis of perfumes. Lc Gc Eur 23(8):430

Balz MK, Schulte E, Thier HP (1993) Simultaneous determination of Alpha-Tocopheryl Acetate, Tocopherols and Tocotrienols by Hplc with fluorescence detection in foods. Fett Wiss Technol 95(6):215–220

Banerjee K, Patil SH, Dasgupta S, Oulkar DP, Patil SB, Savant R, Adsule PG (2008) Optimization of separation and detection conditions for the multiresidue analysis of pesticides in grapes by comprehensive two-dimensional gas chromatography-time-of-flight mass spectrometry. J Chromatogr A 1190(1–2):350–357

Beens J, Adahchour M, Vreuls RJJ, van Altena K, Brinkman UAT (2001) Simple, non-moving modulation interface for comprehensive two-dimensional gas chromatography. J Chromatogr A 919(1):127–132

Beldean-Galea MS, Horga C, Coman MV (2010) Separation and determination of tocopherols in vegetable oils by solid phase extraction on porous polymers SPE cartridges and capillary gas chromatography analysis. Cent Eur J Chem 8(5):1108–1114

Benova B, Adam M, Pavlikova P, Fischer J (2010) Supercritical fluid extraction of piceid, resveratrol and emodin from Japanese knotweed. J Supercrit Fluid 51(3):325–330

Berthod A, Maryutina T, Spivakov B, Shpigun O, Sutherland IA (2009) Countercurrent chromatography in analytical chemistry (Iupac Technical Report). Pure Appl Chem 81(2):355–387

Bertoncini F, Vendeuvre C, Thiebaut D (2005) Interest and applications of multidimensional gas chromatography for trace analysis in the petroleum industry. Oil Gas Sci Technol 60(6):937–950

Bertsch W (2000) Two-dimensional gas chromatography. Concepts, instrumentation, and applications—Part 2: comprehensive two-dimensional gas chromatography. J High Resolut Chromatogr 23(3):167–181

Bicchi C, Binello A, Rubiolo P (2000) Determination of phenolic diterpene antioxidants in rosemary (*Rosmarinus officinalis L.*) with different methods of extraction and analysis. Phytochem Anal 11(4):236–242

Biesaga M, Ochnik U, Pyrzynska K (2009) Fast analysis of prominent flavonoids in tomato using a monolithic column and isocratic HPLC. J Sep Sci 32(15–16):2835–2840

Blanch GP, Reglero G, Herraiz M (1995) Analysis of wine aroma by off-line and online supercritical-fluid extraction gas-chromatography. J Agr Food Chem 43(5):1251–1257

Blasco AJ, Barrigas I, Gonzalez MC, Escarpa A (2005) Fast and simultaneous detection of prominent natural antioxidants using analytical microsystems for capillary electrophoresis with a glassy carbon electrode: a new gateway to food environments. Electrophoresis 26(24):4664–4673

Boselli E, Velazco V, Caboni MF, Lercker G (2001) Pressurized liquid extraction of lipids for the determination of oxysterols in egg-containing food. J Chromatogr A 917(1–2):239–244

Bryson N, Papillard D (2004) An introduction to OPLC—Operation and applications. Lc Gc Eur 17(1):41–47

Buchmabauer G, Jirovetz L, Nikovorov A (1997) The use of GC-FID, FTIR-MS and olfactory of essential oil and plant extracts. Springer, Berlin

Cai JY, Henion J (1995) Capillary electrophoresis mass-spectrometry. J Chromatogr A 703(1–2):667–692

Calderon AI, Wright BJ, Hurst WJ, van Breemen RB (2009) Screening antioxidants using LC-MS: case study with cocoa. J Agr Food Chem 57(13):5693–5699

Calvey EM, Begley TH, Roach JAG (1995) SFC-FTIR spectroscopy and SFC-MS of nonvolatile extractants of microwave susceptor packaging. J Chromatogr Sci 33(2):61–65

Calvey EM, Roach JA, Block E (1994) Supercritical fluid chromatography of garlic (Allium sativum) extracts with mass spectrometric identification of allicin. J Chromatogr Sci 32(3):93–96

CAMAG Research & Development. The new TLC-MS interface. vol CBS 102

Cao J, Murch SJ, O'Brien R, Saxena PK (2006) Rapid method for accurate analysis of melatonin, serotonin and auxin in plant samples using liquid chromatography-tandem mass spectrometry. J Chromatogr A 1134(1–2):333–337
Cao PY, Liu FM, Wang SL, Wang YH, Han LJ (2008) GC-ECD analysis of S-metolachlor (Dual Gold) in cotton plant and soil in trial field. Environ Monit Assess 143(1–3):1–7
Cao XL, Tian Y, Zhang TY, Liu QH, Jia LJ, Ito Y (2003) Separation of dammarane-saponins from notoginseng, root of Panax notoginseng (Burk.) F.H. Chen, by HSCCC coupled with evaporative light scattering detector. J Liq Chromatogr R T 26(9–10):1579–1591
Cesla P, Hajek T, Jandera P (2009) Optimization of two-dimensional gradient liquid chromatography separations. J Chromatogr A 1216(16):3443–3457
Chaimbault P, Petritis K, Elfakir C, Dreux M (1999) Determination of 20 underivatized proteinic amino acids by ion-pairing chromatography and pneumatically assisted electrospray mass spectrometry. J Chromatogr A 855(1):191–202
Chaimbault P, Petritis K, Elfakir C, Dreux M (2000) Ion-pair chromatography on a porous graphitic carbon stationary phase for the analysis of twenty underivatized protein amino acids. J Chromatogr A 870(1–2):245–254
Chan CH, Yusoff R, Ngoh GC, Kung FWL (2011) Microwave-assisted extractions of active ingredients from plants. J Chromatogr A 1218(37):6213–6225
Chang YW, Yao HT, Hsieh SH, Lu TJ, Yeh TK (2007) Quantitative determination of salidroside in rat plasma by on-line solid-phase extraction integrated with high-performance liquid chromatography/electrospray ionization tandem mass spectrometry. J Chromatogr B 857(1):164–169
Chase GW, Akoh CC, Eitenmiller RR (1994) Analysis Ol tocopherols in vegetable-oils by high-performance liquid-chromatography—comparison of fluorescence and evaporative light-scattering detection. J Am Oil Chem Soc 71(8):877–880
Chen ZL, Jin XY, Wang QP, Lin YM, Gan L (2009) Confirmation and determination of sugars in soft drink products by IEC with ESI-MS. Chromatographia 69(7–8):761–764
Chen ZL, Kim KR, Owens G, Naidu R (2008) Determination of carboxylic acids from plant root exudates by ion exclusion chromatography with ESI-MS. Chromatographia 67(1–2):113–117
Cho MJ, Howard LR, Prior RL, Clark JR (2004) Flavonoid glycosides and antioxidant capacity of varous blackberry, blueberry and red grape genotypes determined by high-performance liquid chromatography/mass spectrometry. J Sci Food Agr 84(13):1771–1782
Cicchetti E, Chaintreau A (2009) Comparison of extraction techniques and modeling of accelerated solvent extraction for the authentication of natural vanilla flavors. J Sep Sci 32(11):1957–1964
Cifuentes A (2006) Recent advances in the application of capillary electromigration methods for food analysis. Electrophoresis 27(1):283–303
Claude B, Morin P, Lafosse M, Belmont AS, Haupt K (2008) Selective solid-phase extraction of a triterpene acid from a plant extract by molecularly imprinted polymer. Talanta 75(2):344–350
Co M, Fagerlund A, Engman L, Sunnerheim K, Sjoberg PJR, Turner C (2012) Extraction of antioxidants from spruce (Picea abies) bark using eco-friendly solvents. Phytochem Anal 23(1):1–11
Cole RB (1997) Electrospray ionization mass spectrometry. Wiley, New York
Coleman WM, Dube MF (2005) Headspace solid-phase microextraction analysis of artificial flavors. J Sci Food Agr 85(15):2645–2654
Colombo R, Lancas FM, Yariwake JH (2006) Determination of flavonoids in cultivated sugarcane leaves, bagasse, juice and in transgenic sugarcane by liquid chromatography-UV detection. J Chromatogr A 1103(1):118–124
Corporation L (2004) Comprehensive two-dimensional gas chromatography (GC × GC)—time-of-flight mass spectrometry of metabolic samples obtained from the leaves of a hybrid Aspen (Populus Tremula) Plant, vol 203–821–233. Saint Joseph, Michigan
Corradini C, Cavazza A, Bignardi C (2012) High-performance anion-exchange chromatography coupled with pulsed electrochemical detection as a powerful tool to evaluate carbohydrates of food interest: principles and applications. Int J Carbohydr Chem 2012:13
Cortes-Aguado S, Sanchez-Morito N, Arrebola FJ, Frenich AG, Vidal JLM (2008) Fast screening of pesticide residues in fruit juice by solid-phase microextraction and gas chromatography-mass spectrometry. Food Chem 107(3):1314–1325

Crecelius A, Clench MR, Richards DS (2003) TLC-MALDI in pharmaceutical analysis. Lc Gc Eur 16(4):225–229

Crego A, Ibanez E, Garcia E, de Pablos RR, Senorans FJ, Reglero G, Cifuentes A (2004) Capillary electrophoresis separation of rosemary antioxidants from subcritical water extracts. Eur Food Res Technol 219(5):549–555

Cvek J, Medic-Saric M, Jasprica I, Mornar A (2007) High-performance thin-layer chromatographic analysis of the phenolic acid and flavonoid content of Croatian propolis samples. JPC-J Planar Chromat 20(6):429–435

de Geus HJ, Aidos I, de Boer J, Luten JB, Brinkman UAT (2001) Characterisation of fatty acids in biological oil samples using comprehensive multidimensional gas chromatography. J Chromatogr A 910(1):95–103

De Hoffman E, Charrette J, Stroobant V (1994) Spectrométrie de masse. Masson, Paris

De Medici D, Pieretti S, Salvatore G, Nicoletti M, Rasoanaivo P (2006) Chemical analysis of essential oils of Malagasy medicinal plants by gas chromatography and NMR spectroscopy. Flavour Frag J 7(5):275–281

de Person M, Chaimbault P, Elfakir C (2008) Analysis of native amino acids by liquid chromatography/electrospray ionization mass spectrometry: comparative study between two sources and interfaces. J Mass Spectrom JMS 43(2):204–215

de Villiers A, Kalili KM, Malan M, Roodman J (2010) Improving HPLC separation of polyphenols. Lc Gc Eur 23(9):466

del Rio JC, Gutierrez A, Rodriguez IM, Ibarra D, Martinez AT (2007) Composition of non-woody plant lignins and cinnamic acids by Py-GC/MS, Py/TMAH and FTIR. J Anal Appl Pyrol 79(1–2):39–46

Denery JR, Dragull K, Tang CS, Li QX (2004) Pressurized fluid extraction of carotenoids from *Haematococcus pluvialis* and *Dunaliella salina* and *kavalactones* from *Piper methysticum*. Anal Chim Acta 501(2):175–181

Dhara R, Bhattacharyya DK, Ghosh M (2010) Analysis of sterol and other components present in unsaponifiable matters of mahua, sal and mango kernel oil. J Oleo Sci 59(4):169–176

Donato P, Cacciola F, Dugo P, Mondello L (2011) The secrets of red wine—determination of polyphenols by comprehensive 2D LC with PDA and MS-MS (IT-TOF) detection. LC GC Eur Application Book 3:8–9

Dorich B, Francis E, Murphy B, Richter B, Henderson S (2008) Accelerated solvent extraction with acid pretreatment for improved laboratory productivity. Am Lab 40(8):18

Dost K, Davidson G (2000) Development of a packed-column supercritical fluid chromatography/atmospheric pressure chemical-ionisation mass spectrometric technique for the analysis of atropine. J Biochem Bioph Meth 43(1–3):125–134

Dreux M, Lafosse M (1997) Review of evaporative light-scattering detection for packed-column SFC. LCGC Int 10(6):382–386

Dreux M, Lafosse M, Morin-Allory L (1996) The evaporative light scattering detector—A universal instrument for non-volatile solutes in LC and SFC. LCGC Int 9(3):148–153

Du M, Ahn DU (2002) Simultaneous analysis of tocopherols, cholesterol, and phytosterols using gas chromatography. J Food Sci 67(5):1696–1700

Dugo P, Cacciola F, Herrero M, Donato P, Mondello L (2008) Use of partially porous column as second dimension in comprehensive two-dimensional system for analysis of polyphenolic antioxidants. J Sep Sci 31(19):3297–3308

Eggers R (1996) Supercritical fluid extraction of oilseeds/lipids in natural products. In: King JW, List GR (eds) Supercritical fluid technology in oil and lipid chemistry. AOCS Press, Champaign, pp 35–65

El Deeb S, Abu Iriban M, Gust R (2011) MEKC as a powerful growing analytical technique. Electrophoresis 32(1):166–183

Eskilsson CS, Hartonen K, Mathiasson L, Riekkola ML (2004) Pressurized hot water extraction of insecticides from process dust—Comparison with supercritical fluid extraction. J Sep Sci 27(1–2):59–64

Farina A, Doldo A, Cotichini V, Rajevic M, Quaglia MG, Mulinacci N, Vincieri FF (1995) HPTLC and reflectance mode densitometry of anthocyanins in *Malva silvestris L*: a comparison with gradient-elution reversed-phase HPLC. J Pharm Biomed 14(1–2):203–211

Fellows P (2000) Food processing technology: principles and practice, 3edn. Woodhead Publishing, Witney

Flamini R (2003) Mass spectrometry in grape and wine chemistry. Part I: polyphenols. Mass Spectrom Rev 22(4):218–250

Folch J, Lees M, Stanley GHS (1957) A simple method for the isolation and purification of total lipides from animal tissues. J Biol Chem 226(1):497–509

Fritsche J, Angoelal R, Dachtler M (2002) On-line liquid-chromatography-nuclear magnetic resonance spectroscopy-mass spectrometry coupling for the separation and characterization of secoisolariciresinol diglucoside isomers in flaxseed. J Chromatogr A 972(2):195–203

Fuleki T, Francis FJ (1968) Quantitative methods for anthocyanins. 1. Extraction and determination of total anthocyanin in cranberries. J Food Sci 33(1):72–78

Ganzler K, Salgo A, Valko K (1986) Microwave extraction—a novel sample preparation method for chromatography. J Chromatogr 371:299–306

Gao M, Song BZ, Liu CZ (2006) Dynamic microwave-assisted extraction of flavonoids from Saussurea medusa Maxim cultured cells. Biochem Eng J 32(2):79–83

Garcia-Marino M, Rivas-Gonzalo JC, Ibanez E, Garcia-Moreno C (2006) Recovery of catechins and proanthocyanidins from winery by-products using subcritical water extraction. Anal Chim Acta 563(1–2):44–50

Garcia-Salas P, Morales-Soto A, Segura-Carretero A, Fernandez-Gutierrez A (2010) Phenolic-compound-extraction systems for fruit and vegetable samples. Molecules 15(12):8813–8826

Gentili A, Caretti F, Bellante S, Rocca LM, Curini R, Venditti A (2012) Development and validation of two multiresidue liquid chromatography tandem mass spectrometry methods based on a versatile extraction procedure for isolating non-steroidal anti-inflammatory drugs from bovine milk and muscle tissue. Anal Bioanal Chem 404(5):1375–1388

Gentili A, Caretti F, D'Ascenzo G, Marchese S, Perret D, Di Corcia D, Rocca LM (2008) Simultaneous determination of water-soluble vitamins in selected food matrices by liquid chromatography/electrospray ionization tandem mass spectrometry. Rapid Commun Mass Sp 22(13):2029–2043

Gogus F, Ozel MZ, Kocak D, Hamilton JF, Lewis AC (2011) Analysis of roasted and unroasted Pistacia terebinthus volatiles using direct thermal desorption-GC × GC-TOF/MS. Food Chem 129(3):1258–1264

Gogus F, Ozel MZ, Lewis AC (2005) Superheated water extraction of essential oils of Origanum micranthum. J Chromatogr Sci 43(2):87–91

Gomez-Romero M, Arraez-Roman D, Moreno-Torres R, Garcia-Salas P, Segura-Carretero A, Fernandez-Gutierrez A (2007) Antioxidant compounds of propolis determined by capillary electrophoresis-mass spectrometry. J Sep Sci 30(4):595–603

Goncalves MS, Vieira IJC, Oliveira RR, Braz R (2011) Application of preparative high-speed counter-current chromatography for the separation of two alkaloids from the roots of *Tabernaemontana catharinensis* (Apocynaceae). Molecules 16(9):7480–7487

Gorecki T, Yu XM, Pawliszyn J (1999) Theory of analyte extraction by selected porous polymer SPME fibers. Analyst 124(5):643–649

Gritti F, Guiochon G (2012) The current revolution in column technology: how it began, where is it going? J Chromatogr A 1228:2–19

Groskreutz SR, Swenson MM, Secor LB, Stoll DR (2012) Selective comprehensive multi-dimensional separation for resolution enhancement in high performance liquid chromatography. Part I: Principles and instrumentation. J Chromatogr A 1228:31–40

Grunwald C (1970) Quantitative analysis of free phytosterols by gas chromatography using stationary phase Ov-101. Anal Biochem 34(1):16

Gruz J, Novak O, Strnad M (2008) Rapid analysis of phenolic acids in beverages by UPLC-MS/MS. Food Chem 111(3):789–794

Guiochon G (2007) Monolithic columns in high-performance liquid chromatography. J Chromatogr A 1168(1–2):101–168

Gutzeit D, Winterhalter P, Jerz G (2007) Application of preparative high-speed counter-current chromatography/electrospray ionization mass spectrometry for a fast screening and fractionation of polyphenols. J Chromatogr A 1172(1):40–46

Ha IJ, Kang M, Na YC, Park Y, Kim YS (2011) Preparative separation of minor saponins from Platycodi Radix by high-speed counter-current chromatography. J Sep Sci 34(19):2559–2565

Hakala KS, Laitinen L, Kaukonen AM, Hirvonen J, Kostiainen R, Kotiaho T (2003) Development of LC/MS/MS methods for cocktail dosed Caco-2 samples using atmospheric pressure photoionization and electrospray ionization. Anal Chem 75(21):5969–5977

Hao ZG, Parker B, Knapp M, Yu LL (2005) Simultaneous quantification of alpha-tocopherol and four major carotenoids in botanical materials by normal phase liquid chromatography-atmospheric pressure chemical ionization-tandem mass spectrometry. J Chromatogr A 1094(1–2):83–90

Hasanloo T, Bahmanei M, Sepehrifar R, Kalantari F (2008) Determination of tocopherols and fatty acids in seeds of *Silybum marianum* (L.) gaerth. J Med Plants 7(4):69–76

Hazotte A, Libong D, Matoga A, Chaminade P (2007) Comparison of universal detectors for high-temperature micro liquid chromatography. J Chromatogr A 1170(1–2):52–61

Heldman DR, Hartel RW (1997) Principles of food processing. Springer, New York

Herbreteau B, Lafosse M, Morinallory L, Dreux M (1992) High-performance liquid-chromatography of raw sugars and Polyols using bonded silica-gels. Chromatographia 33(7–8):325–330

Herbreteau B, Salvador A, Lafosse M, Dreux M (1999) SFC with evaporative light-scattering SFC detection and atmospheric-pressure chemical-ionisation mass spectrometry for methylated glucoses and cyclodextrins analysis. Analusis 27(8):706–712

Herrero M, Cifuentes A, Ibanez E (2006) Sub- and supercritical fluid extraction of functional ingredients from different natural sources: plants, food-by-products, algae and microalgae—A review. Food Chem 98(1):136–148

Hilhorst MJ, Somsen GW, de Jong GJ (1998) Potential of capillary electrophoresis for the profiling of propolis. HRC-J High Res Chrom 21(11):608–612

Himbert F, Dreux M, Pennanec R, Chaimbault P, Elfakir C, Lafosse M (2003) Le split en mode actif: Une approche innovante du couplage de la chromatographie liquide en phase normale à la spectrométrie de masse. Spectra Anal 32(235):29–31

Hinshaw JV (2003) Solid-phase microextraction. Lc Gc N Am 21(11):1056–1061

Hirahara Y, Kimura M, Inoue T, Uchikawa S, Otani S, Haganuma A, Matsumoto N, Hirata A, Maruyama S, Iizuka T, Ukyo M, Ota M, Hirose H, Suzuki S, Uchida Y (2005) Validation of multiresidue screening methods for the determination of 186 pesticides in 11 agricultural products using gas chromatography (GC). J Health Sci 51(5):617–627

Hommerson P, Khan AM, de Jong GJ, Somsen GW (2009) Capillary electrophoresis-atmospheric pressure chemical ionization-mass spectrometry using an orthogonal interface: set-up and system parameters. J Am Soc Mass Spectr 20(7):1311–1318

Hompesch RW, Garcia CD, Weiss DJ, Vivanco JM, Henry CS (2005) Analysis of natural flavonoids by microchip-micellar electrokinetic chromatography with pulsed amperometric detection. Analyst 130(5):694–700

Hossain MB, Barry-Ryan C, Martin-Diana AB, Brunton NP (2011) Optimisation of accelerated solvent extraction of antioxidant compounds from rosemary (*Rosmarinus officinalis L.*), marjoram (*Origanum majorana L.*) and oregano (*Origanum vulgare L.*) using response surface methodology. Food Chem 126(1):339–346

Hromadkova Z, Ebringerova A (2003) Ultrasonic extraction of plant materials-investigation of hemicellulose release from buckwheat hulls. Ultrason Sonochem 10(3):127–133

Hurtado-Fernandez E, Gomez-Romero M, Carrasco-Pancorbo A, Fernandez-Gutierrez A (2010) Application and potential of capillary electroseparation methods to determine antioxidant phenolic compounds from plant food material. J Pharm Biomed 53(5):1130–1160

Illes V, Daood HG, Biacs PA, Gnayfeed MH, Meszaros B (1999) Supercritical CO_2 and subcritical propane extraction of spice red pepper oil with special regard to carotenoid and tocopherol content. J Chromatogr Sci 37(9):345–352

Inagaki S, Min JZ, Toyo'oka T (2007) Direct detection method of oligosaccharides by high-performance liquid chromatography with charged aerosol detection. Biomed Chromatogr 21(4):338–342

Ismail SMM, Ali HM, Habiba RA (1993) Gc-Ecd and Gc-Ms analyses of profenofos residues and its biochemical effects in tomatoes and tomato products. J Agr Food Chem 41(4):610–615

Ito Y, Sandlin J, Bowers WG (1982) High-speed preparative countercurrent chromatography with a coil planet centrifuge. J Chromatogr 244(2):247–258

Janicsak G, Toth E, Mathe I (2007) TLC-densitometric investigations of phenylpropanoid glycosides in black horehound (*Ballota nigra L.*). JPC J Planar Chromat 20(6):443–446

Jelen HH, Majcher M, Dziadas M (2012) Microextraction techniques in the analysis of food flavor compounds: a review. Anal Chim Acta 738:13–26

Jenkins AJ, Llosa T, Montoya I, Cone EJ (1996) Identification and quantitation of alkaloids in coca tea. Forensic Sci Int 77(3):179–189

Jerkovic V, Nguyen F, Timmermans A, Collin S (2008) Comparison of procedures for resveratrol analysis in beer: Assessment of stilbenoids stability through wort fermentation and beer aging. J I Brewing 114(2):143–149

Jessome LL, Volmer DA (2006) Ion suppression: a major concern in mass spectrometry. LCGC N Am 24(5):83–89

Jiang CW, Ren QL, Wu PD (2003) Study on retention factor and resolution of tocopherols by supercritical fluid chromatography. J Chromatogr A 1005(1–2):155–164

Jiang LQ, Fang GZ, Zhang Y, Cao GJ, Wang S (2008) Analysis of flavonoids in propolis and Ginkgo biloba by Micellar electrokinetic capillary chromatography. J Agr Food Chem 56(24):11571–11577

Jiao JJ, Zhang Y, Liu CM, Liu J, Wu XQ, Zhang Y (2007) Separation and purification of tricin from an antioxidant product derived from bamboo leaves. J Agr Food Chem 55(25):10086–10092

Jokic S, Velic D, Bilic M, Bucic-Kojic A, Planinic M, Tomas S (2010) Modelling of the process of soli--liquid extraction of total polyphenols from soybeans. Czech J Food Sci 28(3):206–212

Jover E, Adahchour M, Bayona JM, Vreuls RJJ, Brinkman UAT (2005) Characterization of lipids in complex samples using comprehensive two-dimensional gas chromatography with time-of-flight mass spectrometry. J Chromatogr A 1086(1–2):2–11

Justesen U (2000) Negative atmospheric pressure chemical ionisation low-energy collision activation mass spectrometry for the characterisation of flavonoids in extracts of fresh herbs. J Chromatogr A 902(2):369–379

Kalili KM, de Villiers A (2009) Off-line comprehensive 2-dimensional hydrophilic interaction × reversed phase liquid chromatography analysis of procyanidins. J Chromatogr A 1216(35):6274–6284

Karatapanis AE, Fiamegos YC, Stalikas CD (2009) HILIC separation and quantitation of water-soluble vitamins using diol column. J Sep Sci 32(7):909–917

Karuna R, Sashidhar RB (1999) Use of ion-exchange chromatography coupled with TLC-laser scanning densitometry for the quantitation of fumonisin B(1). Talanta 50(2):381–389

Kataoka H, Lord HL, Pawliszyn J (2000) Applications of solid-phase microextraction in food analysis. J Chromatogr A 880(1–2):35–62

Kim TH, Lee SM, Kim YS, Kim KH, Oh S, Lee HJ (2003) Aroma dilution method using GC injector split ratio for volatile compounds extracted by headspace solid phase microextraction. Food Chem 83(1):151–158

Kolouchova-Hanzlikova I, Melzoch K, Filip V, Smidrkal J (2004) Rapid method for resveratrol determination by HPLC with electrochemical and UV detections in wines. Food Chem 87(1):151–158

Konig WA, Hochmuth DH (2004) Enantioselective gas chromatography in flavor and fragrance analysis: strategies for the identification of known and unknown plant volatiles. J Chromatogr Sci 42(8):423–439

Korytar P, Janssen HG, Matisova E, Brinkman UAT (2002) Practical fast gas chromatography: methods, instrumentation and applications. Trac Trend Anal Chem 21(9–10):558–572

Kostiainen R, Kauppila TJ (2009) Effect of eluent on the ionization process in liquid chromatography-mass spectrometry. J Chromatogr A 1216(4):685–699

Kovacevic M, Kac M (2001) Solid-phase microextraction of hop volatiles—potential use for determination and verification of hop varieties. J Chromatogr A 918(1):159–167

Kristo ST, Ganzler K, Apati P, Szoke E, Kery A (2002) Analysis of antioxidant flavonoids from Asteraceae and Moraceae plants by capillary electrophoresis. Chromatographia 56:S121–S126

Kumar V, Mukherjee K, Kumar S, Mal M, Mukherjee PK (2008) Validation of HPTLC method for the analysis of taraxerol in Clitoria ternatea. Phytochem Anal 19(3):244–250

Lamuelaraventos RM, Romeroperez AI, Waterhouse AL, Delatorreboronat MC (1995) Direct HPLC analysis of cis-resveratrol and trans-resveratrol and piceid isomers in Spanish red vitis-vinifera wines. J Agr Food Chem 43(2):281–283

Lang QY, Wai CM (2001) Supercritical fluid extraction in herbal and natural product studies—a practical review. Talanta 53(4):771–782

Larroque V, Desauziers V, Mocho P (2006) Development of a solid phase microextraction (SPME) method for the sampling of VOC traces in indoor air. J Environ Monit 8(1):106–111

Lee I, Boyce MC, Breadmore MC (2010) Quantitative determination of glucoraphanin in Brassica vegetables by micellar electrokinetic capillary chromatography. Anal Chim Acta 663(1):105–108

Leijdekkers AGM, Sanders MG, Schols HA, Gruppen H (2011) Characterizing plant cell wall derived oligosaccharides using hydrophilic interaction chromatography with mass spectrometry detection. J Chromatogr A 1218(51):9227–9235

Lerma-Garcia MJ, Simo-Alfonso EF, Mendez A, Lliberia JL, Herrero-Martinez JM (2010) Fast separation and determination of sterols in vegetable oils by ultraperformance liquid chromatography with atmospheric pressure chemical ionization mass spectrometry detection. J Agr Food Chem 58(5):2771–2776

Lesellier E (2012) Efficiency in supercritical fluid chromatography with different superficially porous and fully porous particles ODS bonded phases. J Chromatogr A 1228:89–98

Lesellier E, Marty C, Berset C, Tchapla A (1989) Optimization of the isocratic non-aqueous reverse phase (NARP) HPLC separation of trans cis alpha-carotenes and beta-carotenes. HRC J High Res Chrom 12(7):447–454

Lesellier E, Tchapla A, Pechard MR, Lee CR, Krstulovic AM (1991) Separation of trans/cis alpha-carotene and beta-carotene by supercritical fluid chromatography. 2. effect of the type of octadecyl-bonded stationary phase on retention and selectivity of carotenes. J Chromatogr 557(1–2):59–67

Li H, Bo C, Zhang Z, Yao S (2004) Focused microwave-assisted solvent extraction and HPLC determination of effective constituents in *Eucommia ulmodies* Oliv. (*E. ulmodies*). Talanta 63(3):659–665

Liang JL, Yang Z, Cao XJ, Wu B, Wu SH (2011) Preparative isolation of novel antioxidant flavonoids of alfalfa by stop-and-go counter-current chromatography and following on-line liquid chromatography desalination. J Chromatogr A 1218(36):6191–6199

Lisa M, Lynen F, Holcapek M, Sandra P (2007) Quantitation of triacylglycerols from plant oils using charged aerosol detection with gradient compensation. J Chromatogr A 1176(1–2):135–142

Liu C, Han JY, Duan YQ, Huang X, Wang H (2007a) Purification and quantification of ginsenoside Rb-3 and Rc from crude extracts of caudexes and leaves of *Panax notoginseng*. Sep Purif Technol 54(2):198–203

Liu HT, Li YF, Luan TG, Lan CY, Shu WS (2007b) Simultaneous determination of phytohormones in plant extracts using SPME and HPLC. Chromatographia 66(7–8):515–520

Longo C, Leo L, Leone A (2012) Carotenoids, fatty acid composition and heat stability of supercritical carbon dioxide-extracted-oleoresins. Int J Mol Sci 13(4):4233–4254

Lopez M, Arce L, Garrido J, Rios A, Valcarcel M (2004) Selective extraction of astaxanthin from crustaceans by use of supercritical carbon dioxide. Talanta 64(3):726–731

Lou ZX, Wang HX, Li J, Zhu S, Lu WP, Ma CY (2011) Effect of simultaneous ultrasonic/microwave assisted extraction on the antioxidant and antibacterial activities of burdock leaves. J Med Plants Res 5(22):5370–5377

Lu YB, Sun CR, Wang Y, Pan YJ (2007) Two-dimensional counter-current chromatography for the preparative separation of prenylflavonoids from Artocarpus altilis. J Chromatogr A 1151(1–2):31–36
Lubda D, Cabrera K, Kraas W, Schaefer C, Cunningham D, Majors RE (2001) New developments in the application of monolithic HPLC columns. LC-GC Europe 14(12):730–732
Lucchesi ME, Chemat F, Smadja J (2004a) An original solvent free microwave extraction of essential oils from spices. Flavour Frag J 19(2):134–138
Lucchesi ME, Chemat F, Smadja J (2004b) Solvent-free microwave extraction of essential oil from aromatic herbs: comparison with conventional hydro-distillation. J Chromatogr A 1043(2):323–327
Luo QL, Andrade JD, Caldwell KD (1998) Thin-layer ion-exchange chromatography of proteins. J Chromatogr A 816(1):97–105
Luque-Garcia JL, Luque de Castro MD (2003) Ultrasound: a powerful tool for leaching. Trac Trend Anal Chem 22(1):41–47
Makino M, Motegi T, Fujimoto Y (2004) Tirucallane-type triterpenes from Juliania adstringens. Phytochemistry 65(7):891–896
Malovana S, Montelongo FJG, Perez JP, Rodriguez-Delgado MA (2001) Optimisation of sample preparation for the determination of trans-resveratrol and other polyphenolic compounds in wines by high performance liquid chromatography. Anal Chim Acta 428(2):245–253
Mandal V, Mohan Y, Hemalatha S (2007) Microwave assisted extraction—An innovative and promising extraction tool for medicinal plant research. Pharmacognosy Rev 1(1):7–18
Manninen P, Laakso P (1997) Capillary supercritical fluid chromatography atmospheric pressure chemical ionization mass spectrometry of triacylglycerols in berry oils. J Am Oil Chem Soc 74(9):1089–1098
Manzano P, Arnaiz E, Diego JC, Toribio L, Garcia-Viguera C, Bernal JL, Bernal J (2011) Comprehensive two-dimensional gas chromatography with capillary flow modulation to separate FAME isomers. J Chromatogr A 1218(30):4952–4959
Maoela MS, Arotiba OA, Baker PGL, Mabusela WT, Jahed N, Songa EA, Iwuoha EI (2009) Electroanalytical determination of catechin flavonoid in ethyl acetate extracts of medicinal plants. Int J Electrochem Sci 4:1497–1510
Marcato B, Cecchin G (1996) Analysis of mixtures containing free fatty acids and mono-, di- and triglycerides by high-performance liquid chromatography coupled with evaporative light-scattering detection. J Chromatogr A 730(1–2):83–90
Marchi I, Rudaz S, Veuthey JL (2009) Atmospheric pressure photoionization for coupling liquid-chromatography to mass spectrometry: a review. Talanta 78(1):1–18
Marioli JM, Sereno LE (1996) Electrochemical Detection of Underivatized Amino Acids with a Ni–Cr alloy electrode. J Liq Chromatogr Relat Technol 19:2505–2515
Marriott PJ, Kinghorn RM (1997) Longitudinally modulated cryogenic system. A generally applicable approach to solute trapping and mobilization in gas chromatography. Anal Chem 69(13):2582–2588
Marston A, Hostettmann K (2006) Developments in the application of counter-current chromatography to plant analysis. J Chromatogr A 1112(1–2):181–194
Martin AJ, Synge RL (1941) A new form of chromatogram employing two liquid phases: a theory of chromatography. 2. Application to the micro-determination of the higher monoamino-acids in proteins. Biochem J 35(12):1358–1368
Matsubara A, Uchikata T, Shinohara M, Nishiumi S, Yoshida M, Fukusaki E, Bamba T (2012) Highly sensitive and rapid profiling method for carotenoids and their epoxidized products using supercritical fluid chromatography coupled with electrospray ionization-triple quadrupole mass spectrometry. J Biosci Bioeng 113(6):782–787
Megoulas NC, Koupparis MA (2005) Twenty years of evaporative light scattering detection. Crit Rev Anal Chem 35(4):301–316
Michel T, Destandau E, Elfakir C (2011a) Evaluation of a simple and promising method for extraction of antioxidants from sea buckthorn (*Hippophae rhamnoides L.*) berries: pressurised solvent-free microwave assisted extraction. Food Chem 126(3):1380–1386

Michel T, Destandau E, Elfakir C (2011b) On-line hyphenation of centrifugal partition chromatography and high pressure liquid chromatography for the fractionation of flavonoids from *Hippophae rhamnoides L.* berries. J Chromatogr A 1218(36):6173–6178
Moini M (2007) Simplifying CE-MS operation. 2. Interfacing low-flow separation techniques to mass spectrometry using a porous tip. Anal Chem 79(11):4241–4246
Mol R, de Jong GJ, Somsen GW (2005) On-line capillary electrophoresis-mass spectrometry using dopant-assisted atmospheric pressure photoionization: setup and system performance. Electrophoresis 26(1):146–154
Mondello L, Casilli A, Tranchida PQ, Costa R, Dugo P, Dugo G (2004) Fast GC for the analysis of citrus oils. J Chromatogr Sci 42(8):410–416
Mondello L, Tranchida PQ, Costa R, Casilli A, Dugo P, Cotroneo A, Dugo G (2003) Fast GC for the analysis of fats and oils. J Sep Sci 26(17):1467–1473
Moreau RA (2006) The analysis of lipids via HPLC with a charged aerosol detector. Lipids 41(7):727–734
Moreau RA, Scott KM, Haas MJ (2008) The identification and quantification of steryl glucosides in precipitates from commercial biodiesel. J Am Oil Chem Soc 85(8):761–770
Moricz AM, Ott PG, Boszormenyi A, Lemberkovics E, Mincsovics E, Tyihak E (2012a) Bioassay-guided isolation and identification of antimicrobial compounds from thyme essential oil by means of overpressured layer chromatography, bioautography and GC-MS. Chromatographia 75(17–18):991–999
Moricz AM, Szarka S, Ott PG, Hethelyi EB, Szoke E, Tyihak E (2012b) Separation and identification of antibacterial chamomile components using OPLC, bioautography and GC-MS. Med Chem 8(1):85–94
Morlock GE, Jautz U (2008) Comparison of two different plunger geometries for HPTLC-MS coupling via an extractor-based interface. JPC-J Planar Chromat 21(5):367–371
Mulbry W, Kondrad S, Buyer J, Luthria DL (2009) Optimization of an oil extraction process for algae from the treatment of manure effluent. J Am Oil Chem Soc 86(9):909–915
Munoz-Gonzalez C, Moreno-Arribas MV, Rodriguez-Bencomo JJ, Cueva C, Alvarez PJM, Bartolome B, Pozo-Bayon MA (2012) Feasibility and application of liquid-liquid extraction combined with gas chromatography-mass spectrometry for the analysis of phenolic acids from grape polyphenols degraded by human faecal microbiota. Food Chem 133(2):526–535
Murayama W, Kobayashi T, Kosuge Y, Yano H, Nunogaki Y, Nunogaki K (1982) A new centrifugal countercurrent chromatograph and its application. J Chromatogr 239:643–649
Ng LK (2002) Analysis by gas chromatography/mass spectrometry of fatty acids and esters in alcoholic beverages and tobaccos. Anal Chim Acta 465(1–2):309–318
Nichols JA, Foster MD (2009) Mass rate attenuator. USA Patent n° 7,575,723 B2
Nkhili E, Tomao V, El Hajji H, El Boustani ES, Chemat F, Dangles O (2009) Microwave-assisted water extraction of green tea polyphenols. Phytochem Anal 20(5):408–415
Nogueira LC, Silva F, Ferreira IMPLVO, Trugo LC (2005) Separation and quantification of beer carbohydrates by high-performance liquid chromatography with evaporative light scattering detection. J Chromatogr A 1065(2):207–210
Novakova L, Matysova L, Solichova D, Koupparis MA, Solich P (2004) Comparison of performance of C18 monolithic rod columns and conventional C18 particle-packed columns in liquid chromatographic determination of Estrogel and Ketoprofen gel. J Chromatogr B 813(1–2):191–197
Novakova L, Vildova A, Mateus JP, Goncalves T, Solich P (2010) Development and application of UHPLC–MS/MS method for the determination of phenolic compounds in Chamomile flowers and Chamomile tea extracts. Talanta 82(4):1271–1280
Nunes C, Rocha SM, Saraiva J, Coimbra MA (2006) Simple and solvent-free methodology for simultaneous quantification of methanol and acetic acid content of plant polysaccharides based on headspace solid phase microextraction-gas chromatography (HS-SPME-GG-FID). Carbohyd Polym 64(2):306–311
Nunez O, Gallart-Ayala H, Martins CPB, Lucci P (2012) New trends in fast liquid chromatography for food and environmental analysis. J Chromatogr A 1228:298–323

Nyiredy S (2001) The bridge between TLC and HPLC: overpressured layer chromatography (OPLC). Trac-Trend Anal Chem 20(2):91–101

Olszewska MA (2012) New validated high-performance liquid chromatographic method for simultaneous analysis of ten flavonoid aglycones in plant extracts using a C18 fused-core column and acetonitrile-tetrahydrofuran gradient. J Sep Sci 35(17):2174–2183

Otsuka K, Terabe S (1996) Micellar electrokinetic chromatography. Capillary electrophoresis guide book. Methods in molecular biology, vol 52. Humana Press Inc, Totowa

Ozel MZ, Gogus F, Lewis AC (2006) Comparison of direct thermal desorption with water distillation and superheated water extraction for the analysis of volatile components of Rosa damascena Mill. using GC × GC-TOF/MS. Anal Chim Acta 566(2):172–177

Ozel MZ, Gogus F, Lewis AC (2008) Composition of Eucalyptus camaldulensis volatiles using direct thermal desorption coupled with comprehensive two-dimensional gas chromatography-time-of-flight-mass spectrometry. J Chromatogr Sci 46(2):157–161

Paolini J, Leandri C, Desjobert JM, Barboni T, Costa J (2008) Comparison of liquid–liquid extraction with headspace methods for the characterization of volatile fractions of commercial hydrolats from typically Mediterranean species. J Chromatogr A 1193(1–2):37–49

Park HS, Lee HJ, Shin MH, Lee KW, Lee H, Kim YS, Kim KO, Kim KH (2007) Effects of cosolvents on the decaffeination of green tea by supercritical carbon dioxide. Food Chem 105(3):1011–1017

Pelusio F, Nilsson T, Montanarella L, Tilio R, Larsen B, Facchetti S, Madsen JO (1995) Headspace solid-phase microextraction analysis of volatile organic sulfur-compounds in black-and-white truffle aroma. J Agr Food Chem 43(8):2138–2143

Peng JY, Fan GR, Hong ZY, Chai YF, Wu YT (2005) Preparative separation of isovitexin and isoorientin from Patrinia villosa Juss by high-speed counter-current chromatography. J Chromatogr A 1074(1–2):111–115

Perez RA, Navarro T, de Lorenzo C (2007) HS-SPME analysis of the volatile compounds from spices as a source of flavour in 'Campo Real' table olive preparations. Flavour Frag J 22(4):265–273

Petritis KN, Chaimbault P, Elfakir C, Dreux M (1999) Ion-pair reversed-phase liquid chromatography for determination of polar underivatized amino acids using perfluorinated carboxylic acids as ion pairing agent. J Chromatogr A 833(2):147–155

Phillips JB, Beens J (1999) Comprehensive two-dimensional gas chromatography: a hyphenated method with strong coupling between the two dimensions. J Chromatogr A 856(1–2):331–347

Pichard H, Caude M, Morin P, Richard H, Rosset R (1990) Identification of pepper essential oil constituents by various chromatography-spectroscopy couplings. Analusis 18(3):167–178

Polesello S, Lovati F, Rizzolo A, Rovida C (1993) Supercritical-fluid extraction as a preparative tool for strawberry aroma analysis. HRC-J High Res Chrom 16(9):555–559

Pongnaravane B, Goto M, Sasaki M, Anekpankul T, Pavasant P, Pavasant P, Shotipruk A (2006) Extraction of anthraquinones from roots of Morinda citrifolia by pressurized hot water: antioxidant activity of extracts. J Supercrit Fluid 37(3):390–396

Poukensrenwart P, Tits M, Wauters JN, Angenot L (1992) Reversed-phase hptlc densitometric evaluation of fraxin in fraxinus-excelsior leaves. J Pharmaceut Biomed 10(10–12):1089–1091

Pruvost A, Becher F, Bardouille P, Guerrero C, Creminon C, Delfraissy JF, Goujard C, Grassi J, Benech H (2001) Direct determination of phosphorylated intracellular anabolites of stavudine (d4T) by liquid chromatography tandem mass spectrometry. Rapid Commun Mass Sp 15(16):1401–1408

Pukalskas A, van Beek TA, de Waard P (2005) Development of a triple hyphenated HPLC-radical scavenging detection-DAD-SPE-NMR system for the rapid identification of antioxidants in complex plant extracts. J Chromatogr A 1074(1–2):81–88

Ramirez-Coronel MAN, Marnet N, Kolli VSK, Roussos S, Guyot S, Augur C (2004) Characterization and estimation of proanthocyanidins and other phenolics in coffee pulp (Coffea arabica) by thiolysis-high-performance liquid chromatography. J Agr Food Chem 52(5):1344–1349

Ramirez P, Senorans FJ, Ibanez E, Reglero G (2004) Separation of rosemary antioxidant compounds by supercritical fluid chromatography on coated packed capillary columns. J Chromatogr A 1057(1–2):241–245
Rather MA, Ganai BA, Kamili AN, Qayoom M, Akbar S, Masood A, Rasool R, Wani SH, Qurishi MA (2012) Comparative GC-FID and GC-MS analysis of the mono and sesquiterpene secondary metabolites produced by the field grown and micropropagated plants of Artemisia amygdalina Decne. Acta Physiol Plant 34(3):885–890
Rauha JP, Vuorela H, Kostiainen R (2001) Effect of eluent on the ionization efficiency of flavonoids by ion spray, atmospheric pressure chemical ionization, and atmospheric pressure photoionization mass spectrometry. J Mass Spectrom 36(12):1269–1280
Remoroza C, Cord-Landwehr S, Leijdekkers AGM, Moerschbacher BM, Schols HA, Gruppen H (2012) Combined HILIC-ELSD/ESI-MSn enables the separation, identification and quantification of sugar beet pectin derived oligomers. Carbohyd Polym 90(1):41–48
Rhourri-Frih B, Chaimbault P, Claude B, Lamy C, Andre P, Lafosse M (2009) Analysis of pentacyclic triterpenes by LC-MS. A comparative study between APCI and APPI. J Mass Spectrom 44(1):71–80
Rhourri-Frih B, Chaimbault P, Dequeral D, Andre P, Lafosse M (2012) Investigation of porous graphitic carbon for triterpenoids and natural resinous materials analysis by high performance liquid chromatography hyphenated to mass spectrometry. J Chromatogr A 1240:140–146
Richter BE, Bornhop DJ, Swanson JT, Wangsgaard JG, Andersen MR (1989) Gas-chromatographic detectors in SFC. J Chromatogr Sci 27(6):303–308
Rodriguez-Amaya DB (2001) A guide to carotenoids analysis in foods ILSI press. International Life Sciences Institute. One Thomas Circle, N.W. Washington, DC
Rosset R, Caude M, Jardy A (1991) Chromatographie en phase liquide et supercritique. Paris
Rostagno MA, Manchon N, D'Arrigo M, Guillamon E, Villares A, Garcia-Lafuente A, Ramos A, Martinez JA (2011) Fast and simultaneous determination of phenolic compounds and caffeine in teas, mate, instant coffee, soft drink and energetic drink by high-performance liquid chromatography using a fused-core column. Anal Chim Acta 685(2):204–211
Rostagno MA, Palma M, Barroso CG (2007) Fast analysis of soy isoflavones by high-performance liquid chromatography with monolithic columns. Anal Chim Acta 582(2):243–249
Ruperez FJ, Martin D, Herrera E, Barbas C (2001) Chromatographic analysis of alpha-tocopherol and related compounds in various matrices. J Chromatogr A 935(1–2):45–69
Sagratini G, Mañes J, Giardiná D, Damiani P, Picó Y (2007) Analysis of carbamate and phenylurea pesticide residues in fruit juices by solid-phase microextraction and liquid chromatography–mass spectrometry. J Chromatogr A 1147(2):135–143
Salvador A, Herbreteau B, Dreux M (1999) Electrospray mass spectrometry and supercritical fluid chromatography of methylated beta-cyclodextrins. J Chromatogr A 855(2):645–656
Sanchez-Avila N, Priego-Capote F, Ruiz-Jimenez J, de Castro MDL (2009) Fast and selective determination of triterpenic compounds in olive leaves by liquid chromatography-tandem mass spectrometry with multiple reaction monitoring after microwave-assisted extraction. Talanta 78(1):40–48
Sannomiya M, Rodrigues CM, Coelho RG, dos Santos LC, Hiruma-Lima CA, Brito ARS, Vilegas W (2004) Application of preparative high-speed counter-current chromatography for the separation of flavonoids from the leaves of Byrsonima crassa Niedenzu (IK). J Chromatogr A 1035(1):47–51
Scampicchio M, Wang J, Mannino S, Chatrathi MP (2004) Microchip capillary electrophoresis with amperometric detection for rapid separation and detection of phenolic acids. J Chromatogr A 1049(1–2):189–194
Schroder M, Vetter W (2011) High-speed counter-current chromatographic separation of phytosterols. Anal Bioanal Chem 400(10):3615–3623
Schurek J, Portoles T, Hajslova J, Riddellova K, Hernandez F (2008) Application of head-space solid-phase microextraction coupled to comprehensive two-dimensional gas chromatography-time-of-flight mass spectrometry for the determination of multiple pesticide residues in tea samples. Anal Chim Acta 611(2):163–172

Seger C, Godejohann M, Tseng LH, Spraul M, Girtler A, Sturm S, Stuppner H (2005) LC-DAD-MS/SPE-NMR hyphenation. A tool for the analysis of pharmaceutically used plant extracts: identification of isobaric iridoid glycoside regioisomers from Harpagophytum procumbens. Anal Chem 77(3):878–885

Seker ME, Celik A, Dost K (2012) Determination of Vitamin E isomers of grape seeds by high-performance liquid chromatography-UV detection. J Chromatogr Sci 50(2):97–101

Seppanen-Laakso T, Laakso I, Hiltunen R (2002) Analysis of fatty acids by gas chromatography, and its relevance to research on health and nutrition. Anal Chim Acta 465(1–2):39–62

Shanker K, Singh SC, Pant S, Srivastava P, Yadav AK, Pandey R, Verma RK, Gupta MM (2008) Quantitative TLC analysis of sterol (24 beta-ethylcholesta-5,22E,25-triene-3 beta-ol) in agnimantha (Clerodendrum phlomidis linn). Chromatographia 67(3–4):269–274

Sharma UK, Sharma N, Gupta AP, Kumar V, Sinha AK (2007) RP-HPTLC densitometric determination and validation of vanillin and related phenolic compounds in accelerated solvent extract of Vanilla planifolia. J Sep Sci 30(18):3174–3180

Shirota O, Nagamatsu K, Sekita S, Komoto N, Kuroyanagi M, Ichikawa M, Ohta S, Ushijima M (2008) Preparative separation of the saponin lancemaside A from Codonopsis lanceolata by centrifugal partition chromatography. Phytochem Anal 19(5):403–410

Shui GH, Leong LP (2004) Analysis of polyphenolic antioxidants in star fruit using liquid chromatography and mass spectrometry. J Chromatogr A 1022(1–2):67–75

Sides A, Robards K, Helliwell S (2000) Developments in extraction techniques and their application to analysis of volatiles in foods. Trac-Trend Anal Chem 19(5):322–329

Soni M, Patidar K, Jain D, Jain S (2010) Ultrasound assisted extraction (UAE): a novel extraction technique for extraction of neutraceuticals from plants. J Pharm Res 3(3):636–638

Srivastava SK (2008) Extraction technologies for medicinal and aromatic plants. International Centre of Science and High Technology, Trieste

Srivastava V, Singh M, Malasonil R, Shanker K, Verma RK, Gupta MM, Gupta AK, Khanuja SPS (2008) Separation and quantification of lignans in Phyllanthus species by a simple chiral densitometric method. J Sep Sci 31(1):47–55

Stecher G, Huck CW, Popp M, Bonn GK (2001) Determination of flavonoids and stilbenes in red wine and related biological products by HPLC and HPLC-ESI-MS-MS. Fresen J Anal Chem 371(1):73–80

Sticher O (2008) Natural product isolation. Nat Prod Rep 25(3):517–554

Stoggl WM, Huck CW, Scherz H, Popp M, Bonn GK (2001) Analysis of vitamin E in food and phytopharmaceutical preparations by HPLC and HPLC-APCI-MS-MS. Chromatographia 54(3–4):179–185

Stoll DR, Li XP, Wang XO, Carr PW, Porter SEG, Rutan SC (2007) Fast, comprehensive two-dimensional liquid chromatography. J Chromatogr A 1168(1–2):3–43

Stolyhwo A, Colin H, Guiochon G (1985) Analysis of triglycerides in oils and fats by liquid-chromatography with the laser-light scattering detector. Anal Chem 57(7):1342–1354

Storm T, Reemtsma T, Jekel M (1999) Use of volatile amines as ion-pairing agents for the high-performance liquid chromatographic-tandem mass spectrometric determination of aromatic sulfonates in industrial wastewater. J Chromatogr A 854(1–2):175–185

Subagio A, Sari P, Morita N (2001) Simultaneous determination of (+)-catechin and (−)-epicatechin in cacao and its products by high-performance liquid chromatography with electrochemical detection. Phytochem Anal 12(4):271–276

Sun P, Armstrong DW (2010) Ionic liquids in analytical chemistry. Anal Chim Acta 661(1):1–16

Suslick KS (1994) Encyclopaedia Britannica Yearbook of science and the future. Britannica, Chicago

Sutherland IA, Fisher D (2009) Role of counter-current chromatography in the modernisation of Chinese herbal medicines. J Chromatogr A 1216(4):740–753

Syage JA, Short LC, Cai SS (2008) Atmospheric pressure photoionization—The second source for LC-MS? Lc Gc N Am 26(3):286

Tangney CC, Driskell JA, Mcnair HM (1979) Separation of Vitamin-E isomers by high-performance liquid-chromatography. J Chromatogr 172:513–515

Taylor LT (2009) Supercritical fluid chromatography for the 21st century. J Supercrit Fluid 47(3):566–573
Tekel J, Hudecova T, Pecnikova K (2001) Isolation and purification techniques for pesticide residue analyses in samples of plant or animal origin. Eur Food Res Technol 213(4–5):250–258
Theodoridis G, Lasakova M, Skerikova V, Tegou A, Giantsiou N, Jandera P (2006) Molecular imprinting of natural flavonoid antioxidants: application in solid-phase extraction for the sample pretreatment of natural products prior to HPLC analysis. J Sep Sci 29(15):2310–2321
Tolstikov VV, Fiehn O (2002) Analysis of highly polar compounds of plant origin: combination of hydrophilic interaction chromatography and electrospray ion trap mass spectrometry. Anal Biochem 301(2):298–307
Toribio A, Destandau E, Elfakir C, Lafosse M (2009) Hyphenation of centrifugal partition chromatography with electrospray ionization mass spectrometry using an active flow-splitter device for characterization of flavonol glycosides. Rapid Commun Mass Sp 23(12):1863–1870
Toribio A, Nuzillard JM, Renault JH (2007) Strong ion-exchange centrifugal partition chromatography as an efficient method for the large-scale purification of glucosinolates. J Chromatogr A 1170(1–2):44–51
Tranchida PQ, Shellie RA, Purcaro G, Conte LS, Dugo P, Dugo G, Mondello L (2010) Analysis of fresh and aged tea tree essential oils by using GC × GC-qMS. J Chromatogr Sci 48 (4):262
Tung YT, Chang WC, Chen PS, Chang TC, Chang ST (2011) Ultrasound-assisted extraction of phenolic antioxidants from Acacia confusa flowers and buds. J Sep Sci 34(7):844–851
Turne C, King JW, Mathiasson L (2001) Supercritical fluid extraction and chromatography for fat-soluble vitamin analysis. J Chromatogr A 936(1–2):215–237
Tyihak E, Mincsovics E, Moricz AM (2012) Overpressured layer chromatography: from the pressurized ultramicro chamber to BioArena system. J Chromatogr A 1232:3–18
Uquiche E, Jerez M, Ortiz J (2008) Effect of pretreatment with microwaves on mechanical extraction yield and quality of vegetable oil from Chilean hazelnuts (*Gevuina avellana Mol*). Innov Food Sci Emerg 9(4):495–500
Urakova IN, Pozharitskaya ON, Shikov AN, Kosman VM, Makarov VG (2008) Comparison of high performance TLC and HPLC for separation and quantification of chlorogenic acid in green coffee bean extracts. J Sep Sci 31(2):237–241
Vaher M, Koel M (2003) Separation of polyphenolic compounds extracted from plant matrices using capillary electrophoresis. J Chromatogr A 990(1–2):225–230
Valcarcel M, Tena MT (1997) Applications of supercritical fluid extraction in food analysis. Fresen J Anal Chem 358(5):561–573
Verhoeven HA, Jonker H, De Vos RC, Hall RD (2012) Solid phase micro-extraction GC-MS analysis of natural volatile components in melon and rice. Methods Mol Biol 860:85–99
Vervoort N, Daemena D, Töröka G (2008) Performance evaluation of evaporative light scattering detection and charged aerosol detection in reversed phase liquid chromatography. J Chromatogr A 1189(1–2):92–100
Viron C, Saunois A, Andre P, Perly B, Lafosse M (2000) Isolation and identification of unsaturated fatty acid methyl esters from marine micro-algae. Anal Chim Acta 409(1–2):257–266
Volpi N (2004) Separation of flavonoids and phenolic acids from propolis by capillary zone electrophoresis. Electrophoresis 25(12):1872–1878
Waksmundzka-Hainos M, Oniszczuk A, Szewczyk K, Wianowska D (2007) Effect of sample-preparation methods on the HPLC quantitation of some phenolic acids in plant materials. Acta Chromatogr 19:227–237
Waksmundzka-Hajnos M, Sherma J, Kowalska T (2008) Thin layer chromatography. In: Phytochemistry, chromatographic science series. CRC Press, Taylor & Francis, Boca Raton
Wanasundara UN, Wanasundara PKJPD, Shahidi F (2005). Bailey's industrial oil and fat products, 6th edn. Wiley, New York
Wang LJ, Weller CL (2006) Recent advances in extraction of nutraceuticals from plants. Trends Food Sci Tech 17(6):300–312

Wang PG, He W (2011) Hydrophilic interaction liquid chromatography (HILIC) and advanced applications. CRC Press, Boca Raton

Weber EJ (1984) High-Performance Liquid-Chromatography of the Tocols in Corn Grain. J Am Oil Chem Soc 61(7):1231–1234

Wichitnithad W, Jongaroonngamsang N, Pummangura S, Rojsitthisak P (2009) A simple isocratic HPLC method for the simultaneous determination of curcuminoids in commercial turmeric extracts. Phytochem Anal 20(4):314–319

Wilson ID (1999) The state of the art in thin-layer chromatography-mass spectrometry: a critical appraisal. J Chromatogr A 856(1–2):429–442

Wu JC, Xie W, Pawliszyn J (2000) Automated in-tube solid phase microextraction coupled with HPLC-ES-MS for the determination of catechins and caffeine in tea. Analyst 125(12):2216–2222

Wu JY, Lin LD, Chau FT (2001) Ultrasound-assisted extraction of ginseng saponins from ginseng roots and cultured ginseng cells. Ultrason Sonochem 8(4):347–352

Xiao Q, Yu CH, Xing J, Hu B (2006) Comparison of headspace and direct single-drop microextraction and headspace solid-phase microextraction for the measurement of volatile sulfur compounds in beer and beverage by gas chromatography with flame photometric detection. J Chromatogr A 1125(1):133–137

Yang Q, Zhang XL, Li XY, Tang WK, Zhang JX, Fang CX, Zheng CY (2007) Coupling continuous ultrasound-assisted extraction with ultrasonic probe, solid-phase extraction and high-performance liquid chromatography for the determination of sodium Danshensu and four tanshinones in Salvia miltiorrhiza bunge. Anal Chim Acta 589(2):231–238

Yu Y, Zhang QW, Li SP (2011) Preparative purification of coniferyl ferulate from Angelica sinensis oil by high performance centrifugal partition chromatography. J Med Plants Res 5(1):104–108

Yue TL, Shao DY, Yuan YH, Wang ZL, Qiang CY (2012) Ultrasound-assisted extraction, HPLC analysis, and antioxidant activity of polyphenols from unripe apple. J Sep Sci 35(16):2138–2145

Zarnowski R, Suzuki Y (2004) Expedient Soxhlet extraction of resorcinolic lipids from wheat grains. J Food Compos Anal 17(5):649–663

Zhang ZS, Wang LJ, Li D, Jiao SS, Chen XD, Mao ZH (2008) Ultrasound-assisted extraction of oil from flaxseed. Sep Purif Technol 62(1):192–198

Zhao X, Xu X, Su R, Zhang H, Wang Z (2012) An application of new microwave absorption tube in non-polar solvent microwave-assisted extraction of organophosphorus pesticides from fresh vegetable samples. J Chromatogr A 1229:6–12

Zhou HY, Liu CZ (2006) Microwave-assisted extraction of solanesol from tobacco leaves. J Chromatogr A 1129(1):135–139

Zhu JH, Qi SD, Zhang HG, Chen XG, Hu ZD (2008) Sample stacking and sweeping in microemulsion electrokinetic chromatography under pH-suppressed electroosmotic flow. J Chromatogr A 1192(2):319–322

Zou J, Li N (2006) Simple and environmental friendly procedure for the gas chromatographic-mass spectrometric determination of caffeine in beverages. J Chromatogr A 1136(1):106–110

Author Biography

Patrick Chaimbault (born 1969) is currently the Dean of the Department of Chemistry in Metz (University of Lorraine). He has been trained as chemist at the University of Orléans (France) and holds an MSc in Chemistry and Physicochemistry of Molecules of Biological Interest. In 2000, he obtained his PhD focusing on the analysis of underivatized amino acids by liquid chromatography coupled to mass spectrometry. After three years of experience as analytical manager in the pharmaceutical industry, he started his academic career. In September 2002, he became Assistant Professor in Analytical Chemistry at the University of Orléans. In 2007, he was habilitated to supervise research (French diploma: Habilitation à Diriger les Recherches). He moved to the University of Metz in November 2008 where he currently holds the position of Professor of Chemistry.

Patrick develops methods concerning the sample preparation and the analyses of complex biological materials (plants, animal cells, body fluids, etc.,) by Mass Spectrometry (MALDI-TOFMS, MALDI-FTICRMS, ESI-MS). He is an expert in coupling separative techniques (GC, LC, Capillary Electrophoresis) with Mass Spectrometry using API sources (ESI, APCI, APPI). To date, he has published over 40 articles in this field. Patrick is currently in charge of the analytical chemistry topic (Identification, Isolation and Analysis) included in the natural products project "NutriOx".

Part III
Connecting Section Between Chapters 3 and 4

At this point, we have already learned a fair amount about various redox active secondary metabolites, their major sources, and how we can identify, purify, and characterize them. We have not yet defined, however, what exactly is meant by 'redox activity'. As this particular term often raises rather different expectations, and frequently also leads to misconceptions, we will now follow the advice provided in Chap. 2 and turn our attention to the 'chemistry' that endows such metabolites with their pronounced biological activities.

The next chapter will therefore provide a brief introduction to redox activity and redox active substances. Here, we will learn about the various types of redox active compounds, from electron donors and acceptors to radicals and electrophiles ready to react with electron-rich molecules. We will see that some of these redox active compounds are very reactive, yet also selective as far as their reaction partners are concerned, while others are less picky and react with virtually any molecule in possession of electrons or—more mysterious—'reducing equivalents'. At the same time, we will see that electron transfer is just one of the possible mechanisms underlying redox activity, while others, such as radical and exchange processes, are perhaps less well known but *de facto* considerably more widespread in cell biology than previously thought.

And, of course, we will touch on the rather sticky issue of pro- and antioxidants. This serious area of research has recently been faced by an equally serious onslaught of pseudo-scientific make-belief, unfounded claims and attempts to make a quick Euro by selling all kinds of reducing agents/antioxidants for all kinds of—often cosmetic or health food related—purposes (see also Chaps. 2 and 5). The widespread uses of preparations, supplements, crèmes, and ointments containing diverse vitamins, the quinone-based coenzyme Q_{10} and the stilbene resveratrol reflect this problem. While some reducing agents indeed act as antioxidants when brought into contact with living cells and organisms, the concept of chemical reductants and oxidants is not simply transferable to Biology. Reductants are not necessarily 'good' antioxidants and oxidants are not necessarily 'bad' pro-oxidants. Indeed, many antioxidants can redox cycle and hence cause significant damage to cells, while more suspect oxidants turn out to be pivotal for intracellular signaling and host defense (see the Chap. 5). Importantly, not all reductants or oxidants unfold their redox activity under physiological conditions. In contrast, other chemical species, such as the zinc ion, can trigger a massive antioxidant response in cells without even being redox active under physiological conditions at all.

The next chapter will therefore consider some of these concepts and misconceptions. It will also showcase several unusual redox active species that only recently have been considered in the context of biological activity (such as polysulfanes, polysulfide ions, diverse electrophiles). Here, we will link the activity of these compounds to specific, redox sensitive intracellular targets, such as the cysteine proteins and enzymes, which together form the 'intracellular thiolstat', an emerging concept in redox signaling and regulation which will be discussed in more detail in some of the following chapters (e.g., in Chaps. 9 and 10).

Chapter 4
A Word on Redox

Torsten Burkholz and Claus Jacob

Keywords Electron transfer • Kinetics • Oxidation state • Redox processes • Thiol/disulfide exchange

1 The Term 'Redox'

Although previous chapters have employed the term 'redox' on numerous occasions, the expression itself is often used with very different meanings and in rather diverse contexts. As part of this chapter, we will therefore take a short detour through the facet-rich landscape of concepts which is associated with the redox notion. By infusing some basic chemistry and, above all, common sense into this matter, we will see that 'redox' is not really the (unhappy) castrated version of 'red-bull', and indeed gives wings, at least as far as the careers of some of our more ambitious colleagues are concerned. On the other hand, we will notice that 'redox' is actually related to "RedCat", our European network considering **red**ox **cat**alysts. Along the way, we will be confronted with some intricate redox chemistry, often in cohort with electrochemical concepts and mechanistic considerations, and also witness some rather shocking events, such as electrons rolling down energy hills in search for a resting place. We will begin this journey at our most preferred parking lot in the green valley of chemistry and then move slowly uphill toward more biochemical and ultimately biological concepts.

Expressions such as 'redox property', 'redox signaling', or 'redox regulation' are frequently used in biochemistry and cell biology to describe the action of particular, 'redox active' intrinsic or extrinsic molecules. While everyone has heard, read, or

T. Burkholz · C. Jacob (✉)
Division of Bioorganic Chemistry, School of Pharmacy, Saarland University, 66041 Saarbruecken, Germany
e-mail: c.jacob@mx.uni-saarland.de

C. Jacob et al. (eds.), *Recent Advances in Redox Active Plant and Microbial Products*,
DOI: 10.1007/978-94-017-8953-0_4,

even used such 'redox' terminology in the past, few of us have actually worried about the exact meaning of this term, and how it may be applied properly in a biological context. In fact, when explaining cellular 'redox processes' to colleagues from physical chemistry—who usually are fairly orthodox when it comes to terminology—one may be faced with utter disbelief. Those colleagues will explain in their perhaps somewhat sarcastic manner (justified when talking to an apparent ignorant), that there may be a red bull, but certainly no such thing as a 'redox'. A 'redox' reaction' in the strict meaning of the word, i.e., as a singular, isolated event, does not exist. It is rather a short form for *two* simultaneously occurring, and firmly linked reactions, one being a 'reduction', the other an 'oxidation' reaction. Formally 'reduction' of a substance involves a decrease of formal oxidation state(s), while the oxidation involves a corresponding increase. Ce^{4+} is reduced to Ce^{3+} and I^- is oxidized to I_2. In most cases, but not always, the reduction reaction therefore consumes electrons which are provided by the associated oxidation reaction. As both reactions are closely connected, and cannot occur without each other, the term 'redox reaction' is therefore rather casually employed to refer to this medley of simultaneous, connected reduction, and oxidation reactions. Yet, it is imprecise, as the 'redox process' at hand actually involves two reactions, not one. As we will see later, this distinction is actually rather important, especially when it comes to reaction mechanisms and to kinetics.

2 The Basics About 'Reduction' and 'Oxidation'

At this point, we briefly need to consider the meaning of 'reduction' and 'oxidation'. As mentioned above, a reduction involves a decrease of the formal oxidation state of one (or several) of the atoms or ions involved. Care should be taken that this 'decrease' also applies to negative oxidation states, which become even more negative during reduction. Hence the reduction of the disulfide H_2S_2 to the sulfide H_2S involves a change of formal oxidation state of sulfur from −1 to −2.[1] An oxidation, on the other hand, involves an increase of formal oxidation state. An Explanatory Box is at hand to explain how such formal oxidation states in complex molecules can be determined (see Explanatory Box 1).

Explanatory Box 1: Determination of formal oxidation states

This chapter discusses at length the more chemical issues surrounding reduction and oxidation processes. One key parameter in such reactions is the (formal) oxidation state of individual atoms in the reaction partners involved. It is therefore important that this particular parameter is determined properly.

[1] As there are no negative Roman numbers, one should refrain from using them to denote oxidation states and employ Arabic numbers instead.

Unfortunately, there are many—often unfounded—mysteries surrounding the oxidation states of particular atoms in complex molecules. On occasion, incorrect numbers are provided which subsequently may result in serious misperceptions. Here, we will consider a few basic rules which should prevent such unnecessary mistakes.

Atoms in elements generally have the oxidation state 0. In single atom ions, such as metal cations or halide anions, the oxidation state corresponds to the charge of the respective ion. Caution should be taken as the charge is given in the form X+ or X−, while the corresponding oxidation state is then +X or −X, respectively. As already mentioned, Roman numbers should be avoided and Arabic ones used instead, as there are also negative and even fractional formal oxidation states (see below). In simple chemical formulae, such as Na_2SO_4 or $KClO_4$, formal oxidation states can be calculated comparably easily for a particular atom as long as one is able to assign oxidation states to the others. In the case of Na_2SO_4, sulfur is therefore +6, assuming Na is +1 and O is −2, as is often (but not always) the case. The same applies to $KClO_4$, where a formal oxidation state of +7 can be calculated for Cl. This simple arithmetic method works well as long as a few basic assumptions are used, such as an oxidation state of −1 for fluorine (also in OF_2, where O is +2), +1 for alkali metal ions and hydrogen (exceptions are hydrides which contain H^- and hydrogen has an oxidation state of −1) and often −2 for oxygen. One should also bear in mind that most atoms try to fulfill the octet rule and hence do not give or take more electrons than necessary to reach the next octet state (exceptions are Noble Gases in their respective compounds). Observing these rules, one can also apply this method to more 'exotic' molecules, such as hydrogen peroxide (H_2O_2), where H is +1 and O is −1 (and not −2), and for the superoxide radical anion ($O_2^{\bullet -}$), where the formal oxidation state of both (equal) oxygen atoms is −0.5 each. Organic molecules are no exception either, and formal oxidation states in methane CH_4 are +1 for H and −4 for C, and in benzene C_6H_6 are +1 for H and −1 for C.

Nonetheless, this method only works well as long as the formal oxidation states of most atoms in a given compound are already known (ideally, only one oxidation state of one particular atom should be in question). The method also fails when the same type of atom is present more than once in a compound, and in different oxidation states. Examples which can easily go wrong include $Na_2S_2O_8$, a peroxo compound which does not contain S as +7 (as one may mistakenly guess and hence break the octet rule). Here, S is +6 and two of the eight oxygen atoms are formally −1, while the other six are −2.

In this case, a more sophisticated method needs to be applied which also requires more extensive chemical knowledge. First, the two-dimensional structural formula of the compound in question needs to be drawn containing all relevant bonding and other outer shell electrons (covalent electron pairs, lone pairs, unpaired electrons). Electronegativity values are then employed to formally assign all electrons (pairs) individually to the atoms they formally belong to, whereby bonding electrons are assigned to the respective, more electronegative atom. For each atom, its formally 'remaining' electrons are then counted and compared to the number of electrons this atom has in its elemental, neutral state (often given by or hidden in the Group number in the Periodic Table). Atoms which have formally 'lost' electrons have positive formal oxidation states, and atoms which have 'gained' electrons have negative ones.

This method can be applied fairly generally. Unlike the simple algorithmic method, this structural formula method does not require any prior knowledge of specific oxidation states and can also discriminate between different atoms of the same element in one compound. It works well in cases such as $Na_2S_2O_8$. It is more difficult to apply, however, for compounds with mesomeric structures, as such structures do not allow the clear assignment of electrons to specific atoms. As the method depends critically on the specific formula used, it is also more complicated to use—but still applicable, of course—in the case of tautomeric structures. In this case, different values are generated for the different tautomers, which is not a shortcoming of the method but a valid reflection of the different 'chemistry' of the individual tautomers. In most instances, this structural formula method therefore provides reliable and comprehensible numbers for formal oxidation states, assuming one agrees on the electronegativity scale used (e.g., according to Pauling or Mulliken or Allred-Rochow or Allen) and pays some extra attention to atoms with almost identical electronegativity values (e.g., sulfur and selenium, carbon and selenium, phosphorous and hydrogen, arsenic and hydrogen, etc.,) where the connecting bond may need to be 'divided'. In those cases, the chemical behavior of the compound in question may provide a hint (e.g., electrophilic centers in substitution reactions, behavior of a hydride versus a typical hydrogen compound). Ultimately, the determination of such formal oxidation states is not an ultimate objective, but a means to an end which subsequently enables the chemist to understand better the reactivity and reactions of specific, redox active compounds.

In many instances, such reduction and oxidation processes proceed via an uptake or donation of one or more electrons, respectively. Here, classical examples include the reaction of Fe^{3+} with Cu^{+}, whereby the copper ion donates one electron to the iron ion and Fe^{2+} and Cu^{2+} are formed. Unfortunately, such reactions involving the 'free flow' of electrons are surprisingly rare in biology. There are no electrons rolling down hills, as they may damage objects on their way. Indeed, the flow of free electrons is a risky business, as they may go astray easily and reduce molecules they are not supposed to, such as dioxygen (O_2), which is commonly

Table 1 Oxidation states of several reactive oxygen and nitrogen species

Reactive species		Ox. state	Occurrence
Singlet oxygen	$^1O_2$	0	UV-radiation, causes skin cancer
Superoxide radical anion	$O_2^{\bullet-}$	−0.5	Inflammation, metabolism
Hydrogen peroxide	H_2O_2	−1	Inflammation, metabolism, cellular signaling
Hydroxyl radical	$HO^\bullet$	−1	Fenton reaction
Nitric oxide	$^\bullet NO$	+2(N)	Nitric oxide synthase
Peroxynitrite	$ONOO^-$	−2(O), +3(N), −1(O, O)	$^\bullet NO + O_2^{\bullet-}$

present in the cell, to the dangerous dioxygen (superoxide) radical anion ($O_2^{\bullet-}$) (Table 1). Here, other, more subtle redox mechanisms are at work, which, among others, involve radicals, atom and ion transfer reactions and nucleophilic/electrophilic substitution reactions. To illustrate the all-importance of such mechanistic aspects, let us briefly consider the reaction of H_2O_2 with a thiol (RSH). While we all know that this reaction is a—please forgive the usage of the term—'redox reaction', resulting in H_2O and (ultimately) a disulfide (RSSR), it does not involve any direct electron transfer. Quite on the contrary, this reaction proceeds in two steps, both of which involve a classical nucleophilic substitution. In the first step, the oxygen–oxygen bond of H_2O_2 is attacked by the thiol(ate), which forms a bond with one of the oxygen atoms and hence releases the second oxygen atom to form an intermediate sulfenic acid RSOH and H_2O. This reaction is usually followed by a second nucleophilic substitution, whereby a second RSH attacks the oxygen–sulfur bond in RSOH to form a sulfur–sulfur bond, i.e., RSSR, and releases a second H_2O. While none of these processes involves a flow of electrons, they nonetheless represent a reduction/oxidation event, as the formal oxidation states of sulfur change from −2 in RSH (assuming R = +1) to 0 in RSOH and −1 in RSSR, while the formal oxidation state of oxygen changes from −1 in H_2O_2 to −2 in H_2O and in RSOH. Please note that in the second step of the overall reaction, the sulfenic acid (RSOH) sulfur, not oxygen, is the oxidant which becomes reduced (0 to −1) while the thiol sulfur is oxidized (−2 to −1). Such a reaction involving the same element in two extreme oxidation states which then form an intermediate oxidation state is called a 'redox comproportionation' (or 'syn-proportionation').

Besides such 'electron-transfer-free' nucleophilic substitutions, which also include the famous thiol/disulfide exchange reactions, there are a few other reaction mechanisms which can result in changes of oxidation state. We will discuss some of these mechanisms using the example of the human form of the enzyme thioredoxin reductase (TrxR). This enzyme contains a flavin, a 2-Cys and a Cys–SeCys redox center and catalyzes the reduction of the protein thioredoxin (Trx) using NADPH as a reductant. When, in a more biological language, considering the 'flow of electrons' or 'reducing equivalents', we first witness the transfer of a hydride (H^-) from NADPH to the flavin redox center of the enzyme. The hydride transferred is indeed reducing the flavin from FAD to $FADH_2$ (a proton

H^+ is taken up simultaneously), while NADPH ends up in its oxidized form, $NADP^+$. Hydride transfer is a common event in biological systems, which indeed results in reduction and oxidation, yet this 'redox' mechanism does not involve the flow of electrons. Other examples include the reduction of a cysteine sulfenic acid by direct H^- transfer from NADH in the enzyme NADH (per)oxidase from *Staphyllococcus faecialis*. Electron transfer does, however, occur in the second step of the TrxR mechanism, whereby two electrons are transferred simultaneously from $FADH_2$ to the Cys–SeCys redox center, which in its oxidized from contains a sulfur–selenium bond (oxidation states of S and Se are both −1). This sulfur–selenium bond becomes reduced to a thiol(ate) and selenol(ate) (oxidation states of S and Se are now both −2). The resulting thiol(ate)/selenol(ate) center then undergoes two nucleophilic exchange reactions with the 2-Cys redox site, reducing the sulfur–sulfur bond at this site to two thiols while itself returning to its oxidized form featuring the sulfur–selenium bond. The two thiols at the 2-Cys site then attack the disulfide bond in Trx, facilitating yet another series of two subsequent nucleophilic substitutions of thiol(ate)s at a sulfur–sulfur bond. Ultimately, the TrxR site is oxidized to the disulfide form, while the disulfide in Trx is reduced to two thiols. In summary, a total of four separate redox events (six if counting each step of the exchange reactions separately) has taken place to oxidize NADPH to $NADP^+$ and simultaneously reduce the disulfide form of Trx to its thiol form. Yet among these reactions, only one of them actually involves a (two) electron transfer.

In fact, there are even more mechanistic options to drive 'redox' reactions, including radical reactions, which we will discuss in more detail in the context of free radical scavenging antioxidants, such as vitamin E and various polyphenols. Here, it should suffice to summarize this discussion by pointing toward the facet-rich chemistry underlying reduction and oxidation reactions and the importance to understand such processes correctly (see also Fig. 1). It is therefore rather silly to describe such processes in metaphors of 'electrons that get stranded at the top of 'energy hills', waiting to roll down the hill toward a low-energy resting place', or by similar phrases indiscriminately mixing and confusing thermodynamic states with particles and skewed or even screwed reaction mechanisms (see Explanatory Box 2).

Explanatory Box 2: Common sense on the rise: The futile battle of Thermodynamics versus Kinetics

According to some reliable sources, Redox Biology is currently on the rise as a field of study, and redox active sulfur species seem to be at the forefront of these developments. This may come as a surprise, as redox processes have been studied since the time of Joseph Priestley (1732–1804) and Antoine Laurent de Lavoisier (1743–1794). Arguably, the adequate description of redox events, especially of 'oxidation' processes was at the heart of the scientific revolution which swept phlogiston and Alchemy from the throne, to be replaced by the 'oxygen theory'. This development helped turn chemistry into a modern science. Hence 'redox' has been on the rise for quite a while, especially in chemistry.

At the same time, redox processes play a major role in Biology, from mitochondrial respiration to host defense and cellular signaling. Many of these issues will be covered in subsequent chapters. From a chemical perspective, redox processes are often described in terms of thermodynamics. Here, the question at hand is "Which direction does a specific redox reaction take?" For instance, does Zn reduce Cu^{2+}, or does Cu rather reduce Zn^{2+}? The answer to such a question tells us, for instance, if a reaction is possible, which products are formed, whether they are stable, and to what extent (e.g., percentage) such a reaction will proceed. Sometimes, the answer to such a thermodynamic question also refers to energy, i.e., to exothermic reactions which proceed spontaneously and to endothermic reactions which require energy from their environment in order to move ahead. Here, thermodynamics employs a wealth of its own descriptors, including the standard normal potential E^0, the potential of specific systems under non-standard conditions E, and the Nernst equation to link E to E^0, the number of electrons transferred, the temperature, and the relative concentrations of oxidized and reduced species. Thermodynamics can be very powerful. It predicts, for instance, that a given thiol becomes oxidized in the presence of hydrogen peroxide, and that the resulting product will mostly be a sulfenic acid which then reacts with excess thiol to form a disulfide (and two equivalents of water).

The one issue thermodynamics paradoxically ignores, however, is the *dynamics* of the system, the speed at which the reaction proceeds. It overlooks, for instance, that certain reactions take time. Time, of course, is often scarce, especially in Biology. Here, kinetics comes into play. Let us return to our example, where H_2O_2 is used to oxidize thiols. While thermodynamics may predict that this oxidation reaction will eventually go to completion, it is ignorant of the time it takes to convert all thiols to disulfides. It may well be that the time available for such a reaction, say, inside a living cell, is far too short for such a process to go to completion. Indeed, some thiols only react slowly, while others are buried inside proteins and inaccessible to reactions, or are locked away in membranes or specific organelles. It therefore seems that thermodynamics as well as kinetics need to be considered in order to understand a biological redox process fully. Or as one may say: "A reaction is efficient once it is fast (kinetics) and goes to completion (thermodynamics)." Thus, attempts to dissociate thermodynamics from kinetics and vice versa, and especially to completely downplay the potential thermodynamic role of GSH as the major cellular thiol because the kinetics of *some* glutathione reactions are slow, seems particularly unproductive and polarizing in the redox field. Both aspects of physicochemical processes will interplay depending on the particular local cellular environment.

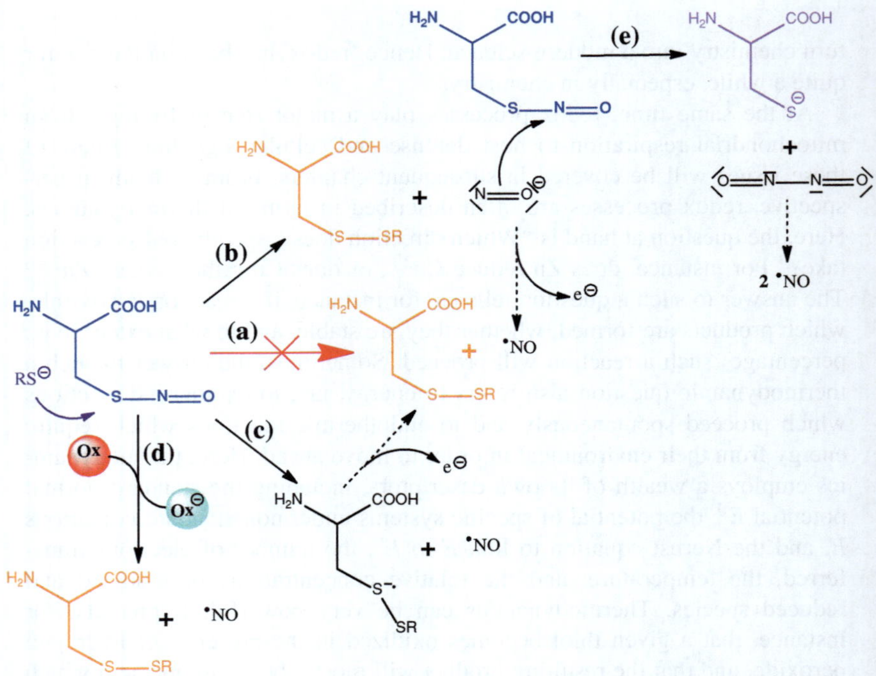

Fig. 1 It is not entirely uncommon that apparently simple biochemical processes cause serious riddles when it comes to their exact 'mechanism'. In those cases, the underlying chemistry needs to be considered in earnest (see also Chap. 2). Here, the release of nitric oxide (•NO) from *S*-nitrosothiols illustrates this point. This kind of release is far from trivial, as a simple attack of a nucleophilic thiol on the cysteine-*S*-nitrosothiol does not simply result in a disulfide and •NO [the electrons do not add up, pathway (*a*)]. From a chemical perspective, an initial release of the nitric oxide (nitroxyl) anion NO^- with subsequent oxidation of NO^- to •NO is more likely [pathway (*b*)]. Alternatively, a three-component reaction involving an intermediate disulfide radical anion formation [pathway (*c*)] or an oxidizing agent [pathway (*d*)] are also possible. Interestingly, chemistry can also propose a more sophisticated, two-step process whereby the nitroxyl anion acts as a nucleophile, attacks another *S*-nitrosothiol at the nitrogen atom and besides a thiol releases dimeric, non-radical N_2O_2, which subsequently decomposes to 2 •NO [pathway (*e*)]. While chemistry may propose and study these mechanisms *in vitro*, their occurrence *in vivo* is considerably more difficult to elucidate

Still, some of these metaphors can be quite useful, especially when it comes to the question as to why Fe^{3+} reacts with Cu^+ to yeild Fe^{2+} and Cu^{2+}, i.e., why the electron is rolling from Cu^+ to Fe^{3+} and not into the reverse direction (i.e., from Fe^{2+} to Cu^{2+} to form Fe^{3+} and Cu^+). In essence, this is a thermodynamic argument related to the energy levels of the individual redox species involved, which can be described in terms of the Gibbs Free Energy *G* and converted into the electrochemical potential *E*. The latter is an indicator of how oxidizing or reducing a particular redox couple, such as the Fe^{2+}/Fe^{3+} couple or the Cu^+/Cu^{2+} couple is. Comparing the two couples, one readily notices that the iron couple is rather oxidizing, while the copper one is more reducing. From this, one may deduce that the electrons are flowing from Cu^+ to Fe^{3+} (and not vice versa) and that the mixture of Fe^{2+} and Cu^{2+} energetically is further 'downhill', and hence more stable thermodynamically, compared to the mixture of Fe^{3+} and Cu^+ (electrochemical normal potentials E^0 can be found in extensive tables for most redox couples assuming they react via electron transfer).

The electrochemical potential E is a rather useful value, not only in electrochemistry but also in biological redox chemistry as it reflects the ability of certain ions or molecules to give or take electrons, or to react as reducing or oxidizing species. Here, one must mention that E is concentration dependent, i.e., the electrochemical potential of a given redox pair in solution (which can be used to estimate the reactivity of the oxidized and reduced forms, or may be compared to other pairs) depends on the ratio of the oxidized and reduced forms present. As a result, E is given by the Nernst equation, which combines the standard potential of a given redox couple E^0 with the concentrations of oxidized and reduced forms and the number of electrons transferred (as well as the temperature). This small but decisive matter is often forgotten in Biology, where E^0 is used in discussions instead of E. While this omission sometimes is not too dramatic (the term accounting for the concentrations is logarithmic), in Biology, such differences in the ratio of oxidized to reduced species often matter considerably. The work of Gary Buettner and Freya Schafer on the electrochemical potential of the glutathione (GSSG/GSH) pair, for instance, has shown that changes in the cytosolic ratio of GSSG to GSH (usually caused by an increase in GSSG) result in a shift of the respective E value by a few tens of millivolts, which is sufficient for the cell to switch between proliferation, differentiation and apoptosis (see Explanatory Box 3 for details on glutathione) (Schafer and Buettner 2001). To avoid any misunderstandings, please note that these major cellular events are not induced by some kind of electric effect; the electrochemical potential posed by the GSSG/GSH pair rather reflects the degree of disulfide formation and *S*-thiolation in the cell, which is a key parameter for cellular signaling, including cell cycle arrest and induction of apoptosis (for a more in-depth discussion of the cellular thiolstat see Chap. 9).

Explanatory Box 3: Glutathione

Glutathione (GSH) is often referred to as the 'redox buffer' of the mammalian cell. GSH is a thiol-containing tripeptide consisting of the amino acids glutamic acid, cysteine, and glycine and containing an unusual γ-linkage between glutamic acid and cysteine. Glutathione carries out many functions in biochemistry. Apart from acting as a redox buffer—it forms a so-called 'redox couple' with its oxidized, disulfide form glutathione disulfide (GSSG)—glutathione is also the substrate of many enzymes, most of which are involved in the control of the intracellular redox status of cysteine proteins and enzymes. Apart from its involvement in redox control processes, GSH is also the substrate of various glutathione *S*-transferases, which are detoxifying enzymes that conjugate GSH to undesired molecules in order to increase their hydrophilicity and hence rate of clearance. While GSH is a reliable reducing agent when it comes to the reduction of a wide variety of (protein-based) disulfides, its oxidized form, GSSG is reduced back to GSH by glutathione disulfide reductase (GR), an enzyme which employs NADPH as the reducing agent. GSSG may also be exported from the cell or locked away in specific cellular compartments. Indeed, the cellular localization and extracellular levels of GSSG represent a hot issue of current redox biology.

3 The Biological Context of 'Redox'

After this rather extensive yet necessary discussion of the 'redox' terminology itself, we will consider 'redox chemistry' in a more biological context. First of all, we will need to address the question of what it means for a molecule to be 'redox active'. After all, this book is about redox active secondary metabolites. Here again, matters are less obvious than they first seem to be. In principle, any atom, ion or molecule (with a few exceptions, such as the H^+ ion) is able to undergo an oxidation *or* a reduction process, it just depends on the partner which simultaneously undergoes reduction and oxidation, respectively. At first sight, this may sound bizarre, as it implies that all molecules are redox active, and hence it would be futile to distinguish 'redox active' secondary metabolites from 'redox inactive' ones. Nonetheless, it is correct that most atoms can donate or accept an electron in the presence of an appropriate, i.e., strong enough, acceptor or donor. This even applies to the noble gases, which have long been considered as (redox) inert, yet can be oxidized by extremely strong oxidants despite their stable octet-like electron configuration. It is, of course, significant whether the products formed during such a forced reduction or oxidation are stable or react further or decompose rather rapidly.

Within a biological context, these considerations provide the most important criteria which can be used to delineate 'redox activity' more narrowly. Here, reduction and oxidation reactions proceed in aqueous media, i.e., in the presence of water. As water itself can be reduced to hydrogen and oxidized to oxygen gas, reduction and oxidation processes able to proceed in water are usually rather limited. It is, for instance, not possible to reduce Na^+ to elemental Na under such *physiological* conditions. This also applies to any other alkali or earth alkali ion. As a rule of thumb, one may therefore only consider redox pairs with an electrochemical potential between, let's say, -1.5 to $+1.5$ V versus the normal hydrogen electrode as relevant in cell biology. These potentials provide, of course, just a rule of thumb based on thermodynamic considerations and, not surprisingly, there are exceptions. The latter mostly result from kinetic complications, such as a hindered reactivity (and hence meta-stable states) or include certain ROS, whose reaction with water would simply regenerate an identical species. This applies, for instance, to compounds such as H_2O_2 and the hydroxyl radical ($HO^{\bullet}$).

Nonetheless, the fact that biological 'redox chemistry' has to proceed under mostly aqueous conditions (membrane environments are an exception) somewhat limits the spectrum of reduction and oxidation processes one may encounter in Biology and enables us to distinguish between 'redox active' and 'redox inactive' metabolites, bearing in mind, of course, that this distinction is instigated primarily from the perspective of biology—and not from the one of chemistry (see also Explanatory Box 4).

Explanatory Box 4: The magic of enzymes

As we have learned as part of this chapter and also in Explanatory Box 2, only reactions sufficiently fast are also relevant within a biochemical context. This has given rise to the idea that enzyme-catalyzed reactions, which often outpace normal, 'chemical' reactions by orders of magnitude, dominate biological processes. In this view, cellular processes resemble a magic flea circus of hundreds of enzymes jumping around quickly catalyzing all kind of reactions and hence are firmly directing biochemistry via kinetic control. Make no mistake, we do indeed find examples of enzymes, such as superoxide dismutase, which work at a speed close to the diffusion limit. Yet in some quarters, this has led to the rather naïve belief that kinetics is more important than thermodynamics. According to this fraternity rate constants are superior to electrochemical potentials, energies and equilibrium constants, and enzymes sway the scepter over intracellular reactions, by deciding which processes occur (fast enough to be relevant) and which are just too slow to be of significance.

Unfortunately, this approach is futile as thermodynamics and kinetics are not mutually exclusive, but complement each other—in chemistry as well as in biology. Even in the presence of the fastest enzymes, we find many reactions which occur in their absence. The reason for this is rather simple. The Achilles Heel of even the most powerful enzymes is their often pronounced selectivity for specific substrates. Their scope (as well as the enormous strength) of action is therefore extremely limited. Most drug–target interactions, for instance, are not catalyzed by enzymes. This is hardly surprising since drugs are mostly not natural and hence unlikely substrates for enzymes. While these reactions are comparably slow compared to enzyme-catalyzed conversions, they still do occur, as numerous mass spectrometry experiments of chemically modified targets have demonstrated. Indeed, reactions taking seconds or minutes to go to completion, such as the reaction of allicin with thiols, are still fast enough to be biologically relevant, even in the absence of any enzymes (see also Chap. 10).

At the same time, enzymes have another major drawback one needs to bear in mind. Enzymes cannot change the direction of a reaction or shift an equilibrium. Here, thermodynamics in the form of energy considerations dominates. Ultimately, spontaneous reactions despite their sluggishness are highly important in biology, medicine, and pharmacy. They enable a myriad of reactions which proceed fast enough to be relevant and without the need of any assistance by enzymes. Such reactions are commonplace in redox biology, as the many cysteine modifications identified by mass spectrometry and Western Blotting illustrate.

Besides this, there is one further issue with the expression 'redox active', which needs to be considered in earnest. In Biology, this expression is used to describe an ion or molecule which is able to undergo either a reduction or oxidation, or both. A thiol (RSH), for instance, is 'redox active' under physiological conditions, and so is a sulfenic acid (RSOH) or the sulfite ion (SO_3^{2-}) (for a discussion, chemical structures and oxidation states of less common Reactive Sulfur Species see Explanatory Box 5 and Fig. 2).

Explanatory Box 5: Reactive Sulfur Species (RSS)

Several chapters of this book refer to natural organic sulfur compounds and to cysteine modifications which contain sulfur in rather unusual oxidation states and 'chemotypes'. Indeed, sulfur is a true 'redox chameleon' which in Biology can occur in over ten different oxidation states (from -2 to $+6$ and including fractional ones), forms part of numerous functional groups and participates in mechanistically rather diverse redox reactions. Here, we will briefly discuss some of the fundamental aspects of this unique biological redox chemistry of sulfur. First of all, we will consider some of the most commonly encountered sulfur modifications. As most of the sulfur species found in Biology are also rather reactive, they are sometimes summarized under the term 'Reactive Sulfur Species' (RSS) in analogy to the better known 'Reactive Oxygen Species' (ROS) and 'Reactive Nitrogen Species' (RNS). RSS are often formed from thiols (RSH, formal oxidation state of sulfur -2, assuming R and H are $+1$ each) by oxidation. They include the better known disulfide RSSR (oxidation state -1), but also various polysulfanes RS_xR' ($R, R' \neq H$ and $x \geq 3$, oxidation state of sulfur -1, $(x-2)$ times $0, -1$). The latter can be reduced to perthiols RSSH (oxidation state -1) and hydropolysulfides RS_xH ($R \neq H$ and $x \geq 2$, oxidation state of sulfur -1, $(x-2)$ times $0, -1$). Oxidation of thiols often also leads to the formation of sulfur-based acids, such as sulfenic acid RSOH (oxidation state 0), sulfinic acid RS(O)OH (oxidation state +2), and sulfonic acid $RS(O)_2OH$ (oxidation state +4). A number of sulfur-centered radicals also exists, of which the thiyl radical $RS^{\bullet}$ (oxidation state −1) is the most commonly encountered. Disulfides can also become oxidized, a chemistry which leads us to disulfide-*S*-oxides, such as the monoxide, known as thiosulfinate RSS(O)R′ (oxidation state −1, +1) and the dioxide, known as thiosulfonate $RSS(O)_2R'$ (oxidation state −1, +3). Interestingly, such disulfide-*S*-oxides can also be formed by condensation reactions. The thiosulfinate allicin, which occurs naturally in garlic and will be discussed in considerable detail in subsequent chapters, is formed from alliin, a sulfoxide RS(O)R′ (oxidation state 0), by an alliinase-catalyzed enzymatic cleavage of a C–S-bond, the formation of a sulfenic acid RSOH, which condenses to RS(O)SR′ and H_2O. It is yet unclear if a similar biological chemistry also exists for sulfones $RS(O)_2R'$ (oxidation state +2), which at least on paper may be cleaved to sulfinic acids RS(O)OH. On the inorganic side, we find a range of biologically relevant compounds, such as the signaling molecule dihydrogen sulfide H_2S, which is in equilibrium with monohydrogen sulfide HS^- and sulfide S^{2-} (oxidation state in each case −2). Between these most reduced forms of sulfur and the most oxidized ones, which include sulfite SO_3^{2-} (oxidation state +4) and sulfate SO_4^{2-} (oxidation state +6), a number of biologically relevant polysulfide anions have recently been emerging. Such polysulfides include various species of the composition H_2S_x (protonated forms, $x \geq 2$), which are classical redox amphoteric molecules and can react as highly reducing as well as fairly oxidizing agents. Other RSS include nitrosothiols RSNO (oxidation state 0) and various inorganic nitrogen–sulfur species which play a role in the chemistry of nitric oxide, $^{\bullet}NO$. Some enzymes also form transient selenium-sulfur RSSeR′ species (oxidation state −1), such as the antioxidant enzyme glutathione peroxidase (GPx) or sulfenylphosphate esters [more details in Jacob et al. (2004)]. The list of biologically relevant sulfur oxidation states and associated chemotypes is still expanding, and only a few of the most apparent species can be presented here.

The same applies to the redox mechanisms underlying the unique redox activity of many of these sulfur species. Depending on the sulfur chemotype, a wide range of reactions and reaction types are possible. Thiols and disulfides, for instance, prefer to react via so-called 'thiol/disulfide exchange' reactions, which involve a nucleophilic attack and subsequent substitution of the thiol at the sulfur–sulfur bond of the disulfide. This redox reaction proceeds without the flow of free electrons (strictly speaking no electron transfer) and is not limited to the disulfide electrophile. It can also be found in the case of polysulfanes, polysulfides, sulfenic and sulfinic acids, thiosulfinates, and thiosulfonates to name just a few. At the same time, perthiols and many hydropolysulfanes and (partially deprotonated) polysulfides can also act as nucleophiles, as can the selenol group (RSeH) present in various proteins and enzymes. While such 'exchange' reactions are clearly at the center of the reactivity of many RSS, other types of reaction are also important. These include radical reactions (especially in the case of the thiyl radical $RS^{\bullet}$), one electron-abstraction and one electron-transfer (in the case of $RS^{\bullet}$), two-electron transfer (in the case of RSOH), hydride transfer (in the case of RSOH) and 'atom transfer' in the case of RS(O)OH and $RS(O)_2OH$ (although this kind of reaction is often based on a sequence of nucleophilic substitiutions). As a consequence, the redox behavior of sulfur (in Biology) is highly complicated and considerable care has to be taken when discussing the different products of the redox reaction, its mechanistic details, energies (thermodynamics), and velocity (kinetics). Sulfur Redox Chemistry and Sulfur Redox Biology are not only for the 'point-hearted', and even the 'professionals' sometimes may be mistaken.

Yet there are some rather significant differences. The thiol may be oxidized under physiological conditions, but it cannot be reduced further. Hence RSH is able to undergo an oxidation reaction, but not a reduction. Is it therefore 'redox active' or only 'ox active'? Surely, one may argue that the term 'redox' does not apply to the thiol in isolation, but to the 'redox' pair it forms part of, such as the RSH/RSOH pair. Indeed, this couple in its entirety is 'redox' active as the thiol can be oxidized to the sulfenic acid which then can be reduced again to the thiol. The sulfenic acid is even both, active as oxidant as well as reductant, i.e., it can be reduced to a thiol (or disulfide) as well as oxidized to a sulfinic (RSO_2H) or sulfonic acid (RSO_3H). Unfortunately, the formation of the sulfonic acid from the sulfinic acid is, as far as we know, irreversible under physiological conditions, hence there is a question if the sulfinic/sulfonic acid couple is indeed a true 'redox couple'. Hence when discussing 'redox active' secondary metabolites, one should always bear in mind that some of these metabolites are active as reducing agents, others as oxidants, and that in some—but certainly not in all—cases, the corresponding redox pair is reversible and hence provides scope for reduction as well as for oxidation. Quinones, for instance, readily redox cycle between the quinone, semiquinone and hydroquinone forms inside cells. In contrast, certain phenolic redox systems, such as vitamin E, can be oxidized, yet their oxidized forms (i.e., radicals) are generally stable or polymerize or decompose, and hence have lost 'redox activity' (to act as oxidants), at least under the conditions they are usually employed in.

This 'loss' of 'redox activity' is actually the key to the antioxidant activity of vitamin E and leads us to the rather foggy and often confused relationships between reducing agents and antioxidants on the one hand, and oxidants and pro-oxidants on the other. Again, this is not a simple or straightforward matter.

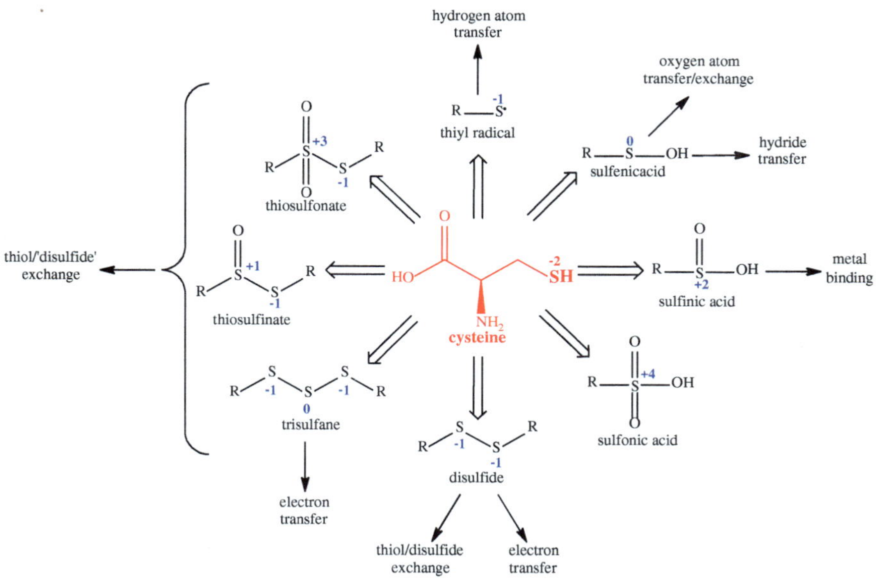

Fig. 2 Chemical structures, oxidation states and main reactivities of selected reactive sulfur species found in numerous plants and also in the human body (oxidation states of R and H are +1)

While it is correct that agents able to donate electrons under physiological conditions are able to reduce certain damaging oxidants, this does neither necessarily nor exclusively turn such 'reducing agents' into 'antioxidants'. In fact, equating reducing agents with antioxidants is another mix-up of terminology at the interface between chemistry and biology. Let us briefly consider two counterexamples to illustrate this point: The semiquinone is, chemically speaking, an excellent one-electron donor; yet it is hardly an antioxidant in Biology, as it donates its one electron to dioxygen, forming the superoxide radical anion ($O_2^{\bullet-}$), a major ingredient of oxidative stress, in the process. Similar considerations apply to the disulfide radical anion ($RSSR^{\bullet-}$), various perthiols (RSSH) and hydropolysulfides (RS_xH, $x \geq 3$) (for Reactive Sulfur Species see Explanatory Box 5). All of these species are electron donors, yet cause the formation of ROS and hence must be considered as potential *pro-oxidants*. In sharp contrast, the Zn^{2+} ion is often seen as an excellent antioxidant. Amazingly, Zn^{2+} is not even redox active under physiological conditions, hence how can it act as antioxidant? Here, biochemical events, such as an activation of certain transcription factors and the subsequent expression of antioxidant proteins and enzymes, explain the antioxidant effects redox inactive Zn^{2+} exerts on the cell.

As the example of Zn^{2+} illustrates, subsequent biochemical events triggered by certain compounds are often considerably more important than the simple reduction or oxidation processes they have initially taken part in. While simple reducing agents are quite useful to sequester radicals (e.g., vitamin C and E, coenzyme Q_{10}

or *N*-acetyl cysteine), the biochemistry associated with these substances is often considerably more complex. Vitamin C, for instance, may cause cellular damage once it is oxidized. Furthermore, excessive supplementation of the human body with vitamin C suppresses the activation of the body's own antioxidant defense, hence assigning an *anti*–antioxidant activation property to (high amounts) of vitamin C (Carr and Frei 1999; Childs et al. 2001; Holloszy 1998; Selman et al. 2006).

Other so-called antioxidants, such as the isothiocyanate sulforaphane or the stilbene xanthohumol, are actually electrophiles which readily modify and hence oxidize cysteine residues in proteins and enzymes.

Isothiocyanates are biologically active molecules which are normally derived from glucosinolate phytochemical precursors. Glucosinolates are *S*-β-thioglucoside *N*-hydroxysulfates that are abundant in cruciferous (*Brassicaceae*) plants. Chemically, there are three different types of glucosinolates, according to the origin of their side chain: (1) aromatic (from phenylalanine or tyrosine); (2) aliphatic (from leucine, isoleucine, methionine, or valine); and (3) indole (from tryptophan) (Fahey et al. 2001; Halkier and Gershenzon 2006). The same plants which contain glucosinolates also have β-thioglucosidase enzymes, known as myrosinases (EC 3.2.3.1), which are physically separated from their glucosinolate substrates. Enzyme and substrate only come into contact when the plant tissue is injured or chewed (by humans or animals). The myrosinase reaction causes a rapid hydrolysis of the glucosinolates which results in a huge variety of reactive compounds. Isothiocyanates represent the major type of agents of the myrosinase reaction and contribute to most of the biological effects that have been associated with glucosinolates. More than 120 natural isothiocyanates have been identified so far (Herr and Buchler 2010; Verkerk et al. 2009).

The isothiocyanate group reacts directly with available sulfur-, nitrogen-, and oxygen-centered nucleophiles. Most common in various plant cells is the reaction of isothiocyanate with cysteine residues in proteins and there especially with glutathione (GSH), forming thiocarbamate products.

The stilbene xanthohumol (2′,4,4′-trihydroxy-3′-prenyl-6′-methoxychalcone,XN) is a prenylated chalcone (chalconoid) found in various plants, e.g., in hops (*Humulus lupulus*, in a concentration around 1 %) and therefore also in beer (in concentrations below 100 ppm) (Stevens et al. 1997; Magalhaes et al. 2009). First studies on hop components described their antiproliferative and cytotoxic activities and identified prenylated flavonoids as antioxidants and modulators of carcinogen metabolism *in vitro* (Miranda et al. 1999, 2000; Henderson et al. 2000). Several recent reports mention that xanthohumol does not only act as an antioxidant, but in fact leads to the induction of reactive oxygen species (ROS). This so-called pro-oxidant activity contributes to its cytotoxic activity and triggers the induction of apoptosis (Strathmann and Gerhauser 2012; Strathmann et al. 2010). Initial studies by Yang et al. (2007) indicate that XN inhibits adipocyte differentiation and induces apoptosis in human preadipocytes via a ROS-mediated mechanism. Xanthohumol treatment of preadipocytes (precursor cells that develop into adipocytes ('fat cells') when fully differentiated) results in a transient increase in 'oxidative stress'.

The apparent antioxidant behavior of the isothiocyanate and stilbene 'oxidants' is then due to the cell's own antioxidant response, which is simply stimulated by the mild form of OS exerted by these 'antioxidants'. There are numerous further examples of chemical reducing agents behaving as pro-oxidants in a biological context, of reducing agents turning nasty upon oxidation, and of entirely redox inactive compounds triggering antioxidant defense mechanisms or, in contrast, causing the up-regulation of intracellular ROS levels.

The matter is extraordinarily complex, and it is generally necessary to study the cellular impact of each suspected pro- or antioxidant separately, and at different concentrations, in order to gain a rough picture of its respective activity. Such studies also need to bear in mind that the human body is not just a cell culture and that bioavailability and metabolic transformations may prevent certain antioxidants from reaching their targets within such a complex organism, as will be discussed later on (see Chap. 16).

4 Conclusions

Ultimately, the dichotomy which exists between the chemical concepts of reductant and oxidant on the one side, and the biochemical concepts of pro- and antioxidant on the other, also explains why simple, and maybe oversimplified attempts to equate those two concepts almost always fail in practice. This is not a reference to the bogged attempts by some gun-slinging chemists who would like to declare every reducing agent a potential antioxidant. The distinction required here is rather between 'antioxidant' properties of substances or natural products *in vitro*, which can be measured, for instance by radical scavenging assays, and expressed in form of ORAC values, total antioxidant capacity (TAC), vitamin C or vitamin E (or Trolox) equivalents, and antioxidant effects caused in the human organism as a whole. As mentioned in the Introduction, this distinction is often blurred and highly delicate from a scientific as well as economic perspective. Only recently, the FDA online database containing ORAC values for a wide range of natural produce and products has been taken off the web because of possible confusion regarding the properties of berries etc., on the one hand, and what these berries may do to promote human wellbeing on the other. Whilst some suspected antioxidants are indeed biologically active, this move nonetheless highlights the complexity of the matter and also the urgent need to conduct serious research in this area.

To explore further how research in this field is currently developing, the next chapter, contributed by Lars-Oliver Klotz from the Friedrich-Schiller-University Jena, Germany, will discuss some of the more recent developments in the field of pro- and antioxidant research. In doing so, Lars-Oliver will not only introduce us to some emerging areas in this field, but also clarify some of the critical issues and clean up some of the myths that may have been associated with certain redox active secondary metabolites in the past.

References

Carr AC, Frei B (1999) Toward a new recommended dietary allowance for Vitamin C based on antioxidant and health effects in humans. Am J Clin Nutr 69(6):1086–1107

Childs A, Jacobs C, Kaminski T, Halliwell B, Leeuwenburgh C (2001) Supplementation with Vitamin C and *N*-acetyl-cysteine increases oxidative stress in humans after an acute muscle injury induced by eccentric exercise. Free Radical Bio Med 31(6):745–753

Fahey JW, Zalcmann AT, Talalay P (2001) The chemical diversity and distribution of glucosinolates and isothiocyanates among plants. Phytochemistry 56(1):5–51

Halkier BA, Gershenzon J (2006) Biology and biochemistry of glucosinolates. Annu Rev Plant Biol 57:303–333

Henderson MC, Miranda CL, Stevens JF, Deinzer ML, Buhler DR (2000) In vitro inhibition of human P450 enzymes by prenylated flavonoids from hops *Humulus lupulus*. Xenobiotica 30(3):235–251

Herr I, Buchler MW (2010) Dietary constituents of broccoli and other cruciferous vegetables: implications for prevention and therapy of cancer. Cancer Treat Rev 36(5):377–383

Holloszy JO (1998) Longevity of exercising male rats: effect of an antioxidant supplemented diet. Mech Ageing Dev 100(3):211–219

Jacob C et al (2004) The sulfinic acid switch in proteins. Org Biomol Chem 2:1953–1956

Magalhaes PJ, Carvalho DO, Cruz JM, Guido LF, Barros AA (2009) Fundamentals and health benefits of xanthohumol, a natural product derived from hops and beer. Nat Prod Commun 4(5):591–610

Miranda CL, Stevens JF, Helmrich A, Henderson MC, Rodriguez RJ, Yang YH, Deinzer ML, Barnes DW, Buhler DR (1999) Antiproliferative and cytotoxic effects of prenylated flavonoids from hops (*Humulus lupulus*) in human cancer cell lines. Food Chem Toxicol 37(4):271–285

Miranda CL, Stevens JF, Ivanov V, McCall M, Frei B, Deinzer ML, Buhler DR (2000) Antioxidant and prooxidant actions of prenylated and nonprenylated chalcones and flavanones in vitro. J Agr Food Chem 48(9):3876–3884

Schafer FQ, Buettner GR (2001) Redox environment of the cell as viewed through the redox state of the glutathione disulfide/glutathione couple. Free Radical Bio Med 30(11):1191–1212

Selman C, McLaren JS, Meyer C, Duncan JS, Redman P, Collins AR, Duthie GG, Speakman JR (2006) Life-long Vitamin C supplementation in combination with cold exposure does not affect oxidative damage or lifespan in mice, but decreases expression of antioxidant protection genes. Mech Ageing Dev 127(12):897–904

Stevens JF, Ivancic M, Hsu VL, Deinzer ML (1997) Prenylflavonoids from *Humulus lupulus*. Phytochemistry 44(8):1575–1585

Strathmann J, Gerhauser C (2012) Anti-proliferative and apoptosis-inducing properties of xanthohumol, a prenylated chalcone from hops. In: Diederich M (ed) Natural compounds as inducers of cell death. Springer, Heidelberg

Strathmann J, Klimo K, Sauer SW, Okun JG, Prehn JHM, Gerhauser C (2010) Xanthohumol-induced transient superoxide anion radical formation triggers cancer cells into apoptosis via a mitochondria-mediated mechanism. Faseb J 24(8):2938–2950

Verkerk R, Schreiner M, Krumbein A, Ciska E, Holst B, Rowland I, De Schrijver R, Hansen M, Gerhauser C, Mithen R, Dekker M (2009) Glucosinolates in Brassica vegetables: the influence of the food supply chain on intake, bioavailability and human health. Mol Nutr Food Res 53:S219–S265

Yang JY, Della-Fera MA, Rayalam S, Baile CA (2007) Effect of xanthohumol and isoxanthohumol on 3T3-L1 cell apoptosis and adipogenesis. Apoptosis 12(11):1953–1963

Authors Biography

Torsten Burkholz (born 1979) has been trained as an inorganic and medicinal chemist at the University of Saarland in Saarbruecken, Germany, graduating with a German "Diplom" in Chemistry. After completing his PhD studies at the University of Saarland in the fields of Chemistry and Pharmacy under the supervision of Prof. Claus Jacob in 2010, he joined the European Marie Curie Initial Training Network "RedCat" as postdoctoral Experienced Researcher, conducting research in the field of Cell Biology in the group of Prof. Paul G. Winyard at the Peninsula College of Medicine and Dentistry, Exeter, UK. In 2012 Torsten moved back to the University of Saarland where he currently holds the position of an "Akademischer Rat" in Bioorganic Chemistry.

As part of this senior position, Torsten is managing the research laboratory of Bioorganic Chemistry, as well as the relevant teaching and the examinations of undergraduate students. Together with Prof. Claus Jacob, he has to date published over 30 publications in the field of oxidative stress, chalcogen-containing natural compounds and their biological activity and, more recently, on the cellular thiolstat. Torsten's ongoing research includes synthetic and analytical chemistry, biological activity studies and 'intracellular diagnostics'.

In 2013, Torsten was appointed as Visiting Professor at the University of Applied Sciences Kaiserslautern, where he is lecturing Pharmacology. In the same year, he established his own small company named "Dr. Burkholz Life Science Consulting UG" which provides scientific consultations for small and medium sized companies in the field of nutrition, food supplements, and natural compounds. His company also offers training and consulting in Inorganic and Analytical Chemistry, as well as in Physics and Pharmacology.

Claus Jacob (born 1969) has been trained as a synthetic (in)organic and biological chemist at the Universities of Kaiserslautern, Leicester, Oxford, and Harvard. He graduated with a 1st class B.Sc. (Hons.) degree from the University of Leicester in 1993, and with a D.Phil. from the University of Oxford in 1997 ("Genetic engineering of redox active enzymes", supervisor Prof. Allen Hill FRS). He subsequently joined the institute of Prof. Bert Vallee at Harvard Medical School as a Feodor Lynen Fellow (Alexander von Humboldt-Foundation) to study processes controlling intracellular zinc homeostasis. During this time, he also obtained a Magister Artium degree in Philosophy, History, and Psychology from the University of

Hagen in Germany (M.A. dissertation on Protochemistry as constructivist foundation of chemistry). He left the US in 1999 to spend some time with Prof. Helmut Sies at the Heinrich-Heine-University in Duesseldorf, Germany, as part of a BASF Research Fellowship from the German Merit Foundation.

Claus started his independent scientific career as lecturer at the University of Exeter in the UK in 1999 and in 2005 moved to the University of Saarland where he currently holds the position of Professor of Bioorganic Chemistry. Claus is an expert in redox active compounds and their impact on biological systems and to date has published over 100 publications in this field. Over the years, his research has focused on Reactive Sulfur Species (RSS) and the cellular thiolstat, terms his team has introduced in 2001 and 2010, respectively. Besides his strong interest in redox active sulfur, Claus has also developed an active research program on synthetic 'sensor/effector' redox modulators based on selenium and tellurium, on redox active plant metabolites and on nanoscopic redox particles. His research includes synthetic and analytical chemistry, biological activity studies and 'intracellular diagnostics' to decipher and map out intracellular events and mode(s) of actions. Claus has coordinated the EU Marie Curie Initial Training Network "RedCat" (2008–2012), has been a partner in the technology transfer project "Corena" (2009–2012) and is currently partially in charge of the natural products project "NutriOx".

Throughout the years, Claus has undertaken many projects to become a highly skilled undertaker, but never a true philosopher, yet his more philosophical and cunning linguistic outpourings are famous and he still maintains a keen interest in various aspects related to the philosophy of chemistry.

Part IV
Connecting Section Between Chapters 4 and 5

After considering the various chemical and biochemical aspects underlying redox activity, we will now turn our attention to more complex biological systems. Not surprisingly, redox activity plays a major part in many intracellular signaling events. Most of the signaling pathways within the mammalian cell contain one or more proteins or enzymes that can be regulated, 'switched on' or 'switched off' by distinct redox processes. Such events may include simple oxidation, mostly of the active site, or more subtle changes, such as the hydroxylation, nitrosation, or thiolation of a particular residue within the protein or enzyme structure. Indeed, many proteins involved in the regulation of the cell cycle, of proliferation, differentiation, and apoptosis, are redox sensitive and hence may be regulated by redox changes. Importantly, such changes may either be 'indogenous', i.e., caused by the cell itself and part of the normal cellular function, or 'exogenous', i.e., caused by an external redox stimulus, such as a redox active secondary metabolite administered to the cell, tissue, or organism. While the overall changes observed in both cases may be similar, the biochemical events often differ: The (mammalian) cell usually employs enzymes, such as specific reductases, to efficiently respond to intrinsic redox stimuli. In contrast, extrinsic stimuli are usually 'alien' to the cell and the result is more complex, with widespread chemical modifications of redox sensitive proteins and enzymes. These modifications often occur rather nonspecifically, even in proteins not related to cellular redox processes. They are also more difficult to control as there are no specific reductases for 'odd' chemical modifications of cysteine side chains.

It is therefore paramount to distinguish between intrinsic redox signaling and control, such as the one provided by specific redox proteins and relays (e.g., Mia40) and extrinsic redox modulation, e.g., triggered by the ingestion of large amounts of thiosulfiantes, polysulfanes, or xanthohumols. This basic difference is frequently forgotten and may cause serious confusion. Here, we will briefly consider redox events intrinsic to the cell first, and in subsequent chapters turn our attention to—extrinsic—redox modulation as a tool in pharmaceutical and agricultural research. In Chap. 5, Lars-Oliver Klotz will therefore open up this discussion with a closer look at specific intracellular signaling pathways and how they may be affected by indogenous and exogenous redox changes.

Chapter 5
Oxidative Stress, Antioxidants, and Chemoprevention: On the Role of Oxidant-Induced Signaling in Cellular Adaptation

Lars-Oliver Klotz

Keywords Antioxidants • Chemoprevention • FoxO • Oxidative stress • Redox signaling

1 Reactive Oxygen Species, Antioxidants, and Oxidative Stress

Reduction/oxidation (redox) processes are at the center of aerobic metabolism. The metabolic generation of molecules with energy-rich bonds that are to be used for biosynthetic purposes requires molecular oxygen, which serves as the ultimate oxidant and electron acceptor, thereby driving numerous redox reactions, including those that are part of the respiratory chain located at the inner mitochondrial membrane. Macronutrient ('fuel') catabolism provides electrons in the form of carrier molecules such as NADH or $FADH_2$ that are reoxidized at the inner mitochondrial membrane, feeding the electron flow along the respiratory chain of protein complexes all the way to molecular oxygen. This flow then drives the build-up of a proton gradient across the inner mitochondrial membrane that is exploited to generate adenosine triphosphate (ATP), the universal energy currency of the living cell that is required for metabolism.

While water is the end product of oxygen reduction at the mitochondrial electron transport chain, ROS result from an incomplete reduction of molecular oxygen: superoxide, a radical anion ($O_2^{\bullet-}$), is the one-electron reduction

L.-O. Klotz (✉)
Department of Nutrigenomics, Institute of Nutrition, Friedrich-Schiller-Universitaet Jena, Jena, Germany
e-mail: lars-oliver.klotz@uni-jena.de

L.-O. Klotz
Faculty of Pharmacy and Pharmaceutical Sciences, University of Alberta, Edmonton, AB, Canada

C. Jacob et al. (eds.), *Recent Advances in Redox Active Plant and Microbial Products*,
DOI: 10.1007/978-94-017-8953-0_5, © Springer Science+Business Media Dordrecht 2014

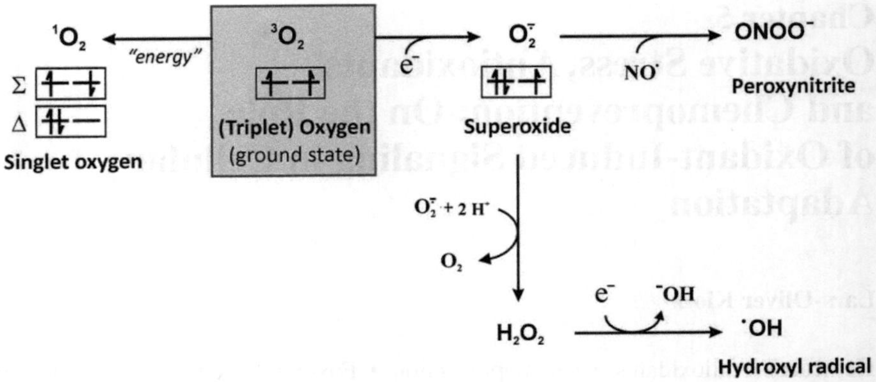

Fig. 1 Overview of the biochemical relationship between reactive oxygen species. Molecules are shown with a depiction of the occupancy of the outermost π* molecular orbitals. Ground state molecular oxygen (with outermost electrons in a triplet state) may be activated to form singlet oxygen by energy transfer, e.g., mediated by photosensitizers that were stimulated by light. Singlet oxygen exists in two forms in solution, $(^1\Delta_g)O_2$, and $\left(^1\Sigma_g^+\right)O_2$; the latter is generally regarded to be too short-lived to be of any significance in biological systems. The term 'singlet oxygen' therefore usually refers to the $(^1\Delta_g)$-form in biological systems. One-electron reduction of molecular oxygen will yield superoxide. Superoxide (or the superoxide anion) either undergoes disproportionation/dismutation to generate hydrogen peroxide, or it may react with nitrogen monoxide to form peroxynitrite. Hydrogen peroxide, in turn, may be reduced (e.g., by redox active metal ions) to result in the formation of hydroxyl radicals

product of O_2. It may rapidly disproportionate ('dismutate') to form water and hydrogen peroxide, which, in turn, may be reduced to form the hydroxyl radical, $^\bullet$OH, a most potent oxidant that immediately reacts with biological molecules wherever generated.

In addition to dismutation (following interaction with another superoxide molecule), superoxide may also react with nitrogen monoxide ('nitric oxide', $^\bullet$NO), another radical generated in biological systems, to form peroxynitrite ($ONOO^-$), a potent oxidant and nitrating species.

Figure 1 provides a schematic overview of the biochemical relations between molecular oxygen and reactive oxygen species (ROS) generated by its step-wise reduction. In addition to reduction (i.e., electron transfer), molecular oxygen may be transformed into a ROS by energy transfer. O_2 is a diradical with its two unpaired electrons in a triplet state, i.e., with parallel spins, in its ground state ($^3O_2$). Energy transfer will result in singlet molecular oxygen ($^1O_2$), an electronically excited version of molecular oxygen, with the two outermost electrons with antiparallel spins—either paired (i.e., occupying the same molecular orbital) or unpaired (Fig. 1; for further detail on the biochemistry of singlet oxygen, see Klotz 2002; Klotz et al. 2003).

In summary, ROS include the oft-cited free radicals, but also non-radicals such as hydrogen peroxide, peroxynitrite, and singlet oxygen. ROS are generated in mammalian cells both under physiological conditions and upon exposure to exogenous stimuli.

1.1 Generation of ROS in Mammalian Cells Under Physiological Conditions

In addition to the somewhat accidental generation of ROS as by-products of mitochondrial respiration, the reduction of oxygen to form superoxide or the formation of H_2O_2 may occur in the vicinity of other electron transport chains, such as those related to xenobiotic metabolism in the endoplasmic reticulum. Furthermore, several enzymatic reactions come with the generation of ROS. For example, NADPH oxidase complexes are a major source of ROS: originally identified in phagocytes as the membrane-bound flavoenzyme responsible for the generation of superoxide and hydrogen peroxide upon cell stimulation (the 'respiratory burst' (Babior 1984)), NADPH oxidase complexes (with five different isoforms) are now known to be present in many non-phagocytes and to be activated by numerous stimuli, including proinflammatory factors (Katsuyama et al. 2012). They are further known to be essential to cellular signaling processes by providing locally generated ROS capable of modulating the activity of signaling molecules.

Further ROS generating enzymes in mammalian cells include amine oxidases, aldehyde oxidases, or oxidases involved in peroxisomal beta-oxidation of fatty acids, to name only a few. In summary, endogenous ROS generating systems contribute to the steady physiological generation of ROS that are required for certain cellular processes to occur.

Several pathophysiological conditions have been demonstrated to come with an enhanced generation of ROS. ROS, being (as the term suggests) reactive, interact with a variety of susceptible biomolecules, leading to their oxidation and to the generation of oxidation products. Accordingly, ROS formation and the oxidative modification of biomolecules have frequently been hypothesized to contribute to the pathogenesis of diseases and conditions such as ischemia/reperfusion injury, side effects of inflammatory processes, diabetes, cardiovascular disease, and carcinogenesis. Interestingly, these contributions to pathogenesis do not necessarily occur by simply destroying cellular components (although at high ROS concentrations this may, of course, be the case), but more frequently appear to be related to inducing confined stress responses and alterations of signaling pathways that are responsible for essential processes such as fuel metabolism, proliferation, and senescence.

1.2 Exogenous Stimuli and the Generation of ROS

The encounter with exogenous stimuli is a major source of ROS for mammalian cells. Such stimuli include the exposure to xenobiotics (including drugs), to physical stimuli (such as UV) or to both, such as particulate matter that serves as a vector for xenobiotics. Although pathogens capable of stimulating the NADPH oxidase response will, of course, also be exogenous stimuli causing the—more indirect—generation of ROS, I will focus on some examples from the field of

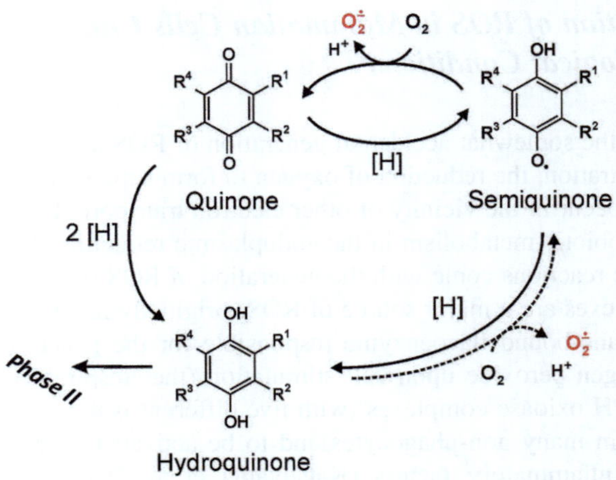

Fig. 2 Redox cycling. Superoxide and other ROS may be generated by cyclic reduction/oxidation of quinones, which may be reduced intracellularly to their corresponding semiquinone, e.g., by microsomal reductases. Semiquinones can either undergo disproportionation to generate quinone and hydroquinone (not shown) or they can be further reduced to the respective hydroquinone. Alternatively, molecular oxygen may reoxidize a semiquinone, regenerating the quinone and producing superoxide. NAD(P)H:quinone oxidoreductase-1 (NQO1, DT diaphorase) catalyzes the two-electron reduction of a quinone to the corresponding hydroquinone at the expense of NADH or NADPH, thereby circumventing the semiquinone stage and allowing the hydroquinone to undergo Phase II xenobiotic metabolism, i.e., to be coupled to water-soluble molecules such as sulfate or glucuronic acid, and to be excreted

toxicology: (a) the exposure to redox cycling drugs, (b) exposure to metal ions, (c) UV radiation, and (d) exposure to particles.

(a) Following their uptake, xenobiotics (i.e., compounds that are, in the broadest sense, strangers to the body or cell) may undergo metabolic steps that come with the cyclic generation of ROS. Xenobiotics may be such redox cycling compounds themselves or may be metabolised to become a redox cycler (Brunmark and Cadenas 1989; Bolton et al. 2000). Certain quinones may undergo redox cycles in cells, i.e., they may be reduced enzymatically to semiquinones at the expense of cellular reducing equivalents such as NADH or NADPH. In the presence of molecular oxygen, which is found physically dissolved and at high micromolar concentrations in most cells, many semiquinones are then easily reoxidized to the respective quinone, with molecular oxygen in turn reduced to superoxide (Fig. 2). This reduction/reoxidation cycle generating superoxide from oxygen may occur repeatedly, provided the supply of reducing equivalents lasts. A second reduction step from semiquinones further on to hydroquinones and subsequent reoxidation may similarly result in the formation of superoxide. The 1,4-naphthoquinone derivatives plumbagin and juglone are found in leadwort (*Plumbago* sp.) and black walnut (*Juglans nigra*), respectively, and are typical examples of potent redox cyclers capable of generating superoxide in mammalian cells.

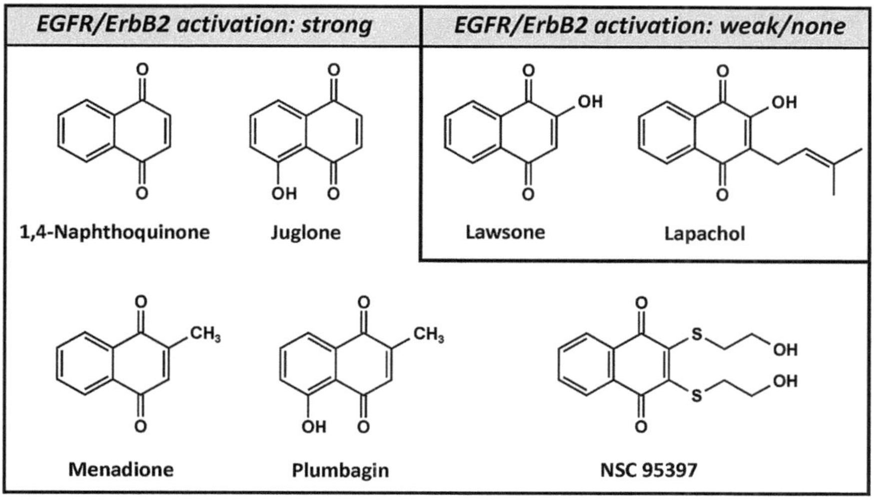

Fig. 3 1,4-Naphthoquinones and RTK activation. Structures of 1,4-naphthoquinone and 1,4-naphthoquinone derivatives sorted by their ability to stimulate the epidermal growth factor receptor (EGFR) and its relative ErbB2 (Klaus et al. 2010; Beier et al. 2006; Melchheier et al. 2005). Some naphthoquinones are derived from natural sources, such as plumbagin (an ingredient of leadwort, *plumbago* sp.), juglone (from various types of walnut, *juglans* sp.), lawsone (found in colorants made from henna, i.e., *lawsonia* sp.) and lapachol (found in lapacho tea). Moreover, the structure of NSC95397 (2,3-bis-[2-hydroxyethylsulfanyl]-1,4-naphthoquinone), an inhibitor of several protein tyrosine phosphatases that causes strong stimulation of EGFR/ErbB2 signaling, is shown

Figure 3 lists some natural and synthetic naphthoquinone derivatives that were shown to stimulate the generation of ROS (and to induce signaling processes that are referred to later in this chapter) (Klaus et al. 2010). Other xenobiotics may be metabolized to form quinoid compounds that may then undergo redox cycling. Prominent examples are the anti-malarial 8-aminoquinoline primaquine or certain pyrimidine glycosides found in fava beans, vicine, and convicine: all of these are metabolized to form products capable of generating ROS through redox cycling (Fig. 4), such as *ortho-/para*-quinones or quinone-imines (Vasquezvivar and Augusto 1992, 1994; Rae et al. 1999).

(b) Redox active metal ions are usually present in cells only in protein-bound and chelated form—for example, virtually no free copper ions are detectable in yeast cells (Lippard 1999; Valko et al. 2005). Nevertheless, the local release of redox active metal ions or metal chelates may cause a detectable generation of ROS. Such metal ions may undergo redox cycling along Eqs. (1–3): in the case of copper ions, high intracellular thiol concentrations will reduce any Cu(II) to cuprous ions that may, in turn, reduce molecular oxygen to generate superoxide. Superoxide may then, if no other target is at hand, disproportionate to generate hydrogen peroxide.

Fig. 4 Metabolism of xenobiotics may generate redox cyclers. Metabolism of vicine and primaquine will generate products that may undergo redox cycling, such as the catechols divicine and 5-hydroxy-6-desmethyl-primaquine (arrow indicates hydroxylation site)

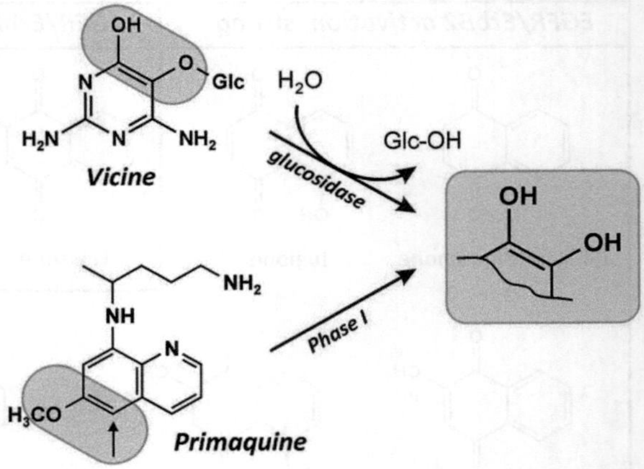

$$2\,Cu^{2+} + 2\,RSH \rightarrow 2\,Cu^{+} + RSSR + 2\,H^{+} \quad (1)$$

$$Cu^{+} + O_2 \rightarrow Cu^{2+} + O_2^{\bullet-} \quad (2)$$

$$2O_2^{\bullet-} + 2H^{+} \rightarrow H_2O_2 + O_2 \quad (3)$$

Cu(I) could also reduce hydrogen peroxide or other peroxides in a Fenton-type reaction to generate hydroxyl/alkoxyl radicals (Eq. 4a/4b; see also Fig. 1).

$$Cu^{+} + H_2O_2 \rightarrow Cu^{2+} + HO^{\bullet} + {}^{-}OH \quad (4a)$$

$$Cu^{+} + ROOH \rightarrow Cu^{2+} + RO^{\bullet} + {}^{-}OH \quad (4b)$$

Similar reactions are known to occur with other redox active metal ions, including iron and cobalt ions (Valko et al. 2005).

(c) Exposure of tissues and cells to ultraviolet (UV) radiation will cause the formation of ROS. The extent of ROS formation depends on the phostosensitizers present, i.e., on the biomolecules available which absorb radiation of appropriate wavelength and are capable of transforming the absorbed energy into reactions that come with the generation of ROS, either directly (i.e., the excited photosensitizer reacts with oxygen) or indirectly (i.e., by interaction of the excited photosensitizer with a reaction partner that, following this interaction, then reacts with oxygen). The extent of ROS formation therefore also depends on the region of wavelengths the tissue is exposed to and that is absorbed by photosensitizers. UV may be roughly divided into UVA (320–400 nm), UVB (280–320 nm) and UVC (<280 nm) portions; sunlight reaching the surface of the earth contains UVA and UVB, as the stratospheric ozone layer hinders UVC from penetrating the atmosphere. UVA penetrates skin most deeply and may reach subepidermal layers (Tyrrell 1996);

its biological actions have been ascribed to an oxidative component due to the photosensitized generation of ROS, such as singlet oxygen. In fact, singlet oxygen (Tyrrell and Pidoux 1989; Klotz et al. 1999, 1997) and hydrogen peroxide (von Montfort et al. 2006a; Mahns et al. 2003) have been demonstrated to mediate several biological effects of UVA, including toxicity and the stimulation of signaling cascades. Porphyrins and flavins are among the photosensitizers that are suitable to generate ROS upon exposure to UVA, and photosensitized peroxide formation upon irradiation of aqueous solutions containing riboflavin and tryptophan was found to be responsible for UVA-induced signaling effects (von Montfort et al. 2006a; Mahns et al. 2003). Of course, these principles of photosensitized ROS generation also apply to other wavelength ranges: for example, visible light is usually employed for photodynamic therapy (PDT), i.e., the local and targeted generation of ROS with the goal of destroying a target tissue that was loaded with a certain photosensitizer.

(d) Formation of ROS may also occur during the interaction of human cells with particles, including ultrafine or nanoparticles (Unfried et al. 2007). While some types of nanoparticles act as photosensitizers, causing the excitation of ground state molecular oxygen to generate both $^1O_2$ and superoxide under the influence of light, the photochemical reactivity of (nano-)particles is usually of little relevance for the interaction with tissues that are not exposed to light. Several other types of reactions are more likely to account for ROS generation, such as the intracellular release of organic matter from combustion-derived (nano-)particles, e.g., polyaromatic hydrocarbons that will undergo Phase I metabolism to yield redox active compounds. Similarly, particulate matter could serve as a vector for redox active transition metals, catalyzing Fenton-type reactions (Eq. 4a, b). In addition to these chemical reactivities, the mere physical interactions of particles with subcellular structures that are involved in redox processes can be envisaged to affect the generation of ROS. Potential (and hypothetical) targets include enzyme complexes in the plasma membrane, mitochondria, and the endoplasmic reticulum. Several plasma membrane-bound enzymes and receptors were identified as structures affected by particles, including integrins (Weissenberg et al. 2010) receptor tyrosine kinases such as the EGFR (Weissenberg et al. 2010; Sydlik et al. 2006) or the ROS generating NADPH oxidase complex (Haberzettl et al. 2008; Mo et al. 2009). Similarly, the enhanced generation of ROS and the damage of mitochondria as well as the presence of ultrafine particulate matter inside mitochondria of exposed cells has been demonstrated (Xia et al. 2006; Li et al. 2003). Mechanistically, two modes of particle action can be envisaged here: (1) the direct interaction of particulate matter with enzyme complexes that leads to a diversion of electron flow to generate ROS, or (2) the interaction with membrane structures that would impair the integrity of membrane-bound complexes or lead to a dysregulation of calcium levels; mitochondria and endoplasmic reticulum are two major cellular Ca^{2+} stores, and an uncontrolled release of Ca^{2+} into the cytosol could cause an activation of calcium-dependent enzymes, such as endothelial or neuronal nitrogen monoxide (•NO) synthases (eNOS or nNOS), resulting in the production of •NO, which may further react with superoxide to form peroxynitrite (see Fig. 1).

1.3 Oxidative Stress and Antioxidants

Obviously, an extensive generation of ROS may be detrimental to cells. Yet cell death is only the ultimate reaction of a cell overwhelmed by ROS; several less drastic reactions of cells will usually be provoked. Hence, it is said that ROS may be stressful in the sense that cells may be forced to adapt to this novel situation of being exposed to ROS. This is implied by a recent re-definition of 'oxidative stress': originally defined by Helmut Sies as "a disturbance in the prooxidant/antioxidant balance in favor of the prooxidants, leading to potential damage" (Sies 1985) it has been redefined by Sies and Jones as "an imbalance between oxidants and antioxidants in favor of the oxidants, leading to a disruption of redox signaling and control and/or molecular damage" (Sies and Jones 2007; Jones 2008). This definition therefore specifically refers to the effect of oxidants on redox signaling, i.e., stress-responsive signaling cascades that are involved in cellular adaptation.

Adaptation is one of the—at least four—major "strategies of antioxidant defense", which also include *prevention, interception* and *repair* (Sies 1993; Klotz and Sies 2003). It may be caused by ROS-induced alterations in signaling cascades that result in changes of cellular levels of antioxidative molecules, such as antioxidative enzymes, which will then enhance the cell's antioxidative capacity at any of the other three antioxidative levels. For example, adaptation may result in higher cellular levels of metal-chelating thioneins, which would be a preventive measure in that the chelation of redox active metal ions will attenuate metal-dependent generation of ROS. Similarly, the stress-induced synthesis of enzymes such as glutathione *S*-transferases (GSTs) is preventive in that the catalyzed coupling of electrophiles to glutathione will prevent them from reacting with crucial protein sites. Finally, an enzyme preventing redox cycles is known to be induced by exposure of cells to stressful stimuli: NAD(P)H:quinone oxidoreductase-1 (or DT-diaphorase) catalyzes a direct two-electron reduction of certain quinones to form the hydroquinone (see Fig. 2, Ernster 1987). This reaction circumvents the semiquinone stage and may be cytoprotective in that certain hydroquinones are directly coupled to glucuronic acid or undergo sulfation to be excreted and thus removed from the system rather than being oxidized to generate superoxide (denoted by 'Phase II' in Fig. 2).

At the level of interception, a battery of low molecular weight antioxidants (water soluble, lipid soluble, cellular, extracellular) and of enzymes with antioxidant activity (such as superoxide dismutases or peroxidases) either quenches, scavenges, or reduces ROS, thereby interrupting any reactions of ROS with biomolecules.

Another line of defense is at the level of repair of oxidative damage to biomolecules. In proteins, sulfur-containing amino acid residues—cysteine and methionine—are particularly susceptible to oxidative modification. Methionine sulfoxide (MetSO) is a major oxidation product detected in proteins exposed to ROS. Such oxidatively damaged proteins can be repaired by methionine sulfoxide reductases (MSR) that catalyze the reduction of MetSO to methionine (Lee et al. 2009). Similarly, oxidative DNA damage may be reverted by enzymes detecting oxidative damage, such as a DNA glycosylase (OGG1) which recognizes 8-hydroxyguanosine residues (Maynard et al. 2009). If oxidative damage of biomolecules is extensive they may also be degraded and disposed of, rather than repaired, in order to avoid accumulation and precipitation inside cells. One such repair and disposal machinery would be the 20S proteasome which recognizes and degrades oxidized proteins (Grune et al. 2003).

2 Oxidant-Induced Damage and Signaling

Exposure of biomolecules to ROS will cause oxidation processes, resulting in oxidized proteins, lipid peroxidation, and oxidative nucleic acid damage, to mention only the major interaction partners. Such damage will have consequences, including (1) inactivation of crucial enzymes and the accumulation of oxidized and insoluble protein, leading to cell death, (2) impairment of membrane integrity and membrane rupture, (3) promutagenic alteration of DNA. It is this very damage, however, that—provided the extent of damage is limited—will stimulate a cellular response by triggering signaling pathways that modulate gene expression and are at the center of an *adaptive* response (see Explanatory Box 1).

Explanatory Box 1: Intracellular Signaling Cascades

This particular chapter considers a few better known intracellular signaling cascades which are pivotal to our understanding of the cell's response to an internal or external stimulus. Such signaling cascades or networks are common in Cell Biology. In fact, there are hardly any processes in the cell which occur in isolation, i.e., devoid of any surrounding network of checks, controls, and responses. Indeed, the living cell uses such networks to ensure the proper functioning of its processes, to become alerted of any undesired events and to be able to respond accordingly, in a swift, effective and well-controlled manner. The idea of intracellular signaling and response networks is not new. Nonetheless, it is sometimes pushed aside or ignored, as it turns the cell into a complex—perhaps even too complex—system to understand and to manipulate. Still, while we may wish to simplify such a system for practical reasons and to rely on the 'one substrate/drug—one target—one response' approach in biochemical and pharmaceutical research as a working hypothesis, we must pay tribute to this complexity. Systems biology as a comparably new branch of biology is inspired by these network ideas and promises many new insights, a more controlled application of 'triggers' (in the form of drugs), personalized medicine and, in the pharmaceutical arena, less side effects. As such networks can be fairly complex, however, any such approach is faced with a considerable number of individual proteins/enzymes involved, and an even greater number of mutual interactions. Bearing in mind that the unawareness of just one player or interaction may seriously compromise the network model, the task at hand is far from simple.

Hence signaling cascades still provide a more realistic, but also more simplistic method. These cascades often involve several key proteins and enzymes which process their signals by a variety of means, including phosphorylation, acetylation, *S*-thiolation, oxidation and reduction, hydrolysis and translocation to name just a few. In this book, redox processes are obviously at the forefront of events. Indeed, many redox active secondary metabolites attack the rather redox sensitive thiol group of cysteine residues in proteins and enzymes, hence leading to oxidation (formation of disulfides, sulfenic and sulfinic acids) or *S*-thiolation (oxidative addition of a thiol, such as a glutathione). These modifications can result in the deactivation (rarely also activation) of the enzyme affected, which leads to a build-up of substrate and decrease of product, i.e., to a 'signal' which is then processed further. To date, it is still unclear which proteins become modified first and to which extent. The concept of the 'cellular thiolstat', which will be introduced in subsequent chapters, tries to address some of these issues. Yet location and accessibility of thiol groups, the variety of oxidation products, kinetic aspects, enzymatic reductions as well as oxidations complicate the emerging picture.

Besides triggering such a signal via the modification of an enzyme, oxidation and *S*-thiolation can also disrupt entire cellular processes. Recent research has illustrated, for instance, that the modification of cysteine residues in tubulin can inhibit cell division, cause cell cycle arrest and induce cell death via apoptosis. Similarly, widespread non-enzymatic oxidation of GSH to GSSG may result in a subsequent, more or less random *S*-glutathiolation of key proteins and enzymes, which in turn would trigger multiple signals via rather diverse signaling pathways.

In any case, when discussing cellular responses to a given signal, the idea of a signaling cascade embedded in an entire network of sensors, checkpoints, controls, response elements, and feedback mechanisms and loops needs to be taken extremely seriously.

2.1 ROS-Induced Signaling: Receptor Tyrosine Kinases and the PI3K/Akt Cascade

In lieu of providing a vast list of cellular signaling cascades known to be affected by ROS, one type of cascade will be focused upon that serves to illustrate principles of the modulation of signaling by ROS as well as (patho-)physiological consequences thereof.

Growth factors such as the epidermal growth factor (EGF), platelet-derived growth factor (PDGF) or insulin act upon their target cells by binding to, and stimulating, their corresponding receptors, the EGF-, PDGF-, or insulin receptor (EGFR, PDGFR, InsR, respectively). These growth factor receptors are members of the receptor tyrosine kinase (RTK) family, with extracellular ligand binding domains. Binding of the respective ligand to the RTK causes (1) dimerization of RTK molecules (provided they are not yet in direct contact, as is the case with InsR family RTKs) (2) stimulation of the integral tyrosine kinase activity which catalyzes (3) auto-tyrosine phosphorylation of the intracellular RTK domain (in *trans*, i.e., one molecule will phosphorylate a second molecule, which—following dimerization—is in close proximity). The resulting phosphotyrosine moieties at the intracellular portion of activated RTK then serve as docking sites for numerous signaling proteins harboring a domain recognizing phosphotyrosine—such as a PTB (phosphotyrosine binding) or an SH2 (Src homology 2) domain. As this receptor tyrosine phosphorylation is therefore critical for the receptor's activity of initiating intracellular signaling processes after ligand binding on the extracellular portion, detection of RTK activity in the laboratory is now usually performed by analysis of RTK tyrosine phosphorylation. Similarly, a large group of low molecular weight RTK inhibitors (tyrphostins) targets the intracellular tyrosine kinase domain.

One of the cascades stimulated upon RTK activation leads to activation of the lipid kinase, phosphoinositide 3′-kinase (PI3K). RTK-dependent activation of PI3K occurs either by direct binding of the PI3K regulatory subunit to phospho-Tyr residues of activated RTK (thereby relieving autoinhibition of the catalytic subunit (Yu et al. 1998)) or indirectly, via adaptor proteins that bind to phospho-Tyr and mediate activation of PI3K (Cully et al. 2006; Vadas et al. 2011). PI3K, once activated, will phosphorylate cell membrane-bound phosphoinositides (phosphatidylinositol phosphates) in their 3′ position, generating, among others, phosphatidylinositol (3′,4′,5′)-trisphosphate (PIP3), which, in turn, constitutes a novel docking site at the membrane for proteins carrying a so-called pleckstrin homology (PH) domain. PH domain-containing proteins are therefore attracted to the membrane and 'concentrated', i.e., brought into each other's vicinity, resulting in phosphorylation and activation of the Ser/Thr kinase Akt (also termed protein kinase B, PKB). Activated Akt will phosphorylate substrate proteins, including nuclear proteins, as activated Akt may migrate into the nucleus (Fig. 5).

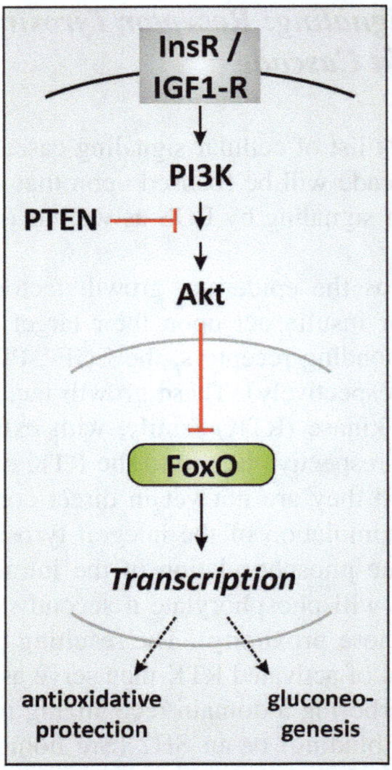

Fig. 5 Schematic representation of insulin signaling to modulate FoxO transcription factors. Stimulation of the mammalian insulin receptor (InsR) or the related insulin-like growth factor-1 receptor (IGF1-R) causes stimulation of the lipid kinase, phosphoinositide 3′-kinase (PI3K), which, by generating phosphatidylinositol (3′,4′,5′)-trisphosphate (PIP_3), triggers phosphorylation and activation of the serine/threonine kinase Akt. Upon activation, Akt may migrate into the nucleus and phosphorylate substrates such as FoxO transcription factors. Following their phosphorylation by Akt, FoxO factors, which regulate the transcription of genes coding for antioxidant enzymes and gluconeogenesis enzymes, are inactivated. The lipid phosphatase PTEN dephosphorylates PIP_3, thereby attenuating Akt activation

Negative regulation of this cascade occurs at several levels, including the dephosphorylation of the RTK (thereby stopping the binding of phospho-tyrosine recognizing proteins) and the dephosphorylation of PIP3 (thereby preventing the recruitment and binding to the membrane of PH domain proteins such as Akt). These inactivating reactions are catalyzed by protein tyrosine phosphatases (PTP) and lipid phosphatases, respectively. One such lipid phosphatase important for the PI3K/Akt pathway is PTEN (phosphatase and tensin homolog) which dephosphorylates PIP3 to phosphatidylinositol (4′,5′-bisphosphate and is frequently found inactivated in human cancers (Knobbe et al. 2002). An inactivation of such negative regulators may cause a net stimulation of the cascade, provided a basal level of substrate (such as PIP3) is present that—as it can no longer be inactivated—will accumulate and stimulate downstream events (Jiang et al. 2010). For a review on the PI3K/Akt cascade, see (Barthel and Klotz 2005).

2.2 Modulation of Signaling by ROS: How?

The oxidation of biomolecules, even if minor, i.e., not yet noticeably interfering with the cellular wellbeing, triggers cellular signaling processes in a variety of ways, four of which are to be discussed briefly.

(a) *Oxidation and inactivation of enzymes, causing a loss of activity and shift in signaling equilibria.*

Many signaling events are initiated at the level of stimulation of a receptor molecule—usually by binding of a receptor ligand. An example of a group of such receptors stimulated by compounds or conditions that come with the formation of ROS, is the group of receptor tyrosine kinases (RTK), some of which have been demonstrated to be stimulated by oxidants; the epidermal growth factor receptor (EGFR), its relative, ErbB2, or platelet-derived growth factor receptors (PDGFR) are among them and were demonstrated to be activated in cells exposed to stimuli such as hydrogen peroxide (Wang et al. 2000), peroxynitrite (Klotz et al. 2000, 2002b), redox cycling agents (Klotz et al. 2002a; Abdelmohsen et al. 2003) or ultraviolet radiation (von Montfort et al. 2006a; Knebel et al. 1996). Knebel et al. then demonstrated that the inactivation of a negative regulator of RTK activation, i.e., a protein tyrosine phosphatase (PTP), mediates this stress-induced RTK stimulation (Knebel et al. 1996). The reasoning for this finding is as follows: RTK are stimulated upon binding of a ligand that causes activation of the receptor's tyrosine kinase activity. Activated RTK will therefore catalyze the phosphorylation of tyrosyl residues, both of other RTK molecules (enhancing their activity) and non-RTK substrates. The resulting phosphotyrosine moieties will serve as docking sites for signaling proteins that will bind and thereby initiate intracellular signaling cascades. As tyrosine phosphorylation therefore is the crucial event in initiating these intracellular signaling processes, reversal of phosphorylation will serve as the off-switch, controlling RTK activity. Dephosphorylation of activated/phosphorylated RTK is catalyzed by PTP, which facilitate the hydrolysis of phosphotyrosine moieties by nucleophilic attack of their active site cysteine thiolate at the pTyr phosphorous—resulting in release of a tyrosyl residue and an intermediate phosphocysteine residue at the PTP's active site, which is then hydrolyzed to release phosphate and the original form of PTP (Kolmodin and Aqvist 2001).

It turns out that PTPs are exquisitely sensitive to oxidants and alkylating agents due to their active site cysteine being present in its deprotonated (thiolate) form (Kolmodin and Aqvist 2001; Ostman et al. 2006). Oxidants, although, of course, reacting with other biomolecules as well, will more rapidly and easily cause oxidation of such sensitive moieties. PTP inactivation has indeed been demonstrated to occur in cells or isolated PTP exposed to oxidants such as H_2O_2 (Knebel et al. 1996; Salmeen et al. 2003; van Montfort et al. 2003), peroxynitrite (Takakura et al. 1999), singlet oxygen (von Montfort et al. 2006b) or alkylating agents that rapidly interact with thiolates (Abdelmohsen et al. 2003; Beier et al. 2006; Iwamoto et al. 2007).

Exposure of cells to oxidants will therefore result in oxidative inactivation of PTPs, implying that tyrosine phosphorylation/dephosphorylation equilibria are shifted to the phosphorylation side, resulting in a net accumulation of phosphotyrosine moieties. If RTK are the substrates of an inactivated PTP, this would cause a net accumulation of phosphorylated and activated RTK.

Interestingly, this mechanism is also exploited naturally, i.e., upon stimulation of RTK with their natural ligands: Stimulation of the insulin receptor with insulin (Meng et al. 2004; Bae et al. 1997), the EGF receptor with EGF (Bae et al. 1997; DeYulia and Carcamo 2005) or the PDGF receptor with PDGF (Sundaresan et al. 1995) all come with a transient and locally confined increase in ROS production and an ROS-induced transient inhibition of PTP associated with the respective RTK—which appears to be required for a significant ligand-induced increase in RTK phosphorylation and therefore a significant intracellular signal to be initiated. Similar findings were reported recently for the NADPH oxidase NOX4, which provides hydrogen peroxide to attenuate EGFR dephosphorylation (Chen et al. 2008).

In essence, the example provided is that of an oxidative inactivation of a susceptible protein (PTP), causing a net stimulation of signaling because the inactivated protein is a negative regulator.

(b) *Oxidative stress and the generation of bioactive oxidation products.*

The oxidation of several biomolecules will yield reactive products that may react further with proteins or other crucial molecules to stimulate signaling events.

Lipid peroxidation, as initiated by exposure of lipid membranes to ROS, will not only result in oxidative breakdown of lipids (and the loss of membrane integrity in affected areas), but will also generate products that may interact with secondary targets not initially affected by ROS. The most prominent examples of such products are reactive aldehydes such as acrolein, malondialdehyde (MDA) and 4-hydroxynonenal (4-HNE). These aldehydes may form adducts with proteins by interacting with amino acid side chains (Fig. 6) (Esterbauer et al. 1991; Uchida 2000; Ishii et al. 2007; LoPachin et al. 2009). It is conceivable that protein functionality may be seriously impaired by such modifications. It has hence been proposed that adduct formation contributes to the pathogenesis of several phenomena, including atherosclerosis and cardiovascular disease (Uchida 2000). As described above, the modulation and impairment of signaling protein activity may entail net activation of a signaling cascade. Likewise, inhibition and alkylation of the lipid and protein dual phosphatase PTEN was demonstrated to coincide with—and postulated to mediate—stimulation of the phosphoinositide 3-kinase (PI3K)/Akt signaling pathway after exposure of cells to 4-HNE (Shearn et al. 2011). In line with such alkylations by lipid peroxidation products inducing pathophysiologically relevant signaling processes, it was demonstrated that exposure of cells to 4-HNE will result in stimulation of expression of the proinflammatory protein cyclooxygenase-2 (COX-2), which appears to be mediated by the stress-responsive kinase $p38^{MAPK}$ (Kumagai et al. 2002). Likewise, exposure of fibroblasts (Chojkier et al. 1989) or hepatic stellate cells (Garcia-Ruiz et al. 2002) to MDA resulted in a stimulation of collagen expression, a crucial step in the pathogenesis of fibrosis.

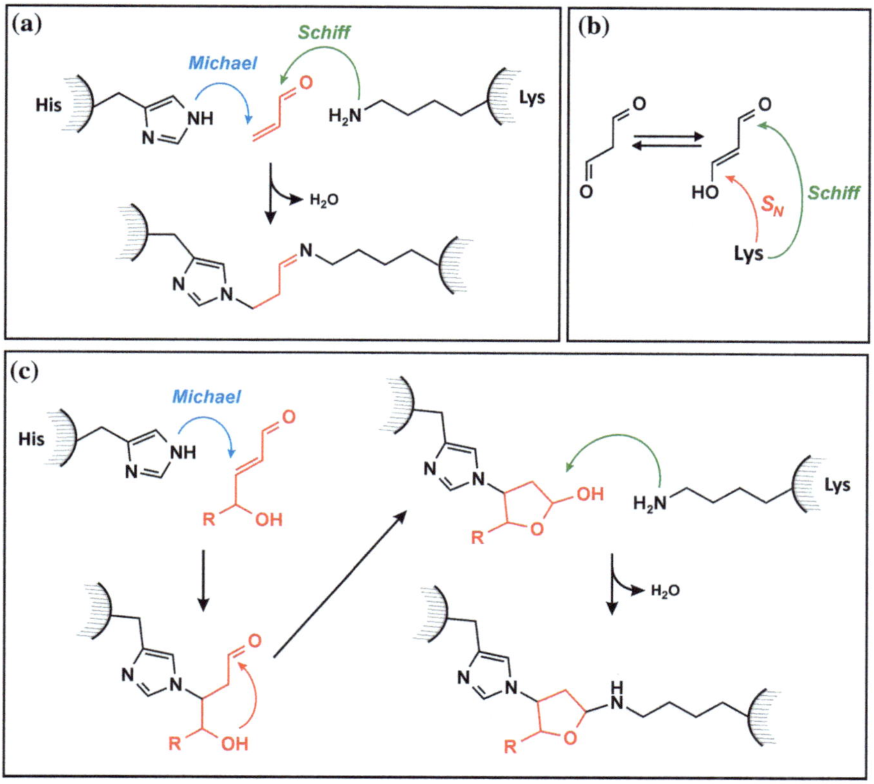

Fig. 6 Covalent modification of proteins by lipid peroxidation products. Lipid peroxidation results in formation of reactive aldehydes such as (**a**) Acrolein (**b**) Malondialdehyde or (**c**) 4-hydroxyalkenals, which may interact with protein amino acid residues and undergo Michael addition reactions, Schiff base formation and other nucleophilic additions to form covalent adducts and cross-links

Glutathione—due to its high intracellular concentrations—is another frequent target of various oxidants. Glutathione oxidation products, such as glutathione sulfenic acid (GSOH), glutathione disulfide (GSSG), GSSG *S*-oxide or *S*-nitroso-(GNSO) and *S*-nitroglutathione ($GSNO_2$) are capable of glutathiolating proteins by interacting with cysteine thiols/thiolates. Vice versa, protein cysteine oxidation products (sulfenic acid, cystine, *S*-oxide, *S*-nitroso) may be attacked by glutathione to form glutathione/protein mixed disulfides (Fig. 7) (Klatt and Lamas 2000; Huang et al. 2007; Hill and Bhatnagar 2012; Jacob 2011; Jacob et al. 2003).

(c) *Gain of activity through protein oxidation.*

While glutathiolation may modulate the activity of proteins via generation of mixed glutathione/protein disulfides, the reversible disulfide formation between a signaling protein and a modifying enzyme and regulator has also been

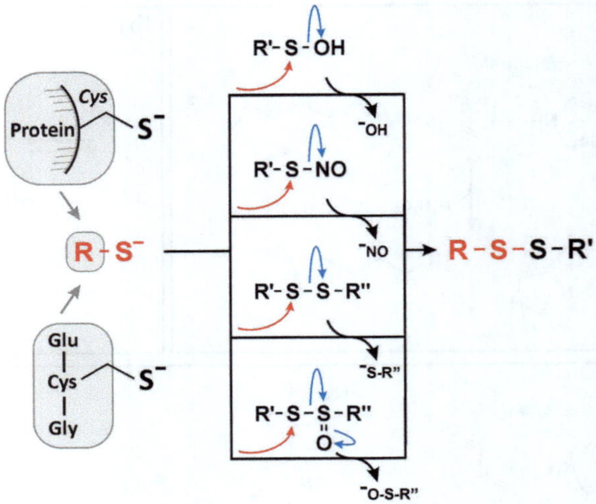

Fig. 7 *S*-Glutathiolation of proteins: mechanisms. *S*-Glutathiolation of proteins, i.e., the formation of a mixed disulfide between glutathione and a protein cysteinyl residue, occurs by nucleophilic attack of a thiol(ate) at an oxidation product of another thiol (resulting from its interaction with a reactive oxygen or nitrogen species), such as (from top to bottom) a sulfenic acid, nitrosothiol, disulfide, or disulfide-*S*-oxide. Depending on which reaction partner is present in the oxidized form, R may be glutathione or a protein cysteine, attacking an oxidized protein cysteine or glutathione moiety R′, respectively. It should be noted that most of these modifications occur spontaneously, i.e., in the absence of enzyme catalysts (see Chaps. 1 and 4)

demonstrated. Dansen et al. reported the modulation of transcription factor FoxO4 by reversible disulfide formation in response to oxidative stress: FoxO4 forms a disulfide with p300/CBP acetyl transferase upon exposure of cells to hydrogen peroxide, resulting in changes of FoxO4 acetylation, which is known to modulate its activity (Dansen et al. 2009).

(d) *Accumulation of oxidation products, resulting in modulation of signaling processes.*

Recently, the oxidation of proteins and the degradation of oxidized proteins were related to enhanced signaling. Oxidized proteins are preferably degraded by the proteasome (Grune et al. 2003) which does not appear to require ubiquitination of the target proteins (Kastle and Grune 2011). Extensive oxidation and cross-linking of oxidized proteins, however, may cause the inhibition of the proteasomal system, with consequences for the turnover of other cellular proteins, including proteins involved in the regulation of gene expression: proteasomal inhibition in skin fibroblasts induced by exposure to UVA and/or singlet oxygen subsequently led to an accumulation of c-Jun and phosphorylated c-Jun and the activation of activator protein-1, i.e., transcription factors known to respond to stressful stimuli (Catalgol et al. 2009).

3 Oxidant-Induced Signaling and Chemoprevention

3.1 Antioxidants and Chemoprevention

Oxidative stress may cause damage to crucial cellular molecules. It may also induce signaling pathways—in part through those very oxidative modifications to biomolecules that occupy central positions in the control of signaling pathways. At first sight, damage to biomolecules clearly appears as an effect of oxidants that needs to be prevented in order to uphold cellular integrity—hence the notion that antioxidants might be a proper remedy to interfere with such degradative processes (see also Chap. 2). In this context, the rather broad term, 'chemoprevention', would therefore refer to the prevention of oxidative stress-induced damage by application of antioxidants.

The outcome of stress-induced signaling, however, is not necessarily negative: assuming that stimulation of a stress-responsive signal cascade has beneficial effects—e.g., in terms of inducing adaptation—it might be undesired to block these processes by blocking the initial stress, i.e., by scavenging ROS. Indeed, such effects were observed in studies with rats and humans, where vitamin C supplementation was demonstrated to *decrease* training efficiency, apparently because it prevents cellular adaptations to exercise (Gomez-Cabrera et al. 2008, 2012). Similarly, vitamin C/vitamin E supplementation decreases exercise-induced insulin sensitivity and attenuates the induction of the body's own ROS defense systems (Ristow et al. 2009).

It therefore seems that chemoprevention of stress-induced damage needs to target processes induced by ROS rather than to target ROS themselves. Proper antioxidants would selectively attenuate adverse effects of ROS but leave positive consequences unaffected.

3.2 FoxO Transcription Factors in ROS-Induced Signaling

The aforementioned stimulation of receptor tyrosine kinase signaling leading to activation of the protein kinase Akt is an excellent example of stress-induced signaling and its consequences for cell survival and adaptation.

Akt is regarded as antiapoptotic, i.e., as attenuating processes leading to apoptotic cell death. This is due to its numerous substrates that are known to play an important role in regulating apoptosis and that, upon phosphorylation by Akt, are either activated (such as IKK, the inhibitor of kappa B kinase, leading to stimulation of the NF-κB pathway which regulates the expression of antiapoptotic proteins) or inactivated (in the case of proapoptotic proteins such as caspase-9, Bad, and others) (Duronio 2008; Sale and Sale 2008). FoxO (forkhead box, class O) transcription factors are Akt substrates that not only control apoptotic processes, but also play an important role both in the regulation of fuel metabolism and in the cellular response to (oxidative) stress. This versatility is due to the target genes controlled by FoxOs (Greer and Brunet 2005; Barthel et al. 2005; Walter et al. 2008). FoxO transcription factors control apoptosis by regulating the expression of proapoptotic factors such as Fas-Ligand. They further regulate gluconeogenesis by controlling expression of crucial enzymes, such as glucose 6-phosphatase and phosphoenol pyruvate carboxykinase, and they respond to insulin by being phosphorylated in a phosphoinositide 3′-kinase (PI3K)—and Akt-dependent manner (Greer and Brunet 2005). In addition to this crucial role in glucose homeostasis and fuel metabolism, FoxO factors play an important role in the cellular stress response, e.g., by controlling the expression of antioxidant enzymes, such as catalase, manganese-superoxide dismutase (MnSOD) (Kops et al. 2002), selenoprotein P (Speckmann et al. 2008; Walter et al. 2008) and ceruloplasmin (Leyendecker et al. 2011).

Four FoxO isoforms are known to date, with FoxO1a, FoxO3a and FoxO4 ubiquitously expressed in humans, and FoxO6 expressed in confined areas (Greer and Brunet 2005). Transcriptional activity of FoxOs 1a, 3a, 4 is controlled by phosphorylation by Akt, which results in FoxO inactivation as well as sequestration to the cytosol. A *stimulation* of Akt—e.g., by stimulation of cells with growth factors, such as EGF or insulin—will therefore result in an *inactivation* of FoxOs and a downregulation of FoxO-dependent transcription. Thus, insulin—essentially signaling glucose sufficiency to hepatocytes—will attenuate the expression of hepatocyte genes coding for key gluconeogenesis enzymes (Fig. 5).

FoxOs being regulated by signaling cascades emanating from RTKs implies that there is regulation by stressful stimuli that affect RTK-dependent signaling. As receptor tyrosine kinases react to xenobiotics and ROS generating systems, including quinones (Abdelmohsen et al. 2003, 2007; Beier et al. 2006), ultraviolet radiation (von Montfort et al. 2006a; Mahns et al. 2003; Knebel et al. 1996) and heavy metal ions (Eckers et al. 2009; Barthel et al. 2007; Walter et al. 2006; Ostrakhovitch et al. 2002), PI3K-dependent signaling is activated by such stimuli, resulting in stimulation of the Ser/Thr kinase Akt. FoxO transcription factors were then identified as phosphorylated and/or inactivated upon exposure of mammalian liver cells to stressful stimuli such as copper or zinc ions (Barthel et al. 2007; Walter et al. 2006). Thus, these metal ions initiated insulin-mimetic signaling. This stress response appears to be highly conserved, as the FoxO ortholog, DAF-16, was also inactivated in *Caenorhabditis elegans* worms when grown in the presence of Cu^{2+} (Walter et al. 2006).

FoxO activity is extensively regulated by posttranslational modifications, including modifications other than phosphorylation, such as acetylation and (mono- and poly-) ubiquitination (Barthel et al. 2005). Several kinases other than Akt contribute to inactivation of FoxOs, such as serum/glucocorticoid responsive kinase (SGK) or "dual-specificity, tyrosine-phosphorylated and regulated kinase" (DYRK1A) (Calnan and Brunet 2008). Moreover, activating phosphorylation was described, e.g., by c-Jun *N*-terminal kinases (JNK), and demonstrated to mediate an adaptive response to oxidative stress that causes FoxO-dependent expression of MnSOD (Essers et al. 2004). Acetylation (by p300/CBP) and deacetylation (by sirtuins) of FoxO in response to (oxidative) stress were described, although there are contradictory results on whether deacetylation results in activation (Frescas et al. 2005) or inactivation (Motta et al. 2004) of FoxOs. As mentioned above, FoxO4 activity was demonstrated to be regulated by reversible disulfide formation between FoxO4 and p300/CBP acetyl transferase upon exposure of cells to hydrogen peroxide (Dansen et al. 2009). In summary, there is both activation of FoxOs by oxidative stress (e.g., via JNK-dependent phosphorylation) and inactivation (via stress-induced Akt activation). It has not yet been fully elucidated, however, which effect prevails under which cellular conditions, and to what effect.

The spectrum of genes regulated by FoxOs and the fine-tuning of FoxO activity is diversified significantly by interaction with transcriptional coregulators (van der Vos and Coffer 2008). Two major xenosensors are among these coregulators modulating FoxO activity (and vice versa), i.e., constitutive androstane receptor (CAR) and pregnane X-receptor (PXR), allowing for an indirect modulation of FoxOs by CAR/PXR-specific xenobiotics, such as phenobarbital (Kodama et al. 2004). An interaction between the FoxO and aryl hydrocarbon receptor systems has also been hypothesized recently (Eckers et al. 2011).

3.3 FoxOs in Chemoprevention: The Cases of Green Tea Flavonoids and Selenium

As outlined above, FoxO activity is affected by ROS and xenobiotics, and FoxOs, in turn, modulate the expression of antioxidant proteins and of other proteins involved in the stress response. This would render FoxO transcription factors excellent targets for chemoprevention, aiming at an upregulation of a FoxO-dependent endogenous defense against noxious stimuli. Such approaches are based on the assumption that stimulation of FoxOs is desired—which, of course, cannot always be the case, as FoxO transcription factors also stimulate the production of pro-apoptotic factors (see Sect. 3.2).

There appear to be ways of stimulating FoxO activity using naturally occurring compounds—irrespective of whether FoxO stimulation is beneficial, and under which circumstances. The following paragraphs will first give one example of such a group of compounds and will then provide an example of the kind of consequences an overactivation of FoxOs might have, thereby illustrating a chemopreventive approach targeting FoxOs and some potential pitfalls.

(a) *FoxO activation by green tea flavonoids*

Green tea flavonoids, epicatechin (EC), EC gallate, epigallocatechin, and epigallocatechin gallate (EGCG) (Fig. 8) were tested for their effect on FoxO activity in different cell types, including human hepatoma cells (HepG2) and human fibroblasts. EGCG is of particular interest in *in vitro* studies as it is the major green tea flavonoid (Cabrera et al. 2003) and as *in vivo* a high percentage of consumed

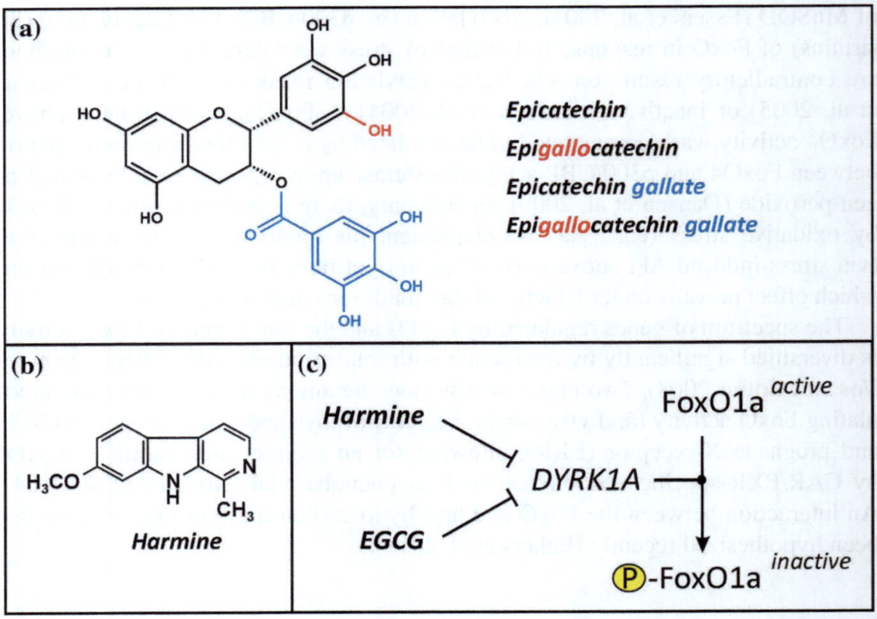

Fig. 8 Green tea flavonoids as modulators of FoxO activity. Structures of (**a**) Green tea flavonoids and of (**b**) Harmine (**c**) Hypothesis on the mode of stimulation of FoxO activity by epigallocatechin gallate (EGCG) and harmine: Stimulation of FoxO activity by inhibition of an inhibitor of FoxO proteins, "dual-specificity, tyrosine-phosphorylated and regulated kinase" (DYRK)-1A

EGCG reaches the circulation unmetabolised (Unno et al. 1996). Interestingly, phosphorylation, nuclear exclusion and inactivation of FoxOs were observed in these cells using EGCG, but this effect was found only at supraphysiological concentrations of above 10 µM: H_2O_2 was generated upon incubation of cells in the presence of epigallocatechin gallate (EGCG) (Bartholome et al. 2010). Lower concentrations of EGCG (1 µM) did not cause significant H_2O_2 generation, and (rather than not affecting FoxO activity) elicited the opposite effect, i.e., FoxO stimulation (Bartholome et al. 2010). Therefore, EGCG is a typical example of a hormetic substance that does have an effect at low concentrations which is actually *inverted* at higher concentrations.

Regarding the mechanism of action of EGCG that led to FoxO stimulation, it is currently assumed that the FoxO kinase, DYRK1A, may be the molecular target leading to the effects observed (Bartholome et al. 2010). Phosphorylation of FoxOs by DYRK1A was demonstrated to attenuate FoxO transcriptional regulation (Woods et al. 2001). DYRK1A, in turn, was demonstrated to be efficiently inhibited by EGCG (Bain et al. 2003). In support of this hypothesis, a known inhibitor of DYRK1A, i.e., harmine (Fig. 8b), had the same effect on FoxO activity as EGCG (Bartholome et al. 2010). In summary, the mode of action of EGCG is hypothesised to be the inhibition of an endogenous inhibitor of the effector to result in the effector's net stimulation (Fig. 8c)—a mode of action that reminds of the above-mentioned inhibition of PTPs by ROS to result in RTK stimulation (see Sect. 2.2). The concentrations of EGCG employed for these experiments are in a region of concentrations achievable *in vivo* (approx. 0.2 μM to approx. 5 μM in human serum after consumption of green tea or green tea extract) (Unno et al. 1996; Nakagawa et al. 1997; Maiani et al. 1997). One should note that these actions of EGCG are fairly specific and not simply due to random redox reactions of this redox active polyphenolic compounds (see also Chap. 2).

(b) *FoxOs and selenium homeostasis*

Expression of the major plasma selenium transporter, selenoprotein P (SelP), was demonstrated to be controlled by FoxOs (Walter et al. 2008) and FoxO effects were enhanced by peroxisomal proliferator activated receptor coactivator-1α (PGC1α) (Speckmann et al. 2008). SelP *per se* has antioxidant properties (Traulsen et al. 2004) and is required for selenium supply to extrahepatic tissues in order to allow for biosynthesis of selenium-dependent antioxidant enzymes, such as glutathione peroxidases (Steinbrenner et al. 2006b).

While this appears as a beneficial FoxO effect *per se*, three findings place this into a different context: (1) exposure of hepatoma cells to insulin downregulates SelP production (as expected for a FoxO target gene) (Walter et al. 2008; Speckmann et al. 2008) (2) High SelP levels were found to be correlated with, and to cause, insulin resistance (Misu et al. 2010) (3) Overexpression of cytosolic glutathione peroxidase (GPx-1) in mice caused insulin resistance (McClung et al. 2004).

Again, it seems that too much of an antioxidant may have adverse effects (see Sect. 3.1): FoxO activation might result in enhanced production of SelP and supply of selenium to extrahepatic tissues. Selenium supply via SelP will induce production of GPx-1 (Steinbrenner et al. 2006a, b), an antioxidant selenoprotein that catalyzes the reduction of peroxides at the expense of glutathione. Elevation of GPx-1 levels may therefore interfere with insulin signaling—which, in turn, relies on locally confined generation of H_2O_2 (see Sect. 2.1).

These data may help to explain recent epidemiological data linking high selenium intake to Type II diabetes (see (Steinbrenner et al. 2011) for review): the identified control of SelP expression by insulin would thus be part of a regulatory circuit counteracting selenium saturation (and insulin resistance) of peripheral tissues.

3.4 Conclusion: Modulation of Signaling: Antioxidant and Chemopreventive Approach of the Future

The identification of novel antioxidant approaches that are based upon pharmacological modulation of signaling cascades rather than direct scavenging of ROS is based upon the hypothesis that stressful stimuli, including ROS generated endogenously under exposure of cells to xenobiotics, initiate adaptive processes that support cellular stress resistance.

Considering the fact that several common 'antioxidants' and compounds with beneficial effects reported in the literature—such as green tea flavonoids or similar plant secondary metabolites—show only limited bioavailability and low *in vivo* concentrations, it is suggested that they may act as perceived antioxidants not by directly scavenging ROS, but rather in an indirect manner, i.e., by modulating cellular signaling cascades—which would require only minute amounts of active agent, as opposed to a direct reaction with ROS (see also Chap. 2).

Some signaling cascades suitable as potential targets for chemopreventive approaches are known (including the PI3K/Akt/FoxO cascade), but measures will have to be identified to modulate signaling in a beneficial or desired direction, avoiding adverse consequences or side effects resulting from that very modulation of the signaling cascade of interest. These measures include the identification or development of novel pharmacological modulators.

Acknowledgments The author gratefully acknowledges support by the Canada Research Chairs (CRC) program, the Canada Foundation for Innovation (CFI), Alberta Advanced Education and Technology (AET), the Natural Sciences and Engineering Research Council of Canada (NSERC, Discovery Grant RGPIN 402228-2011), and the Faculty of Pharmacy and Pharmaceutical Sciences, University of Alberta.

References

Abdelmohsen K, Gerber PA, von Montfort C, Sies H, Klotz LO (2003) Epidermal growth factor receptor is a common mediator of quinone-induced signaling leading to phosphorylation of connexin-43 role of glutathione and tyrosine phosphatases. J Biol Chem 278(40):38360–38367

Abdelmohsen K, Sauerbier E, Ale-Agha N, Beier J, Walter P, Galban S, Stuhlmann D, Sies H, Klotz LO (2007) Epidermal growth factor- and stress-induced loss of gap junctional communication is mediated by ERK-1/ERK-2 but not ERK-5 in rat liver epithelial cells. Biochem Biophys Res Commun 364(2):313–317

Babior BM (1984) The respiratory burst of phagocytes. J Clin Invest 73(3):599–601

Bae YS, Kang SW, Seo MS, Baines IC, Tekle E, Chock PB, Rhee SG (1997) Epidermal growth factor (EGF)-induced generation of hydrogen peroxide role in EGF receptor-mediated tyrosine phosphorylation. J Biol Chem 272(1):217–221

Bain J, McLauchlan H, Elliott M, Cohen P (2003) The specificities of protein kinase inhibitors: an update. Biochem J 371:199–204

Barthel A, Klotz LO (2005) Phosphoinositide 3-kinase signaling in the cellular response to oxidative stress. Biol Chem 386(3):207–216

Barthel A, Schmoll D, Unterman TG (2005) FoxO proteins in insulin action and metabolism. Trends Endocrinol Metab 16(4):183–189

Barthel A, Ostrakhovitch EA, Walter PL, Kampkötter A, Klotz LO (2007) Stimulation of phosphoinositide 3-kinase/Akt signaling by copper and zinc ions: mechanisms and consequences. Arch Biochem Biophys 463(2):175–182

Bartholome A, Kampkötter A, Tanner S, Sies H, Klotz LO (2010) Epigallocatechin gallate-induced modulation of FoxO signaling in mammalian cells and *C. elegans*: FoxO stimulation is masked via PI3K/Akt activation by hydrogen peroxide formed in cell culture. Arch Biochem Biophys 501(1):58–64

Beier JI, von Montfort C, Sies H, Klotz LO (2006) Activation of ErbB2 by 2-methyl-1,4-naphthoquinone (menadione) in human keratinocytes: role of EGFR and protein tyrosine phosphatases. FEBS Lett 580(7):1859–1864

Bolton JL, Trush MA, Penning TM, Dryhurst G, Monks TJ (2000) Role of quinones in toxicology. Chem Res Toxicol 13(3):135–160

Brunmark A, Cadenas E (1989) Redox and addition chemistry of quinoid compounds and its biological implications. Free Radical Biol Med 7(4):435–477

Cabrera C, Gimenez R, Lopez MC (2003) Determination of tea components with antioxidant activity. J Agric Food Chem 51(15):4427–4435

Calnan DR, Brunet A (2008) The FoxO code. Oncogene 27(16):2276–2288

Catalgol B, Ziaja I, Breusing N, Jung T, Höhn A, Alpertunga B, Schroeder P, Chondrogianni N, Gonos ES, Petropoulos I, Friguet B, Klotz LO, Krutmann J, Grune T (2009) The Proteasome is an integral part of solar ultraviolet a radiation-induced gene expression. J Biol Chem 284(44):30076–30086

Chen K, Kirber MT, Xiao H, Yang Y, Keaney JF (2008) Regulation of ROS signal transduction by NADPH oxidase 4 localization. J Cell Biol 181(7):1129–1139

Chojkier M, Houglum K, Solisherruzo J, Brenner DA (1989) Stimulation of collagen gene-expression by ascorbic-acid in cultured human-fibroblasts. A Role for Lipid-Peroxidation. J Biol Chem 264(28):16957–16962

Cully M, You H, Levine AJ, Mak TW (2006) Beyond PTEN mutations: the PI3K pathway as an integrator of multiple inputs during tumorigenesis. Nat Rev Cancer 6(3):184–192

Dansen TB, Smits LMM, van Triest MH, de Keizer PLJ, van Leenen D, Koerkamp MG, Szypowska A, Meppelink A, Brenkman AB, Yodoi J, Holstege FCP, Burgering BMT (2009) Redox-sensitive cysteines bridge p300/CBP-mediated acetylation and FoxO4 activity. Nature Chemical Biology 5 (9):664-672

DeYulia GJ, Carcamo JM (2005) EGF receptor-ligand interaction generates extracellular hydrogen peroxide that inhibits EGFR-associated protein tyrosine phosphatases. Biochem Biophys Res Commun 334(1):38–42

Duronio V (2008) The life of a cell: apoptosis regulation by the PI3K/PKB pathway. Biochem J 415:333–344

Eckers A, Reimann K, Klotz LO (2009) Nickel and copper ion-induced stress signaling in human hepatoma cells: analysis of phosphoinositide 3'-kinase/Akt signaling. Biometals 22(2):307–316

Eckers A, Sauerbier E, Anwar-Mohamed A, Hamann I, Esser C, Schroeder P, El-Kadi AOS, Klotz LO (2011) Detection of a functional xenobiotic response element in a widely employed FoxO-responsive reporter construct. Arch Biochem Biophys 516(2):138–145

Ernster L (1987) Dt-diaphorase: a historical review. Chemica Scripta 27A:1–13

Essers MAG, Weijzen S, de Vries-Smits AMM, Saarloos I, de Ruiter ND, Bos JL, Burgering BMT (2004) FOXO transcription factor activation by oxidative stress mediated by the small GTPase Ral and JNK. EMBO J 23(24):4802–4812

Esterbauer H, Schaur RJ, Zollner H (1991) Chemistry and biochemistry of 4-Hydroxynonenal, Malonaldehyde and related aldehydes. Free Radical Biol Med 11(1):81–128

Frescas D, Valenti L, Accili D (2005) Nuclear trapping of the forkhead transcription factor FoxO1 via Sirt-dependent deacetylation promotes expression of glucogenetic genes. J Biol Chem 280(21):20589–20595

Garcia-Ruiz I, de la Torre P, Diaz T, Esteban E, Fernandez I, Munoz-Yague T, Solis-Herruzo JA (2002) Sp1 and Sp3 transcription factors mediate malondialdehyde-induced collagen alpha 1(I) gene expression in cultured hepatic stellate cells. J Biol Chem 277(34):30551–30558

Gomez-Cabrera MC, Domenech E, Romagnoli M, Arduini A, Borras C, Pallardo FV, Sastre J, Vina J (2008) Oral administration of vitamin C decreases muscle mitochondrial biogenesis and hampers training-induced adaptations in endurance performance. Am J Clin Nutr 87(1):142–149

Gomez-Cabrera MC, Ristow M, Vina J (2012) Antioxidant supplements in exercise: worse than useless? American J Physiol-Endocrinol Metab 302(4):E476–E477

Greer EL, Brunet A (2005) FOXO transcription factors at the interface between longevity and tumor suppression. Oncogene 24(50):7410–7425

Grune T, Merker K, Sandig G, Davies KJA (2003) Selective degradation of oxidatively modified protein substrates by the proteasome. Biochem Biophys Res Commun 305(3):709–718

Haberzettl P, Schins RPF, Hohr D, Wilhelmi V, Borm PJA, Albrecht C (2008) Impact of the Fc gamma II-receptor on quartz uptake and inflammatory response by alveolar macrophages. American J Physiol-Lung Cell Mol Physiol 294(6):L1137–L1148

Hill BG, Bhatnagar A (2012) Protein S-glutathiolation: redox-sensitive regulation of protein function. J Mol Cell Cardiol 52(3):559–567

Huang KP, Huang FL, Shetty PK, Yergey AL (2007) Modification of protein by disulfide S-monoxide and disulfide S-dioxide: distinctive effects on PKC. Biochemistry 46(7):1961–1971

Ishii T, Yamada T, Mori T, Kumazawa S, Uchida K, Nakayama T (2007) Characterization of acrolein-induced protein cross-links. Free Radical Res 41(11):1253–1260

Iwamoto N, Sumi D, Ishii T, Uchida K, Cho AK, Froines JR, Kumagai Y (2007) Chemical knockdown of protein-tyrosine phosphatase 1B by 1,2-naphthoquinone through covalent modification causes persistent transactivation of epidermal growth factor receptor. J Biol Chem 282(46):33396–33404

Jacob C (2011) Redox signalling via the cellular thiolstat. Biochem Soc Trans 39:1247–1253

Jacob C, Giles GL, Giles NM, Sies H (2003) Sulfur and selenium: the role of oxidation state in protein structure and function. Angewandte Chemie-Int Ed 42(39):4742–4758

Jiang XN, Chen S, Asara JM, Balk SP (2010) Phosphoinositide 3-kinase pathway activation in phosphate and tensin homolog (PTEN)-deficient prostate cancer cells is independent of receptor tyrosine kinases and mediated by the p110 beta and p110 delta catalytic subunits. J Biol Chem 285(20):14980–14989

Jones DP (2008) Radical-free biology of oxidative stress. American J Physiol-Cell Physiol 295(4):C849–C868

Kastle M, Grune T (2011) Proteins bearing oxidation-induced carbonyl groups are not preferentially ubiquitinated. Biochimie 93(6):1076–1079

Katsuyama M, Matsuno K, Yabe-Nishimura C (2012) Physiological roles of NOX/NADPH oxidase, the superoxide-generating enzyme. J Clin Biochem Nutr 50(1):9–22

Klatt P, Lamas S (2000) Regulation of protein function by S-glutathiolation in response to oxidative and nitrosative stress. Eur J Biochem 267(16):4928–4944

Klaus V, Hartmann T, Gambini J, Graf P, Stahl W, Hartwig A, Klotz LO (2010) 1,4-Naphthoquinones as inducers of oxidative damage and stress signaling in HaCaT human keratinocytes. Arch Biochem Biophys 496(2):93–100

Klotz LO (2002) Oxidant-induced signaling: effects of peroxynitrite and singlet oxygen. Biol Chem 383(3–4):443–456

Klotz LO, Sies H (2003) Defenses against peroxynitrite: selenocompounds and flavonoids. Toxicol Lett 140:125–132

Klotz LO, Briviba K, Sies H (1997) Singlet oxygen mediates the activation of JNK by UVA radiation in human skin fibroblasts. FEBS Lett 408(3):289–291

Klotz LO, Pellieux C, Briviba K, Pierlot C, Aubry JM, Sies H (1999) Mitogen-activated protein kinase (p38-, JNK-, ERK-) activation pattern induced by extracellular and intracellular singlet oxygen and UVA. Eur J Biochem 260(3):917–922

Klotz LO, Schieke SM, Sies H, Holbrook NJ (2000) Peroxynitrite activates the phosphoinositide 3-kinase/Akt pathway in human skin primary fibroblasts. Biochem J 352:219–225

Klotz LO, Patak P, Ale-Agha N, Buchczyk DP, Abdelmohsen K, Gerber PA, von Montfort C, Sies H (2002a) 2-Methyl-1,4-naphthoquinone, vitamin K-3, decreases gap-junctional intercellular communication via activation of the epidermal growth factor receptor/extracellular signal-regulated kinase cascade. Cancer Res 62(17):4922–4928

Klotz LO, Schroeder P, Sies H (2002b) Peroxynitrite signaling: receptor tyrosine kinases and activation of stress-responsive pathways. Free Radical Biol Med 33(6):737–743

Klotz LO, Kröncke KD, Sies H (2003) Singlet oxygen-induced signaling effects in mammalian cells. Photochem Photobiol Sci 2(2):88–94

Knebel A, Rahmsdorf HJ, Ullrich A, Herrlich P (1996) Dephosphorylation of receptor tyrosine kinases as target of regulation by radiation, oxidants or alkylating agents. EMBO J 15(19):5314–5325

Knobbe CB, Merlo A, Reifenberger G (2002) Pten signaling in gliomas. Neurooncol 4(3):196–211

Kodama S, Koike C, Negishi M, Yamamoto Y (2004) Nuclear receptors CAR and PXR cross talk with FOXO1 to regulate genes that encode drug-metabolizing and gluconeogenic enzymes. Mol Cell Biol 24(18):7931–7940

Kolmodin K, Aqvist J (2001) The catalytic mechanism of protein tyrosine phosphatases revisited. FEBS Lett 498(2–3):208–213

Kops GJPL, Dansen TB, Polderman PE, Saarloos I, Wirtz KWA, Coffer PJ, Huang TT, Bos JL, Medema RH, Burgering BMT (2002) Forkhead transcription factor FOXO3a protects quiescent cells from oxidative stress. Nature 419(6904):316–321

Kumagai T, Nakamura Y, Osawa T, Uchida K (2002) Role of p38 mitogen-activated protein kinase in the 4-hydroxy-2-nonenal-induced cyclooxygenase-2 expression. Arch Biochem Biophys 397(2):240–245

Lee BC, Dikiy A, Kim HY, Gladyshev VN (2009) Functions and evolution of selenoprotein methionine sulfoxide reductases. Biochimica Et Biophysica Acta-General Subjects 1790(11):1471–1477

Leyendecker M, Korsten P, Reinehr R, Speckmann B, Schmoll D, Scherbaum WA, Bornstein SR, Barthel A, Klotz LO (2011) Ceruloplasmin expression in rat liver cells is attenuated by insulin: role of FoxO transcription factors. Horm Metab Res 43(4):268–274

Li N, Sioutas C, Cho A, Schmitz D, Misra C, Sempf J, Wang MY, Oberley T, Froines J, Nel A (2003) Ultrafine particulate pollutants induce oxidative stress and mitochondrial damage. Environ Health Perspect 111(4):455–460

Lippard SJ (1999) Biochemistry: free copper ions in the cell? Science 284(5415):748–749

LoPachin RM, Gavin T, Petersen DR, Barber DS (2009) Molecular Mechanisms of 4-hydroxy-2-nonenal and acrolein toxicity: nucleophilic targets and adduct formation. Chem Res Toxicol 22(9):1499–1508

Mahns A, Melchheier I, Suschek CV, Sies H, Klotz LO (2003) Irradiation of cells with ultraviolet-A (320–400 nm) in the presence of cell culture medium elicits biological effects due to extracellular generation of hydrogen peroxide. Free Radical Res 37(4):391–397

Maiani G, Serafini M, Salucci M, Azzini E, FerroLuzzi A (1997) Application of a new high-performance liquid chromatographic method for measuring selected polyphenols in human plasma. J Chromatogr B 692(2):311–317

Maynard S, Schurman SH, Harboe C, de Souza-Pinto NC, Bohr VA (2009) Base excision repair of oxidative DNA damage and association with cancer and aging. Carcinogenesis 30(1):2–10

McClung JP, Roneker CA, Mu WP, Lisk DJ, Langlais P, Liu F, Lei XG (2004) Development of insulin resistance and obesity in mice overexpressing cellular glutathione peroxidase. Proc Natl Acad Sci USA 101(24):8852–8857

Melchheier I, von Montfort C, Stuhlmann D, Sies H, Klotz LO (2005) Quinone-induced Cdc25A inhibition causes ERK-dependent connexin phosphorylation. Biochem Biophys Res Commun 327(4):1016–1023

Meng TC, Buckley DA, Galic S, Tiganis T, Tonks NK (2004) Regulation of insulin signaling through reversible oxidation of the protein-tyrosine phosphatases TC45 and PTP1B. J Biol Chem 279(36):37716–37725

Misu H, Takamura T, Takayama H, Hayashi H, Matsuzawa-Nagata N, Kurita S, Ishikura K, Ando H, Takeshita Y, Ota T, Sakurai M, Yamashita T, Mizukoshi E, Yamashita T, Honda M, Miyamoto K, Kubota T, Kubota N, Kadowaki T, Kim HJ, Lee IK, Minokoshi Y, Saito Y, Takahashi K, Yamada Y, Takakura N, Kaneko S (2010) A liver-derived secretory protein, selenoprotein P, causes insulin resistance. Cell Metab 12(5):483–495

Mo YQ, Wan R, Chien SF, Tollerud DJ, Zhang QW (2009) Activation of endothelial cells after exposure to ambient ultrafine particles: the role of NADPH oxidase. Toxicol Appl Pharmacol 236(2):183–193

Motta MC, Divecha N, Lemieux M, Kamel C, Chen D, Gu W, Bultsma Y, McBurney M, Guarente L (2004) Mammalian SIRT1 represses forkhead transcription factors. Cell 116(4):551–563

Nakagawa K, Okuda S, Miyazawa T (1997) Dose-dependent incorporation of tea catechins, (-)-epigallocatechin-3-gallate and (-)-epigallocatechin, into human plasma. Biosci Biotechnol Biochem 61(12):1981–1985
Östman A, Hellberg C, Böhmer FD (2006) Protein-tyrosine phosphatases and cancer. Nat Rev Cancer 6(4):307–320
Ostrakhovitch EA, Lordnejad MR, Schliess F, Sies H, Klotz LO (2002) Copper ions strongly activate the phosphoinositide-3-kinase/Akt pathway independent of the generation of reactive oxygen species. Arch Biochem Biophys 397(2):232–239
Rae TD, Schmidt PJ, Pufahl RA, Culotta VC, O'Halloran TV (1999) Undetectable intracellular free copper: the requirement of a copper chaperone for superoxide dismutase. Science 284(5415):805–808
Ristow M, Zarse K, Oberbach A, Kloting N, Birringer M, Kiehntopf M, Stumvoll M, Kahn CR, Blüher M (2009) Antioxidants prevent health-promoting effects of physical exercise in humans. Proc Natl Acad Sci USA 106(21):8665–8670
Sale EM, Sale GJ (2008) Protein kinase B: signalling roles and therapeutic targeting. Cell Mol Life Sci 65(1):113–127
Salmeen A, Andersen JN, Myers MP, Meng TC, Hinks JA, Tonks NK, Barford D (2003) Redox regulation of protein tyrosine phosphatase 1B involves a sulphenyl-amide intermediate. Nature 423(6941):769–773
Shearn CT, Smathers RL, Stewart BJ, Fritz KS, Galligan JJ, Hail N, Petersen DR (2011) Phosphatase and tensin homolog deleted on chromosome 10 (PTEN) inhibition by 4-hydroxynonenal leads to increased Akt activation in hepatocytes. Mol Pharmacol 79(6):941–952
Sies H (1985) Oxidative stress: introductory remarks. In: Sies H (ed) Oxidative stress. Academic Press, London, pp 1–8
Sies H (1993) Strategies of antioxidant defense. Eur J Biochem 215(2):213–219
Sies H, Jones DP (2007) Oxidative stress. In: Fink G (ed) Encyclopedia of stress, vol 2. Elsevier, San Diego, pp 45–48
Speckmann B, Walter PL, Alili L, Reinehr R, Sies H, Klotz LO, Steinbrenner H (2008) Selenoprotein P expression is controlled through interaction of the coactivator PGC-1 alpha with FoxO1a and hepatocyte nuclear factor 4 alpha transcription factors. Hepatology 48(6):1998–2006
Steinbrenner H, Alili L, Bilgic E, Sies H, Brenneisen P (2006a) Involvement of selenoprotein P in protection of human astrocytes from oxidative damage. Free Radical Biol Med 40(9):1513–1523
Steinbrenner H, Bilgic E, Alili L, Sies H, Brenneisen P (2006b) Selenoprotein P protects endothelial cells from oxidative damage by stimulation of glutathione peroxidase expression and activity. Free Radical Res 40(9):936–943
Steinbrenner H, Speckmann B, Pinto A, Sies H (2011) High selenium intake and increased diabetes risk: experimental evidence for interplay between selenium and carbohydrate metabolism. J Clin Biochem Nutr 48(1):40–45
Sundaresan M, Yu ZX, Ferrans VJ, Irani K, Finkel T (1995) Requirement for generation of H_2O_2 for platelet-derived growth-factor signal-transduction. Science 270(5234):296–299
Sydlik U, Bierhals K, Soufi M, Abel J, Schins RPF, Unfried K (2006) Ultrafine carbon particles induce apoptosis and proliferation in rat lung epithelial cells via specific signaling pathways both using EGF-R. American J Physiol-Lung Cell Mol Physiol 291(4):L725–L733
Takakura K, Beckman JS, MacMillan-Crow LA, Crow JP (1999) Rapid and irreversible inactivation of protein tyrosine phosphatases PTP1B, CD45, and LAR by peroxynitrite. Arch Biochem Biophys 369(2):197–207
Traulsen H, Steinbrenner H, Buchczyk DP, Klotz LO, Sies H (2004) Selenoprotein P protects low-density lipoprotein against oxidation. Free Radical Res 38(2):123–128

Tyrrell RM (1996) Activation of mammalian gene expression by the UV component of sunlight—from models to reality. BioEssays 18(2):139–148
Tyrrell RM, Pidoux M (1989) Singlet oxygen involvement in the inactivation of cultured human-fibroblasts by UVA (334 nm, 365 nm) and near-visible (405 nm) radiations. Photochem Photobiol 49(4):407–412
Uchida K (2000) Role of reactive aldehyde in cardiovascular diseases. Free Radical Biol Med 28(12):1685–1696
Unfried K, Albrecht C, Klotz LO, Von Mikecz A, Grether-Beck S, Schins RPF (2007) Cellular responses to nanoparticles: target structures and mechanisms. Nanotoxicology 1(1):52–71
Unno T, Kondo K, Itakura H, Takeo T (1996) Analysis of (-)-epigallocatechin gallate in human serum obtained after ingesting green tea. Biosci Biotechnol Biochem 60(12):2066–2068
Vadas O, Burke JE, Zhang XX, Berndt A, Williams RL (2011) Structural basis for activation and inhibition of class I phosphoinositide 3-kinases. Sci Signal 4(195): re2
Valko M, Morris H, Cronin MTD (2005) Metals, toxicity and oxidative stress. Curr Med Chem 12(10):1161–1208
van der Vos KE, Coffer PJ (2008) FOXO-binding partners: it takes two to tango. Oncogene 27(16):2289–2299
van Montfort RLM, Congreve M, Tisi D, Carr R, Jhoti H (2003) Oxidation state of the active-site cysteine in protein tyrosine phosphatase 1B. Nature 423(6941):773–777
Vasquezvivar J, Augusto O (1992) Hydroxylated metabolites of the antimalarial drug primaquine. Oxidation and redox cycling. J Biol Chem 267(10):6848–6854
Vasquezvivar J, Augusto O (1994) Oxidative activity of primaquine metabolites on rat erythrocytes in vitro and in vivo. Biochem Pharmacol 47(2):309–316
von Montfort C, Fernau NS, Beier JI, Sies H, Klotz LO (2006a) Extracellular generation of hydrogen peroxide is responsible for activation of EGF receptor by ultraviolet A radiation. Free Radical Biol Med 41(9):1478–1487
von Montfort C, Sharov VS, Metzger S, Schöneich C, Sies H, Klotz LO (2006b) Singlet oxygen inactivates protein tyrosine phosphatase-1B by oxidation of the active site cysteine. Biol Chem 387(10–11):1399–1404
Walter PL, Kampkötter A, Eckers A, Barthel A, Schmoll D, Sies H, Klotz LO (2006) Modulation of FoxO signaling in human hepatoma cells by exposure to copper or zinc ions. Arch Biochem Biophys 454(2):107–113
Walter PL, Steinbrenner H, Barthel A, Klotz LO (2008) Stimulation of selenoprotein P promoter activity in hepatoma cells by FoxO1a transcription factor. Biochem Biophys Res Commun 365(2):316–321
Wang XT, McCullough KD, Franke TF, Holbrook NJ (2000) Epidermal growth factor receptor-dependent Akt activation by oxidative stress enhances cell survival. J Biol Chem 275(19):14624–14631
Weissenberg A, Sydlik U, Peuschel H, Schroeder P, Schneider M, Schins RPF, Abel J, Unfried K (2010) Reactive oxygen species as mediators of membrane-dependent signaling induced by ultrafine particles. Free Radical Biol Med 49(4):597–605
Woods YL, Rena G, Morrice N, Barthel A, Becker W, Guo SD, Unterman TG, Cohen P (2001) The kinase DYRK1A phosphorylates the transcription factor FKHR at Ser(329) in vitro, a novel in vivo phosphorylation site. Biochem J 355:597–607
Xia T, Kovochich M, Brant J, Hotze M, Sempf J, Oberley T, Sioutas C, Yeh JI, Wiesner MR, Nel AE (2006) Comparison of the abilities of ambient and manufactured nanoparticles to induce cellular toxicity according to an oxidative stress paradigm. Nano Lett 6(8):1794–1807
Yu JH, Zhang YT, McIlroy J, Rordorf-Nikolic T, Orr GA, Backer JM (1998) Regulation of the p85/p110 phosphatidylinositol 3'-kinase: Stabilization and inhibition of the p110 alpha catalytic subunit by the p85 regulatory subunit. Mol Cell Biol 18(3):1379–1387

Author Biography

Lars-Oliver Klotz received his Diploma (MSc equivalent) in biochemistry from the University of Tuebingen, Germany, in 1995, followed by a PhD in Biochemistry from the University of Duesseldorf, Germany, in 1998 (supervisor: Prof. Helmut Sies). He then worked as a postdoctoral fellow at the National Institute on Aging in Baltimore, MD, USA, and returned to the University of Duesseldorf in 2000 to establish his own laboratory at the Department of Biochemistry and Molecular Biology and to obtain his lecturer's qualification ("Habilitation") in 2001. From 2007 to 2010, Lars-Oliver was Head of the "Environment and Stress Signaling" section at the Leibniz Research Institute for Environmental Medicine (IUF) in Duesseldorf. In 2010, he moved to the University of Alberta, Edmonton, AB, Canada, where he was an Associate Professor at the Faculty of Pharmacy and Pharmaceutical Sciences and held the Canada Research Chair in Pharmaceutical Sciences. In September 2013, he assumed his current position as Chair of the Department of Nutrigenomics at the University of Jena, Germany. In 2006, Lars-Oliver was recipient of the Catherine Pasquier Award of the European Society of Free Radical Research (SFRR-Europe). His research interests include the biochemistry of oxidative stress and stress-induced signal transduction.

Part V
Connecting Section Between Chapters 5 and 6

Redox active secondary metabolites have long been considered as extrinsic redox modulators that may exert certain beneficial effects on the (human) body. Indeed, we are almost daily confronted with the simplistic idea of 'good antioxidants' fighting 'bad free radicals'. Not surprisingly, many natural materials rich in 'antioxidants' have been turned into commercial products. Redox active plant products are all around us, starting from antioxidants in fruit juices and particular 'health foods' specially designed for eternal life to anti-aging, anti-wrinkle, rejuvenating crèmes used in cosmetics to mummify the generation 60+. Many of these antioxidant effects are minor, belittled by the body's own antioxidant defenses, and hence more of a sales pitch than anything else (see also Chap. 2). Nonetheless, recent nutritional and biochemical research has provided mounting evidence that some redox active plant secondary metabolites may indeed impact on the body at the level of cellular redox processes. The resulting health benefits are often diverse and sometimes even rather unexpected. They include, for instance, a positive impact on the cardiovasculature, possibly cancer preventive actions, a strengthening of antimicrobial (e.g., antibacterial) defense, a reduction in nutritional uptake, an improved lipid metabolism, and even some distinct epigenetic changes.

As part of the following chapter, we will turn our attention to the vast and chemically diverse families of flavonoids and stilbenes. These families include some of the better known antioxidants, such as the epicatechins, taxifolin, quercetin, anthocyanidins, and resveratrol. Some of these substances are nowadays on the market as 'miracle cures', yet the scientific basis for such claims is often rather shaky. Here, one frequently cited phenomenon in support of the antioxidant activity of such compounds is the "French Paradox", which will be discussed, together with the various compounds implicated in this phenomenon, their chemistry, reactivity, and emerging biochemical mode(s) of action. As many of these compounds are well known to chemistry and accessible by chemical synthesis, this particular chapter will also consider how—partial or total—chemical synthesis can unlock the full potential of such natural products and specially tailored derivatives.

Chapter 6
The French Paradox at Tea Time: From Antioxidant Flavonoids and Stilbenes Toward Bio-inspired Synthetic Derivatives

Oualid Talhi, Diana C. G. A. Pinto and Artur M. S. Silva

Keywords Epicatechin • Flavonoids • Frenchparadox • Resveratrol • Stilbenes

1 Introduction

Nowadays, science has become more immerged into daily life. Who has not heard or at least observed the word 'antioxidant', which is frequently advertised on many beverages and dairy products of the supermarket? Although most people ignore the real significance of an antioxidant, many of them do use the term 'antioxidant' while discussing nutrition, food, and health; hence, the question must be posed "What is an antioxidant?"

In a general aspect, an antioxidant is a substance able to prevent the oxidation reaction mainly caused by oxygen. A concrete example is commonly observed in the well-known phenomena of metal corrosion. Iron is oxidized by oxygen from the air. Anticorrosive agents, which in fact are antioxidants, are used to protect metallic surfaces from corrosive damage. From a chemical point of view, the oxidation reaction is a redox reaction involving electron transfer from a substance toward the oxidant agent. This reaction can produce free radicals that are highly reactive species that attack molecules by capturing electrons, thus modifying their molecular structures. In other words, any substance that reduces the oxidative damage due to oxygen (and/or free radicals) is called an 'antioxidant'. It is therefore capable to stop the destructive oxidation process by reacting with free radicals and stopping their activity (Valko et al. 2007; Hermes-Lima 2005). These antioxidant properties are encountered in specific families of natural and synthetic organic compounds; the most common are phenolic compounds.

O. Talhi · D. C. G. A. Pinto · A. M. S. Silva (✉)
Department of Chemistry and QOPNA, University of Aveiro, 3810-193 Aveiro, Portugal
e-mail: artur.silva@ua.pt

C. Jacob et al. (eds.), *Recent Advances in Redox Active Plant and Microbial Products*,
DOI: 10.1007/978-94-017-8953-0_6,

In biology, despite the crucial role of oxidation reactions for metabolism and organism functioning, they can also be highly destructive. A biological paradox is eventually noticed for most beings (animals and plants) requiring oxygen to ensure life, while this molecule is extremely reactive and able to induce degradation for many organisms. Under such circumstances, antioxidation systems are set up in the form of antioxidant agents acting together with enzymes to prevent the formation of highly reactive species or even to eliminate them just before they damage cell-components, like DNA, lipids, and proteins. Plants and animals utilize and produce various antioxidants to protect themselves, such as glutathione, vitamin C, and vitamin E, or enzymes like catalase, peroxidase, and superoxide dismutase. Similarly, our body itself produces, involving the amino acid cysteine, a powerful antioxidant, α-lipoic acid. A deficiency or total absence of antioxidants causes oxidative stress that can damage or destroy cells. Oxidative stress has been implicated in the pathogenesis of many human diseases, mainly cancer. So far, the application of antioxidants in pharmacology has been studied in order to treat several pathologies namely cardio, cerebral, vascular, atherosclerosis, neoplasia, and neurodegenerative diseases. However, it still remains unclear whether oxidative stress is the cause or the consequence of these diseases (Valko et al. 2007).

Great attention was focused on the naturally occurring antioxidant phytochemicals as possible therapy for cardiovascular diseases. These natural occurring components are considered as important nutritional ingredients acting as protectors for the health maintenance and preventing certain diseases like cancer or heart failure. Although studies suggest that nutritional antioxidants are beneficial to health, extensive clinical trials did not reveal a very clear *in vivo* biological action on humans and have even suggested that excessive intake of these substances could sometimes have negative effects (Barbosa 2007; Galati and O'Brien 2004; Gerhaeuser 2001).

A huge amount of organic compounds from both natural and synthetic sources are recognized as potent antioxidant agents. The flavonoid family, widespread in the plant kingdom is associated with antioxidant capacities, especially flavones. Great attention is due to the stilbene family in which resveratrol, the main grape skin and red wine active component, is the chief leader of this family, displaying a broad spectrum of biological effects, also followed by a queue of numerous biological active synthetic analogs.

This chapter deals with a brief structural introduction to the flavonoid and the stilbene compounds. A word is given on the chemistry behind the drive to synthesize more effective derivatives of the previously mentioned natural products, including flavones, 2-styrylchromones, and stilbenes. A discussion on nutritional polyphenols is presented emphasizing some biologically impactful molecules mainly from the flavonoid and stilbene families. Flavonoids such as quercetin, anthocyanins, catechins, and resveratrol, along with other small phenolic molecules constitute the subject of this chapter. These compounds are frequently encountered in vegetables, fruits, and beverages. Red wine, tea, especially green, and chocolate are, without any doubt, the most fascinating examples considered in people's eating habits and most important sources of the above highlighted

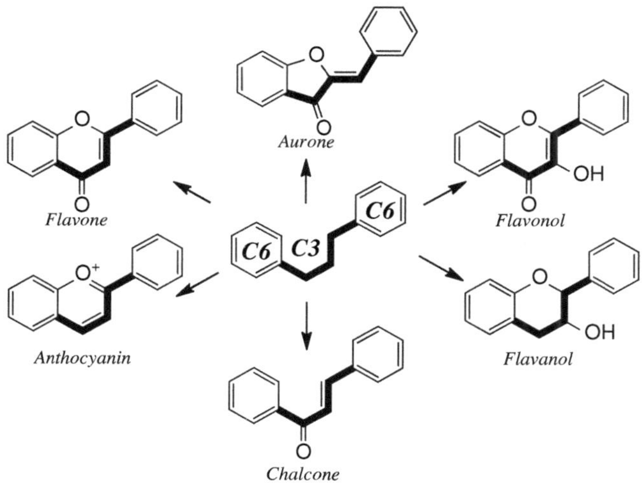

Fig. 1 Flavonoid main basic structures

nutritional polyphenols. We will revisit the French Paradox which relates the observation that mortality rates due to coronary heart disease are relatively low in some regions of France despite a diet rich in saturated fats. A similar concept will be treated regarding tea drinks and their potential cancer prevention. Chocolate is a tasty source of polyphenols which is taking part in our survey on food-nutrients and health effects. Aspects of biology, recent evidences on beneficial effects, *in vitro*, *in vivo* and clinical trials, for antioxidant assays of these polyphenol-rich foods are outlined. Some social circumstances end up with different opinions on whether these foods are with or without risk for human health.

2 Flavonoids and Related Compounds

Flavonoids form a family of phenolic secondary metabolites of plants. All the classes of this family of compounds share the same basic structure formed by two aromatic rings attached together via a three carbon chain giving rise to a C_6–C_3–C_6 system. Usually their structure contains a pyrano- or furano-heterocycle with a ketone function. They constitute one of the most numerous and widespread families of natural plant phytoconstituents with more than 4,000 structures identified, and categorized into several classes, namely the six member containing form (pyrano-) including flavones, flavanols (catechins), flavanones, flavonols, isoflavones, anthocyanins, and procyanidins; the five member containing form (furano-) is represented by aurones. In addition to the open forms like chalcones and dihydrochalcones exist (Fig. 1) (Harborne and Grayer 1993; Andersen and Markham 2006).

Flavonoids are low-molecular-weight substances extracted from plants by various methods (Harborne and Grayer 1993; Andersen and Markham 2006). They have primarily been identified as pigments responsible for the autumnal shades (yellow, orange, and red) of many kinds of flowers and plants. Flavonols (quercetin, myricetin, and kaempferol) and flavones (apigenin and luteolin) are the most common phenolics in plant-based foods. Quercetin is a predominant component in onions, apples, and berries (Tereschuk et al. 2004). Flavonoids like the colorful anthocyanins are found in vegetables and fruits, such as red cabbage (McDougall et al. 2007) and red grapes (Orak 2007). They are also found in red wine phytochemicals (Perez-Magarino and Gonzalez-San Jose 2006; Cliff et al. 2007; Crozier et al. 2010). More colorful flavonoids are prominent components of citrus fruits and other food sources. Flavanones, like narrigin are typically present in citrus fruits, and flavanols, particularly catechin, are present as catechin gallate in beverages such as green or black tea (Amarowicz and Shahidi 1996; Armoskaite et al. 2011; Atoui et al. 2005; Gao et al. 2008) and red wine (Rosenkranz et al. 2002).

Flavonoids not only give to food its colors but play a crucial protective role in human health. Consequently, many structures are established as potential biologically active nutrients (Harborne and Grayer 1993; Andersen and Markham 2006; Martino 2000). They have also been credited with many diverse key functions in plant growth and development, including stress protection, reproduction, signaling, and protection from insect and mammalian consumption (Harborne and Grayer 1993; Andersen and Markham 2006). The daily intake of flavonoids in humans can reach an approximate value of 25 mg/day, an average amount which qualifies a pharmacological level to human body fluids and tissues, guaranteeing a good absorption from the gastrointestinal tract. In 1938, Szent-Gÿorgyi has first initiated the biological activity of flavonoids, in his study on citrus peel flavonoids which provide an efficient activity in preventing the capillary bleeding and fragility associated with scurvy (Tereschuk et al. 2004). Certain individual members of the flavonoid family displayed a multiplicity of biological activities, and therefore this most promising family of biologically active compounds becomes the key title of several recent research works. Among the authors, Morton et al. (2000) has published a review on distribution, bioavailability, and biological activity of the flavonoid compounds suggesting that they may have a physiological role as antioxidants.

Actually, it is accepted that natural flavonoids present in fruits and plant-derived-foods are relevant, not only for technological reasons and organoleptic properties, but also because of their potential health-promoting effects, as suggested by the available experimental and epidemiological data. Human trials on the antioxidant effects of beverages rich in such polyphenolic compounds, like red wine, fruit juice, and tea, have been limited and results are, at present, inconclusive. This fact is particularly due to poor inconvincible methodologies available to measure oxidative damage *in vivo*, and that is why still further research efforts are being required. The use of appropriate biomarkers of oxidant damage *in vivo* measure is primordial in order to prove that these compounds can be conclusively considered as dietary antioxidants with nutritional benefit. In contrast, the beneficial biological effects of these food components may be depicted by two of their characteristic properties, their

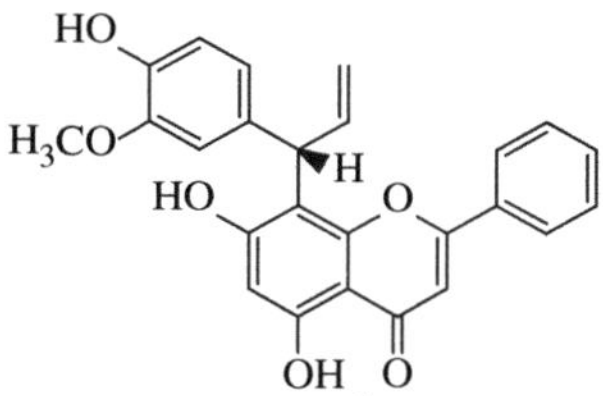

1

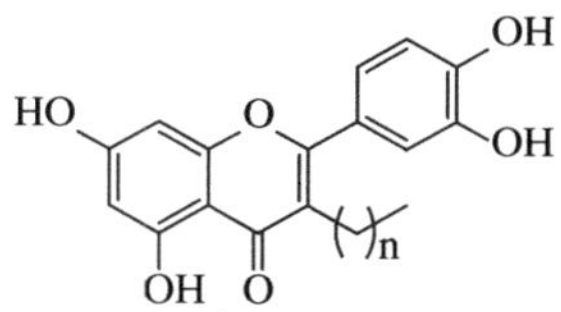

2

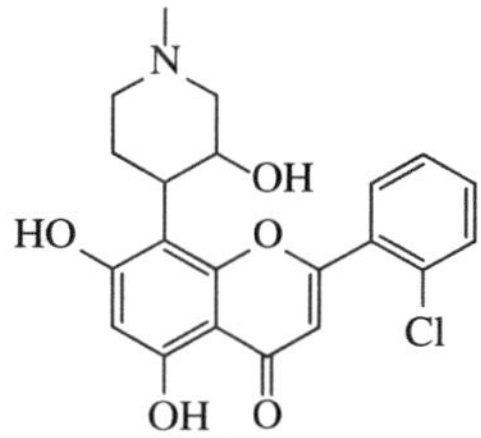

3

Fig. 2 Molecular structures of (7″*R*)-8-[1-(4-hydroxy-3-methoxyphenyl)prop-2-en-1-yl]galangin **1**, 3-alkyl-3′,4′,5,7-tetrahydroxyflavones **2**, flavopiridol [2-(2-chlorophenyl)-5,7-dihydroxy-8-(3-hydroxy-1-methylpiperidin-4-yl)-4*H*-chromen-4-one] **3**

affinity for proteins and their antioxidant activity. Over the last 15 years, numerous publications have demonstrated that besides their *in vitro* antioxidant capacity (measured by DPPH, ORAC, and other techniques) (Salucci et al. 2002) and *in vivo* evaluation (Dai et al. 2004), certain flavonoids, such as anthocyanins, catechins, proanthocyanidins encountered in our daily food, may regulate different signaling pathways involved in cell survival, growth, and differentiation. These compounds act differently and selectively in various models as far as their antioxidant capacity is concerned, suggesting that multi-models should be utilized in order to evaluate an antioxidant from natural sources (de Pascual-Teresa et al. 2010).

Flavones (2-arylchromones) are a group of flavonoids which gained great attention over the last decade due to their potential biological and medicinal utilities. These compounds can be characterized as ‘privileged structures’ for their ability to interact with a number of different receptors in the body, thereby precipitating a wide range of biological responses (Verma and Pratap 2010). Among the naturally occurring flavones and their synthetic analogs, several derivatives displayed important biological properties, such as anticancer (Cummings et al. 1989; Cardenas et al. 2006; Lin et al. 2007; Zhu et al. 2007), anti-inflammatory (Park et al. 1999; Nagaoka et al. 2003; Moscatelli et al. 2006) and antioxidant (Beyer and Melzig 2003; Vaya et al. 2003) activities. These first promising results explain the increasing interest in this class of the most abundant naturally occurring compounds and the continuous isolation of new biologically active derivatives, such as (7″*R*)-8-[1-(4-hydroxy-3-methoxyphenyl)prop-2-en-1-yl]galangin **1** having a cytotoxic activity against PANC-1 human pancreatic cancer cells (Li et al. 2010). Considerable attention was paid to the synthesis of flavone derivatives especially those with biological activity predictions, for instance, we underline some molecules like 3-alkyl-3′,4′,5,7-tetrahydroxyflavones **2** proved as potent active antioxidant evaluated in various biological systems, including *in vitro* assays (Filipe et al. 2009; Gomes et al. 2009b). Flavopiridol **3** shows a cyclin-dependent kinase inhibitory effect which is actually under phase II clinical trials for a number of different malignancies (Fig. 2) (Murthi et al. 2000; Kosmider and Osiecka 2004; Teillet et al. 2008; Ahmadi et al. 2009; Diaz-Padilla et al. 2009; Hallek and Pflug 2011).

2-Styrylchromones constitute a further important biological active chromone-based structure, but very scarce as naturally occurring compounds (Gomes et al. 2010).

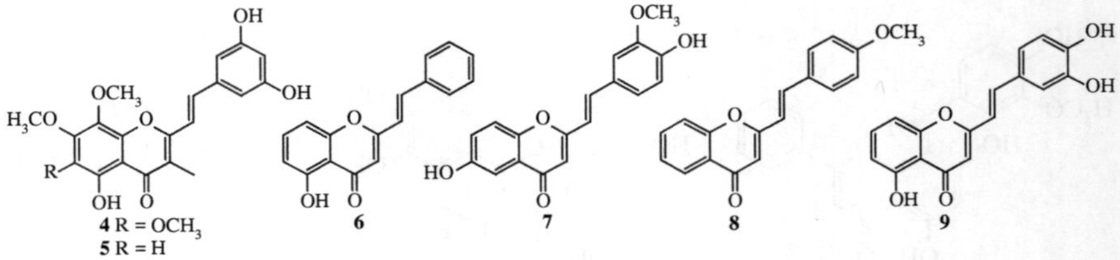

Fig. 3 Structures of hormothamnione **4**, 6-demethoxyhormothamnione **5**, (*E*)-5-hydroxy-2-styrylchromone **6**, (*E*)-6,4′-dihydroxy-3′-methoxy-2-styrylchromone **7**, (*E*)-3′methoxy-2-styrylchromone **8** and (*E*)-5,3′,4′-trihydroxy-2-styrylchromone **9**

Only four derivatives have been isolated from nature, namely hormothamnione **4** and 6-demethoxyhormothamnione **5** from the marine blue-green algae *Chrysophaem taylori* (Gerwick et al. 1986; Gerwick 1989), (*E*)-5-hydroxy-2-styrylchromone **6** from the rhizomes of *Imperata cylindrical* (Yoon et al. 2006) and, more recently, (*E*)-6,4′-dihydroxy-3-methoxy-2-styrylchromone **7** isolated from the tree *Aquilaria sinensis* (Yang et al. 2012) (Fig. 3). Natural derivatives have only demonstrated cytotoxic activity against leukemia cells (Gerwick et al. 1986; Gerwick 1989) while a range of biological effects has been evidenced in synthetic derivatives, such as antiviral (Desideri et al. 2000), antitumor (Brion et al. 1991), antimitotic (Marinho et al. 2008), anti-inflammatory (Gomes et al. 2009a) and antioxidant (Gomes et al. 2007, 2008, 2010) activities. Some of the biologically active synthetic derivatives present simple structures; for example, the antimitotic (*E*)-4′-methoxy-2-styrylchromone **8** also considered as a potent anti-norovirus agent (Marinho et al. 2008), along with (*E*)-5-hydroxy-2-styrylchromone **6** (Rocha-Pereira et al. 2010). Moreover, 2-styrylchromones with a catechol pattern such as derivative **9**, present considerable anti-inflammatory and antioxidant activity (Fig. 3) (Gomes et al. 2007, 2008).

In light of the biological significance of the mentioned chromone-based compounds, many researchers dedicate their work to develop efficient synthetic methodologies for such compounds. The most used synthetic routes include the Baker–Venkataraman method and the cyclodehydrogenation of 2′-hydroxychalcones and 2′-hydroxycinnamylideneacetophenones.

The Baker–Venkataraman route is one of the oldest approaches drawn to the synthesis of flavone derivatives and still being one of the most used efficient routes for 2-styrylchromones production (Baker 1933; Mahal and Venkataraman 1934; Price et al. 1993; Silva et al. 2004; Pinto et al. 1998, 1999). It involves a three-step sequence where the final step consists in the cyclization of β-diketones **11** (obtained from ester **10** via Baker–Venkataraman rearrangement), which exists in equilibrium with its enolic form, into chromone derivatives **12**. Several conditions can be employed to perform this cyclization, mostly under acidic conditions. Extensive studies performed by Silva et al. indicate that molecular iodine leads to a successful manner of flavones and (*E*)-2-styrylchromones synthesis (Scheme 1) (Pinto et al. 2000a, b; Barros and Silva 2006, 2009).

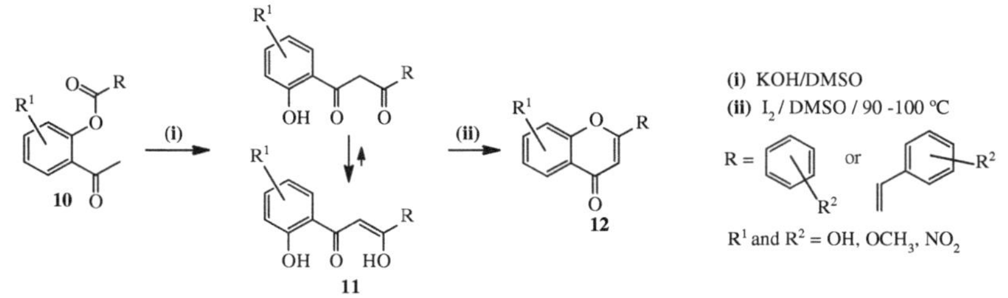

Scheme 1 Synthesis of flavones (R = aryl) and (*E*)-2-styrylchromones (R = styryl) by the Baker–Venkataraman method

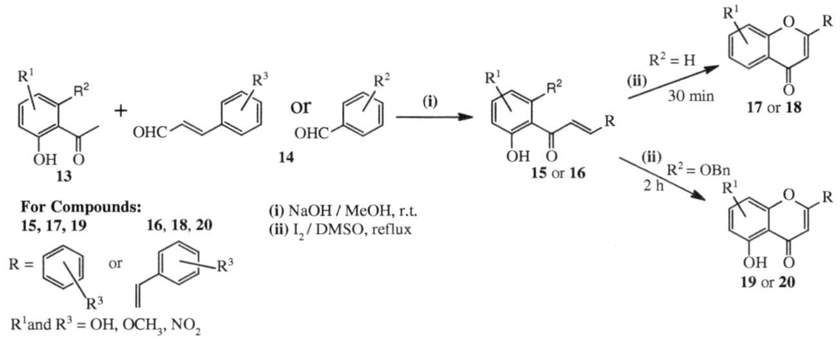

Scheme 2 Synthesis of flavones **17**, **19** and (*E*)-2-styrylchromones **18**, **20** by cyclodehydrogenation of the corresponding 2′-hydroxychalcones **15** and 2′-hydroxy-2-cinnamylideneacetophenones **16**

As mentioned previously, the oxidative ring closure of the appropriate 2′-hydroxychalcones and 2′-hydroxy-2-cinnamylideneacetophenones is another important well-documented approach toward flavones and (*E*)-2-styrylchromones synthesis. Various reagent systems are known for the oxidative cyclization of the 2′-hydroxychalcones **15** to the corresponding flavones **17**, namely disulfides (Hoshino et al. 1986), sodium periodate (Hans and Grover 1993), hypervalent iodine reagents (Gulacsi et al. 1998), DDQ (Chan et al. 2006), oxalic acid (Zambare et al. 2009), Wacker–Cook-related oxidation (Lorenz et al. 2010), and finally selenium dioxide (Gupta et al. 2000). Nevertheless, the use of molecular iodine/DMSO system seems to be more flexible for its lower toxicity and cost, leading to better yields and shorter reaction times. Further systematic studies have been conducted by Silva et al. disclosing scopes and limits of this method (Silva et al. 1997; Patonay et al. 1997). Yet, the most important aspect of their work was the successful application of this methodology to the synthesis of flavones **17** and (*E*)-2-styrylchromones **18**, via the oxidative ring closure of the 2′-hydroxychalcones **15** and 2′-hydroxy-2-cinnamylideneacetophenones **16** in 30 min. Also 5-hydroxyflavones **19** and (*E*)-5-hydroxy-2-styrylchromones **20** have been similarly produced but lasting a longer reaction time (Scheme 2) (Silva et al. 1994, 1998).

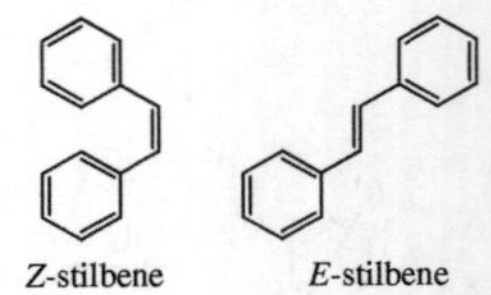

Fig. 4 Structures of stilbene diastereomers

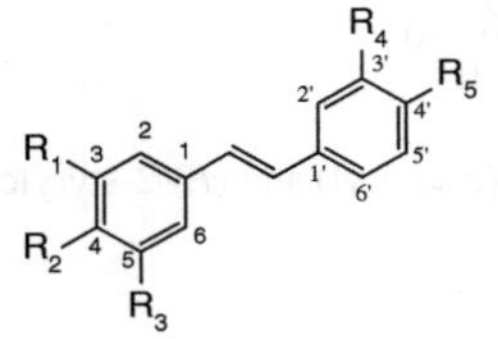

21. $R^1 = R^3 = R^5 = OH, R^2 = R^4 = H$
22. $R^1 = R^3 = R^4 = R^5 = OH, R^2 = H$
23. $R^1 = R^4 = OH, R^2 = R^3 = R^5 = H$
24. $R^1 = R^2 = OH, R^3 = R^4 = R^5 = H$
25. $R^2 = R^5 = OH, R^1 = R^3 = R^4 = H$

Fig. 5 Resveratrol **21** and other biologically active synthetic stilbenes **22**–**25**

3 Stilbenes, Resveratrol, and More Active Synthetic Derivatives

The stilbene skeleton consists in a C=C double bond linking two aromatic rings and presents two diastereomeric forms, *E*-1,2-diphenylethylene (*E*-stilbene) and *Z*-1,2-diphenylethylene (*Z*-stilbene) (Fig. 4). The name '*stilbene*' is also used to design the class of its poly-substituted hydroxy- and alkoxy-derivatives which are natural occurring phenolic compounds present in many families of plants. The most fascinating example is resveratrol **21** (Fig. 5), isolated from red fruits and different plants. Resveratrol (*Z*-3, 5, 4′-trihydroxystilbene) **21** is an abundant phytochemical of grape skins and therefore present in red wine (Likhtenshtein 2010). In the mid-1990s, resveratrol **21** was identified as one of the possible factors responsible for the "French Paradox". Numerous biomedical studies have been carried out on the pharmacological properties of **21** and its health benefits on humans. The discovery of David Sinclair's research group (Howitz et al. 2003; Wood et al. 2004) at Harvard University in 2003, that resveratrol **21** extends lifespan in yeast, was a step forward making resveratrol a sole biological active agent and chief leader of polyphenolic compounds. An overview of resveratrol **21** and its biological properties is presented in Sect. 4.1.4.

On the other side, the great attention devoted to resveratrol **21** due to its unique biological properties, does not really make it so exceptional because several of its synthetic analogs demonstrate better biological applications. Various operative challenges are faced in achieving the isolation of **21** in sufficient quantity from natural sources. Furthermore, its limited bioavailability in the blood circulation is seen as the main obstacle for therapeutic use. For such reasons, many researchers directed their attention toward the synthetic derivatives aiming the production of biologically more active agents. The elaboration of new stilbene derivatives was mainly based on modifying some structural features, for example, the number and position of the hydroxyl and other functional groups on the stilbene skeleton. Thakkar et al. was the first to describe the synthesis of piceatannol **22** and other polyhydroxylated analogs, which presented antiproliferative action on leukemic cells. Other synthetic substituted (hydroxy, methoxy, fluoro) stilbenes have been evaluated on tumor cell lines [HCT-56 (colon), MDA-MB-468 (breast)], revealing better antiproliferative effect of the 3,3′-dihydroxystilbene **23** ($GI_{50} = 2.7\ \mu M$) compared to resveratrol **21** ($GI_{50} = 41\ \mu M$) (Fig. 5) (Thakkar et al. 1993). An established structure–antioxidant activity relationship of polyhydroxystilbenes showed great dependency on the number and relative position of their hydroxyl groups (Lion et al. 2005; Zhou and Liu 2005). Thus, 3,4-dihydroxystilbene **24** proves to have better antioxidant properties than *trans*-resveratrol **21**, however, in the stilbenes' case, the presence of a catechol moiety does not justify properly the antioxidant capacity of **24** since 4,4′-dihydroxystilbene **25** is even more powerful (Fig. 5). Besides, a clear observation was made on the presence of the *para*-hydroxyl function which greatly influences the antioxidant activity of stilbenes as confirmed by comparative results evaluated on the non-*para*-substituted derivatives (Fan et al. 2009; Petralia et al. 2004). Methoxy- and ethoxy-stilbenes have been tested on both NF-κB and TPA-induced activation of AP-1 pathways, revealing higher inhibitions than resveratrol **21**. In terms of structure–biological activity relationship, the absence of the *para*-hydroxyl function, in this case, does not explain the pro-apoptotic activity and its association with an inexistent antioxidant potential (Deck et al. 2008).

The construction of organic hybrids between resveratrol and other scaffolds consist of a new strategy toward new biological target design. As resveratrol **21** bioavailability is low in the organism and promptly metabolized, Hauss et al. (2007) focused on the increasing its presence in the blood circulation. Hence, the combination of the neuroregenerative properties of fatty alcohols with the neuroprotective properties of resveratrol **21** resulted in effective series of compounds with a dual biological activity and high metabolic resistance. The antiproliferative effects of phosphoric amino acids coupled to stilbenes were found 15 times more potent than **21**, as a result of CNE-1 and CNE-2 cell lines evaluation (Liu et al. 2008a). **21** was also functionalized by triphenylphosphonium iodide in order to confer a better solubility to the whole molecule in the mitochondrial medium and therefore a better inhibition of several tumor cell lines was observed (Biasutto et al. 2008). Resveratrol oligomers have been largely ignored despite their high biological activity. A recent report on a programmable, controlled, and potentially scalable synthesis of the resveratrol family via a three-stage design has been published (Snyder et al. 2011).

The C=C double bond of the stilbene skeleton is stabilized by conjugation electronic effects, making it more resistant to hydrolysis or other addition-type reaction and oxidative cleavages. Several strategies have been reported to build this relevant C=C double bond. The Julia olefination, for instance, is one of the C–C simple-bond formation reaction implicating the reaction of phenyl sulfones **26** with aldehydes (or ketones) **27** (Julia and Paris 1973). Under basic conditions, the obtained sulfinate **28** yield the alkene structure after a Smiles rearrangement and spontaneous reductive elimination of the phenolate **29** accompanied by sulfur

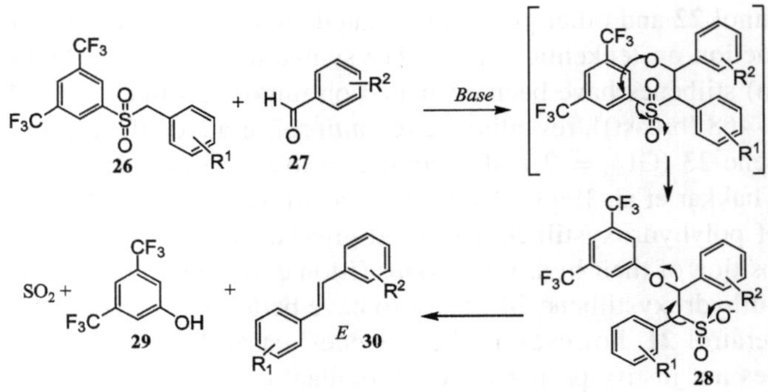

Scheme 3 Synthesis of stilbenes by the Julia olefination

dioxide release (Scheme 3). This transformation highly favors the formation of the *E*-alkene **30**. Thus different stilbenes have been diastereoselectively obtained by this method (Alonso et al. 2004).

The Wittig reaction is largely known to produce a C=C double bond by coupling an aldehyde (or ketone) with a phosphorous ylide giving rise to an alkene and triphenylphosphine oxide. In this regard, stilbenes are usually synthesized via Wittig coupling reaction of aldehydes with substituted benzylic ylides, which tends to afford the thermodynamically stable *E*-products (Vedejs et al. 1993). The Wittig reaction is usually carried out in basic media, therefore, the presence of hydroxyl groups on both of the starting materials (aldehyde **33** and phosphine ylides **34**) can induce some problems due to the labile protons. In case of resveratrol **21** or similar bioactive polyhydroxylated derivatives (e.g., piceatannol **22**), the synthetic procedure start by an initial hydroxyl-protection step of the Wittig reaction partners **31**, **32** using appropriate protecting groups, such as methyl, trimethylsilyl, acetyl, etc. The required cleavage of these protecting groups at the end of the synthesis can cause some further problems. Pettit et al. have used *t*-butyldimethysilyl (TBDMS) as a protecting group (air-stable and easily cleaved with Bu_4NF) to achieve the production of a series of stilbenes **35** (mixture of separable *E* and *Z* isomers) (Scheme 4). These compounds have been assessed for their therapeutic effects in some cell line targets and allowed to establish a structure-activity relationship (Pettit et al. 1995, 2002). Interestingly, a set of antimicrobial resveratrol derivatives have also been elaborated using the Wittig olefination (Albert et al. 2011).

A multi-component strategy was developed by Hilt et al. involving a cobalt(I)-catalyzed Diels–Alder reaction of propargylic phosphonium salts **36** (or higher homologs) with 1,3-dienes **37** leading to dihydroaromatic phosphonium salt intermediates which underwent *in situ* a one-pot Wittig-type olefination reaction with aldehydes **38**. Subsequent oxidation led to stilbene-type products **39**. The semi-stabilized dihydroaromatic phosphonium ylides intermediates are predominantly leading to the *E*-configured products (Scheme 5) (Hilt and Hengst 2007).

The Horner–Wadsworth–Emmons reaction is an alternative of the Wittig coupling reaction using stabilized phosphonate carbanions derived from **40** brought

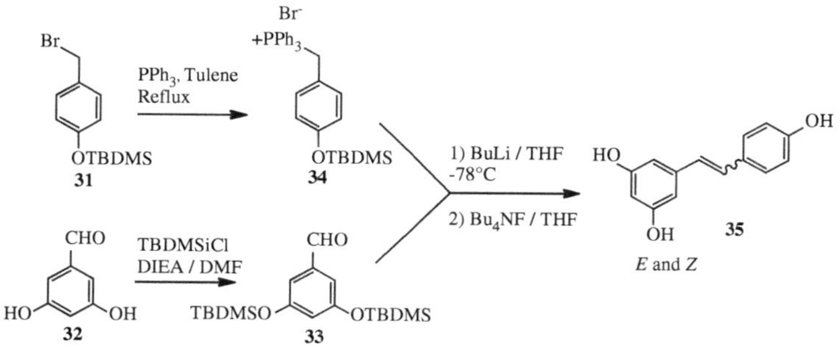

Scheme 4 Synthesis of stilbenes through the Wittig reaction

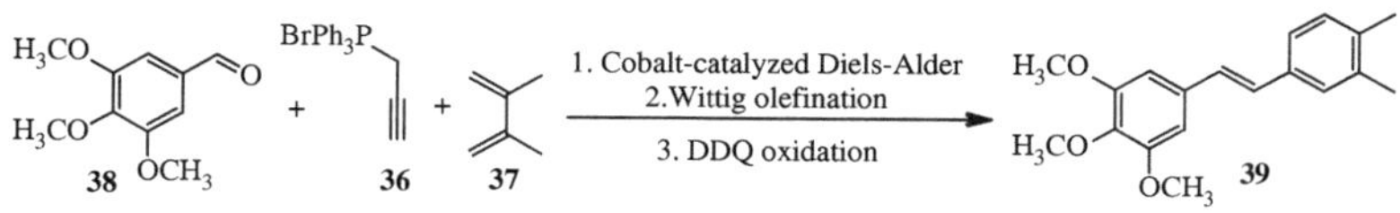

Scheme 5 A concise synthesis of substituted stilbenes by a cobalt(I)-catalyzed Diels-Alder/Wittig olefination reaction sequence starting from propargylic phosphonium salts

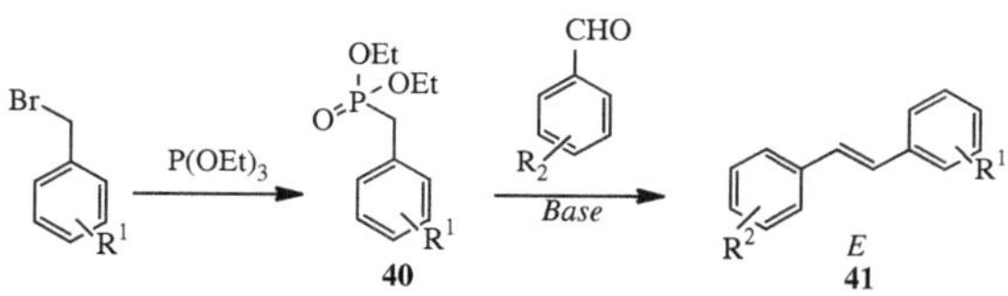

Scheme 6 Synthesis of stilbenes through the Horner–Wadsworth–Emmons reaction

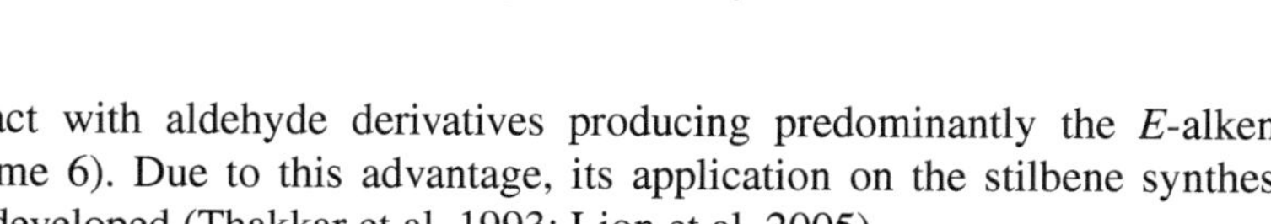

Scheme 7 Synthesis of stilbenes through the McMurry reaction

to react with aldehyde derivatives producing predominantly the *E*-alkenes **41** (Scheme 6). Due to this advantage, its application on the stilbene synthesis has been developed (Thakkar et al. 1993; Lion et al. 2005).

More C=C bond formation procedures have been used in the synthesis of stilbenes. In the McMurry reaction, two aromatic aldehydes or ketones are coupled to give the corresponding alkene using titanium(III) chloride and a reducing agent. The *Z*-stilbene skeleton **42** is predominantly obtained when coupling benzaldehyde derivatives using McMurry's conditions (Scheme 7) (McMurry and Fleming 1974; Rele et al. 2008).

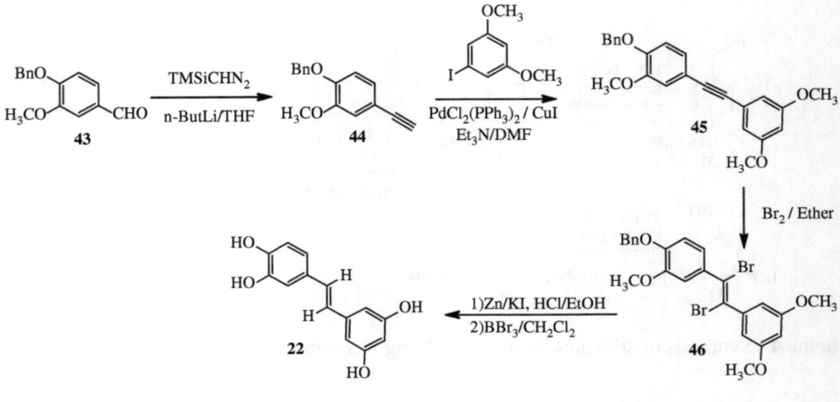

Scheme 8 Synthesis of stilbenes involving a Sonogashira cross-coupling reaction

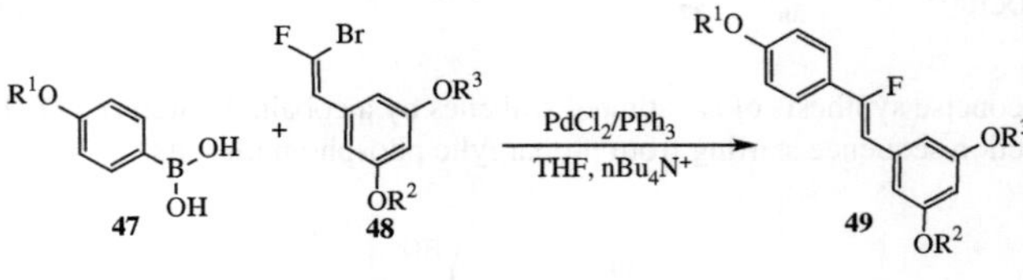

Scheme 9 Synthesis of stilbenes involving a Suzuki cross-coupling reaction

The Sonogashira cross-coupling reaction aims the formation of carbon–carbon bonds by the reaction of a terminal alkyne and an aryl or vinyl halide using a palladium catalyst. The resulting substituted alkyne is potential precursors for the synthesis of stilbenes. An example for the *E*-piceatannol **22** synthesis was described starting from vanillin. A one-step transformation of the *para*-protected-vanillin **43** to the corresponding terminal-alkyne **44**, was followed by the Sonogashira cross-coupling reaction with the appropriate 3,5-dimethoxyiodobenzene to afford the desired alkyne **45**. This latter compound was subjected to bromination to give selectively the *E*-dibromostilbene **46** and finally transformed to *E*-piceatannol **22** by a reductive debromination and deprotection of the phenolic functions (Scheme 8) (Han et al. 2008).

The Suzuki cross-coupling reaction is another palladium catalyzed transformation of aryl boronic acids **47** (or vinylboronic acids) into alkenes by reacting with vinyl halides **48** (or aryl halides). It is widely used to synthesize poly-olefins like styrenes **49** and substituted biphenyls, therefore, its application on the stilbene synthesis can be of a great interest. Some synthetic derivatives of resveratrol have been elaborated by this method (Scheme 9) (Eddarir et al. 2001).

The Heck reaction is another important synthetic access to substituted alkenes. It consists in the coupling of an aryl halide (or triflate) **50** with an alkene **51** catalyzed by palladium reagents under basic conditions. Comparing to the other

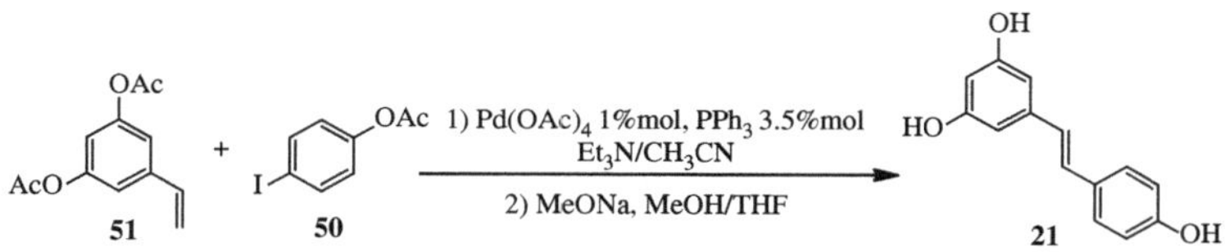

Scheme 10 Synthesis of stilbenes involving a Heck reaction

palladium-catalyzed cross-coupling reactions (Sonogashira and Suzuki), this reaction allows a concrete diastereoselective elaboration of resveratrol **21** (Scheme 10) (Guiso et al. 2002).

4 Nutritional Antioxidant Phenolic Compounds in Daily Life

Exceptive epidemiological studies agree that human diet rich in fruit, vegetables, and plant-derived foods decrease diseases incidence, like cardiovascular diseases (CVD) (Valko et al. 2007; de Pascual-Teresa et al. 2010; Loke et al. 2010; Berthelot-Garcias et al. 2009) diabetes (Ferruzzi 2010), cancer (Serafini 2004), and stroke (Bejot et al. 2009), which are the most common and the main worldwide death causes recorded in the past few years. According to some statistics of 2003, 44 % of worldwide deaths were related to acute myocardial infarction, 33 % to stroke, and 23 % to arterial hypertension or to other CVDs, such as pulmonary embolism and causes of heart failure. Age, male sex, smoking, increase in LDL-cholesterol, and type 2 diabetes, together with familial antecedents, lack of physical training, android obesity, and menopause are the main risk factors predisposed to CVD (Berthelot-Garcias et al. 2009).

Many reports associated the benefit of fruits, vegetables, and plants with their antioxidant polyphenolic content (Gerhaeuser 2001; Loke et al. 2010; Zloch 1996; Visioli et al. 2000). Polyphenolic compounds are ubiquitous in all plant organs and constitute, therefore, an integral part of the human diet. Despite their lifestyle and food fashion differences, most people have the tendency toward processed drinks and dairy products rather than taking raw fruits and vegetables. Many examples of our actual eating habits support this social observation, like the open day coffee or tea. Usually, lunch or dinner time is accompanied by wine, spirits, and beers, whereas at any time abnormal quantities of various attractive colored and tasty products are consumed like chocolates, cakes, and candies (Holdsworth 2008). In fact, although these products are submitted to food processing that definitely causes modifications in the initial raw material (for example, production of black tea, roasted coffee, matured wines, production of chocolate, jam, etc.,) (Ferruzzi 2010; Dominguez-Perles et al. 2011; Kaack and Christensen 2010; Nishiyama et al. 2010; Lee et al. 2008; Negukhula et al. 2011) they play an enormous role as major human dietary polyphenolic sources. The daily intake of polyphenolic compounds (up to 1 g) has shown a relationship with reduced risk of CVDs and cancer prevention. Also, recent epidemiologic data further support the association of polyphenols to their antioxidant action but very restrictive and even without providing clear evidence of the contradictory properties of these compounds widely spread in our diet (Barbosa 2007; Galati and O'Brien 2004; Gerhaeuser 2001; Serafini 2004).

Polyphenolic compounds display multiple structure conditioned interactions with various biomolecules, namely the activity modulation of various enzyme systems. In the diet they act mostly as health promoting factors during various chemical and physical stresses of the organism. They are antiatherogenic and anticarcinogenic, on the principle of inhibition of oxidative destruction of various biological structures, inhibition of processes of bioactivation of carcinogens, blocking LDL oxidation, and stimulating the activity of antioxidant and detoxication enzymes. Some of them have shown some mutagenic properties in genotoxicity tests. However, results of animal experiments and epidemiological studies do not confirm the risk of neoplastic disease in subjects with a normal intake of these substances. The use of the health promoting properties of polyphenols isolated from plants and their administration in a pure state is not foreseen. However, under certain conditions it is desirable to increase the consumption of foods which are important sources of these substances (Zloch 1996).

Initially, the protective effect of dietary polyphenolic compounds was thought to be due to their antioxidant properties which result in a lowering of the free radicals levels within the body (Fernandez-Panchon et al. 2008). There is now emerging evidence that the metabolites of dietary polyphenolic compounds, which appear in the circulatory system in low concentrations, exert modulatory effects in cells through selective actions on different components of the intracellular signaling cascades vital for cellular functions, such as growth, proliferation, and apoptosis. In addition, the intracellular concentrations required to affect cell signaling pathways are considerably lower than those required to impact on antioxidant capacity (Crozier et al. 2009).

"An increased intake of dietary antioxidant polyphenols may protect against CVDs!" This possibility has always been considered in many research works due to the antioxidant property of polyphenols in one hand, and the oxidative events *in vivo* which may play a role in the pathogenesis of atherosclerosis in the other hand. In this regard, vitamins E and C are mostly known as potent antioxidant agents, while a slight to progressive knowledge has been reported about the similar antioxidant role of plant-derived polyphenolic compounds, especially, flavonoids and stilbenes. These two well-known families have been our previous case of discussion because of their bioavailability in our daily food. They are mostly reported to be members of red wines and teas compositions and therefore these common beverages should be exciting examples to be treated due to people's preferences. French Paradox at a relaxed tea time is a social circumstance that can be interpreted to a scientific dialogue on health promoting food and nutrition. The reader of this chapter can probably be without any tendency toward these two popular beverages and, therefore, other tasty dietary sources are brought up in this subject in order to satisfy not only the food desire but also to enrich the scientific background.

4.1 A Red Wine Desire or in a Tea Mood?

At the pub, maybe the choice between ordering a cup of red wine or tea can be without great importance because this depends only on people's minds. However, in a bioanalytical laboratory, scientists cannot choose between them since both offer valuable phytochemical compositions taking into account their uncountable biological properties. In this part, a major discussion is presented on the French Paradox concept while having a relaxed tea time. These two popular beverages have been subjects of many recent works treating their health promotion and antioxidant effects, based on different new clinical trials and ancient epidemiological data. A knowledge update on the cardiovascular effects and cancer prevention of pure anthocyanins, proanthocyanidins, catechins, and stilbenes, as the main phytochemicals of red wine and tea, is being described. Basic information, some key molecule structures, *in vitro* and *in vivo* biological evaluation and clinical studies on both of the two beverages including their isolated phytochemicals are also indicated. The exceptional stilbene resveratrol **21** will take a great part of our discussion with an up-to-date overview on its biological manifestations contributing to the French Paradox explanation. Necessary data are mentioned in order to understand these polyphenols' role in reducing risk factors and preventing cardiovascular health problems through different aspects of their bioefficacy on vascular parameters such as platelet aggregation, atherosclerosis, antioxidant status (Sparwel et al. 2009; Cooper et al. 2004; de Lange et al. 2007) blood pressure (Kappagoda et al. 2000), inflammation, myocardial conditions, and whole-body metabolism (Dixon et al. 2002). Better designed clinical studies are strongly required to improve the current knowledge on the potential health benefits of these polyphenols on cardiovascular and metabolic health (Sun et al. 2002).

4.1.1 The French Paradox, History, and Actuality

An old friendship and cultural history are known between red wine and mankind dating back thousands of years. It was known and very common in many ancient civilizations having an important role in religions as well. For a long time, people admire drinking a lot of wine ignoring that its abuse has serious effects on physical and mental health and causes acute and chronic damage. But, in fact, light to moderate intake of red wine can be beneficial in healthy individuals, which was also observed in antiquity. Red wine was even used in ointments for its disinfectant effects and maybe due to its alcohol content (Feher et al. 2007). In the Muslim society, although drinking wine is absolutely inadmissible and considered a sin for socio-religious reasons, a meaningful sentence in the Quran (the religious Islamic book) refers to its health effects on humans. The phenomena established in the French Paradox can also explain the positive effects of red wine, notably on high fatty diet French society. Behind these social, historical, and philosophical circumstances, enormous research efforts have been conducted toward a real scientific explanation.

The French Paradox was coined in the 1990s explaining the reason for the relatively low incidence of cardiovascular disease in the French population, despite a relatively high dietary intake of saturated fats. This was potentially attributable to moderate consumption of red wine (Lippi et al. 2010; Yoo et al. 2010). Statistical observations are leading to some explanations of the French Paradox (Bejot et al. 2009; Tunstall-Pedoe 2008). An evident inverse relationship between moderate wine consumption and coronary artery disease mortality was observed in France with an approximate mortality rate of 50 % compared to other European countries and the United States (Rosenkranz et al. 2002). Over the last 20 years, considerable amount of studies investigated the crucial biological actions and clinical associations of red wine consumption with cardiovascular disease and mortality. The putative cardioprotective effects of alcohol and other substances in alcoholic beverages have been discussed taking into account the French Paradox. Accordingly, our literature survey is quoting various hypotheses that explain the protective effects of moderate intake of red wine which offers a polyvalence of biological effects targeting all phases of the atherosclerotic process. Taking into account its polyphenolic profile, red wine favors a decrease of oxidative stress, enhance cholesterol efflux from vessel and inhibit lipoproteins LDL oxidation, macrophage cholesterol accumulation, and foam-cell formation (De Gaetano et al. 2005; Vidavalur et al. 2006; Zern and Fernandez 2005; Szmitko and Verma 2005; Raghava 1993; Howard et al. 2002; Shrikhande 2000) increase antioxidant capacity in humans (Logan et al. 2008; Pinzani et al. 2010; de Gaulejac et al. 1999) and reduce susceptibility of human plasma to lipid peroxidation (Barbaste et al. 2002; Belleville 2002). Red wine may also increase nitric oxide bioavailability, thereby antagonizing the development of endothelial dysfunction and improved endothelial function, decrease blood viscosity, improve insulin sensitivity, counteract platelet hyperactivity, inhibit platelet aggregation and platelet adhesion to fibrinogen-coated surfaces, and decrease plasma levels of von Willebrand factor, fibrinogen, and coagulation factor VII (Providencia 2006). Many reports also consider reactive oxygen species or free radical oxidations to be responsible for the accompanying disorders of most pathologies including CVDs, aging, and cancer. Hence, it is conceivable that natural plant metabolites such as polyphenols are likely to play an important role in ensuring this protection. Indeed, not only their presence, particularly high amounts and varieties in our daily food, especially in red wine, but also, their very potent antioxidant or radical scavenging properties are making polyphenols the best contributing for the paradoxical part of the French Paradox (Chong-Han 2010).

Some recent opinions are against the French Paradox concept since it was relatively observed that French coronary heart disease rates are not so low, nor fat diet intake so high, nor the diet-heart concept so unique, as to support the French Paradox any further, except as cultural whimsy or a marketing stratagem (Ducimetière 2008). In fact, many other countries worldwide record low coronary heart disease rate as red wine consumption is not so popular. Overweight and increased cardiovascular mortality are some of the negative results attributed to red wine consumption and therefore, other opinions disagree totally with the idea of the healthy cup of red wine (Hu 2005). Of course, evidence is present for and against the French Paradox hypothesis, while strong epidemiological data favor the specific cardiovascular benefit of red wine which could at least explain it, while, epidemiological and mechanistic evidence proved that the alcohol intake is not without a degree of risk on healthy humans. More convincing evidence is that human studies with de-alcoholized red wine show short-term cardiovascular benefits. The specific components of the de-alcoholized wine that are active on cardiovascular endpoints are the polyphenols found in red wine (Vinson et al. 2001).

Further, Paradox states are linked to alcohol consumption, which, on one side, was associated with incidence of type 2 diabetes and cardiovascular disease in type 2 diabetes, while at the opposite side, a number of reports consistently suggest that the acute effect of alcohol induces a state of insulin resistance and improvement in insulin sensitivity (Zilkens and Puddey 2003). Difficulties are encountered to explain the effect of alcohol on risk factors associated with CVDs by a uniform biochemical mechanism. Moreover, its protective effects are counterbalanced by its addictive properties (Barbaste et al. 2002; Belleville 2002; Brenner et al. 2001; Iijima et al. 2000, 2002).

Although the majority of the scientific reports are in favor of beneficial cardiovascular effects related to the moderate red wine consumption, no one has yet considered the limit scale of red wine consumption for healthy individuals (Lippi et al. 2010). The limit line between 'moderate' and 'excessive' consumption is literary understandable but since red wine displays positive and negative effects, this limit line becomes a mathematical equation to be solved in order to determine whether or not one cup (or more, or less) is the appropriate amount for healthy people (Mudry 2010; Goldfinger 2003). In connection, Cordova et al. suggested to prescribe one or two drinks of red wine every day with meals for patients. This may translate to a longer, healthier, and better quality of life (Cordova and Sumpio 2009). This scientific prescription was based on the polyphenolic compounds responsible for these biological effects, including flavonoids and stilbenes, which are more abundant in red wine than in other beverages. However, this conclusion seems to be premature because no significant bioactive evidence of wine polyphenols has been shown in humans so far. Available data, justify the results of reduced cardiovascular risk and moderate consumption of red wine, but it was also associated with detrimental effects in pregnant women, in children, and in patients with various organic diseases, particularly hepatic, as well as in the case of regular administration of certain medicines (Feher et al. 2007; van de Wiel and de Lange 2008). International comparisons, starting from 1819, shows that a cup of red wine has a main role on individual's health without ignoring that a bottle should not be healthier (Yoo et al. 2010;

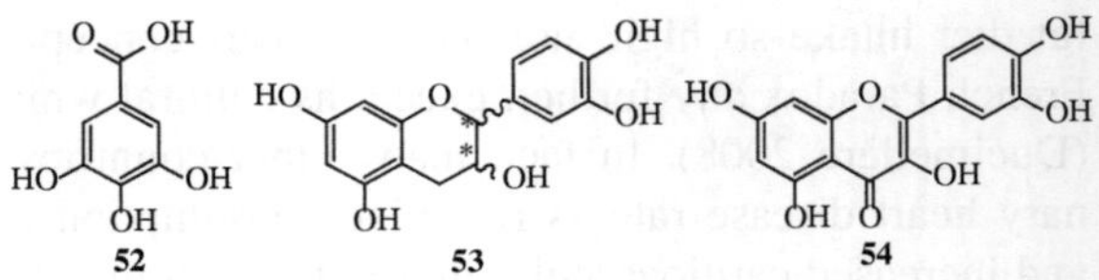

Fig. 6 Structure of gallic acid **52**, catechin base-structure **53**, and quercetin **54**

Chetreanu and Iliescu 2009). The French Paradox may have its basis within a milieu containing several key molecules exhibiting several biological actions coupled with probable or visible side effects. That is why its favorable cardiovascular benefits might be attributable to combined, additive, or perhaps synergistic effects of ethanol and other wine ingredients such as resveratrol **21**.

We have generalized the beneficial actions related to red wine as a whole beverage, but in fact a cup of red wine is composed of several individual components and each of the component display its specific biological role. *In vivo* evidences of red wine isolated and purified products have only been recently published consisting of a new background which may help to a better understanding of the French Paradox.

The oxidation of human Low Density Lipoproteins (or LDL) is responsible for atherosclerosis and arterial damage increase. In fact, molecules absorption through cell membrane (such as LDL and other proteins) has to be strictly regulated. This process known as endocytosis used by all cells of the body, can be enhanced in case of oxidised LDL which are not capable to pass through the hydrophobic plasma or cell membrane and cause accumulations in favor of atheroma and platelet aggregates formation. These are considered as primary risk factors of cardiovascular complications (Berthelot-Garcias et al. 2009; Zern and Fernandez 2005; Raghava 1993). Lectin-like oxidized LDL receptor-1 (LOX-1) is an endothelial receptor for oxidized LDL (ox-LDL) and plays multiple roles in the development of cardiovascular diseases. A chronic administration of purified oligomeric procyanidins from grapes and apples inhibit lipid accumulation in vascular wall in hypertensive rats fed with high fat diet. These results show the selective LOX-1 inhibition by procyanidins but not any other polyphenols (Howard et al. 2002; Nishizuka et al. 2011). This potent inhibition capacity can be particular evidence underlying the mechanism of the vascular action of red wine procyanidins as derived from red grapes.

Actual findings attribute the low incidence of cardiovascular disorders especially in Mediterranean countries to the antioxidant capacity of red wine polyphenols. Conceivably, other anti-inflammatory pathways may contribute to at least a similar extent to the atheroprotective activity of these polyphenols. Investigations have confirmed that gallic acid **52** (Fig. 6), an abundant red wine polyphenol, modulates the activity of P-selectin, an adhesion molecule that is critically involved in the recruitment of inflammatory cells to the vessel wall and thus in atherosclerosis, by binding and antagonizing this protein (Appeldoorn et al. 2005).

The protective effect of red wine on thrombosis is clinical evidence toward the concept of the French Paradox. Alcohol-free red wine supplementation almost completely reverted the prothrombotic effect of the cholesterol-rich-diet in experimental animals, supporting the concept of the French Paradox that regular consumption of wine (rather than alcohol) was able to prevent arterial thrombosis associated with dietary-induced hypercholesterolemia, an effect mediated by down regulation of platelet function (De Curtis et al. 2005).

Rosenkranz et al., have comparatively demonstrated that pre-incubation of vascular smooth muscle cells (VSMCs) with red wine, but not white wine, inhibits ligand binding and the subsequent tyrosine phosphorylation of the platelet-derived growth factor beta receptor (beta-PDGFR), which plays a critical role in the pathogenesis of atherosclerosis. Analytical data revealed flavonoids of the catechin base-structure **53** (Fig. 6) as major constituents of red wine and potent inhibitors of beta-PDGFR signaling. Importantly, the concentrations of red wine/catechins necessary to inhibit the PDGFR *in vitro* correlate with the serum levels after red wine consumption in humans. It was then concluded that non-alcoholic constituents of red wine, which accumulate during the mash fermentation, inhibit beta-PDGFR activation and PDGF-dependent cellular responses in VSMCs. Hence, catechin-mediated inhibition of beta-PDGFR signaling constitutes one of the molecular explanations for the French Paradox (Rosenkranz et al. 2002).

Several animal and epidemiological studies suggest that red wine polyphenol constituents possess antioxidant activities that favor protection against cardiovascular and, probably, central nervous system disorders, such as Alzheimer's disease (AD) and ischemia. Bastianetto et al. studied the potential of the three major red wine derived-polyphenols resveratrol **21**, quercetin **54**, and (+)-catechin **53** to protect cultured rat hippocampus cells against toxicity induced by the nitric oxide free radical donors, sodium nitroprusside (SNP), and 3-morpholinosydnonimine (SIN-1) (Bastianetto et al. 2000). Among the phenolic compounds tested, only the flavonoids afforded significant protection against SIN-1-induced toxicity (5 mM). The effects of phenolic constituents were shared by Trolox (100 μM), a vitamin E analog, but not by selective inhibitors of cyclooxygenases (COX) and lipoxygenases (LOX). These results suggest that the neuroprotective abilities of quercetin **54**, resveratrol **21**, and (+)-catechin **53** result from their antioxidant properties rather than from their supposed inhibitory effects on intracellular enzymes such as COX, LOX, or nitric oxide synthase. Quercetin **54** (Fig. 6), however, may also act via protein kinase C (PKC) to produce its protective effects (Bastianetto et al. 2000; Rendig et al. 2001).

Rabai et al. (2010) showed that red wine and alcohol-free red wine have some beneficial effects on hemorheological parameters. These effects may play a role in the pathophysiology of the French Paradox regarding the cardiovascular protective impacts of red wine. The opening of mitochondrial KATP channels was obtained by a non-alcoholic red wine extract in guinea pigs, Therefore, this effect was prevented by the mitochondrial KATP channel blocker 5-hydroxydecanoate, confirming this subcellular mechanism as underlying the French Paradox (Aiello and Cingolani 2011).

A recent report demonstrated that red wine anthocyanins are rapidly absorbed in humans and affect monocyte chemoattractant protein 1 levels and antioxidant capacity of plasma (Garcia-Alonso et al. 2009). These interesting red wine phythochemicals have been previously proved to have a great participation in the whole antioxidant activity (Rivero-Perez et al. 2008).

4.1.2 The Tea Society

Diseases incidence and diet modes are geographically distributed in the world. We always focus on the diet mode of a population in order to explain certain low rate disease incidences in specific geographical parts. A similar situation, like in France, can be observed in countries like China, India, and most of Middle East Arabian countries, in which, the cancer incidence is relatively low comparing to the Europe and Western countries. If we want to mimic the French Paradox, in this case, the cup of tea replaces red wine. Tea drinks are diverse and very popular in the world, usually prepared by infusion of different tea herbs; green and black teas (*Camellia sinensis* family: (L.) Kuntze *Theaceae*) (Dominguez-Perles et al. 2011; Yao et al. 2006; Wang and Ho 2009). A wide occurrence of structurally diverse polyphenols is known throughout the plant kingdom. Recent interest on tea varieties from different geographical parts has increased greatly because of the antioxidant and free radical-scavenging abilities associated with some phenolics and their potential effects on human health (Armoskaite et al. 2011; Atoui et al. 2005; Black et al. 2011; Wu et al. 2010; Thuong et al. 2009; Terasawa and Yamazaki 2002; Viveros-Valdez et al. 2008; Jayasekera et al. 2011; Nuengchamnong et al. 2011; Omar et al. 2011; Wu et al. 2012; Etoh et al. 2004; Komes et al. 2010; Lee et al. 2006; Liu et al. 2009; Zhu et al. 2002; Unachukwu et al. 2010; Rechner et al. 2002; Rusak et al. 2008; Su et al. 2007; de Mejia et al. 2010; McKay and Blumberg 2006; Kulisic-Bilusic et al. 2008; Dominguez-Perles et al. 2012; Fale et al. 2011; Banerjee et al. 2010; AlGamdi et al. 2011; Ariffin et al. 2011; Bueyuekbalci and El 2008; Chen et al. 2012).

Catechins are the major class of polyphenols found in great amounts in green tea and, therefore, a scientific attention was paid to the regular consumption of green tea and its relation with atherosclerosis and cancer prevention (Salucci et al. 2002; de Mejia et al. 2010; Ishii et al. 2010; Carvalho et al. 2010). Green tea polyphenols have been proposed to exert protective effects against several types of cancer, based on preclinical and clinical trial data. Green tea extracts strongly inhibited the growth of renal carcinoma cell lines in a concentration-dependent manner. This is the first report showing that green tea is likely to be an effective anticancer agent for renal cell carcinoma (Carvalho et al. 2010). Tea catechins and tea catechin metabolites/catabolites are bioavailable in the systemic circulation after oral intake of green tea or green tea catechins. The metabolites/catabolites identified in humans include glucuronide/sulfate conjugates, methylated tea catechin conjugates, and microflora-mediated ring fission products and phenolic acid catabolites. Plasma levels of unchanged tea catechins in humans are mostly in the sub-μM or nM concentration range, which is much lower than the effective concentrations determined in most *in vitro* studies. However, some of the catechin metabolites/catabolites are present in the systemic circulation at levels much higher than those of the parent catechins. The contribution of catechin derived metabolites/catabolites to the biological effects associated with green tea is not yet defined. A limited number of chemoprevention trials of green tea or green tea catechins have been conducted to date and have shown potential preventive activity for oral, prostate, and colorectal cancer (Chow and Hakim 2011).

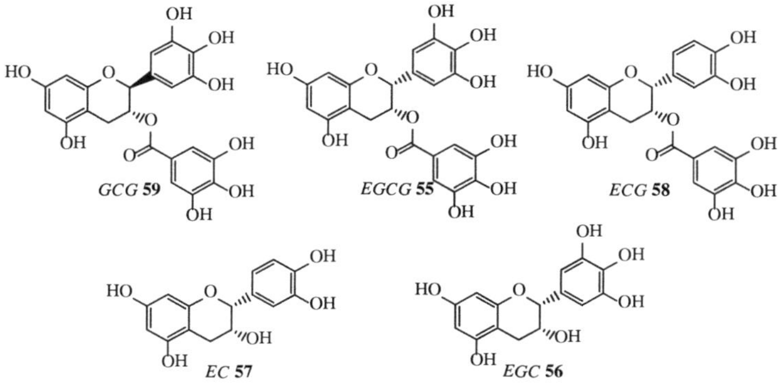

Fig. 7 Structures of catechin derivatives found in tea

The amounts of phenolics and flavonoid in the herbal green tea infusions were highly correlated with their anti-glycation activity (Bueyuekbalci and El 2008; Azevedo et al. 2011; Ho et al. 2010; Ankolekar et al. 2011). Thus, these compounds, like other that inhibit the formation of advanced glycation end-products, are supposed to have therapeutic potentials in patients with diabetes or age-related diseases.

Green tea might also be utilized as a natural antimicrobial agent to inactivate *Vibrio parahaemolyticus* in oysters and extend the shelf life during refrigeration storage (Xi et al. 2012). The green tea polyphenolic compounds treatment may be a useful method for preserving the human saphenous vein and could be exploited to craft strategies for the long-term preservation of other tissues under physiological conditions (Han et al. 2005). The ingestion of either green tea or black tea results in a major increase in the excretion of hippuric acid into urine (Mulder et al. 2005). Both black and green tea extracts may have synergistic or antagonistic effects on certain anti-streptococcal antibiotics. These effects are more prominent with black tea (Neyestani et al. 2007). Also, black tea extract has selective pro-inflammatory cytokine-suppressing effects on human peripheral blood mononuclear cells (Neyestani et al. 2009). Phytochemicals from Chinese herb teas showed antioxidant activity and inhibition of hepatoma cell proliferation *in vitro* assays, exhibiting a great potential as new nutraceutical agents (Li et al. 2009).

Besides their flavor fingerprint composed of terpenoids (Pripdeevech and Machan 2011) many kinds of teas display a variety of antioxidant flavonoids, especially, the flavanol subclass represented by catechins, namely epigallocatechin gallate (EGCG) **55**, epigallocatechin (EGC) **56**, epicatechin (EC) **57**, epicatechin gallate (ECG) **58**, and gallocatechin gallate (GCG) **59** (Fig. 7) (Amarowicz and Shahidi 1996; Williamson et al. 2011; Wu et al. 2011; Song et al. 2011; Dalluge and Nelson 2000; Karori et al. 2007; Kodama et al. 2010; Zuo et al. 2002; Zhu et al. 2009; Galati et al. 2006). According to a new phytochemical report on

Taiwan's teas (Wu et al. 2011; Wang et al. 2006) green tea drinks are found to contain the highest level of these major components comparing to other types, like oolong and black teas. The scavenging abilities against 1,l-diphenyl-2-picrylhydrazyl (DPPH) and 2,2′-azino-bis(3-ethylbenzthiazoline-6-sulphonic acid) (ABTS) radicals, used to determine the antioxidant potential of tea drinks, resulting in a higher free radical scavenging activity of the green tea. Total phenolics, total catechins, and EGC **56** of tea drinks were positively and significantly correlated ($r > 0.8$) to the scavenging abilities against DPPH and ABTS radicals (Wu et al. 2011). Total antioxidant capacities of herbal green tea extracts were further confirmed by other different assays including, trolox equivalent antioxidant capacity (TEAC) and oxygen radical absorbance capacity (ORAC) (Tsai et al. 2008). Complimentary studies suggest that iron may modify the antioxidant properties of phenolic compounds when adding to green tea, being noticed a decrease of its antioxidant capacity in rats after an oral dose of the mixture. However, further studies on the effect of iron on the bioavailability and the antioxidant capacity of phenolic compounds are required (Kapsokefalou et al. 2006).

Further recent knowledge was published on the antioxidant capability of green tea polyphenols, underlying the mechanism of antioxidant antagonism in peroxidising liposomes with β-carotene radical cation addition. The previously mentioned green tea catechins, including EC, EGC, ECG, and EGCG, all showed antioxidant effect in liposomes for lipid oxidation initiated in the lipid phase (antioxidant efficiency EC > EGCG > ECG > EGC) or in the aqueous phase (EC ≫ EGC > EGCG > ECG) as monitored by the formation of conjugated dienes. For initiation in the lipid phase, it is suggested that the β-carotene radical cation is rather reacting with the tea polyphenols through addition, in effect preventing regeneration of β-carotene as an active lipid phase antioxidant and leading to the observed antagonism with the polyphenols (EC > ECG > EGCG > EGC) (Song et al. 2011).

Tea phenolic acids along with the some previously highlighted catechins containing gallic acid moieties display further medical benefits. The cytotoxicity of these tea phenolic components toward isolated rat hepatocytes have been evaluated (epigallocatechin-3-gallate > propyl gallate > epicatechin-3-gallate > gallic acid, epigallocatechin > epicatechin). Using gallic acid as a model tea phenolic and comparing it with the tea catechins and gallic acid-derivative food supplements, the major cytotoxic mechanism found with hepatocytes was mitochondrial membrane potential collapse and ROS formation. Epigallocatechin-3-gallate was the most effective at collapsing the mitochondrial membrane potential and inducing ROS formation. Liver injury was also observed *in vivo* when these tea phenolics were administered into mice. In contrast, GSH conjugation, methylation, metabolism by NAD(P)H:quinone oxidoreductase 1, and formation of an iron complex were important in detoxifying the gallic acid (Galati et al. 2006).

Green and black tea have been considered as candidates for a chemopreventive evaluation in prostate cancer due to their valuable antioxidant activity. The polyphenol compositions of green and black tea are different due to post-harvest processing. As far as green tea contains higher concentrations of monomeric polyphenols (catechins), it affects numerous intracellular signaling pathways involving prostate cancer (CaP) development. Black tea polymeric polyphenols, on the other hand, are poorly absorbed and are converted to phenolic acids by the colonic microflora. Therefore, after consumption of green tea, higher concentrations of polyphenols are found in the circulation, whereas in the case of black tea consumption, the phenolic acid levels in the circulation are higher. The majority of *in vitro* cell culture, *in vivo* animal, and clinical intervention tests of green tea extracts (or purified EGCG) on prostate carcinogenesis, provide strong evidence supporting a chemopreventive effect of green tea. While the evidence for a chemopreventive effect of black tea is much weaker, there are several animal black tea intervention studies demonstrating the inhibition of CaP growth (Henning et al. 2011).

Olech et al. (2012) demonstrated that *Rosa rugosa* Thunb. teas possess high antiradical activity and their polyphenols constituents showed a considerable impact in the anticancer activities against ovarian (TOV-112D), cervical (HeLa), breast (T47D), and lung cancer (A549) cell lines.

First convincing evidence showed that green tea polyphenols are effective in reducing tributyltin (TBT)-induced oxidative damage both *in vivo* and *in vitro*. The possible protective mechanism may be due to the powerful ability of polyphenols to scavenge ROS, nitric oxide and prevent DNA breaks (Liu et al. 2008b; Tsai et al. 2007). Thus, green tea could be an effective agent or food supplement to reduce the cytotoxicity of TBT (Liu et al. 2008b). Further studies reveal that tea polyphenols are able to act as pro-oxidants to cause a response to oxidative stress in yeasts under certain conditions (Maeta et al. 2007).

Anti-cholinesterase and antioxidant active constituents of *Plectranthus barbatus* Andrews (Indian coleus) aqueous extract were found in plasma of rats after its administration. The *Plectranthus barbatus* Andrews herbal tea extract components also inhibit lysozyme activity with IC_{50} values around 100 μM. This inhibition activity may be an additional mechanism for the anti-inflammatory activity of their polyphenolic constituents (Fale et al. 2011).

According to a recent pharmacokinetic study on healthy humans, a consumption of an average cup of green tea (200 mL) containing 112 mg of (−)-epigallocatechin gallate **55**, 51 mg of EGC **56** and 15 mg of EC **57**, gave rise to predicted C_{max} values (total free and sulfate/glucuronide conjugates) in plasma of 125, 181, and 76 nM, respectively, together with 94 nM methyl-EGC and 51 nM methyl-EC. Most studies with chlorogenic acids report a very low amount of intact molecules in plasma (Williamson et al. 2011).

Interestingly, the methanol extract of fresh tea leaves of *Camellia sinensis* (L.) Kuntze (Theaceae) inhibited enzymes with hydrolytic activity in snake [*Naja naja kaouthia* (L.) Kuntze (Elapidae) and *Calloselasma rhodostoma K.* (Viperidae)] venoms, by *in vitro* neutralization and *in vivo* inhibition of the hemorrhagic and the dermonecrotic activities. These snake venom enzymes are responsible for the early effects of envenomation, such as local tissue damage and inflammation. It is suggested that the inhibitory potential of the *Camellia*

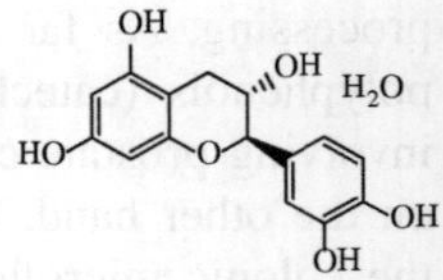

Fig. 8 Structure of catechin hydrate **60**

sinensis (L.) Kuntze extract against local tissue damage induced by snake venom may be attributed to complexation between the venom proteins and the phenolic contents of the extract (Pithayanukul et al. 2010).

4.1.3 Tea and Red Wine Common Factors

Several common biological activities approved for both red wine and tea may be due to their similar polyphenolic background. Catechin hydrate **60** (Fig. 8), a strong antioxidant that scavenges free radicals, is evidenced as an abundant phyto-constituent of both green tea and red wine. Catechin hydrate **60** possesses anticancer potential and effectively kills 100 % MCF-7 cells after 72 h of exposure, inducing apoptosis, which was confirmed by terminal deoxynucleotidyl transferase-mediated dUTP nick end labeling (TUNEL) and real-time PCR assays. The induction of apoptosis by catechins hydrate is affected by its ability to increase the expression of pro-apoptotic genes such as caspase-3, -8, and -9 and TP53 (Alshatwi 2010).

Moreover, a comparative study shows that water extracts of black tea had the highest α-glucosidase inhibitory activity than several selected red wines. The α-glucosidase inhibitory activity of the examined teas and wines correlated with the phenolic content, antioxidant activity, and phenolic profile of the extracts (Kwon et al. 2008).

Dietary polyphenols in teas and wines have been associated with beneficial health effects. After ingestion, most polyphenols are metabolized by the colonic microbiota to a limited number of key metabolites. The metabolic profile depends on the individual and on the polyphenol sources. Varying metabolite pathways among individuals result in different metabolome profiles and therefore related health effects are hypothesized to differ between subjects (Gross et al. 2010; Roowi et al. 2010).

4.1.4 Resveratrol, the Difference Between Red Wine and Tea

Polyphenols, among them resveratrol **21**, have generated a great amount of scientific interests due to its *in vivo* and *in vitro* antioxidant capabilities. Since **21** has been evidenced in red wine, the birth of a new French Paradox key molecule has been noticed (Zhuang et al. 2003; Gusman et al. 2001; De Leiris and Boucher 2008; Das and Das 2007). The efficient *in vitro* cardioprotection effect and *in vivo* on different animal models of **21** have been well documented (Zhuang et al. 2003; Liu et al. 2007). For the last few years, an outstanding focus on the therapeutic properties of **21** has been reported describing its antioxidant actions on one hand and its selectivity to some cell-targets on the other hand (Nikolova 2007; Poussier et al. 2005).

Resveratrol **21** is present in grapevine as constitutive compound of the woody organs, and as induced substance in leaves and fruit acting like phytoalexin in the mechanisms of grapevine resistance toward some pathogens (Bavaresco et al. 2000). As the main ingredient of red wine was thought to be its active principle involved in heart diseases prevention, it can reduce ischemic damage in heart ischemia reperfusion injury and also in brain ischemia/reperfusion in rodent models. Most of the protective biological actions associated with **21** have been associated with its intrinsic radical scavenger properties. It was shown that it exerts neuroprotective actions on primary neuronal cultures (Zhuang et al. 2003). It is a potent peroxidase-dependent mechanism-based inactivator of COX-1, a desired target for antiplatelet agents, but no similar effects have been noticed on COX-2. These findings imply that resveratrol **21** is not the sole agent responsible for the antiplatelet activity of red wine and suggest that all dietary meta-hydroquinones should be examined for cardioprotective effects (Szewczuk and Penning 2004).

A growing body of evidence supports the protecting role of resveratrol **21** in the cardiovascular system in a multidimensional way. The most important point is that at a very low concentration it inhibits apoptotic cell death, thereby providing protection from various diseases including myocardial ischemic reperfusion injury, atherosclerosis, and ventricular arrhythmias. **21** when used at higher doses facilitates apoptotic cell death and behaves as a chemopreventive alternative (Das and Das 2007).

In vivo animal tests of daily oral administration demonstrate that resveratrol **21** failed to protect against environmental tobacco smoke (ETS) exposure impaired endothelial function and increased oxidative stress, which was associated with pulmonary and systemic inflammation, in juvenile male pigs. However, it exerted a positive effect on left ventricular function which may help explain the French Paradox (Al-Dissi and Weber 2011).

A number of resveratrol **21** valuable properties have been attributed to its intrinsic antioxidant capabilities, although their potential level in the circulation is likely not enough to neutralize free radical scavenging. The brain and the heart are uniquely vulnerable to hypoxic conditions and oxidative stress injuries. Increased heme oxygenase activity, stimulated by resveratrol **21,** has led to significant protection against models of *in vitro* and *in vivo* oxidative stress injury (Dore 2005).

Resveratrol **21** acts as reactive oxygen species inhibitor, which together with the accumulation of the reactive oxygen intermediate (ROI) produced from cell antioxidant self-defense (enzymes), are also responsible for cell tissue damage, aging, and carcinogenesis. ROS and ROI lead to oxidative stress phenomena responsible for the development of cardiovascular diseases and oxidation of different macromolecules (DNA, lipids, and proteins) (Alcaraz et al. 2009). **21** is

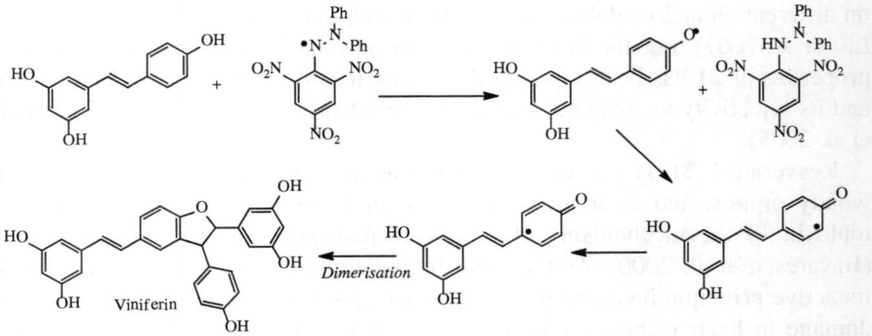

Scheme 11 DPPH radical-induced dimerization of trans-resveratrol

transformed itself into a stabilized free radical upon reacting with DPPH radical leading to viniferin dimer (Scheme 11) (Wang et al. 1999).

Oxidative stress can also decrease the bioavailability of nitric oxide radical in vessels, which is highly associated with endothelial dysfunction. One of the mechanisms involved in beneficial effects of **21,** is its capacity to maintain sufficient nitric oxide radical bioavailability in vascular endothelium (Frombaum et al. 2012).

A huge amount of studies have shown that phenolic compounds contained in red wine inhibit the susceptibility of low-density lipoproteins (LDLs) to oxidation, thereby potentially reducing their atherogenicity (Iijima et al. 2000, 2002). The effects of **21** on isolated tissues or organs are well described and include molecular mechanisms leading to decrease arterial damage, decrease activity of angiotensin-II, increase nitric oxide, and decrease platelet aggregation. Anti-ischemic effects include stimulation of prosurvival paths, decrease LDL-oxidation, atheroma, and on the ischemic-beneficial metabolic changes. Most recently, the agonist effect of **21** on the anti-senescence factor sirtuin has lessened cell death in myocytes from failing hearts. Mechanistic feasibility strengthens the case for prospective therapeutic trials of alcohol *vs* red wine *vs* resveratrol, for example in those with heart failure (Opie and Lecour 2007).

Recently, **21** was discovered to be a putative activator of SIRT1 which can partially mimic the physiological effects of calorie restriction, such as the life span extension of model organisms. It is important to notice that SIRT1 activation is a promising new therapeutic approach for treating diseases of aging such as type 2 diabetes (Hu et al. 2011; Lekli et al. 2010).

4.2 Polyphenols from Other Tasty Sources

Few people do not like it, other few people like but avoid it, however most people adore chocolate. It was only recently published that cocoa-derived products (dark chocolate, milk chocolate, and cocoa powder) are great sources of polyphenols particularly, catechins (flavan-3-ols), and procyanidins. However, the data vary remarkably due to the quantity of cocoa liquor used in the recipe of the cocoa products but also due to the analytical procedure employed. In 1994, the per head consumption of chocolate and chocolate confectionery in the European Union ranged from 1.3 kg/year in Portugal to 8.8 kg/year in Germany. In general, consumers in the northern countries consume on average more than people in the south. Thus, chocolate can be seen as a relevant source for phenolic antioxidants for some European populations. However, this alone does not imply that chocolate could be beneficial to human health. Some epidemiological evidence suggests a beneficial effect to human health by following a polyphenol-rich diet, namely rich in fruits and vegetables and to a less obvious extent an intake of tea and wine having a similar polyphenol composition as cocoa. In many experiments cellular targets have been identified and molecular mechanisms of disease prevention proposed, in particular for the prevention of cancer and cardiovascular diseases as well as for alleviating the response to inflammation reactions. However, it has to be demonstrated whether polyphenols exert these effects *in vivo*.

One prerequisite is that the polyphenols are absorbed from the diet. For monomeric flavonoids such as the catechins, there is increasing evidence for their absorption, while for complex phenols and tannins (procyanidins) these questions have to be addressed for the future. Another open question is related to polyphenol metabolism. For example, much effort has been invested to show antioxidant effects of free unbound polyphenols, especially of catechins and the flavonol quercetin. However, only a very small part can be found in plasma in the free form, but they occur as conjugated or even metabolized to several phenolic acids and other ring scission products. From the papers reviewed, it is too early to give an answer to the question, whether chocolate and/or other sources rich in catechins and procyanidins are beneficial to human health. Even though some data are promising and justify further research in the field, it has to be shown in future whether the intake of these functional compounds and/or their sources is related to measurable effects on human health and/or the development of diseases (Wollgast and Anklam 2000).

Flavanol-rich foods, i.e., wines, chocolates, and teas, and of purified flavonoids inhibited angiotensin converting enzyme (ACE) activity; red wines being more effective than white wine, and green tea more effective than black tea. When isolated polyphenols were assayed, procyanidins (dimer and hexamer) and epigallocatechin significantly inhibited enzyme activity. Similar concentrations of (+)-catechin (−)-epicatechin, gallic acid, chlorogenic acid, caffeic acid, quercetin, kaempferol, and resveratrol were ineffective. The ACE inhibition activity of rat kidney membranes in the presence of chocolate extracts or purified procyanidins depend on the chocolate content of flavanols and the number of flavanol units constituting the procyanidin. These experiments demonstrate that flavanols either isolated or present in foods could inhibit ACE activity. The occurrence of such inhibition *in vivo* needs to be determined, although it is supported by the association between the consumption of flavanol-rich foods and reductions in blood pressure observed in several experimental models (Actis-Goretta et al. 2006).

Lee et al. prefer cocoa, not for the delicious taste of chocolate, but in fact, for its high phenolic content and higher antioxidant capacity than teas and red wine. A comparative report between phenolic and flavonoid contents and total antioxidant capacities of cocoa, black tea, green tea, and red wine revealed that the cocoa contained much higher levels of total phenolics and flavonoids. Cocoa exhibited the highest antioxidant activity, and the relative total antioxidant capacities were as follows in decreasing order: cocoa, red wine, green tea, black tea. The total antioxidant capacities were highly correlated with phenolic and flavonoid contents. These results suggest that cocoa is more beneficial to health than teas and red wine in terms of its higher antioxidant capacity (Lee et al. 2003).

5 Conclusions

The first part of the chapter focused on the move from the natural antioxidant nutrients to synthetic agents which are mostly bio-inspired from natural models like flavonoid-type compounds and stilbenes. The result of the chemistry laboratory demonstrated better activity of synthetic models. However nature remains the first supplier of nutritional dietary components while further research efforts are needed to prove the reliability of the synthetic ones. The second part of the chapter was devoted to some examples of people's eating habits according to their geographical and cultural belongings including, the so-called French Paradox or red wine in EU and USA, tea in Far East and Middle East, and finally chocolate, the worldwide tendency. By knowing the beneficial actions of each food, it was possible to establish a comprehensive relationship between health and nutrition mode underlying the limit of excessive intake and risk factors. The phytochemical knowledge of these daily consumed food is such important to assess their selective beneficial effects on humans and also some side effects almost related to an over intake. Individual biological evaluations of these food-isolated polyphenols provide a better understanding of the overall biological benefit and related negative effects. The case of resveratrol, catechins, quercetin, and other polyphenolic compounds is already quoted but further examinations are required for other key molecules and their metabolites. To sum up, natural or synthetic ingredients of our food should be promotable for our health but we always have to consider that the positive biological effect ensured by these phytochemicals and/or additive dietary synthetic chemicals can be transformed to negative effect after a certain limit line and therefore this is the line that has to be evaluated or at least estimated by scientists in order to avoid this negative side.

References

Actis-Goretta L, Ottaviani JI, Fraga CG (2006) Inhibition of angiotensin converting enzyme activity by flavanol-rich foods. J Agric Food Chem 54(1):229–234

Ahmadi T, Maniar T, Schuster S, Stadtmauer E (2009) Chronic lymphocytic Leukemia: new concepts and emerging therapies. Curr Treat Option Oncol 10(1–2):16–32

Aiello EA, Cingolani HE (2011) A possible subcellular mechanism underlying the "French paradox": the opening of mitochondrial K-ATP channels. Appl Physiol Nutr Metab 36(5):768–772

Al-Dissi AN, Weber LP (2011) Resveratrol preserves cardiac function, but does not prevent endothelial dysfunction or pulmonary inflammation after environmental tobacco smoke exposure. Food Chem Toxicol 49(7):1584–1591

Albert S, Horbach R, Deising HB, Siewert B, Csuk R (2011) Synthesis and antimicrobial activity of (E) stilbene derivatives. Bioorgan Med Chem 19(17):5155–5166

Alcaraz M, Acevedo C, Castillo J, Benavente-Garcia O, Armero D, Vicente V, Canteras M (2009) Liposoluble antioxidants provide an effective radioprotective barrier. Brit J Radiol 82(979):605–609

AlGamdi N, Mullen W, Crozier A (2011) Tea prepared from *Anastatica hierochuntica* seeds contains a diversity of antioxidant flavonoids, chlorogenic acids and phenolic compounds. Phytochemistry 72(2–3):248–254

Alonso DA, Najera C, Varea M (2004) 3,5-bis(trifluoromethyl)phenyl sulfones in the modified Julia olefination: application to the synthesis of resveratrol. Tetrahedron Lett 45(3):573–577

Alshatwi AA (2010) Catechin hydrate suppresses MCF-7 proliferation through TP53/Caspase-mediated apoptosis. J Exp Clin Canc Res 29

Amarowicz R, Shahidi F (1996) A rapid chromatographic method for separation of individual catechins from green tea. Food Res Int 29(1):71–76

Andersen ØM, Markham KR (2006) Flavonoids: chemistry, biochemistry and applications. CRC publication, Taylor & Francis, Boca Raton

Ankolekar C, Terry T, Johnson K, Johnson D, Barbosa ACL, Shetty K (2011) Anti-hyperglycemia properties of Tea (*Camellia sinensis*) bioactives using in vitro assay models and influence of extraction time. J Med Food 14(10):1190–1197

Appeldoorn CCM, Bonnefoy A, Lutters BCH, Daenens K, van Berkel TJC, Hoylaerts MF, Biessen EAL (2005) Gallic acid antagonizes P-selectin-mediated platelet-leukocyte interactions—Implications for the French paradox. Circulation 111(1):106–112

Ariffin F, Chew SH, Bhupinder K, Karim AA, Huda N (2011) Antioxidant capacity and phenolic composition of fermented *Centella asiatica* herbal teas. J Sci Food Agric 91(15):2731–2739

Armoskaite V, Ramanauskiene K, Maruska A, Razukas A, Dagilyte A, Baranauskas A, Briedis V (2011) The analysis of quality and antioxidant activity of green tea extracts. J Med Plants Res 5(5):811–816

Atoui AK, Mansouri A, Boskou G, Kefalas P (2005) Tea and herbal infusions: their antioxidant activity and phenolic profile. Food Chem 89(1):27–36

Azevedo MF, Lima CF, Fernandes-Ferreira M, Almeida MJ, Wilson JM, Pereira-Wilson C (2011) Rosmarinic acid, major phenolic constituent of Greek sage herbal tea, modulates rat intestinal SGLT1 levels with effects on blood glucose. Mol Nutr Food Res 55:S15–S25

Baker W (1933) Molecular rearrangement of some o-acyloxyacetophenones and the mechanism of the production of 3-acylchromones. J Chem Soc 1381–1389

Banerjee D, Hassarajani SA, Maity B, Narayan G, Bandyopadhyay SK, Chattopadhyay S (2010) Comparative healing property of kombucha tea and black tea against indomethacin-induced gastric ulceration in mice: possible mechanism of action. Food Funct 1(3):284–293

Barbaste M, Berke B, Dumas M, Soulet S, Delaunay JC, Castagnino C, Arnaudinaud V, Cheze C, Vercauteren J (2002) Dietary antioxidants, peroxidation and cardiovascular risks. J Nutr Health Aging 6(3):209–223

Barbosa DS (2007) Green tea polyphenolic compounds and human health. J Verbrauch Lebensm 2(4):407–413
Barros AIRNA, Silva AMS (2006) Efficient synthesis of nitroflavones by cyclodehydrogenation of 2′-hydroxychalcones and by the Baker-Venkataraman method. Monatsh Chem 137(12):1505–1528
Barros A, Silva A (2009) Synthesis and structure elucidation of three series of nitro-2-styrylchromones using 1D and 2D NMR spectroscopy. Magn Reson Chem 47(10):885–896
Bastianetto S, Zheng WH, Quirion R (2000) Neuroprotective abilities of resveratrol and other red wine constituents against nitric oxide-related toxicity in cultured hippocampal neurons. Brit J Pharmacol 131(4):711–720
Bavaresco L, Fregoni M, Trevisan M (2000) Resveratrol: from grapevine to wine and relationship with human health. Acta Medica Romana 38(4):466–471
Bejot Y, Osseby GV, Aboa-eboule C, Durier J, Lorgis L, Cottin Y, Moreau T, Giroud M (2009) Dijon's vanishing lead with regard to low incidence of stroke. Eur J Neurol 16(3):324–329
Belleville J (2002) The French paradox: possible involvement of ethanol in the protective effect against cardiovascular diseases. Nutrition 18(2):173–177
Berthelot-Garcias E, Meuleman C, Dufaitre G, Ederhy S, Haddour N, Boccara F, Baudin B, Cohen A (2009) Cardiovascular diseases: epidemiology and medical care by both the cardiologist and the biologist. Rev Francophone Lab 39(409):27–39
Beyer G, Melzig MF (2003) Effects of selected flavonoids and caffeic acid derivatives on hypoxanthine-xanthine oxidase-induced toxicity in cultivated human cells. Planta Med 69(12):1125–1129
Biasutto L, Mattarei A, Marotta E, Bradaschia A, Sassi N, Garbisa S, Zoratti M, Paradisi C (2008) Development of mitochondria-targeted derivatives of resveratrol. Bioorg Med Chem Lett 18(20):5594–5597
Black P, Saleem A, Dunford A, Guerrero-Analco J, Walshe-Roussel B, Haddad P, Cuerrier A, Arnason JT (2011) Seasonal variation of phenolic constituents and medicinal activities of Northern Labrador Tea, *Rhododendron tomentosum* ssp subarcticum, an Inuit and cree first nations traditional medicine. Planta Med 77(14):1655–1662
Brenner H, Rothenbacher D, Bode G, Marz W, Hoffmeister A, Koenig W (2001) Moderate alcohol consumption and coronary heart disease: case-control study of a prominantly beer-drinking population. Gesundheitswesen 63(7):469
Brion D, Le Baut G, Zammatio F, Pierre A, Atassi G, Belachm L (1991) Preparation of 2-styryl-4-chromanones as anticancer agents
Bueyuekbalci A, El SN (2008) Determination of In vitro antidiabetic effects, antioxidant activities and phenol contents of some herbal teas. Plant Foods Hum Nutr 63(1):27–33
Cardenas M, Marder M, Blank VC, Roguin LP (2006) Antitumor activity of some natural flavonoids and synthetic derivatives on various human and murine cancer cell lines. Bioorgan Med Chem 14(9):2966–2971
Carvalho M, Jeronimo C, Valentao P, Andrade PB, Silva BM (2010) Green tea: a promising anticancer agent for renal cell carcinoma. Food Chem 122(1):49–54
Chan KF, Zhao YZ, Burkett BA, Wong ILK, Chow LMC, Chan TH (2006) Flavonoid dimers as bivalent modulators for P-glycoprotein-based multidrug resistance: synthetic apigenin homodimers linked with defined-length poly(ethylene glycol) spacers increase drug retention and enhance chemosensitivity in resistant cancer cells. J Med Chem 49(23):6742–6759
Chen HX, Zhang Y, Lu XM, Qu ZS (2012) Comparative studies on the physicochemical and antioxidant properties of different tea extracts. J Food Sci Tech 49(3):356–361
Chetreanu L, Iliescu ML (2009) About wine intake and cardiovascular diseases. Rev Med Chir Soc Med Nat Iasi 113(3):845–848
Chong-Han K (2010) Dietary lipophilic antioxidants: implications and significance in the aging process. Crit Rev Food Sci Nutr 50(10):931–937
Chow HHS, Hakim IA (2011) Pharmacokinetic and chemoprevention studies on tea in humans. Pharmacol Res 64(2):105–112

Cliff MA, King MC, Schlosser J (2007) Anthocyanin, phenolic composition, colour measurement and sensory analysis of BC commercial red wines. Food Res Int 40(1):92–100
Cooper KA, Chopra M, Thurnham DI (2004) Wine polyphenols and promotion of cardiac health. Nutr Res Rev 17(1):111–130
Cordova AC, Sumpio BE (2009) Polyphenols are medicine: Is it time to prescribe red wine for our patients? Int J Angiol Official Publ Int Coll Angiol, Inc 18(3):111–117
Crozier A, Borges G, Ryan D (2010) The glass that cheers—Phenolic and polyphenolic constituents and the beneficial effects of moderate red wine consumption. Biochemist 32(6):4–6
Crozier A, Jaganath IB, Clifford MN (2009) Dietary phenolics: chemistry, bioavailability and effects on health. Nat Prod Rep 26(8):1001–1043
Cummings J, Double JA, Bibby MC, Farmer P, Evans S, Kerr DJ, Kaye SB, Smyth JF (1989) Characterization of the major metabolites of flavone acetic-acid and comparison of their disposition in humans and mice. Cancer Res 49(13):3587–3593
Dai D-Z, Chen S-H, Feng Y (2004) Tea polyphenols and quercetin preventing the heart, brain and liver from the injury by free radicals in comparison with ascorbic acid. Chin J Nat Med 2(4):223–231
Dalluge JJ, Nelson BC (2000) Determination of tea catechins. J Chromatogr A 881(1–2):411–424
Das S, Das DK (2007) Resveratrol: a therapeutic promise for cardiovascular diseases. Recent Pat Cardiovasc Drug Discovery 2(2):133–138
De Curtis A, Murzilli S, Di Castelnuovo A, Rotilio D, Donati MB, De Gaetano G, Iacoviello L (2005) Alcohol-free red wine prevents arterial thrombosis in dietary-induced hypercholesterolemic rats: experimental support for the 'French paradox'. J Thromb Haemost 3(2):346–350
De Gaetano G, Di Castelnuovo A, Rotondo S (2005) Cardiovascular protective effect of moderate wine consumption: evidence after the French Paradox. Sang Thrombose Vaiss 17(1):47–60
de Gaulejac NS, Glories Y, Vivas N (1999) Free radical scavenging effect of anthocyanins in red wines. Food Res Int 32(5):327–333
de Lange DW, Verhoef S, Gorter G, Kraaijenhagen RJ, van de Wiel A, Akkerman JWN (2007) Polyphenolic grape extract inhibits platelet activation through PECAM-1: an explanation for the French paradox. Alcohol Clin Exp Res 31(8):1308–1314
De Leiris J, Boucher F (2008) Does wine consumption explain the French paradox? Dialogues Cardiovasc Med 13(3):183–192
de Mejia EG, Song YS, Heck CI, Ramirez-Mares M (2010) Yerba mate tea (*Ilex paraguariensis*): phenolics, antioxidant capacity and in vitro inhibition of colon cancer cell proliferation. J Funct Food 2(1):23–34
de Pascual-Teresa S, Moreno DA, Garcia-Viguera C (2010) Flavanols and anthocyanins in cardiovascular health: a review of current evidence. Int J Mol Sci 11(4):1679–1703
Deck LM, Hunsaker LA, Gonzales AM, Orlando RA, Vander Jagt DL (2008) Substituted trans-stilbenes can inhibit or enhance the TPA-induced up-regulation of activator protein-1. BMC Pharmacol 8:19
Desideri N, Conti C, Mastromarino P, Mastropaolo F (2000) Synthesis and anti-rhinovirus activity of 2-styrylchromones. Antivir Chem Chemother 11(6):373–381
Diaz-Padilla I, Siu LL, Duran I (2009) Cyclin-dependent kinase inhibitors as potential targeted anticancer agents. Invest New Drug 27(6):586–594
Dixon JB, Dixon ME, O'Brien PE (2002) Reduced plasma homocysteine in obese red wine consumers: a potential contributor to reduced cardiovascular risk status. Eur J Clin Nutr 56(7):608–614
Dominguez-Perles R, Moreno DA, Carvajal M, Garcia-Viguera C (2011) Composition and antioxidant capacity of a novel beverage produced with green tea and minimally-processed byproducts of broccoli. Innov Food Sci Emerg 12(3):361–368
Dominguez-Perles R, Moreno DA, Garcia-Viguera C (2012) Analysis of the tumoral cytotoxicity of green tea-infusions enriched with broccoli. Food Chem 132(3):1197–1206

Dore S (2005) Unique properties of polyphenol stilbenes in the brain: More than direct antioxidant actions; Gene/protein regulatory activity. Neurosignals 14(1–2):61–70

Ducimetière P (2008) Coronary heart disease in France and in Europe: Where are the facts? Dialogues Cardiovasc Med 13(3):193–199

Eddarir S, Abdelhadi Z, Rolando C (2001) Fluorinated resveratrol and pterostilbene. Tetrahedron Lett 42(52):9127–9130

Etoh H, Murakami K, Yogoh T, Ishikawa H, Fukuyama Y, Tanaka H (2004) Anti-oxidative compounds in barley tea. Biosci Biotechnol Biochem 68(12):2616–2618

Fale PLV, Ascensao L, Serralheiro MLM, Haris PI (2011) Interaction between *Plectranthus barbatus* herbal tea components and human serum albumin and lysozyme: binding and activity studies. Spectrosc Biomed Appl 26(2):79–92

Fan GJ, Liu XD, Qian YP, Shang YJ, Li XZ, Dai F, Fang JG, Jin XL, Zhou B (2009) 4,4′-Dihydroxy-trans-stilbene, a resveratrol analogue, exhibited enhanced antioxidant activity and cytotoxicity. Bioorgan Med Chem 17(6):2360–2365

Feher J, Lengyel G, Lugasi A (2007) The cultural history of wine—Theoretical background to wine therapy. Cent Eur J Med 2(4):379–391

Fernandez-Panchon MS, Villano D, Troncoso AM, Garcia-Parrilla MC (2008) Antioxidant activity of phenolic compounds: from in vitro results to in vivo evidence. Crit Rev Food Sci 48(7):649–671

Ferruzzi MG (2010) The influence of beverage composition on delivery of phenolic compounds from coffee and tea. Physiol Behav 100(1):33–41

Filipe P, Silva AMS, Seixas RSGR, Pinto DCGA, Santos A, Patterson LK, Silva JN, Cavaleiro JAS, Freitas JP, Maziere JC, Santus R, Morliere P (2009) The alkyl chain length of 3-alkyl-3′,4′,5,7-tetrahydroxyflavones modulates effective inhibition of oxidative damage in biological systems: Illustration with LDL, red blood cells and human skin keratinocytes. Biochem Pharmacol 77(6):957–964

Frombaum M, Le Clanche S, Bonnefont-Rousselot D, Borderie D (2012) Antioxidant effects of resveratrol and other stilbene derivatives on oxidative stress and center dot NO bioavailability: potential benefits to cardiovascular diseases. Biochimie 94(2):269–276

Galati G, Lin A, Sultan AM, O'Brien PJ (2006) Cellular and in vivo hepatotoxicity caused by green tea phenolic acids and catechins. Free Radical Biol Med 40(4):570–580

Galati G, O'Brien PJ (2004) Potential toxicity of flavonoids and other dietary phenolics: significance for their chemopreventive and anticancer properties. Free Radical Biol Med 37(3):287–303

Gao DF, Zhang YJ, Yang CR, Chen KK, Jiang HJ (2008) Phenolic antioxidants from green tea produced from *Camellia taliensis*. J Agric Food Chem 56(16):7517–7521

Garcia-Alonso M, Minihane AM, Rimbach G, Rivas-Gonzalo JC, de Pascual-Teresa S (2009) Red wine anthocyanins are rapidly absorbed in humans and affect monocyte chemoattractant protein 1 levels and antioxidant capacity of plasma. J Nutr Biochem 20(7):521–529

Gerhaeuser C (2001) Flavonoide und andere pflanzliche Wirkstoffe. Aktuelle Ernahrungsmedizin 26(4):137–143

Gerwick WH (1989) 6-Desmethoxyhormothamnione, a New Cyto-Toxic Styrylchromone from the Marine Cryptophyte Chrysophaeum-Taylori. J Nat Prod 52(2):252–256

Gerwick WH, Lopez A, Vanduyne GD, Clardy J, Ortiz W, Baez A (1986) Hormothamnione, a Novel Cytotoxic Styrlchromone from the Marine Cyanophyte Hormothamnion-Enteromorphoides Grunow. Tetrahedron Lett 27(18):1979–1982

Goldfinger TM (2003) Beyond the French paradox: the impact of moderate beverage alcohol and wine consumption in the prevention of cardiovascular disease. Cardiol Clin 21(3):449–457

Gomes A, Fernandes E, Garcia MBQ, Silva AMS, Pinto DCGA, Santos CMM, Cavaleiro JAS, Lima JLFC (2008) Cyclic voltammetric analysis of 2-styrylchromones: relationship with the antioxidant activity. Bioorgan Med Chem 16(17):7939–7943

Gomes A, Fernandes E, Silva AMS, Pinto DCGA, Santos CMM, Cavaleiro JAS, Lima JLFC (2009a) Anti-inflammatory potential of 2-styrylchromones regarding their interference with arachidonic acid metabolic pathways. Biochem Pharmacol 78(2):171–177

Gomes A, Fernandes E, Silva AMS, Santos CMM, Pinto DCGA, Cavaleiro JAS, Lima JLFC (2007) 2-Styrylchromones: novel strong scavengers of reactive oxygen and nitrogen species. Bioorgan Med Chem 15(18):6027–6036

Gomes A, Freitas M, Fernandes E, Lima JLFC (2010) Biological activities of 2-Styrylchromones. Mini-Rev Med Chem 10(1):1–7

Gomes A, Neuwirth O, Freitas M, Couto D, Ribeiro D, Figueiredo AGPR, Silva AMS, Seixas RSGR, Pinto DCGA, Tome AC, Cavaleiro JAS, Fernandes E, Lima JLFC (2009b) Synthesis and antioxidant properties of new chromone derivatives. Bioorgan Med Chem 17(20):7218–7226

Gross G, Jacobs DM, Peters S, Possemiers S, van Duynhoven J, Vaughan EE, van de Wiele T (2010) In vitro bioconversion of polyphenols from Black Tea and Red Wine/Grape Juice by human intestinal microbiota displays strong interindividual variability. J Agric Food Chem 58(18):10236–10246

Guiso M, Marra C, Farina A (2002) A new efficient resveratrol synthesis. Tetrahedron Lett 43(4):597–598

Gulacsi K, Litkei G, Antus S, Gunda TE (1998) A short and facile synthetic route to prenylated flavones. Cyclodehydrogenation of prenylated 2′-hydroxychalcones by a hypervalent iodine reagent. Tetrahedron 54(45):13867–13876

Gupta M, Paul S, Gupta R, Loupy A (2000) A rapid method for the cyclization of 2′-hydroxychalcones into flavones. Org Prep Proced Int 32(3):280–283

Gusman J, Malonne H, Atassi G (2001) A reappraisal of the potential chemopreventive and chemotherapeutic properties of resveratrol. Carcinogenesis 22(8):1111–1117

Hallek M, Pflug N (2011) State of the art treatment of chronic lymphocytic leukaemia. Blood Rev 25(1):1–9

Han DW, Park YH, Kim JK, Jung TG, Lee KY, Hyon SH, Park JC (2005) Long-term preservation of human saphenous vein by green tea polyphenol under physiological conditions. Tissue Eng 11(7–8):1054–1064

Han SY, Bang HB, Lee HS, Hwang JW, Choi DH, Yang DM, Jun JG (2008) A new synthesis of stilbene natural product piceatannol. B Korean Chem Soc 29(9):1800–1802

Hans N, Grover SK (1993) An efficient conversion of 2′-Hydroxychalcones to flavones. Synthetic Commun 23(7):1021–1023

Harborne JB, Grayer RJ (1993) The flavonoids, advances in research since 1986. Chapman and Hall, London

Hauss F, Liu JW, Michelucci A, Coowar D, Morga E, Heuschling P, Luu B (2007) Dual bioactivity of resveratrol fatty alcohols: Differentiation of neural stem cells and modulation of neuroinflammation. Bioorg Med Chem Lett 17(15):4218–4222

Henning SM, Wang PW, Heber D (2011) Chemopreventive effects of tea in prostate cancer: Green tea versus black tea. Mol Nutr Food Res 55(6):905–920

Hermes-Lima M (2005) Oxygen in biology and biochemistry: role of free radicals. Wiley-Liss, New York

Hilt G, Hengst C (2007) A concise synthesis of substituted Stilbenes and Styrenes from propargylic phosphonium salts by a cobalt-catalyzed Diels-Alder/Wittig olefination reaction sequence. J Org Chem 72(19):7337–7342

Ho SC, Wu SP, Lin SM, Tang YL (2010) Comparison of anti-glycation capacities of several herbal infusions with that of green tea. Food Chem 122(3):768–774

Holdsworth M (2008) How important are differences in national eating habits in France? Dialogues Cardiovasc Med 13(3):200–208

Hoshino Y, Oohinata T, Takeno N (1986) The direct preparation of flavones from 2′-Hydroxychalcones using disulfides. B Chem Soc Jpn 59(7):2351–2352

Howard A, Chopra M, Thurnham DI, Strain JJ, Fuhrman B, Aviram M (2002) Red wine consumption and inhibition of LDL oxidation: what are the important components? Med Hypotheses 59(1):101–104

Howitz KT, Bitterman KJ, Cohen HY, Lamming DW, Lavu S, Wood JG, Zipkin RE, Chung P, Kisielewski A, Zhang LL, Scherer B, Sinclair DA (2003) Small molecule activators of sirtuins extend *Saccharomyces cerevisiae* lifespan. Nature 425(6954):191–196

Hu FB (2005) Overweight and increased cardiovascular mortality—No French paradox. Hypertension 46(4):645–646
Hu Y, Liu J, Wang JF, Liu QS (2011) The controversial links among calorie restriction, SIRT1, and resveratrol. Free Radical Bio Med 51(2):250–256
Iijima K, Yoshizumi M, Hashimoto M, Kim S, Eto M, Ako J, Liang YQ, Sudoh N, Hosoda K, Nakahara K, Toba K, Ouchi Y (2000) Red wine polyphenols inhibit proliferation of vascular smooth muscle cells and downregulate expression of cyclin A gene. Circulation 101(7):805–811
Iijima K, Yoshizumi M, Ouchi Y (2002) Effect of red wine polyphenols on vascular smooth muscle cell function-molecular mechanism of the 'French paradox'. Mech Ageing Dev 123(8):1033–1039
Ishii T, Mori T, Ichikawa T, Kaku M, Kusaka K, Uekusa Y, Akagawa M, Aihara Y, Furuta T, Wakimoto T, Kan T, Nakayama T (2010) Structural characteristics of green tea catechins for formation of protein carbonyl in human serum albumin. Bioorgan Med Chem 18(14):4892–4896
Jayasekera S, Molan AL, Garg M, Moughan PJ (2011) Variation in antioxidant potential and total polyphenol content of fresh and fully-fermented Sri Lankan tea. Food Chem 125(2):536–541
Julia M, Paris JM (1973) Syntheses using sulfones. 5. Method for general synthesis of doubles. Tetrahedron Lett 49:4833–4836
Kaack K, Christensen LP (2010) Phenolic acids and flavonoids in tea processed from flowers of black elder (*Sambucus nigra* L.) stored in different packing materials. Eur J Hortic Sci 75(5):214–220
Kappagoda CT, Karim M, McCormick K, Kandaswami C (2000) Unraveling the French paradox. Chem Innov 30(9):26–31
Kapsokefalou M, Zhu L, Miller DD (2006) Adding iron to green tea may decrease its antioxidant capacity in rats after an oral dose of the mixture. Nutr Res 26(9):480–485
Karori SM, Wachira FN, Wanyoko JK, Ngure RM (2007) Antioxidant capacity of different types of tea products. Afr J Biotechnol 6(19):2287–2296
Kodama DH, Goncalves AEDS, Lajolo FM, Genovese MI (2010) Flavonoids, total phenolics and antioxidant capacity: comparison between commercial green tea preparations. Ciencia Tecnol Alime 30(4):1077–1082
Komes D, Horzic D, Belscak A, Ganic KK, Vulic I (2010) Green tea preparation and its influence on the content of bioactive compounds. Food Res Int 43(1):167–176
Kosmider B, Osiecka R (2004) Flavonoid compounds: a review of anticancer properties and interactions with cis-diamminedichloroplatinum(II). Drug Develop Res 63(4):200–211
Kulisic-Bilusic T, Katalinic V, Dragovic-Uzelac V, Ljubenkov I, Krisko A, Dejanovic B, Jukic M, Politeo O, Pifat G, Milos M (2008) Antioxidant and acetylcholinesterase inhibiting activity of several aqueous tea infusions in vitro. Food Technol Biotechnol 46(4):368–375
Kwon YI, Apostolidis E, Shetty K (2008) Inhibitory potential of wine and tea against α-amylase and α-glucosidase for management of hyperglycemia linked to type 2 diabetes. J Food Biochem 32(1):15–31
Lee HC, Jenner AM, Low CS, Lee YK (2006) Effect of tea phenolics and their aromatic fecal bacterial metabolites on intestinal microbiota. Res Microbiol 157(9):876–884
Lee KW, Kim YJ, Lee HJ, Lee CY (2003) Cocoa has more phenolic phytochemicals and a higher antioxidant capacity than teas and red wine. J Agric Food Chem 51(25):7292–7295
Lee SC, Jeong SM, Lee JM, Jang A, Kim DH, Jo C (2008) Effect of irradiation on total phenol and catechins contents and radical scavenging activity of green tea leaf and stem extract. J Food Biochem 32(6):782–794
Lekli I, Ray D, Das DK (2010) Longevity nutrients resveratrol, wines and grapes. Genes Nutr 5(1):55–60
Li F, Awale S, Tezuka Y, Esmi H, Kadota S (2010) Study on the constituents of Mexican propolis and their cytotoxic activity against PANC-1 human pancreatic cancer cells. J Nat Prod 73(4):623–627

Li JH, Chang RL, Wu SC (2009) The effects of antioxidative activity and human hepatoma cells of Chinese herbal teas. Taiwanese J Agric Chem Food Sci 47(3):145–154

Likhtenshtein G (2010) Stilbenes, applications in chemistry, life sciences and materials science. Wiley-VCH Verlag GmbH & Co, KGaA, Germany

Lin YP, Hsu FL, Chen CS, Chern JW, Lee MH (2007) Constituents from the Formosan apple reduce tyrosinase activity in human epidermal melanocytes. Phytochemistry 68(8):1189–1199

Lion CJ, Matthews CS, Stevens MFG, Westwell AD (2005) Synthesis, antitumor evaluation, and apoptosis-inducing activity of hydroxylated (E)-stilbenes. J Med Chem 48(4):1292–1295

Lippi G, Franchini M, Favaloro EJ, Targher G (2010) Moderate red wine consumption and cardiovascular disease risk: beyond the "French Paradox". Semin Thromb Hemost 36(1):59–70

Liu BL, Zhang X, Zhang W, Zhen HN (2007) New enlightenment of French Paradox: resveratrol's potential for cancer chemoprevention and anti-cancer therapy. Cancer Biol Ther 6(12):1833–1836

Liu HC, Dong AJ, Gao CM, Tan CY, Liu HX, Zu XY, Jiang YY (2008a) The design, synthesis, and anti-tumor mechanism study of N-phosphoryl amino acid modified resveratrol analogues. Bioorgan Med Chem 16(23):10013–10021

Liu HG, Gu ZL, Xu LH, Hsu S (2008b) Protective effect of green tea polyphenols on tributyltin-induced oxidative damage detected by in vivo and in vitro models. Environ Toxicol 23(1):77–83

Liu Q, Zhang YJ, Yang CR, Xu M (2009) Phenolic antioxidants from Green Tea Produced from Camellia crassicolumna Var. multiplex. J Agric Food Chem 57(2):586–590

Logan BA, Hammond MP, Stormo BM (2008) The French paradox—Determining the superoxide-scavenging capacity of red wine and other beverages. Biochem Mol Biol Edu 36(1):39–42

Loke WM, Proudfoot JM, Hodgson JM, McKinley AJ, Hime N, Magat M, Stocker R, Croft KD (2010) Specific dietary polyphenols attenuate atherosclerosis in apolipoprotein E-Knockout Mice by alleviating inflammation and endothelial dysfunction. Arterioscl Throm Vas 30(4):749–757 (U 243)

Lorenz M, Kabir MS, Cook JM (2010) A two step synthesis of BzR/GABAergic active flavones via a Wacker-related oxidation. Tetrahedron Lett 51(7):1095–1098

Maeta K, Nomura W, Takatsume Y, Izawa S, Inoue Y (2007) Green tea polyphenols function as prooxidants to activate oxidative-stress-responsive transcription factors in yeasts. Appl Environ Microb 73(2):572–580

Mahal HS, Venkataraman K (1934) Synthetical experiments in the chromone group. Part XIV. The action of sodamide on 1-acyloxy-2-acetonaphthones. J Chem Soc 1767–1769

Marinho J, Pedro M, Pinto DCGA, Silva AMS, Cavaleiro JAS, Sunkel CE, Nascimento MSJ (2008) 4′-Methoxy-2-styrylchromone a novel microtubule-stabilizing antimitotic agent. Biochem Pharmacol 75(4):826–835

Martino VS (2000) Los flavonoides como promisorios agentes preventivos y terapéuticos. Acta Farmacéutica Bonaerense 19(4):303–308

McMurry JE, Fleming MP (1974) A new method for the reductive coupling of carbonyls to olefins. Synthesis of beta-carotene. J Am Chem Soc 96(14):4708–4709

McDougall GJ, Fyffe S, Dobson P, Stewart D (2007) Anthocyanins from red cabbage—Stability to simulated gastrointestinal digestion. Phytochemistry 68(9):1285–1294

McKay DL, Blumberg JB (2006) A review of the bioactivity and potential health benefits of chamomile tea (*Matricaria recutita* L.). Phytother Res 20(7):519–530

Morton LW, Caccetta RA, Puddey IB, Croft KD (2000) Chemistry and biological effects of dietary phenolic compounds: relevance to cardiovascular disease. Clin Exp Pharmacol Physiol 27(3):152–159

Moscatelli V, Hnatyszyn O, Acevedo C, Megias J, Alcaraz MJ, Ferraro G (2006) Flavonoids from Artemisia copa with anti-inflammatory activity. Planta Med 72(1):72–74

Mudry J (2010) The Poison is in the dose the french paradox, the healthy drinker and the medicalization of virtue. Food Cult Soc 13(1):91–114

Mulder TP, Rietveld AG, van Amelsvoort JM (2005) Consumption of both black tea and green tea results in an increase in the excretion of hippuric acid into urine. Am J Clin Nutr 81(1):256s–260s

Murthi KK, Dubay M, McClure C, Brizuela L, Boisclair MD, Worland PJ, Mansuri MM, Pal K (2000) Structure-activity relationship studies of flavopiridol analogues. Bioorg Med Chem Lett 10(10):1037–1041

Nagaoka T, Banskota AH, Tezuka Y, Midorikawa K, Matsushige K, Kadota S (2003) Caffeic acid phenethyl ester (CAPE) analogues: potent nitric oxide inhibitors from the Netherlands propolis. Biol Pharm Bull 26(4):487–491

Negukhula S, Mudau FN, Mariga IK, Liphadzi KB (2011) Effects of soaking conditions on total phenolic and antioxidant activity of black tea and black tea combined (50:50) with bush tea. Indian J Hortic 68(1):91–95

Neyestani TR, Gharavi A, Kalayi A (2009) Selective effects of tea extract and its phenolic compounds on human peripheral blood mononuclear cell cytokine secretions. Int J Food Sci Nutr 60:79–88

Neyestani TR, Khalaji N, Gharavi A (2007) Black and green teas may have selective synergistic or antagonistic effects on certain antibiotics against *Streptococcus pyogenes* in vitro. J Nutr Environ Med 16(3–4):258–266

Nikolova V (2007) Resveratrol: a crossroad of enology and biomedicine. Biomed Rev 18:89–101

Nishiyama MF, Costa MAF, da Costa AM, de Souza CGM, Boer CG, Bracht CK, Peralta RM (2010) Brazilian green tea (*Camellia sinensis* var assamica): effect of infusion time, mode of packaging and preparation on the extraction efficiency of bioactive compounds and on the stability of the beverage. Ciencia Tecnol Alime 30:191–196

Nishizuka T, Fujita Y, Sato Y, Nakano A, Kakino A, Ohshima S, Kanda T, Yoshimoto R, Sawamura T (2011) Procyanidins are potent inhibitors of LOX-1: a new player in the French Paradox. Proc Jpn Acad B-Phys 87(7):104, 431

Nuengchamnong N, Krittasilp K, Ingkaninan K (2011) Characterisation of phenolic antioxidants in aqueous extract of Orthosiphon grandiflorus tea by LC-ESI-MS/MS coupled to DPPH assay. Food Chem 127(3):1287–1293

Olech M, Nowak R, Los R, Rzymowska J, Malm A, Chrusciel K (2012) Biological activity and composition of teas and tinctures prepared from *Rosa rugosa* Thunb. Cent Eur J Biol 7(1):172–182

Omar MH, Mullen W, Crozier A (2011) Identification of proanthocyanidin dimers and trimers, Flavone C-Glycosides, and antioxidants in ficus deltoidea, a Malaysian herbal tea. J Agric Food Chem 59(4):1363–1369

Opie LH, Lecour S (2007) The red wine hypothesis: from concepts to protective signalling molecules. Eur Heart J 28(14):1683–1693

Orak HH (2007) Total antioxidant activities, phenolics, anthocyanins, polyphenoloxidase activities of selected red grape cultivars and their correlations. Sci Hortic-Amsterdam 111(3):235–241

Park KY, Lee SH, Min BK, Lee KS, Choi JS, Chung SR, Min KR, Kim Y (1999) Inhibitory effect of luteolin 4′-O-glucoside from Kummerowia striata and other flavonoids on interleukin-5 bioactivity. Planta Med 65(5):457–459

Patonay T, Cavaleiro JAS, Levai A, Silva AMS (1997) Dehydrogenation by iodine-dimethylsulfoxide system: a general route to substituted chromones and thiochromones. Heterocycl Commun 3(3):223–229

Perez-Magarino S, Gonzalez-San Jose ML (2006) Polyphenols and colour variability of red wines made from grapes harvested at different ripeness grade. Food Chem 96(2):197–208

Petralia S, Spatafora C, Tringali C, Foti MC, Sortino S (2004) Hydrogen atom abstraction from resveratrol and two lipophilic derivatives by tert-butoxyl radicals. A laser flash photolysis study. New J Chem 28 (12):1484-1487

Pettit GR, Grealish MP, Jung MK, Hamel E, Pettit RK, Chapuis JC, Schmidt JM (2002) Antineoplastic agents. 465. Structural modification of resveratrol: Sodium resverastatin phosphate. J Med Chem 45(12):2534–2542

Pettit GR, Singh SB, Boyd MR, Hamel E, Pettit RK, Schmidt JM, Hogan F (1995) Antineoplastic Agents. 291. Isolation and Synthesis of Combretastatins a-4, a-5, and a-6. J Med Chem 38(15):1666, 2994

Pinto DCGA, Silva AMS, Almeida LMPM, Cavaleiro JAS, Levai A, Patonay T (1998) Synthesis of 4-aryl-3-(2-chromonyl)-2-pyrazolines by the 1,3-dipolar cycloaddition of 2-styrylchromones with diazomethane. J Heterocyclic Chem 35(1):217–224

Pinto DCGA, Silva AMS, Cavaleiro JAS (2000a) A convenient synthesis of new (E)-5-hydroxy-2-styrylchromones by modifications of the Baker-Venkataraman method. New J Chem 24(2):85–92

Pinto DCGA, Silva AMS, Cavaleiro JAS (2000b) Synthesis of 3-(2-benzyloxy-6-hydroxyphenyl)-1-methylpyrazoles by the reaction of chromones with methylhydrazine. J Heterocyclic Chem 37(6):1629–1634

Pinto DCGA, Silva AMS, Cavaleiro JAS, Foces-Foces C, Llamas-Saiz AL, Jagerovic N, Elguero J (1999) Synthesis and molecular structure of 3-(2-benzyloxy-6-hydroxyphenyl)-5-styrylpyrazoles. Reaction of 2-styrylchromones and hydrazine hydrate. Tetrahedron 55(33):10187–10200

Pinzani P, Petruzzi E, Magnolfi SU, Malentacchi F, De Siena G, Petruzzi I, Motta M, Malaguarnera M, Marchionni N, Pazzagli M (2010) Red or white wine assumption and serum antioxidant capacity. Arch Gerontol Geriatr 51(3):e72–e74

Pithayanukul P, Leanpolchareanchai J, Bavovada R (2010) Inhibitory effect of tea polyphenols on local tissue damage induced by snake venoms. Phytother Res 24:S56–S62

Poussier B, Cordova AC, Becquemin JP, Sumpio BE (2005) Resveratrol inhibits vascular smooth muscle cell proliferation and induces apoptosis. J Vasc Surg 42(6):1190–1197

Price WA, Silva AMS, Cavaleiro JAS (1993) 2-Styrylchromones—Biological action, synthesis and reactivity. Heterocycles 36(11):2601–2612

Pripdeevech P, Machan T (2011) Fingerprint of volatile flavour constituents and antioxidant activities of teas from Thailand. Food Chem 125(2):797–802

Providencia R (2006) Cardiovascular protection from alcoholic drinks: scientific basis of the French Paradox. Revista portuguesa de cardiologia : orgao oficial da Sociedade Portuguesa de Cardiologia = Portuguese journal of cardiology : an official journal of the Portuguese Society of Cardiology 25(11):1043–1058

Rabai M, Toth A, Kenyeres P, Mark L, Marton Z, Juricskay I, Toth K, Czopf L (2010) In vitro hemorheological effects of red wine and alcohol-free red wine extract. Clin Hemorheol Microcirc 44(3):227–236

Raghava GS (1993) Inhibition of LDL oxidation by phenolic substances in red wine: a clue to the French paradox?. Nutr Rev 51(6):185–187

Rechner AR, Wagner E, Van Buren L, Van de Put F, Wiseman S, Rice-Evans CA (2002) Black tea represents a major source of dietary phenolics among regular tea drinkers. Free Radical Res 36(10):1127–1135

Rele SM, Nayak SK, Chattopadhyay S (2008) Salt/ligand-activated low-valent titanium formulations: the 'salt effect' on diastereoselective carbon-carbon bond forming SET reactions. Tetrahedron 64(30–31):7225–7233

Rendig SV, Symons JD, Longhurst JC, Amsterdam EA (2001) Effects of red wine, alcohol, and quercetin on coronary resistance and conductance arteries. J Cardiovasc Pharm 38(2):219–227

Rivero-Perez MD, Muniz P, Gonzalez-Sanjose ML (2008) Contribution of anthocyanin fraction to the antioxidant properties of wine. Food Chem Toxicol 46(8):2815–2822

Rocha-Pereira J, Cunha R, Pinto DCGA, Silva AMS, Nascimento MS (2010) (E)-2-Styrylchromones as potential anti-norovirus agents. Bioorgan Med Chem 18(12):4195–4201

Roowi S, Stalmach A, Mullen W, Lean MEJ, Edwards CA, Crozier A (2010) Green Tea Flavan-3-ols: colonic degradation and urinary excretion of catabolites by humans. J Agric Food Chem 58(2):1296–1304

Rosenkranz S, Knirel D, Dietrich H, Flesch M, Erdmann E, Bohm M (2002) Inhibition of the PDGF receptor by red wine flavonoids provides a molecular explanation for the "French paradox". Faseb J 16(12):1958–+

Rusak G, Komes D, Likic S, Horzic D, Kovac M (2008) Phenolic content and antioxidative capacity of green and white tea extracts depending on extraction conditions and the solvent used. Food Chem 110(4):852–858

Salucci M, Stivala LA, Maiani G, Bugianesi R, Vannini V (2002) Flavonoids uptake and their effect on cell cycle of human colon adenocarcinoma cells (Caco2). Brit J Cancer 86(10):1645–1651

Serafini M (2004) Overview of polyphenols. Agro Food Ind Hi Tec 15(2):10–11

Shrikhande AJ (2000) Wine by-products with health benefits. Food Res Int 33(6):469–474

Silva AMS, Pinto DCGA, Cavaleiro JAS (1994) 5-Hydroxy-2-(Phenyl or Styryl) Chromones—One-Pot Synthesis and C-6, C-8 C-13 Nmr assignments. Tetrahedron Lett 35(32):5899–5902

Silva AMS, Pinto DCGA, Cavaleiro JAS, Levai A, Patonay T (2004) Synthesis and reactivity of styrylchromones. Arkivoc 106–123

Silva AMS, Pinto DCGA, Tavares HR, Cavaleiro JAS, Jimeno ML, Elguero J (1998) Novel (E)- and (Z)-2-styrylchromones from (E, E)-2′-hydroxycinnamylideneacetophenones—Xanthones from daylight photooxidative cyclization of (E)-2-styrylchromones. Eur J Org Chem 9:2031–2038

Silva AMS, Tavares HR, Barros AINRA, Cavaleiro JAS (1997) NMR and structural and conformational features of 2′-hydroxychalcones and flavones. Spectrosc Lett 30(8):1655–1667

Snyder SA, Gollner A, Chiriac MI (2011) Regioselective reactions for programmable resveratrol oligomer synthesis. Nature 474(7352):461–466

Song LL, Liang R, Li DD, Xing YD, Han RM, Zhang JP, Skibsted LH (2011) Beta-carotene radical cation addition to green tea polyphenols. Mechanism of antioxidant antagonism in peroxidizing liposomes. J Agric Food Chem 59(23):12643–12651

Sparwel J, Vantler M, Caglayan E, Kappert K, Fries JW, Dietrich H, Bohm M, Erdmann E, Rosenkranz S (2009) Differential effects of red and white wines on inhibition of the platelet-derived growth factor receptor: impact of the mash fermentation. Cardiovasc Res 81(4):758–770

Su XG, Duan J, Jiang YM, Duan XW, Chen F (2007) Polyphenolic profile and antioxidant activities of oolong tea infusion under various steeping conditions. Int J Mol Sci 8(12):1196–1205

Sun AY, Simonyi A, Sun GY (2002) The "French paradox" and beyond: neuroprotective effects of polyphenols. Free Radical Biol Med 32(4):314–318

Szewczuk LM, Penning TM (2004) Mechanism-based inactivation of COX-1 by red wine m-hydroquinones: a structure-activity relationship study. J Nat Prod 67(11):1777–1782

Szmitko PE, Verma S (2005) Antiatherogenic potential of red wine: clinician update. Am J Physiol-Heart C 288(5):H2023–H2030

Teillet F, Boumendjel A, Boutonnat J, Ronot X (2008) Flavonoids as RTK inhibitors and potential anticancer agents. Med Res Rev 28(5):715–745

Terasawa N, Yamazaki N (2002) Radical scavenging activity of Japanese black tea, Toyama Kurocha. Food Sci Technol Res 8(3):218–220

Tereschuk ML, Baigori MD, De Figueroa LI, Abdala LR (2004) Flavonoids from argentine tagetes (Asteraceae) with antimicrobial activity. Methods Mol Biol 268:317–330

Thakkar K, Geahlen RL, Cushman M (1993) Synthesis and protein-tyrosine kinase inhibitory activity of polyhydroxylated stilbene analogs of piceatannol. J Med Chem 36(20):2950–2955

Thuong PT, Su ND, Ngoc TM, Hung TM, Dang NH, Thuan ND, Bae K, Oh WK (2009) Antioxidant activity and principles of Vietnam bitter tea Ilex kudingcha. Food Chem 113(1):139–145

Tsai PJ, Tsai TH, Yu CH, Ho SC (2007) Comparison of NO-scavenging and NO-suppressing activities of different herbal teas with those of green tea. Food Chem 103(1):181–187

Tsai TH, Tsai TH, Chien YC, Lee CW, Tsai PJ (2008) In vitro antimicrobial activities against cariogenic streptococci and their antioxidant capacities: a comparative study of green tea versus different herbs. Food Chem 110(4):859–864

Tunstall-Pedoe H (2008) The French paradox. Dialogues Cardiovasc Med 13(3):159–182

Unachukwu UJ, Ahmed S, Kavalier A, Lyles JT, Kennelly EJ (2010) White and green teas (*Camellia sinensis* var. sinensis): variation in phenolic, methylxanthine, and antioxidant profiles. J Food Sci 75(6):C541–C548

Valko M, Leibfritz D, Moncol J, Cronin MTD, Mazur M, Telser J (2007) Free radicals and antioxidants in normal physiological functions and human disease. Int J Biochem Cell B 39(1):44–84

van de Wiel A, de Lange DW (2008) Cardiovascular risk is more related to drinking pattern than to the type of alcoholic drinks. Neth J Med 66(11):467–473

Vaya J, Mahmood S, Goldblum A, Aviram M, Volkova N, Shaalan A, Musa R, Tamir S (2003) Inhibition of LDL oxidation by flavonoids in relation to their structure and calculated enthalpy. Phytochemistry 62(1):89–99

Vedejs E, Cabaj J, Peterson MJ (1993) Wittig ethylidenation of Ketones—Reagent control of Z/E selectivity. J Org Chem 58(23):6509–6512

Verma AK, Pratap R (2010) The biological potential of flavones. Nat Prod Rep 27(11):1571–1593

Vidavalur R, Otani H, Singal PK, Maulik N (2006) Significance of wine and resveratrol in cardiovascular disease: French paradox revisited. Exp Clin Cardiol 11(3):217–225

Vinson JA, Teufel K, Wu N (2001) Red wine, dealcoholized red wine, and especially grape juice, inhibit atherosclerosis in a hamster model. Atherosclerosis 156(1):67–72

Visioli F, Borsani L, Galli C (2000) Diet and prevention of coronary heart disease: the potential role of phytochemicals. Cardiovasc Res 47(3):419–425

Viveros-Valdez E, Rivas-Morales C, Carranza-Rosales P, Mendoza S, Schmeda-Hirschmann G (2008) Free radical scavengers from the Mexican herbal tea "Poleo" (*Hedeoma drummondii*). Zeitschrift Fur Naturforschung Section C-a Journal of Biosciences 63(5–6):341–346

Wang HF, Tsai YS, Lin ML, Ou ASM (2006) Comparison of bioactive components in GABA tea and green tea produced in Taiwan. Food Chem 96(4):648–653

Wang MF, Jin Y, Ho CT (1999) Evaluation of resveratrol derivatives as potential antioxidants and identification of a reaction product of resveratrol and 2,2-diphenyl-1-picryhydrazyl radical. J Agric Food Chem 47(10):3974–3977

Wang Y, Ho CT (2009) Polyphenols chemistry of tea and coffee: a century of progress. J Agric Food Chem 57(18):8109–8114

Williamson G, Dionisi F, Renouf M (2011) Flavanols from green tea and phenolic acids from coffee: critical quantitative evaluation of the pharmacokinetic data in humans after consumption of single doses of beverages. Mol Nutr Food Res 55(6):864–873

Wollgast J, Anklam E (2000) Polyphenols in chocolate: is there a contribution to human health? Food Res Int 33(6):449–459

Wood JG, Rogina B, Lavu S, Howitz K, Helfand SL, Tatar M, Sinclair D (2004) Sirtuin activators mimic caloric restriction and delay ageing in metazoans. Nature 431(7004):107, 430:686

Wu CY, Xu HR, Heritier J, Andlauer W (2012) Determination of catechins and flavonol glycosides in Chinese tea varieties. Food Chem 132(1):144–149

Wu JJA, Chiang MT, Chang YW, Chen JY, Yang HT, Lii CK, Lin JH, Yao HT (2011) Correlation of major components and radical scavenging activity of commercial tea drinks in Taiwan. J Food Drug Anal 19(3):289–300

Wu LC, Jou AFJ, Chen SH, Tien CY, Cheng CF, Fan NC, Ho JAA (2010) Antioxidant, anti-inflammatory and anti-browning activities of hot water extracts of oriental herbal teas. Food Funct 1(2):200–208

Xi DY, Liu CC, Su YC (2012) Effects of green tea extract on reducing Vibrio parahaemolyticus and increasing shelf life of oyster meats. Food Control 25(1):368–373

Yang L, Qiao LR, Xie D, Yuan YH, Chen NH, Dai JG, Guo SX (2012) 2-(2-Phenylethyl)chromones from Chinese eaglewood. Phytochemistry 76:92–97

Yao LH, Jiang YM, Caffin N, D'Arcy B, Datta N, Liu X, Singanusong R, Xu Y (2006) Phenolic compounds in tea from Australian supermarkets. Food Chem 96(4):614–620

Yoo YJ, Saliba AJ, Prenzler PD (2010) Should red wine be considered a functional food? Compr Rev Food Sci F 9(5):530–551

Yoon JS, Lee MK, Sung SH, Kim YC (2006) Neuroprotective 2-(2-phenylethyl)chromones of *Imperata cylindrica*. J Nat Prod 69(2):290–291
Zambare AS, Sangshettia JN, Kokare ND, Shinde DB (2009) Development of mild and efficient method for synthesis of substituted flavones using oxalic acid catalyst. Chinese Chem Lett 20(2):171–174
Zern TL, Fernandez ML (2005) Cardioprotective effects of dietary polyphenols. J Nutr 135(10):2291–2294
Zhou B, Liu ZL (2005) Bioantioxidants: from chemistry to biology. Pure Appl Chem 77(11):1887–1903
Zhu JTT, Choi RCY, Chu GKY, Cheung AWH, Gao QT, Li J, Jiang ZY, Dong TTX, Tsim KWK (2007) Flavonoids possess neuroprotective effects on cultured pheochromocytoma PC12 cells: a comparison of different flavonoids in activating estrogenic effect and in preventing beta-amyloid-induced cell death. J Agric Food Chem 55(6):2438–2445
Zhu Q, Liang CP, Cheng KW, Peng XF, Lo CY, Shahidi F, Chen F, Ho CT, Wang MF (2009) Trapping effects of green and black tea extracts on peroxidation-derived carbonyl substances of seal blubber oil. J Agric Food Chem 57(3):1065–1069
Zhu QY, Hackman RM, Ensunsa JL, Holt RR, Keen CL (2002) Antioxidative activities of oolong tea. J Agric Food Chem 50(23):6929–6934
Zhuang H, Kim YS, Koehler RC, Dore S (2003) Potential mechanism by which resveratrol, a red wine constituent, protects neurons. Ann NY Acad Sci 993:276–286
Zilkens RR, Puddey IB (2003) Alcohol and type 2 diabetes—Another paradox? J Cardiovasc Risk 10(1):25–30
Zloch Z (1996) The role of dietary plant polyphenols in health maintenance. Casopis lekaru ceskych 135(3):84–88
Zuo YG, Chen H, Deng YW (2002) Simultaneous determination of catechins, caffeine and gallic acids in green, Oolong, black and pu-erh teas using HPLC with a photodiode array detector. Talanta 57(2):307–316

Authors Biography

Oualid Talhi (born 1985) received his first degree of graduate studies in Chemistry in 2007 at the University of Science and Technology Houari Boumediene (USTHB), Algiers, Algeria. Afterwards, he joined the laboratory of heterocyclic compounds and organic synthesis at USTHB to conduct his Master thesis research on solid catalysis and Knoevenagel reactions, successfully defending his Master thesis in 2009. In November 2009, he joined the laboratory of organic chemistry and natural products QOPNA at the University of Aveiro, Portugal, as an early stage researcher under the supervision of Prof. Artur Silva. He conducted his research project on natural-type polyphenols, their synthesis and biological applications within the European Initial Training Network "RedCat", obtaining a PhD degree in chemistry from the University of Aveiro by the end of 2012.

Oualid currently holds a postdoctoral position at the QOPNA group of the University of Aveiro, and is involved in organo-synthetic research aiming at new strategies against neuropathological disorders. Oualid's expertise is mainly in the field of naturally occurring polyphenol dyads, their total synthesis, characterization with sophisticated techniques (2D-NMR and X-ray crystallography), and their application to biological models.

Diana C. G. A. Pinto studied chemistry at the University of Aveiro, Portugal, graduating from this university in Analytical Chemistry in 1991. In 1996 she received her PhD in Chemistry from Aveiro University. She then joined the Department of Chemistry at Aveiro University where she is currently Assistant Professor of Organic Chemistry.

Diana is an expert in organic synthesis, including the development of new strategies toward synthesis of nitrogen- and oxygen-containing heterocyclic compounds that can be used as new drugs. Over the years, her research has also focused on the application of environmentally friendly methodologies in organic synthesis, with a certain focus on the application of microwave irradiation. Besides her strong interest in organic synthesis, Diana is also developing an active research program in the isolation and characterization of natural products, focusing on medicinal plants.

Artur M. S. Silva is Full Professor at the University of Aveiro in Portugal. He obtained both his B.Sc. (1987) and PhD (1993) degrees from the University of Aveiro. He joined the Department of Chemistry of the same University in 1987 and was appointed to Auxiliary Professor in 1996, Associate Professor in 1999 and Full Professor in 2001. Artur has published over 410 SCI-listed papers and 15 book chapters and has delivered more than 30 lectures at scientific meetings. His research interests involve synthetic organic chemistry (especially the development of new synthetic methods for oxygen- and nitrogen-containing heterocyclic compounds and organo-catalyzed transformations), natural products identification, and structural characterization by NMR.

Part VI
Connecting Section Between Chapters 6 and 7

While the previous chapter has considered the "French Paradox" and some of the redox active secondary metabolites associated with it in more general terms, we will now consider one of the compounds often associated with this apparent paradox in more detail. Resveratrol is found primarily but not exclusively in grapes and hence forms part of the human diet, especially in the form of red wine. Many health benefits have been associated with resveratrol, among them a protection from cardiovascular diseases.

The following chapter will take a closer look at resveratrol, and the current state of the art as far as its biological activity, biochemical modes of action, and possible practical applications are concerned. Not surprisingly, this chapter has been provided by colleagues from Burgundy, one of the prime wine producing regions in France, where the university even owns its own vineyard. And, *nomen est omen*, it has been edited very carefully, of course, by Paul Winyard, who himself rather curiously prefers a pint of xanthohumol over a glass of resveratrol.

Chapter 7
Bioreactivity of Resveratrol Toward Inflammation Processes

Norbert Latruffe, Allan Lançon, Emeric Limagne and Jean-Jacques Michaille

Keywords Cardiovascular diseases • French paradox • Functional food • Red wine • Sirtuins

1 Dietary Polyphenols

To adapt to or mount defenses against their often unfavorable environment, plants produce many non-energy compounds called secondary metabolites (e.g., flavonoids, polyphenols), numbering between 5,000 and 8,000 of such currently known substances. They protect against radiation, microbial infections, oxidizing stress, hydric, or chemical stress and even, through pigments and odorant molecules, enhance pollination, or protect against predators. Similarly, these plant micro-constituents often provide valuable bioactive properties in humans and animals for essential physiological function (signaling, gene regulation, acquired or infectious disease prevention, etc.). The essential biochemical processes put in place by sometimes primitive organisms have been selected through evolution and are generally preserved in all living beings. With hindsight, this can be exemplified with the substance called resveratrol, the well-known polyphenol from grapes that plays an essential role in wine as an elicitor of the natural defenses, which, interestingly, has been shown to be a protector of health in humans. For some researchers, this is an anti-infectious agent against pathogenic microorganisms such as *Botritis cinerea.* In humans, it can delay, or even block, the appearance of predominant diseases such as atherosclerosis, diabetes, cancer and inflammation. At the same time, it is considered that regular consumption of green vegetables, fruits,

N. Latruffe (✉) · A. Lançon · E. Limagne · J.-J. Michaille
Laboratoire de Biochimie (Bio-peroxIL), INSERM IFR100—Faculté des Sciences, Université de Bourgogne, 6, Bd Gabriel, 21000 Dijon, France
e-mail: Norbert.Latruffe@u-bourgogne.fr

C. Jacob et al. (eds.), *Recent Advances in Redox Active Plant and Microbial Products*,
DOI: 10.1007/978-94-017-8953-0_7,

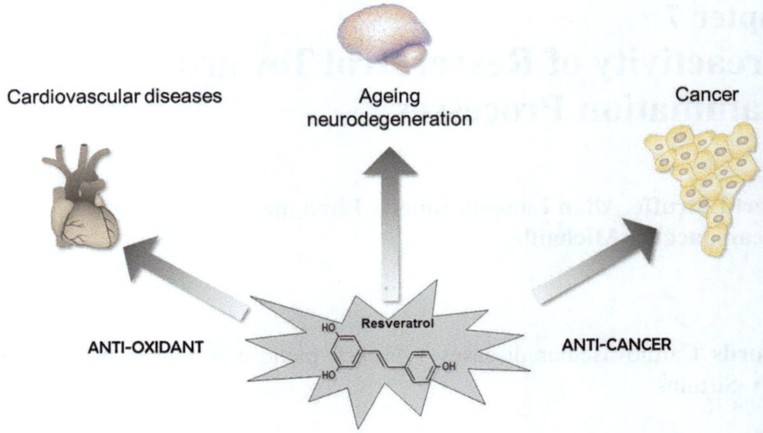

Fig. 1 Resveratrol, a beneficial molecule for human health

fiber, and fish proteins, accompanied by daily physical exercise has a protective effect against the appearance of disease and is consequently a factor of longevity. Grapes, like tea and coffee, soy, peanuts, cacao, apples, onions, cabbage, broccoli, tomato, almonds, olive oil, pomegranates, and red berries (blueberries, black currants, raspberries), etc., are rich in polyphenols (both colored and uncolored) and in vitamins possessing powerful antioxidant properties.

2 Resveratrol: A Unique Polyphenol from Vine

Resveratrol (or *trans*-3′, 4, 5′-trihydroxystilbene) (Fig. 1), as far as we know today, is the grape vine's main defense molecule (so-called phytoalexin) and is most particularly massively produced in response to a fungal attack. Although other plants belonging to around 20 other species also synthesize resveratrol including nonedible plants such as *Polygonum cuspidatum*, known to be rich in resveratrol, and *Veratrum album* (European White Hellebore or White Veratrum, found for instance in the plateaux of the Haut-Doubs region near the Jura mountain in France, from which the name resveratrol originates) (Aggarwal and Shishodia 2006). A few are edible (except for peanut plants in which resveratrol is found in the seeds, or in blueberries). Historically, Asian civilizations did not commonly cultivate grape vines and therefore were not familiar with resveratrol. Nonetheless, their pharmacopeia included extracts of *Polygonum cuspidatum* roots as a vasorelaxant and preparations based on *Yucca schidigera* for their antimutagenic properties. These two medicinal plants have been identified over the past few years as rich in resveratrol. Langcake and Pryce detected this new molecule in grapes and wine after infection of the grapevines by *Botrytis cinerea* (Langcake and Pryce 1976). The *trans (E)* isomer of resveratrol is the most abundant and active form of resveratrol as compared to the *cis (Z)* isomer. In grapes resveratrol mainly accumulates in a glycosylated conjugated state (piceid). Some di-methoxylated derivatives are also present (pterostilbene) as well as resveratrol oligomers (ε-viniferin, a dimer, and hopeaphenol, Renaud et al., showed that a large cohort of moderate consumers of wine presented lower cancer mortality (Renaud et al. 1998). Interestingly, over the past few years new properties of resveratrol have been discovered, at least in laboratory mammals, such as its possible beneficial role in longevity, (Howitz et al. 2003) prevention of neurodegenerative disease, (Parker et al. 2005) delay of cerebral aging, (Chan et al. 2008; Ritz et al. 2008) maintenance of a high level of physical activity in mice subjected to a diet including resveratrol, (Baur et al. 2006; Lagouge et al. 2006) and the prevention of oxidative stress (OS) in ischemia-reperfusion during organ transplantation (see Explanatory Box 1. For resveratrol, Sirtuins and aging) (Hassan-Khabbar et al. 2008).

Explanatory Box 1: Aging, Epigenetics, and SIRT (Human Sirtuin)

During the last couple of decades, many beneficial effects have been ascribed to resveratrol. These include not only antioxidant properties but also various chemopreventive, anticancer properties, a beneficial influence on the cardio-vasculature and diverse antimicrobial activities to mention just a few. This chapter will address some of these activities and their underlying cellular mechanisms in more detail.

Recently, resveratrol has also fuelled a rather different debate. It seems that this compound is able to slow down aging and increase the lifespan of some mammalian test animals. Not surprisingly, these findings have stirred up a rather intense debate, given the implication that it might be possible to delay aging in humans as well, and hence to achieve longevity by taking certain natural products, either as food or food supplements or even as anti-aging drugs. Here, the debate goes well beyond the more traditional 'anti-aging' crèmes which are commonly used in cosmetics to protect against skin damage by UV-radiation or free radicals. It appears that substances such as resveratrol not only simply protect the organism from external stresses, but retard the natural aging process of cells and the organism as a whole.

Nonetheless, such 'anti-aging' pills are not just part of science-fiction or a clever decoy to transfer money down the age pyramid. There is some quite convincing scientific evidence which points towards epigenetic effects associated with resveratrol (and also other natural products, including xanthohumol from hop). In brief, such compounds interfere with key epigenetic processes. Xanthohumol, for instance, may chemically modify relevant lysine and/or arginine residues of specific histones and hence cause a state resembling (hyper-)acetylation, a detachment of DNA and an (over-)expression of certain proteins. These proteins may, for instance, assist the cell in functioning normally, to differentiate and also to enter apoptosis if any serious damage has occurred. Indeed, an increase of histone acetylation is often desired and there are certain drugs, such as the hydroxamic acid vorinostat, which cause this state by inhibiting the enzymes responsible for the controlled removal of such acetyl groups, i.e., the histone deacetylases (HDACs). Vorinostat belongs to the SAHA-type HDAC inhibitors and is used in the treatment of cutaneous T cell lymphoma.

Resveratrol, in contrast, seems to act more indirectly by activating a specific class of HDACs, namely (some of) the sirtuins (SIRT enzymes). These enzymes remove acetyl-groups from histones and hence decrease acetylation. In contrast to the more common HDAC inhibitors, SIRT *activators* therefore decrease the acetylation status of specific histones. This results in a tighter binding of DNA and a *reduced* expression of specific proteins. As some of the proteins down-regulated by these processes actually promote aging, the sirtuins seem to delay aging (and promote DNA repair). Taken together, the activation of sirtuins by compounds such as resveratrol may therefore delay aging and hence indeed increase the lifespan of the organism affected. The notion of longevity drugs is therefore not just a pipedream, but may indeed possess a rather solid biochemical basis related to epigenetics.

3 Antioxidant Properties of Resveratrol

Antioxidants, both endogenous and supplied by the diet, are essential in the vital processes because cell aging is directly related to the presence of free radicals, oxygenated or others, presenting a lone electron that is chemically very reactive. Thus, one of the mechanisms of action of an antioxidant is to scavenge oxygen free radicals (Fig. 2). The other mechanism that an antioxidant uses is to stimulate the cell's antioxidant defenses (e.g., enzymes detoxifying free radicals). Given their content of hydroxyl chemical functional groups related to their benzene nuclei (or phenols), phytophenols have essential antioxidant properties. It should be remembered that living mammalian cells naturally produce oxidant compounds, such as some types of free radicals that present a highly reactive single electron (e.g., superoxide radical anions, etc.,) (Fig. 2). These free radicals have dual roles, one defending the body with bactericidal or antiviral effects (produced by macrophages), the other producing harmful effects by altering the essential macromolecules of life: DNA breaks, peroxidation of lipids, or oxidation of proteins. These free radicals are for the most part produced by the mitochondria in which the oxygen from breathing is corrupted to superoxide radical anion. Their toxic effect is the source of the transformation of healthy cells into cancerous cells as well as cell aging. Polyphenols therefore trap single electrons by making them mobile within the polyphenol molecule and therefore much less reactive to neighboring molecules. Concomitantly the polyphenols oxidize, however, with the phenol groups

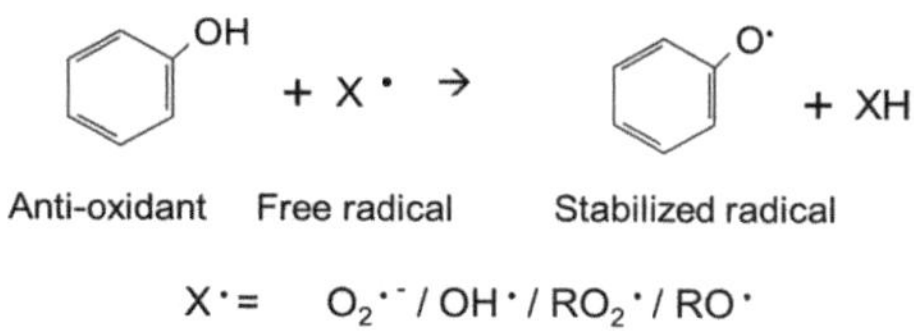

Fig. 2 Anti-oxidative properties of (poly-) phenols

becoming quinone groups, which in some cases (when polyphenols are in excess) can also become pro-oxidants. In conjunction with polyphenols, vitamins C and E also contribute to the antioxidant potential brought by fruit and vegetables. Resveratrol has been established as a powerful antioxidant with a direct impact on oxidative stress. Many tests are available to measure the antioxidant potential of a fluid or an extract, e.g., the measurement of malondialdehyde, isoprostanes, the occurrence of 8-hydroxydesoxyguanine in DNA, etc.

4 Bioavailability

In nutri-pharmacological potency or in toxicology, the notion of bioavailability is essential. This concerns the processes of absorption, transformation (metabolism), elimination (excretion), and the pharmacokinetics. It is known that resveratrol, which is found mostly in the glycosylated form in grapes and wine, undergoes deglycosylation by the intestinal flora and by glycosidases at the surface of enterocytes and is then absorbed in this form (called aglycone). Its rapid transfer through the cell membrane is mediated by a passive diffusion phenomenon accompanied by a facilitated diffusion process because resveratrol is amphiphilic (soluble in both hydrophobic medium, such as membrane phospholipids, and hydrophilic medium such as extracellular or cytoplasmic spaces) (Lancon et al. 2004). Resveratrol (all or in part) is then transformed (metabolized) by conjugating enzymes (UDP-glucuronyl-transferases, sulfotransferases) to turn it more hydrosoluble, e.g., in a glucuronide or sulfate form (Lancon et al. 2007). Resveratrol is also converted by a hydroxylated form, the piceatannol or a hydrogenated form at the conjugated double bond between the two phenolic groups. The elimination of resveratrol and its by-products by the intestinal cells, and therefore their passage in the bloodstream, involves the intervention of ATP-dependent efflux pumps called MDRs (multidrug resistance proteins) located in the cell's plasmic membrane. The passage of these by-products through the liver accentuates their metabolism and part of the conjugated forms is recycled back to the aglycone (the active form), which is distributed throughout the body. From a pharmacokinetic point of view, resveratrol is rapidly absorbed with a plasma peak between 15 and 30 min and a concentration depending on the quantity ingested, which is on the order of the micromolar (Colin, Ph.D thesis, University Bourgogne, Dijon, France, 2008). Conjugated resveratrol is found eliminated in the feces and urine.

A general, recurrent, and complex question in this research area is "can resveratrol concentrations inducing an *in vitro* effect be reached *in vivo*?" The current knowledge is as follows. (1) the plasmatic resveratrol concentrations can reach micromolar levels in animal and humans receiving pharmacological doses of resveratrol in resveratrol-supplemented diet. Moreover, the plasma level of polyphenols represents just a part of the blood content since these molecules largely accumulate in blood cells (Ginsburg et al. 2011) and (2) the plasmatic resveratrol concentration does not reflect tissue concentrations since several papers report accumulation of resveratrol in the liver (Bertelli et al. 1998). In addition, we have shown that resveratrol can accumulate in hepatic cells not only through diffusion, but also through active carrier-mediated uptake (Lancon et al. 2004). In colon intestine cells, raise up to 40 micromolar (Patel et al. 2010). This concentration is compatible with those required for resveratrol binding to and inhibition of enzymes such as COX1 (cyclo-oxygenase 1) and COX2 or for stimulating the integrin alpha V beta 3 receptor (Calamini et al. 2010; Lin et al. 2006).

5 Bioactivity of Resveratrol

Resveratrol has been established as a powerful antioxidant with a direct impact on oxidative stress. Indeed, in 1995 it was shown that the powerful antioxidant properties of resveratrol were capable of preventing the oxidation of LDL cholesterol and therefore to protect the arteries against atherosclerosis (Fig. 1) (Goldberg et al. 1995).

Resveratrol has also been shown to inhibit lipoxygenases and cyclo-oxygenases (that synthesize pro-inflammatory mediators from arachidonic acid), protein kinases (such as PKCs and PKD), receptor tyrosine kinases and lipid kinases, as well as IKKα, an activator of the pro-inflammatory NF-κB pathway (Delmas et al. 2011). In addition, resveratrol regulates apoptosis (Colin et al. 2011) and cell cycle progression and down-regulates the MAP kinase signaling pathway, the NF-κB pathway, and the AP-1 (Activator Protein 1) pathway (Delmas et al. 2002). Resveratrol interferes with many other cell functions such as phosphorylation signaling and gene regulation. This requires that mechanisms of action also include activation of membrane proteins, such as recruitment of death receptors to set off apoptosis (Delmas et al. 2003), activation of kinases, such as AMP-kinase and CDKs (cyclin-dependent kinases) (Delmas et al. 2002), or activate nuclear receptors to estrogens regulating the transcription of target genes. Recent data showed that resveratrol monosulfate and bisulfate derivatives display biological effects, such as the inhibition of COX1, COX2, •NO production and iNOS expression, or the activation of Sirtuin 1 (SIRT1) which are compatible with anticancer effects. Recently, we have discovered a resveratrol-dependent new regulatory pathway through the regulation of microRNA activities (see further below) (Hoshino et al. 2010).

6 Anti-inflammatory Properties of Resveratrol

6.1 Resveratrol and Inflammation; Systemic Effect

Inflammation is the result of a complex immune response to pathogens, allergens, damaged cells, tissue injury, or toxic molecules (Fig. 3). For the body, this inflammation is beneficial and self-contained, yet may become chronic. Chronic inflammation has been linked to many pathologies such as vascular alterations, neurodegenerative diseases, rheumatoid arthritis, chronic asthma, multiple sclerosis, psoriasis, inflammatory bowel disease, and various types of cancers. For instance, it has been established that inflammation is associated with the induction or the aggravation of more then 25 percent of cancers (Colotta et al. 2009).

The inflammation process is the result of signaling the emission of molecules and capitation of so-called chemokines. The chemokines are small and chemoattractive proteins which will mobilize leucocytes from the lymphea/plasma to the site of inflammation which is marked by chemokine emission responsible for the production of pro-inflammatory compounds (e.g., prostaglandins and leukotrienes) (Bureau et al. 2008). These chemokines (including interleukins) will bind to receptors at the membrane surface of monocytes, a process which will result in macrophage activation, consequently eliminating damaged tissues. These events are usually accompanied by pain. The inflammatory process will end when chemokines are enzymatically degraded. Numerous pathologies are linked to such an inflammatory process.

One way to limit inflammation is to inhibit chemokines production, which can be achieved by employing steroid anti-inflammatory drugs or non steroid anti-inflammatory drugs. Interestingly, resveratrol, as well as curcumin, have also been shown to exert a variety of anti-inflammatory effects through the inhibition of

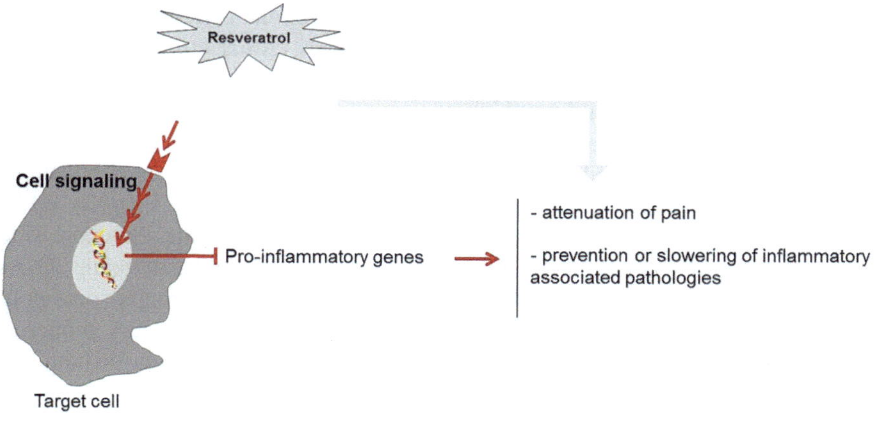

Fig. 3 Resveratrol anti-inflammatory properties

lipoxygenases and cyclo-oxygenases that synthesize pro-inflammatory mediators from arachidonic acid (Csaki et al. 2009). Inhibition of protein kinases such as PKCs and PKD, receptor tyrosine kinases and lipid kinases, as well as IKKα, an activator of the pro-inflammatory NF-κB pathway also provides some relief (Delmas et al. 2011).

6.2 Resveratrol-Dependent Control of Inflammation Through MicroRNA Modulation

MicroRNA (miRNA) function in the cell is an expanding new field of research. The first noncoding small regulatory RNA (*lin4*) was identified by Lee et al. as a developmental regulator in *C. elegans.* miRNAs were rapidly shown to be present not only in animals but also in plants and viruses (Lee et al. 1993). Since then, miRNAs have been implicated in the regulation of cell proliferation, differentiation and homeostasis, as well as in the innate and adaptive immune response. To date around 1,500 miRNAs have been identified in humans. miRNA misexpression has been linked to major pathologies such as cancer or cardiovascular, neurodegenerative and autoimmune diseases (Tili et al. 2007). Finally miRNAs have recently been found in blood and other body fluids. They are transported from cell to cell either through the gap junction or through blood secretion and exert their targeting capabilities in recipient cells. In blood miRNAs have been found either in microvesicles, exosomes, HDLs, or associated with RNA-binding proteins such as Ago2 or nucleophosmin 1 (Kosaka and Ochiya 2011). MiRNAs are capable of delivering an effect to distant cells, and may even be responsible for the induction of metastases at a distant location of the original tumor (Kosaka and Ochiya 2011). In contrast it is probable that some pharmaceutical compounds, including resveratrol, may possibly exert wide anti-inflammatory and antitumor effects in the body by causing the secretion of anti-inflammatory and antitumor miRNAs into the bloodstream. Excellent reviews have recently described the effects of resveratrol in animal models (Athar et al. 2007; Tili and Michaille 2011).

Despite a number of studies which have recently investigated several signaling and transcriptional pathways, the mechanisms of pleiotropic action of resveratrol is presently still poorly understood (Delmas et al. 2011). Some recent publications, however, have established that one reason resveratrol can affect so many different regulatory pathways might be due to its ability to modulate the expression, and consequently the regulatory effects, of a number of small noncoding RNAs, namely microRNAs (miRNAs) (Tili and Michaille 2011). Interestingly some polyphenols, including resveratrol, are known to exhibit anti-inflammatory properties and we recently showed that resveratrol can regulate the expression of both pro- and anti-inflammatory miRNAs (Tili et al. 2010). In human THP-1 monocytic cells as well as in human blood monocytes, for instance, resveratrol upregulates *miR-663*, an anti-inflammatory and tumor-suppressor miRNA that decreases AP-1 transcriptional activity and impairs its up-regulation by lipopolysaccharides (LPS) at least in part by targeting *JunB* and *JunD* transcripts. In contrast, resveratrol impairs the upregulation of pro-inflammatory and oncogenic *miR-155* by LPS in a *miR-663*-dependent manner. These results open the perspective of manipulating *miR-663* levels to potentiate anti-inflammatory and antitumor effects of resveratrol in pathologies associated with elevated levels of *miR-155*. In contrast to 'classical' coding transcripts noncoding RNAs have been generally much less conserved during evolution.

6.3 Resveratrol and Osteoarthritis

Osteoarthritis is a chronic and 'wear-and-tear'-associated pathology of articulations. This age-linked disease is very handicapping and painful and is characterized anatomically by the lack of articular cartilage (collagen, chondroitin sulfate) regeneration. This disregulation of cartilage production results into pain (mechanical and diurne) as well as difficulties to move the articulations. The disease can evolve into a sub-chondral bone fissuring. The osteophyte formation (bone extension), accompanied or not, of a synovite is characterized by immune cell infiltration (macrophage, neutrophiles) and accute inflammation at the synovial cavity. Currently no curative treatment is available, an inhibition of disease progression is equally difficult. The only approach to delay the handicap is to maintain regular and very moderate physical exercise, and the supplementation with chondroitin derivatives. Local and heavy pain and inflammation can be attenuated by anti-inflammatory drugs. These drugs, however, show often undesirable side effect. In contrast, some polyphenols such as resveratrol are known to be good natural anti-inflammatory molecules (Shakibaei et al. 2007, 2008; Wang et al. 2011) and interesting analgesic substances (Pham-Marcou et al. 2008).

While the anti-inflammatory effects of resveratrol as well as of other phytophenols are known the knowledge of their impact on chondrocyte model is so far limited (Shakibaei et al. 2007, 2008; Sharma et al. 2007; Wang et al. 2011). The mechanisms of action may involve signaling pathways where NF-κB and AP-1 become inhibited (Fig. 4). Alterations of chondrocytes are mainly responsible for arthritis accompanied by inflammation and pain. Resveratrol (RSV) shows anti-inflammatory properties by inhibiting IL-6, IL-8 secretion in LPS- treated cultured human chondrocytes (Ragot et al. unpublished)

Thus resveratrol a natural and safe polyphenol appears to be a good anti-inflammatory compound which could substitute partially or even completely for classical steroid anti-inflammatory drugs and non steroidal anti-inflammatory drugs. A recent review has been published by Shen et al. (2012) These new data open interesting perspectives for further studies, and aim at the prevention and the treatment (possibly co-treatment with glucocorticoids) of inflammation linked to arthritis.

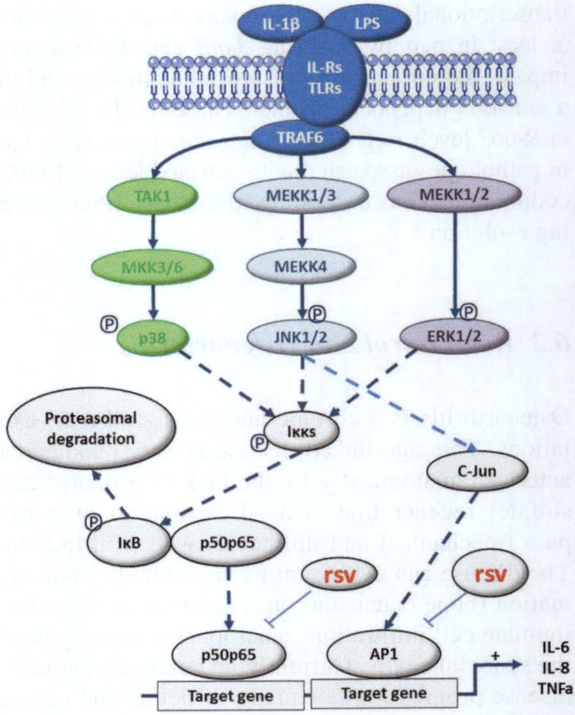

Fig. 4 Signaling pathway of the inflammatory process and possible interference(s) of resveratrol

6.4 Resveratrol and ALMD (Age-Linked Macular Degeneration)

A few years ago, it has been considered that inflammatory processes were also associated with retinal disorders such as diabetic retinopathy and ALMD (Ambati et al. 2003; Joussen et al. 2004). Later one, an *in vivo* study on mice has reported the ocular inflammation may be induced by endotoxins, and relieved in part by treatment with resveratrol (Kubota et al. 2009). This paper also demonstrated that a 5-day pretreatment with an oral resveratrol supplementation leads to the inhibition of ICAM-1 and MCP-1, two important proteins in the inflammatory process. MCP-1 (Monocyte Chimoattractant Protein 1) is a chemokine expressed by endothelial cells covering the vascular wall. Its role is to attract immune cells such as leukocytes to the inflammatory site. ICAM-1 (Inter-Cellular Adhesion Molecule 1) is expressed at the endothelial surface. Its role is to extract leukocytes from the bloodstream to allow them to diffuse at the tissue target.

In diabetes, the sustained high level of blood glucose leads to a chronic inflammation accompanied by a slow but regular degradation of Retine Pigmentar Epithelial cells (RPE) leading to the alteration of the blood-retinal

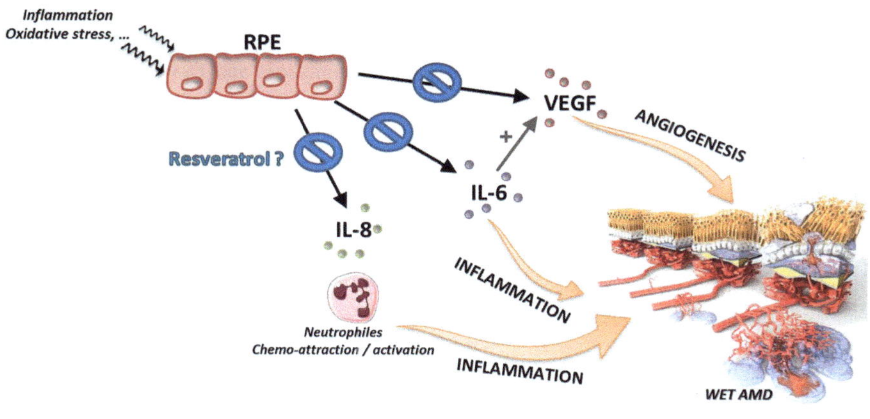

Fig. 5 Hypothetical inhibition effect(s) of resveratrol in the context of the pathogenesis of wet AMD

barrier and the loss of the central vision. Recently, an *in vitro* study on retinal pigmented cells analyzed the inflammatory phenomena to hyperglycemia conditions (Losso et al. 2010). The authors have shown that cells submitted to the diabetes test are producing pro-inflammatory cytokines like interleukin 6 and interleukin 8 and that resveratrol was able to inhibit this reaction in a dose-dependent manner. At the same time cyclo-oxygenase-2 (COX2) activity, which is responsible for the pro-inflammatory prostaglandin production, is also inhibited by resveratrol while the expression of Connexin 43 and Gap-junction, two proteins involved in cell–cell interaction is conserved. The cell cohesion is maintained thus preventing retinal-blood barrier degradation. A tentative explanation of the hypothetical inhibition effect of resveratrol in the context of the pathogenesis of wet AMD is presented in Fig. 5.

It has been shown that resveratrol inhibits ROS production leading to a protection of trabecular net cells which are submitted to OS following hyperoxygenation, a factor which can initiate glaucoma. A similar study shows that resveratrol is able to decrease the expression of interleukin-6 (IL-6), interleukin-8 (IL-8), messenger of interleukin-1α (IL-1α) as well as selectin-E, all of which are markers of inflammation. Selectin-E, also called ELAM-1 (endothelial-leukocyte adhesion molecule-1) is involved in the recruitment of leucocytes on the inflammation site, similar to ICAM-1 (Hua et al. 2011).

Resveratrol exhibits *in vitro* and *in vivo* anti-inflammatory capabilities at the molecular level by limiting the expression of pro-inflammatory factors such us interleukins and prostaglandins, but also at the cellular dimension by decreasing the chemoattraction and the recruitment on cells of the immune system to the inflammatory site.

Acknowledgments The team was supported in part by the Regional Council of Burgundy; contracts with Merck MF Co (Dijon), Théa Co (Clermont), and APICIL foundation (Lyon). Thanks to Kevin Ragot for initial experiments on inflammation, and Drs Gérard Lizard and Patrick Dutartre for helpful discussions.

References

Aggarwal BB, Shishodia S (2006) Molecular targets of dietary agents for prevention and therapy of cancer. Biochem Pharmacol 71(10):1397–1421

Ambati J, Anand A, Fernandez S, Sakurai E, Lynn BC, Kuziel WA, Rollins BJ, Ambati BK (2003) An animal model of age-related macular degeneration in senescent Ccl-2-or Ccr-2-deficient mice. Nat Med 9(11):1390–1397

Athar M, Back JH, Tang X, Kim KH, Kopelovich L, Bickers DR, Kim AL (2007) Resveratrol: a review of preclinical studies for human cancer prevention. Toxicol Appl Pharm 224(3):274–283

Baur JA, Pearson KJ, Price NL, Jamieson HA, Lerin C, Kalra A, Prabhu VV, Allard JS, Lopez-Lluch G, Lewis K, Pistell PJ, Poosala S, Becker KG, Boss O, Gwinn D, Wang MY, Ramaswamy S, Fishbein KW, Spencer RG, Lakatta EG, Le Couteur D, Shaw RJ, Navas P, Puigserver P, Ingram DK, de Cabo R, Sinclair DA (2006) Resveratrol improves health and survival of mice on a high-calorie diet. Nature 444(7117):337–342

Bertelli A, Bertelli AAE, Gozzini A, Giovannini L (1998) Plasma and tissue resveratrol concentrations and pharmacological activity. Drug Exp Clin Res 24(3):133–138

Bureau G, Longpre F, Martinoli MG (2008) Resveratrol and quercetin, two natural polyphenols, reduce apoptotic neuronal cell death induced by neuroinflammation. J Neurosci Res 86(2):403–410

Calamini B, Ratia K, Malkowski MG, Cuendet M, Pezzuto JM, Santarsiero BD, Mesecar AD (2010) Pleiotropic mechanisms facilitated by resveratrol and its metabolites. Biochem J 429:273–282

Chan SL, Tabellion A, Bagrel D, Perrin-Sarrado C, Capdeville-Atkinson C, Atkinson J (2008) Impact of chronic treatment with red wine polyphenols (RWP) on cerebral arterioles in the spontaneous hypertensive rat. J Cardiovasc Pharm 51(3):304–310

Colin D, Limagne E, Jeanningros S, Jacquel A, Lizard G, Athias A, Gambert P, Hichami A, Latruffe N, Solary E, Delmas D (2011) Endocytosis of resveratrol via lipid rafts and activation of downstream signaling pathways in cancer cells. Cancer Prev Res 4(7):1095–1106

Colotta F, Allavena P, Sica A, Garlanda C, Mantovani A (2009) Cancer-related inflammation, the seventh hallmark of cancer: links to genetic instability. Carcinogenesis 30(7):1073–1081

Csaki C, Mobasheri A, Shakibaei M (2009) Synergistic chondroprotective effects of curcumin and resveratrol in human articular chondrocytes: inhibition of IL-1 beta-induced NF-kappa B-mediated inflammation and apoptosis. Arthritis Res Ther 11 (6)

Delmas D, Passilly-Degrace P, Jannin B, Cherkaoui-Malki M, Latruffe N (2002) Resveratrol, a chemopreventive agent, disrupts the cell cycle control of human SW480 colorectal tumor cells. Int J Mol Med 10(2):193–199

Delmas D, Rebe C, Lacour S, Filomenko R, Athias A, Gambert P, Cherkaoui-Malki M, Jannin B, Dubrez-Daloz L, Latruffe N, Solary E (2003) Resveratrol-induced apoptosis is associated with Fas redistribution in the rafts and the formation of a death-inducing signaling complex in colon cancer cells. J Biol Chem 278(42):41482–41490

Delmas D, Solary E, Latruffe N (2011) Resveratrol, a phytochemical inducer of multiple cell death pathways: apoptosis, autophagy and mitotic catastrophe. Curr Med Chem 18(8):1100–1121

Ginsburg I, Kohen R, Koren E (2011) Quantifying oxidant-scavenging ability of blood. New Engl J Med 364(9):883–885

Goldberg DM, Yan J, Ng E, Diamandis EP, Karumanchiri A, Soleas G, Waterhouse AL (1995) A global survey of trans-resveratrol concentrations in commercial wines. Am J Enol Viticult 46(2):159–165

Hassan-Khabbar S, Cottart CH, Wendum D, Vibert F, Clot JP, Savouret JF, Conti M, Nivet-Antoine V (2008) Postischemic treatment by trans-resveratrol in rat liver ischemia-reperfusion: a possible strategy in liver surgery. Liver transpl 14(4):451–459

Hoshino J, Park EJ, Kondratyuk TP, Marler L, Pezzuto JM, van Breemen RB, Mo S, Li Y, Cushman M (2010) Selective synthesis and biological evaluation of sulfate-conjugated resveratrol metabolites. J Med Chem 53(13):5033–5043

Howitz KT, Bitterman KJ, Cohen HY, Lamming DW, Lavu S, Wood JG, Zipkin RE, Chung P, Kisielewski A, Zhang LL, Scherer B, Sinclair DA (2003) Small molecule activators of sirtuins extend Saccharomyces cerevisiae lifespan. Nature 425(6954):191–196

Hua J, Guerin KI, Chen J, Michan S, Stahl A, Krah NM, Seaward MR, Dennison RJ, Juan AM, Hatton CJ, Sapieha P, Sinclair DA, Smith LE (2011) Resveratrol inhibits pathologic retinal neovascularization in Vldlr(-/-) mice. Invest Ophthalmol Vis Sci 52(5):2809–2816

Joussen AM, Poulaki V, Le ML, Koizumi K, Esser C, Janicki H, Schraermeyer U, Kociok N, Fauser S, Kirchhof B, Kern TS, Adamis AP (2004) A central role for inflammation in the pathogenesis of diabetic retinopathy. FASEB J 18(12):1450–1452

Kosaka N, Ochiya T (2011) Unraveling the mystery of cancer by secretory microRNA: horizontal microRNA transfer between living cells. Front genet 2:97

Kubota S, Kurihara T, Mochimaru H, Satofuka S, Noda K, Ozawa Y, Oike Y, Ishida S, Tsubota K (2009) Prevention of ocular inflammation in endotoxin-induced uveitis with resveratrol by inhibiting oxidative damage and nuclear factor-kappaB activation. Invest Ophthalmol Vis Sci 50(7):3512–3519

Lagouge M, Argmann C, Gerhart-Hines Z, Meziane H, Lerin C, Daussin F, Messadeq N, Milne J, Lambert P, Elliott P, Geny B, Laakso M, Puigserver P, Auwerx J (2006) Resveratrol improves mitochondrial function and protects against metabolic disease by activating SIRT1 and PGC-1 alpha. Cell 127(6):1109–1122

Lancon A, Delmas D, Osman H, Thenot JP, Jannin B, Latruffe N (2004) Human hepatic cell uptake of resveratrol: involvement of both passive diffusion and carrier-mediated process. Biochem Bioph Res Co 316(4):1132–1137

Lancon A, Hanet N, Jannin B, Delmas D, Heydel JM, Lizard G, Chagnon MC, Artur Y, Latruffe N (2007) Resveratrol in human hepatoma HepG2 cells: Metabolism and inducibility of detoxifying enzymes. Drug Metab Dispos 35(5):699–703

Langcake P, Pryce RJ (1976) Production of resveratrol by vitis-vinifera and other members of vitaceae as a response to infection or injury. Physiol Plant Pathol 9(1):77–86

Lee RC, Feinbaum RL, Ambros V (1993) The C-elegans heterochronic gene lin-4 encodes small Rnas with antisense complementarity to lin-14. Cell 75(5):843–854

Lin HY, Lansing L, Merillon JM, Davis FB, Tang HY, Shih A, Vitrac X, Krisa S, Keating T, Cao HJ, Bergh J, Quackenbush S, Davis PJ (2006) Integrin alphaVbeta3 contains a receptor site for resveratrol. FASEB J 20(10):1742–1744

Losso JN, Truax RE, Richard G (2010) Trans-resveratrol inhibits hyperglycemia-induced inflammation and connexin downregulation in retinal pigment epithelial cells. J Agric Food Chem 58(14):8246–8252

Parker JA, Arango M, Abderrahmane S, Lambert E, Tourette C, Catoire H (2005) Resveratrol rescues mutant polyglutamine cytotoxicity in nematode and mammalian neurons. Nat Genet 37(5):349–350

Patel KR, Brown VA, Jones DJL, Britton RG, Hemingway D, Miller AS, West KP, Booth TD, Perloff M, Crowell JA, Brenner DE, Steward WP, Gescher AJ, Brown K (2010) Clinical pharmacology of resveratrol and its metabolites in colorectal cancer patients. Cancer Res 70(19):7392–7399

Pham-Marcou TA, Beloeil H, Sun X, Gentili M, Yaici D, Benoit G, Benhamou D, Mazoit JX (2008) Antinociceptive effect of resveratrol in carrageenan-evoked hyperalgesia in rats: prolonged effect related to COX-2 expression impairment. Pain 140(2):274–283

Renaud SC, Gueguen R, Schenker J, d'Houtaud A (1998) Alcohol and mortality in middle-aged men from Eastern France. Epidemiology 9(2):184–188

Ritz MF, Ratajczak P, Curin Y, Cam E, Mendelowitsch A, Pinet F, Andriantsitohaina R (2008) Chronic treatment with red wine polyphenol compounds mediates neuroprotection in a rat model of ischemic cerebral stroke. J Nutr 138(3):519–525

Shakibaei M, Csaki C, Nebrich S, Mobasheri A (2008) Resveratrol suppresses interleukin-1 beta-induced inflammatory signaling and apoptosis in human articular chondrocytes: potential for use as a novel nutraceutical for the treatment of osteoarthritis. Biochem Pharmacol 76(11):1426–1439

Shakibaei M, John T, Seifarth C, Mobasheri A (2007) Resveratrol inhibits IL-1 beta-induced stimulation of caspase-3 and cleavage of PARP in human articular chondrocytes in vitro. Ann Ny Acad Sci 1095:554–563

Sharma S, Chopra K, Kulkarni SK (2007) Effect of insulin and its combination with resveratrol or curcumin in attenuation of diabetic neuropathic pain: participation of nitric oxide and TNF-alpha. Phytother Res 21(3):278–283

Shen CL, Smith BJ, Lo DF, Chyu MC, Dunn DM, Chen CH, Kwun IS (2012) Dietary polyphenols and mechanisms of osteoarthritis. J Nutr Biochem 23(11):1367–1377

Tili E, Michaille JJ (2011) Resveratrol, MicroRNAs, Inflammation, and Cancer. J nucleic acids 2011:102431

Tili E, Michaille JJ, Adair B, Alder H, Limagne E, Taccioli C, Ferracin M, Delmas D, Latruffe N, Croce CM (2010) Resveratrol decreases the levels of miR-155 by upregulating miR-663, a microRNA targeting JunB and JunD. Carcinogenesis 31(9):1561–1566

Tili E, Michaille JJ, Gandhi V, Plunkett W, Sampath D, Calin GA (2007) miRNAs and their potential for use against cancer and other diseases. Future oncol 3(5):521–537

Wang KT, Chen LG, Tseng SH, Huang JS, Hsieh MS, Wang CC (2011) Anti-inflammatory effects of resveratrol and oligostilbenes from Vitis thunbergii var. taiwaniana against lipopolysaccharide-induced arthritis. J Agric Food Chem 59(8):3649–3656

Authors Biography

Norbert Latruffe studied at the University of Besançon and the University of Lyon I and obtained his PhD in 1977. He was appointed as Professor in 1989 and is currently Full Professor of Biochemistry at the University of Burgundy. He has established and, until 2006, headed the Laboratory of Molecular and Cellular Biology. He was in charge of the group of Biochemistry of Metabolism and Nutrition at the INSERM research center, UMR 866 of Dijon until 2011. Early in his career, Norbert has been interested in the following research topics: energetic metabolism of lipids (UA CNRS 531, Besançon); phospholipid-dependent membrane enzymes (Postdoc at Vanderbilt University, Nashville TN, USA); toxicology of peroxisome proliferators (at Dijon). He has also collaborated in various projects as visitor at different international universities (e.g., Stockholm, Bern, Himeji). Starting in 1998, he launched a new challenge on the preventing role of resveratrol, a well-known phytophenol, against age-related pathologies, such as cancer, inflammation, and cardiovascular diseases. With his collaborators, he was one of the first to explore resveratrol metabolism (2004), its pro-apoptotic properties (2004), and recently discovered a new resveratrol signaling pathway through the modulation of micro RNAs (2010). To date, Norbert is the author of more than 150 international, peer-reviewed publications and of more than 120 lectures. He is a current (or past) expert member of several national evaluation councils (e.g., CNRS, AFSSA, AERES, CNU) and at international level (EU). He has been awarded several distinctions, such as the Prize at the 16th Oncology and Molecular Medicine Meeting at Rhodes, is a laureate of the APICIL foundation, and recipient of the Academic Medal.

Allan Lançon holds a PhD in Biochemistry, Molecular and Cell Biology from the University of Burgundy, Dijon, France, which he obtained in 2006 with a thesis on a "Study of the tvransport and the metabolism of resveratrol in human hepatic cells". At this time, he also explored the cellular uptake of compounds, the biological activities of polyphenols and endocrine disruptors From 2007 until 2013, Allan has carried out several postdoctoral projects at the Laboratory of Nutritional and Metabolic Biochemistry (Dijon). These projects have been concerned with the prevention of age-related macular degeneration, the fight against metabolic syndrome, the fight against Type 2 diabetes and associated inflammation, to limit joint inflammation in osteoarthritis and to strengthen the antioxidant capacity of the body. In 2009, Allan also received training in the field of Management, especially for the creation and running of companies. Allan is currently a co-author of 11 publications.

Emeric Limagne is currently a PhD student at the INSERM research center number 866 in Dijon where he conducts his studies in the field of cancer research. Emeric previously worked as a research technician in a project on the prevention of inflammation in osteoarthritis using chondrocyte–monocyte co-cultures. Emeric has already published several manuscripts in international journals.

Jean-Jacques Michaille is Full Professor in Cell Biology at the University of Burgundy. He graduated from the University of Lyon and subsequently was appointed as Assistant Professor at the University of Grenoble and Lyon I, before being recruited as Professor at the University of Burgundy. Initially, Jean-Jacques' prime research interest was the Biology of Development. Currently, he is an expert in the field of molecular biology of mircoRNAs. Jean-Jacques collaborates closely with Professor Carlo Croce at the University of Columbus, Ohio, USA, where he regularly spends several months each year to conduct his studies in this emerging field of research. Jean-Jacques is the author or co-author of numerous publications in internationally leading journals.

Part VII
Connecting Section Between Chapters 7 and 8

Anyone believing that plant-derived polyphenols are simply electron-rich, low molecular weight antioxidants is terribly mistaken. Research conducted during the last decade or two has shown that many flavone-based polyphenols do not just occur as monomeric structures or esterified with gallic acid, but may also appear in oligomeric and even polymeric forms. Indeed, such *proanthocyanidins* and *tannins* represent a separate class of biopolymers, which is fairly unique and from the perspective of biological activity should be considered on par with other common biopolymers, such as DNA, RNA, proteins, and enzymes.

Some physicochemical properties and biological activities of these compounds, as well as potential practical applications, have been known for a while, while others are only now emerging. The astringency of tannins, for instance, has been considered for centuries and has been employed in 'tanning' processes since ancient times and around the world. Astringency is also used to 'spice up' certain beverages, and it is hardly surprising that in most fruit juices, it seems to go hand in hand with a high 'antioxidant' power (e.g., in blueberry juice, pomegranate juice). The notion that astringency is related to antioxidant activity and hence somehow also to health benefits is not new and finds its chemical basis in the accumulation of hydroxyl groups and hydroquinone redox centers in such polymeric structures.

Unfortunately, this kind of collective redox activity of numerous redox centers has long overshadowed a number of additional physicochemical properties which emerge when moving from the monomeric to the oligomeric and subsequently to the polymeric forms of such molecules. Indeed, astringency itself often is not the result of redox processes but rather reflects a strong noncovalent interaction between the oligomeric secondary metabolites on the one hand, and proteins and enzymes on the other. Since these interactions tend to denature and hence inactive the proteins affected, astringent compounds have many potential applications, for instance as antimicrobial and antidigestive agents.

The following chapter will present some of the most commonly found astringent polymers based on flavonoids (e.g., proanthocyanidins), together with their chemistry, biochemical action, and biological activities and a brief look at their monomeric building blocks (i.e., flavanols). As such, polymeric secondary metabolites are omnipresent and may be extracted or processed and subsequently used in everyday life rather readily (e.g., in the form of grape seed flour); this chapter will also consider possible (health) benefits associated with these unique substances. Not surprisingly, perhaps, the chapter is first-authored by Hadi Ebrahimnejad from *Iran*, a country that is especially rich in astringent fruits, and is home to the pomegranate.

Chapter 8
Flavanols and Proanthocyanidins

Hadi Ebrahimnejad, Torsten Burkholz and Claus Jacob

Keywords Enzyme inhibition • Flavanols • Polyphenols • Oligomeric proanthocyanidins • Tannins

1 Introduction

Amelioration of health and lowering the risk of many diseases is associated with legions of studies tied to the plant food intake. At the same time, approaches in chemistry and biology have propelled the interdisciplinary fields of natural product chemistry research and pharmacognosy to yield information about many classes of dietary phytochemicals (Kelm et al. 2005).

Flavanol-based oligomers/polymers were discovered by the French researcher Jacques Masquelier in the late 1940s. Flavanols (or flavan-3-ols) have a 15-carbon structural backbone designated as C6–C3–C6, in which three carbons connect the aromatic A and B rings (Fig. 1). The heterocyclic benzopyran ring is admitted as the 'C' ring, the fused aromatic ring as the 'A' ring and the phenyl constituent as the 'B' ring. These compounds have a saturated C3 bridge in the heterocyclic C ring. Flavanols can occur in different states such as monomers, oligomeric or polymeric forms, whereby the latter are referred to proanthocyanidins. Proanthocyanidins represent a

H. Ebrahimnejad
Department of Food Hygiene, Faculty of Veterinary Medicine,
Shahid Bahonar University, Kerman, Iran
e-mail: Ebrahimnejad.email@gmail.com

H. Ebrahimnejad · T. Burkholz · C. Jacob (✉)
Division of Bioorganic Chemistry, School of Pharmacy,
University of Saarland, 66041 Saarbruecken, Germany
e-mail: c.jacob@mx.uni-saarland.de

C. Jacob et al. (eds.), *Recent Advances in Redox Active Plant and Microbial Products*,
DOI: 10.1007/978-94-017-8953-0_8, © Springer Science+Business Media Dordrecht 2014

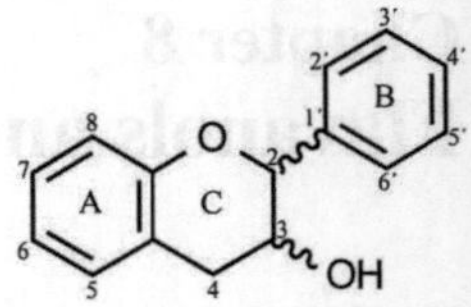

Fig. 1 Structural backbone of flavan-3-ols

unique class of secondary metabolites with high-molecular weight that despite their limited bioavailability, exhibit a wide range of sometimes astonishing bioactivities in the body, including anti-oxidative, chemopreventive, anticancer, vasorelaxing, anti-adhesive, antimicrobial, and even anti-nutritional effects (Xu et al. 2012; de la Iglesia et al. 2010; Aron and Kennedy 2008).

Proanthocyanidins—also called condensed tannins—have a retentive chronicle of acting as skin tanning agents, moreover, they possess a significant role in flavor and astringency of foods and beverages (Dixon et al. 2005). Nonetheless, assigning proanthocyanidins to the tannins is also more or less inauspicious. Tannins embrace water-soluble and insoluble (highly polymeric) substances with a broad spectrum of biological properties. The category of 'tannins' clearly refers to a rather diffuse group of natural products which often have very little in common (Xu et al. 2012).

Not surprisingly, flavanols have recently experienced a certain evolution in nutritional and pharmaceutical research. During the last decade, a number of research articles have addressed the different aspects of these exceptional substances, and now there are several excellent reviews available on this topic. Nonetheless, certain aspects of these phytochemicals in biochemistry still remain obscure (Xu et al. 2012).

2 Structure

Flavan-3-ols are non-planar by virtue of their saturated C3 element. Structurally, flavanols belong to the most complex group of flavonoids, ranging from the simple monomers to the complex structures of oligomeric and polymeric proanthocyanidins (Crozier et al. 2009). Proanthocyanidin structures differ depending upon the stereochemistry and hydroxylation of the flavan-3-ol subunits, the degree of polymerization (DP), the position of the interflavanyl bond and its stereochemistry and eventually, with regard to different modifications, for instance esterification, of a particular flavan-3-ol (Dixon et al. 2005).

2.1 Monomeric Forms

(+)-Catechin (1) and its isomer (−)-epicatechin (2) represent the simple monomers and 'building blocks' for most of the more complex proanthocyanidins (Fig. 2). Markedly, (+)-epicatechin has a 2,3-*trans* stereochemistry while (−)-catechin possesses 2,3-*cis* stereochemistry. The hydroxylated forms of these monomers

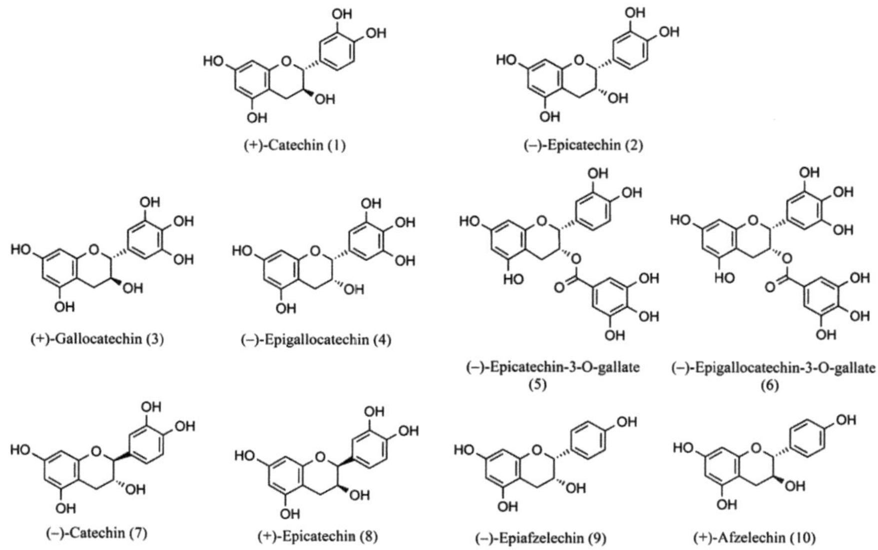

Fig. 2 Chemical structures of selected catechin monomers

are (+)-gallocatechin and (−)-epigallocatechin (3, 4) and respectively galloylated flavan-3-ols emanate commonly from the esterification of the 3-OH group with gallic acid (Crozier et al. 2009; Dixon et al. 2005).

Despite the extensive distribution of (+)-catechin and (−)-epicatechin in nature (−)-catechin (7) and (+)-epicatechin (8), which are the two other isomers resulting from the chiral centers at C2 and C3 of the flavan-3-ols, are comparatively rare. Furthermore, (−)-epiafzelechin (9) and (+)-afzelechin (10) are additional structural subunits discovered in the less common proanthocyanidins (Crozier et al. 2009).

The proanthocyanidin IUPAC system of naming—although troublesome in the case of larger molecules—is assumed useful in cases that involve defining the absolute stereochemistry at all chiral centers. Here, the stereochemistry and location of the interflavanyl bond can be referred with bracketed symbols α and β. Furthermore, since most of the common flavan-3-ols exist as 2*R* isomers, less common 2*S* isomers are tagged with '*ent*' at the beginning; such as *ent*-catechin [(−)-catechin] (Aron and Kennedy 2008) (Fig. 2).

2.2 Dimers, Oligomers, and Polymers

Many flavanols such as proanthocyanidins are composed of more than one monomer. Generally, proanthocyanidins consist of successive monomeric units linked through carbon–carbon and ether linkages.

In the prominent proanthocyanidins of plant origin and consumed as part of daily (human) nutrition, the monomeric units are primarily linked through single

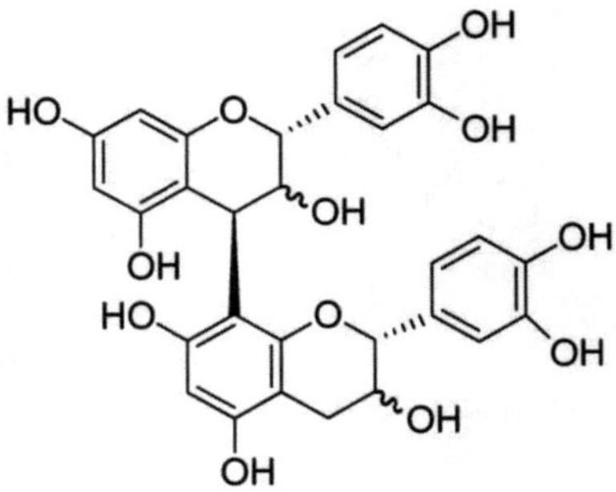

C4→C8 linkage (11)

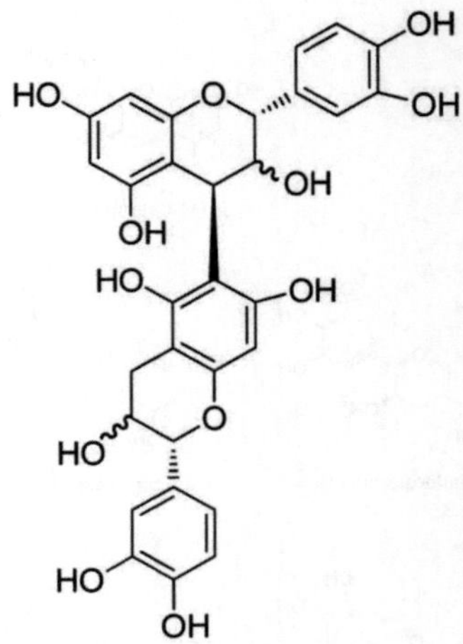

C4→C6 linkage (12)

'B'-type Dimers

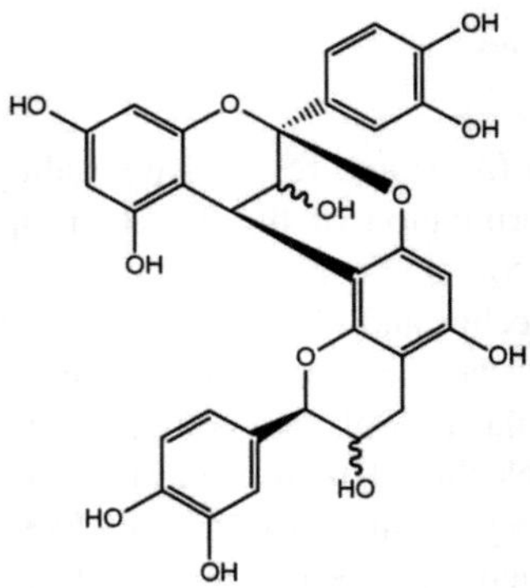

C4→C8 linkage (13)

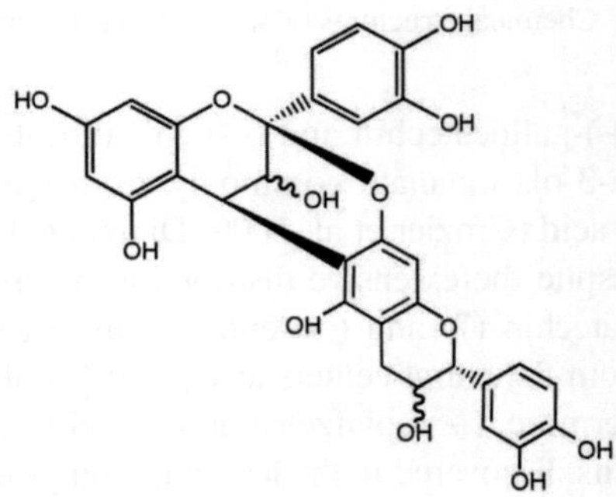

C4→C6 linkage (14)

'A'-type Dimers

Fig. 3 'A'-type and 'B'-type dimeric proanthocyanidins composed of flavanol monomers

C4 → C8 or, to a lesser extent, C4 → C6 carbon–carbon bonds (B linkages); or through C4 → C8 or C4 → C6 carbon–carbon linkages and an additional ether bond between C2 and C7 positions (A linkages) (Beecher 2003). Dimers with B linkages are termed 'B'-type with the oxidative coupling of C4 → C8 (11) or C4 → C6 (12). Generally, the diversity of this type of flavanols is merely due to the (+)-catechin and (−)-epicatechin arrangement as the starter and extension units. 'A'-type proanthocyanidin dimers are more rigid in conformation than the representatives of B category due to the possession of A linkages; one C4 → C8 (13) or C4 → C6 (14) bond and an extra C → O bond (Dixon et al. 2005; Aron and Kennedy 2008; Kalili and de Villiers 2009; Ricardo da Silva et al. 1991) (Fig. 3).

Proanthocyanidins that comprise only (epi)catechin units, are called procyanidins, and are the most abundant type of proanthocyanidins in plants. The less common proanthocyanidins are prodelphinidins and propelargonidins. Prodelphinidins contain gallocatechin and epigallocatechin units, while propelargonidins consist of afzelechin and epiafzelechin monomers (Crozier et al. 2009; Kalili and de Villiers 2009).

The degree of polymerization (DP) in 'classical' oligomeric proanthocyanidins can reach up to 10 units, while more 'tannin' like proanthocyanidin can occur as polymers of up to 50 units. Soluble proanthocyanidin polymers typically possess molecular weight (MW) averages of 1,000–6,000 Da, but sometimes even reach 20,000 Da (Crozier et al. 2009; Xu et al. 2012; Aron and Kennedy 2008).

3 Basic Considerations

Flavanols provide plenty of opportunities for health research, which ultimately may result in certain health benefits in future. Here, unveiling the more realistic mechanisms of proanthocyanidin interactions with biomolecules and biopolymers may also unravel the complexity of their biological activities.

3.1 Proanthocyanidin–Protein Interactions

One of the most stunning properties of proanthocyanidins is their ability to bind rather strongly to proteins. The interactions of proanthocyanidins with proteins are mainly due to hydrogen bonding, van der Waals, and electrostatic interactions, but they may also include covalent bond formation. These interactions influence the function of the proteins and enzymes affected. The complexation of proteins with these bio-oligomers is probably responsible for most of the biological activities associated with proanthocyanidins. This property of the proanthocyanidins is only emerging when the oligomer embraces 4 or 5 monomers, and often 'vanishes' again at chain lengths above 10 or 11, due to the fact that most of them are no longer soluble in aqueous media (Xu et al. 2012). Furthermore, although the esterified form of (−)-epigallocatechin with gallic acid (EGCG) shows a direct interaction with some plasma proteins, such as fibrinogen and fibronectin (Sazuka et al. 1996), in comparison to proanthocyanidin oligomers, the monomers lack certain properties, such as truly strong binding to proteins, which only occurs at an oligomeric state, once a certain 'cooperative synergy' is reached, i.e., usually from the tetramer upward.

As highlighted, most of the beneficial and therapeutic properties associated with proanthocyanidins appear to be due to effects on proteins and enzymes. The effects proanthocyanidins exert on proteins and enzymes in the human Gastrointestinal Tract (GIT) can even be felt in everyday life: astringency of certain foodstuffs is a property associated closely with proanthocyanidin oligomers and tannins, and is a direct result of binding—and subsequently denaturing—of salivary, proline-rich proteins

(PRPs). This property, which is actually a characteristic of plant defense against pathogens, also spices up our food and beverages with a distinct taste, and is notably absent in the corresponding monomers (Xu et al. 2012; Aron and Kennedy 2008). Here, the importance of proline in proteins is presumably due to its inability to fit into the α-helix, which leads to a loose, open, and accessible structure and hence to the formation of hydrogen bonds with these oligomers (Mehansho et al. 1987).

Interestingly, besides the proline content of proteins, the glycosylation status of some proteins may also affect their affinity to bind to proanthocyanidins (see also proanthocyanidin–carbohydrate interactions) (Santos-Buelga and Scalbert 2000). Furthermore, proanthocyanidin–protein interactions depend on both chain length and presence of galloyl groups in the proanthocyanidin structure (Aron and Kennedy 2008). Hydroxyl groups of the oligomeric proanthocyanidins—whose number rather increases with an increasing chain length—are available for extensive hydrogen and electrostatic bonding with the carbonyl groups of the peptide linkages (Xu et al. 2012; Oh et al. 1980). Furthermore, covalent cross-linking of proanthocyanidins to proteins enables them to interact with proteins without the need for hydrogen bonding, electrostatic, or van der Waals interactions. Intriguingly, this covalent cross-linking does not require large oligomeric proanthocyanidin chains but rather requires the prior oxidation of (some of) the B-rings, i.e., there may be a redox 'activation' step for this particular proanthocyanidin interaction (Xu et al. 2012). Under oxidative conditions, hydroquinones are readily transformed to quinones which might then react with nucleophilic groups (e.g., SH, NH_2) of the protein (Beart et al. 1985). These covalent interactions of (oxidized) proanthocyanidins with proteins ultimately result in protein–protein crosslinking as most anthocyanidins possess two or more reactive (hydro-)quinone moieties. Such interactions therefore include (several) subsequent Michael addition reactions at various B-rings, and also Schiff base formation. The prime targets for this type of chemistry are cysteine and lysine residues in peptides, proteins, and enzymes as well as certain DNA bases.

From a chemical point of view, such crosslinking of proteins by galloylated proanthocyanidin dimers, trimers, and oligomers is indeed possible, since the (oxidized) gallic acid residues are able to participate readily in Michael addition reactions and imine formation with amine groups of proteins (Xu et al. 2012).

Apart from those linkages, the presence of aromatic nuclei of proanthocyanidins and the aliphatic and aromatic side chains of the protein amino acids brings to mind that hydrophobic interactions might also, be relevant to the formation of protein-proanthocyanidin complexes (Oh et al. 1980).

3.2 *Proanthocyanidin–Nucleic Acid Interactions*

Scientist have noticed colored nucleic acids in catechin-treated cells, and hence have theorized that catechins also directly interact with nucleic acids (Kuzuhara et al. 2006). The red color produced is also indicative of hydrolysis of polymeric proanthocyanidins to anthocyanidins (Wang and Vodkin 1994).

Indeed, it has been illustrated that (−)-epigallocatechin gallate (EGCG) can bind to single-stranded DNA, RNA, and also to double-stranded AG:CT oligomers of various nucleotide lengths (Kuzuhara et al. 2006).

The results obtained by surface plasmon resonance and cold spray ionization-mass spectrometry indicate that there are multiple binding sites for EGCG on DNA and RNA oligomers. Moreover, the accumulation of catechins in both double-stranded DNA and RNA suggests that DNA and RNA molecules may even act as biological reservoirs for EGCG (Kuzuhara et al. 2006, 2007). Indeed, it is likely that EGCG plays an important role in the protection of double-stranded DNA, for instance preventing it from dissociating ('melting') into the single-stranded DNA (Kuzuhara et al. 2007).

It has been speculated that the spatial arrangement of phenolic groups optimizes a conformation that could interact with nucleic acids, due to hydrogen bonding to the phosphate backbone (Todd and Vodkin 1993). Furthermore, since both galloyl and catechol groups of EGCG are essential for DNA binding, both groups seem to hold the strands of DNA via their branching structure (Kuzuhara et al. 2006). Notably, while the EGCG interacts with 20-mer double-stranded DNA, but other catechins do not, it seems that one hydroxyl group of the trihydroxyphenyl ring in EGCG is essential for binding to double-stranded DNA (Kuzuhara et al. 2007).

The proanthocyanidin reactions in cells are likely to be affected by EGCG binding to DNA and RNA. Furthermore, the accumulation of EGCG at nucleic acids plays a significant role for cancer prevention in humans. Nonetheless, it remains unclear if such effects are due to direct interactions of proanthocyanidins with signaling cascades, regulator proteins (such as transcription factors) or possibly with DNA itself, for instance by protecting DNA from mutagenic agents (Xu et al. 2012).

3.3 Proanthocyanidin–Carbohydrate Interactions

Proanthocyanidin heterosides are not frequently reported. Nonetheless, their occurrence seems to be somewhat underestimated as sugars are often associated with purified proanthocyanidin polymers (Santos-Buelga and Scalbert 2000).

Polysaccharides, like proteins, can bind to proanthocyanidins reversibly and through non-covalent interactions, such as hydrogen bonding and van der Waals forces. Such interactions may also be irreversible, for instance by formation of covalent bonds between the polysaccharide and the phenolic ring system of the proanthocyanidin.

Several glycosylated flavanols have been identified. Flavan-3-ol subunits for instance can carry a glycosyl substituent on the C3, C5, C7, C3′ or C4′ positions through an *O*-glycosidic bond. Likewise *C*-glycosides have been recognized with a glycosyl moiety at either C6 or C8 positions. Furthermore, some flavan-3-ol subunits have been characterized with two glycosyl moieties attached through di-*O*-glycosidic bonds (Fig. 4) (Morimoto et al. 1986b; Kashiwada et al. 1986). Interestingly, several proanthocyanidin dimers with *O*-glycosidic or *C*-glycosidic bonds have also been identified through hydrolysis or degradation processes (Kashiwada et al. 1986; Morimoto et al. 1986a).

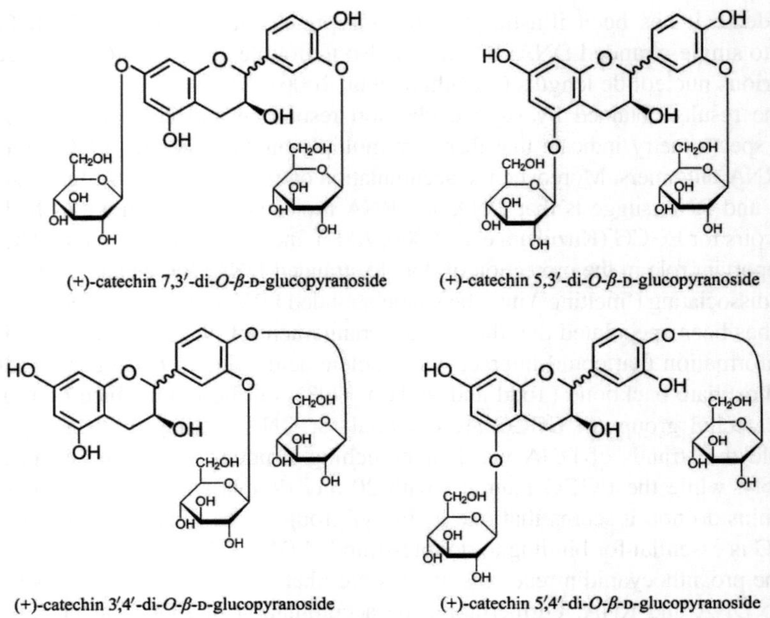

Fig. 4 Putative structures of glycosylated catechins with two glycosyl moieties

Importantly, glycosylation of flavanols might change their chemical properties and subsequently their interactions and functional applications. Here, glycosylation of catechin leads to decreased antioxidant activity in both lipid and aqueous phase (Plumb et al. 1998).

Glycosylation might also affect the binding, affinity, and specificity of proanthocyanidins to proteins. Not surprisingly, the carbohydrate residues in salivary glycoproteins may enhance this interaction with proteins and ultimately also interfere with the astringency in the mouth. (Santos-Buelga and Scalbert 2000). It is likely that the linkage between proanthocyanidins and carbohydrates results from their biosynthesis in the plant cell or is derived subsequently by processes analog to the acid-catalyzed degradation (Beart et al. 1985). The reaction of proanthocyanidins with polysaccharides and proteins may also increase the amount of insoluble proanthocyanidins and therefore affects their bioavailability (Plumb et al. 1998).

3.4 Proanthocyanidin Interactions with Cellular Membranes

During the last two decades, the interaction of proanthocyanidins with cell membranes has attracted considerable attention as this interaction might affect many of the important roles of proanthocyanidins, including their cytoprotective, anti-inflammatory, anticarcinogenic, antibacterial, and antiviral effects. Apart from bioactivity, the solubility of proanthocyanidins in lipids is a critical determinant of their bioavailability. At the same time proanthocyanidins can crosslink membranes to produce biocompatible membranes which can be applied as tissue-engineering scaffolds and drug-delivery vehicles (Kim et al. 2005).

Notably, most research articles in this area are traditionally based on model membranes such as liposomes, vesicles or lipid monolayers, while more recently, some researchers have begun to asses membrane proteins. Investigations on the distribution of proanthocyanidins within cells are hindered by the difficulty of detection of the flavan-3-ol. Fluorescence-labeling and confocal fluorescent microscopy have been applied recently for this purpose (Revesz et al. 2011).

At this point it has been demonstrated that the affinity of tea catechins for lipid bilayers is affected by the number of B-ring hydroxyl groups [e.g., (−)-epicatechin binds stronger than (−)-epigallocatechin, and (−)-epicatechin gallate stronger than (−)-epigallocatechin gallate], the presence of a galloyl moiety [e.g., (−)-epicatechin gallate stronger than (−)-epicatechin, and (−)-epigallocatechin gallate stronger than (−)-epigallocatechin] and the stereochemistry of the catechin [e.g., (−)-epicatechin gallate stronger than (−)-catechin gallate, and (−)-epigallocatechin gallate stronger than (−)-gallocatechin gallate]. Besides, synthetic (+)-catechin derivatives containing chains with chain lengths of three carbons or more showed very strong affinity for membranes and derivatives with chain lengths of four carbons or more caused leakage of the liposome (Kajiya et al. 2002, 2004).

The strength of such an interaction also seems to be inversely related to the polarity of the compound. In another study, Lázaro et al. showed that bulky and hydrophobic conjugates of (−)-epicatechin clearly interact with the phospholipids and may have a tendency to penetrate into the hydrophobic core of the vesicles. In contrast, the smaller, more polar cationic derivatives of epicatechin may be located at the outer interface of the lipid membrane (Lazaro et al. 2007). Consistently, a very polar proanthocyanidin trimer composed of catechin-(4→8)-catechin-(4→8)-catechin adsorbed on the membrane surface without disturbing the membrane head groups (Yu et al. 2011).

Regarding external factors, the affinity of the catechins for lipid bilayers strongly depended on the salting out effect of the aqueous medium, hence increased concentrations of salt in aqueous medium result in more incorporation of catechins into the lipid bilayers. Furthermore, the negative electric charge of the lipid bilayers provides a converse effect on the incorporation of catechins (Kajiya et al. 2002).

The integrity of the apical plasma membrane and the conservation of tight junctions of the enterocyte are essential for maintaining the intestinal epithelial barrier. This barrier is necessary for preventing the penetration of undesired agents and microorganisms. Here, large procyanidins may exert beneficial effects, such as alterations in intestinal barrier permeability. Indeed, interactions of procyanidins with synthetic membranes can protect them from disruption by detergents and also form free radical mediated lipid oxidation (Erlejman et al. 2006).

It is also possible to produce films with improved structural properties from natural membranes with the aid of proanthocyanidins. The presence of hydroxyl-groups in the proanthocyanidin structure enables crosslinking with chitosan and gelatin polymers through ester linkages resulting in an elastic and flexible film. The stabilizing effect of proanthocyanidin on this film might be due to the high number of covalent crosslinks. Such a film is a biocompatible membrane with greatly improved mechanical properties (Kim et al. 2005).

Despite the recent studies concerning this subject many questions remain, including the precise mechanism of interaction and transport of proanthocyanidins through the different membranes and polymers.

3.5 Metal Chelating Activity

Interactions of proanthocyanidins with metal ions are important for a number of reasons. Such interactions have a pronounced impact on the activity of the proanthocyanidins and bioavailability of the metal ions.

Hydroxyl groups of proanthocyanidins or their monomers play a basic role in the interaction of such molecules with iron ions most likely through specific 'iron binding motifs' present in the flavanol structure. Here *O*-dihydroxyphenyl/catechol groups of the B-ring are supposed to be the prime metal-chelation sites (Aron and Kennedy 2008; Fraga et al. 2010). The degree of polymerization and metabolism, which ultimately affects the hydroxyl content of proanthocyanidins, and might impact on the functionality of their catechol groups, also has an impact on the metal chelating activity of these phytochemicals.

At the same time, the bioavailability of the metals is also affected. Nonetheless, it seems that Fe^{2+}, Zn^{2+} and even Cu^{2+} are not particularly well chelated by proanthocyanidins. It is also likely that vitamin C prevents iron chelation, due to the reduction of Fe^{3+} to Fe^{2+}, and subsequently causes a higher bioavailability of iron ions (Santos-Buelga and Scalbert 2000).

3.6 Pro- and Antioxidant Activity

It is likely that antioxidant efficiency of proanthocyanidins—established repeatedly in various studies—plays an important role. Flavan-3-ols seem to develop their antioxidant activity through different mechanisms, including free radical scavenging, transition metal chelation and interference with specific enzymes (Aron and Kennedy 2008).

The flavan-3-ol structure—which includes several hydroquinone moieties—obviously facilitates radical scavenging by simple chemical reductions. Antioxidant efficiency is therefore affected by the degree of polymerization (also the position of hydroxyl groups), by galloylation, methoxylation and glycosylation (Xu et al. 2012; Aron and Kennedy 2008).

In contrast, semiquinone radicals show the tendency to chemically create oligomeric structures. This coupling through a radical mechanism of addition effectively maintains the scavenging activity of proanthocyanidins by retaining the catechol/pyrogallol moieties. The relative stability of radicals produced when such structures act as antioxidants ultimately results in higher antioxidant activity compared to Vitamin C and E (Aron and Kennedy 2008). A noteworthy theory implies that such a redox active system may sustain redox cycling which ultimately may result in a pronounced ROS generating activity. Furthermore, proanthocyanidins may interact with enzymes and bind to metal ions, and hence may also indirectly cause the generation of other pro- or antioxidants. Ultimately a more or less unintelligible behavior of proanthocyanidins—sometimes paradoxical—might ensue (Xu et al. 2012).

3.7 Anti-inflammatory Activity

Inflammation is an elaborated process which results in the activation of signaling pathways and transcription factors and also the release of various proinflammatory molecules (such as TNF and IL-1) and oxidants in cells. These lead to detrimental effects on normal tissues and finally result in systemic disorders (for instance cardiovascular and neurodegenerative diseases, various types of cancer and diabetes) (Pan et al. 2010).

Although the anti-inflammatory mechanisms of flavan-3-ols are not yet clearly defined or fully understood, an inhibition of the NF-κB (nuclear factor-kappa B) signaling cascade and suppression of AP-1 (activator protein-1) activation processes are among the anti-inflammatory mechanisms currently discussed in the context of flavan-3-ols. Furthermore, inhibition of proinflammatory mediators and leukocyte infiltration are also possible actions associated with these phytochemicals in the context of preventing inflammation (Crozier et al. 2009; Serafini et al. 2010; Sharma and Katiyar 2010).

Nonetheless, conclusive anti-inflammatory trials in humans with the focus on chemically pure flavan-3-ols—instead of studies on flavanol-rich foods—clinically are still required in order to clarify the exact nature and extent of such anti-inflammatory properties, if any (Serafini et al. 2010).

3.8 Degradation of Proanthocyanidins

From a chemical point of view, the 'B'-type C–C linkage between proanthocyanidin monomers is not particularly stable and may be cleaved easily by nucleophilic agents such as thiols (resulting in 'thiolysis') under acidic conditions at room or elevated temperatures (Xu et al. 2012; de la Iglesia et al. 2010; Chen et al. 2009).

In contrast, these acid-sensitive linkages show stability under moderate basic conditions, and moreover, 'A'-type proanthocyanidins are resistant against thiolytic degradation (Xu et al. 2012; Kelm et al. 2005).

Nucleophiles like phloroglucinol and various thiol containing reagents (such as benzyl mercaptan, thiophenol, mercaptoacetic acid, and cysteamine) capture

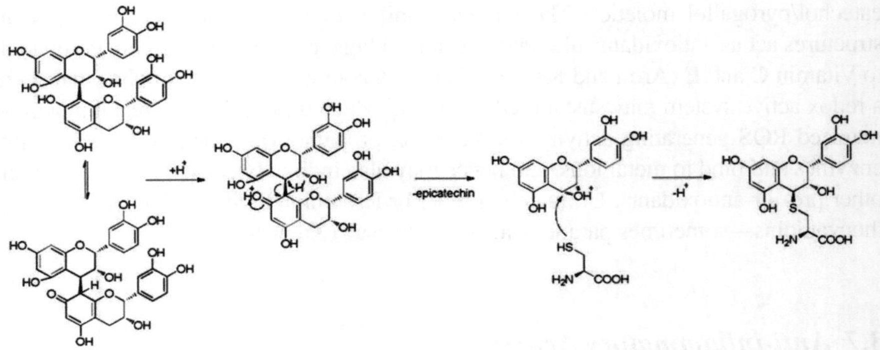

Fig. 5 Degradation of a proanthocyanidin in the presence of L-cysteine with release of epicatechin and the formation of thioether ('thiolysis') (Xu et al. 2012)

carbocations while reacting with proanthocyanidins and subsequently result in sulfur-containing adducts, such as different thioethers (Fig. 5).

Interestingly, these semisynthetic adducts themselves also possess medical importance in the context of chemoprevention and activity against cancer cells (Xu et al. 2012; de la Iglesia et al. 2010; Chen et al. 2009; Iglesias et al. 2012). Furthermore, phloroglucinolysis and thiolysis are also used as analytical tools in the determination of the mean degree of polymerization (mDP) and also to define the monomeric composition of such proanthocyanidins (Xu et al. 2012).

Despite their value as an analytical and 'semi-synthetic' tool in proanthocyanidin degradation, the application of thiolytic methods has some limitations, including epimerization of monomers produced, low yields due to instability of product, side reactions and an unbearable laboratory environment because of the unpleasant odour of some of the thiolysation reagents (e.g., thiophenol and benzyl mercaptan) (Kelm et al. 2005; Chen et al. 2009).

3.9 Synthesis of Proanthocyanidins

Proanthocyanidins are secondary metabolites present in tissues of many higher plants, including seeds, leaves, bark, and fruits (Paolocci et al. 2007; Dixon et al. 2005). In some plants, like strawberries, most of the structural genes related to the proanthocyanidin biosynthesis have been characterized and today, genetic control plays an important role in the production of value-added engineered plants (Schaart et al. 2013).

The first attempts to synthesize proanthocyanidins chemically were based on their biosynthetic pathways and included the condensation of a flavan-3,4-diol (leucoanthocyanidin)—an electrophile—and a flavan-3-ol, which acts as a nucleophile (Ferreira and Coleman 2011).

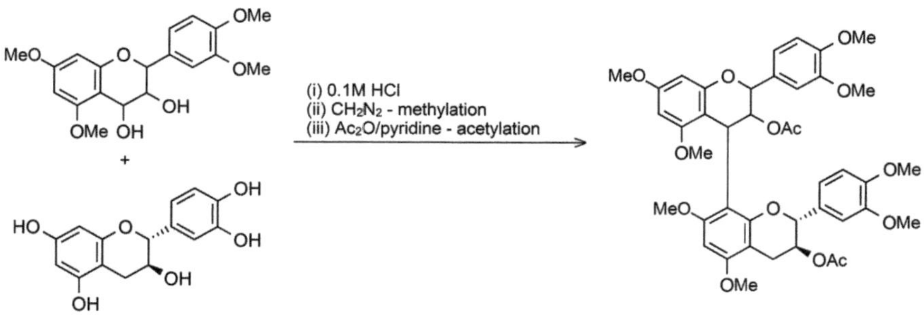

Fig. 6 Initial synthesis of a proanthocyanidin dimer via acid-catalyzed condensation

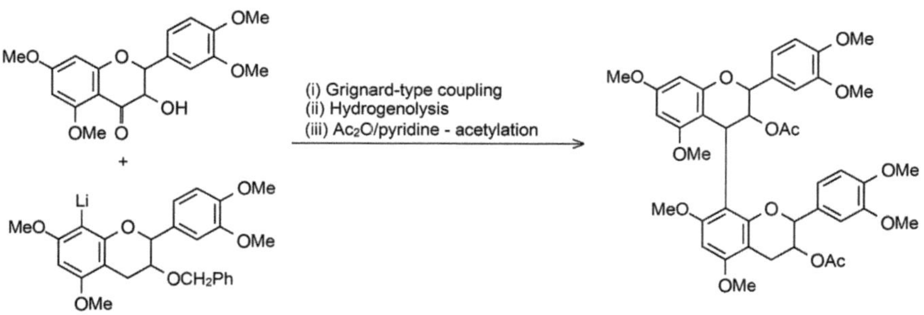

Fig. 7 Procyanidin synthesis via a Grignard-type coupling

Here Geissman and Yoshimura (1966) elucidated the synthesis of a procyanidin derivative through a mild acid-catalyzed condensation of a flavan-3,4-diol analog and a flavan-3-ol followed by a methylation and acetylation process (Fig. 6).

Thereafter an alternative method has been developed using a Grignard-type coupling reaction between a derivative of catechin and a taxifolin derivative, followed by hydrogenolysis and acetylation to produce a procyanidin B-3 derivative (Fig. 7).

Acid-catalyzed thiolysis was first used by Haslam and his collaborators for free phenolic procyanidin synthesis (Haslam 1974; Fletcher et al. 1977).

Indeed, most of the synthetic protocols classically depict interflavanyl bond formation. In the recent past, such reactions were modified toward high yield and performance (Ferreira and Coleman 2011).

Finally, besides the fact that most of the naturally occurring free phenolic analogs are now synthetically accessible, an easy-to-conduct chemical synthesis of proanthocyanidins is still a wishful dream, not least because of the presence of additional chiral centers. Ultimately, the lack of straightforward methods for the synthesis of oligomers containing more than 5 monomers and which hence exert biological activity, is highly disappointing from a biological point of view (Xu et al. 2012; Ferreira and Coleman 2011).

3.10 Bioavailability of Proanthocyanidins

Bioavailability is a limiting factor affecting the biological activity of proanthocyanidins, and hence also limits their impact on human health (Monagas et al. 2010). Processes such as digestion, absorption, metabolism, and elimination affect the bioavailability of flavan-3-ols. After digestion, flavonoids are being solubilized for uptake through intestinal epithelial cells (Smith 2013). Here, variables such as the degree of polymerization (DP) and galloylation have a large impact on flavan-3-ol bioavailability. Hence, monomers are readily absorbed in the small intestine while galloylation considerably reduces their absorption. In contrast, hydrolytic degradation of polymeric proanthocyanidins to monomers under the acidic condition of the stomach may increase bioavailability yet may 'dramatically' reduce activity (Monagas et al. 2010). Although low molecular weight flavanols (DP ≤ 3) are easily absorbed intact in the gastrointestinal tract, proanthocyanidins are considered to be absorbed 10–100-fold less than their corresponding monomers which might be due to their lower paracellular permeability and their binding capacity to the mucosal and luminal proteins (Manach et al. 2005; Aron and Kennedy 2008).

To date, most interest in human health effects of proanthocyanidins is attributed to the flavanol metabolites rather than the original proanthocyanidine forms present in foods (Monagas et al. 2010). Notably, flavanols are predominantly present as conjugated forms in plasma (Aron and Kennedy 2008). Here, extensive metabolism of monomeric flavanols is likely to occur in the enterocytes or hepatocytes via Phase II enzymes resulting in conjugates. Later, these metabolites can enter systemic circulation or are eliminated via bile.

Proanthocyanidins with DP > 3 pass unaltered through the small intestine and enter the colon, where they become metabolized by the *gut microflora*. Microbial biotransformation of flavanols before absorption in the colon results in microbial-derived metabolites which are later on mostly excreted via the urine (Fig. 8) (Monagas et al. 2010).

Despite the many unknowns involved in these microbial pathways, a large percentage of microbial-derived metabolites ultimately results in considerable bioavailability of such proanthocyanidin products.

Research in this area is cumbersome, and proanthocyanidin analysis may be hampered due to the fact that such compounds can readily associate with proteins in blood and tissues resulting in a notorious underestimation of proanthocyanidin content in biological samples (Xu et al. 2012).

Intriguingly, several studies have illustrated that diverse phytochemical targets in the gastrointestinal tract, respiratory system or on the skin capitalize on the remarkable potential of flavanols without the necessity of entering into the blood circulation (Xu et al. 2012).

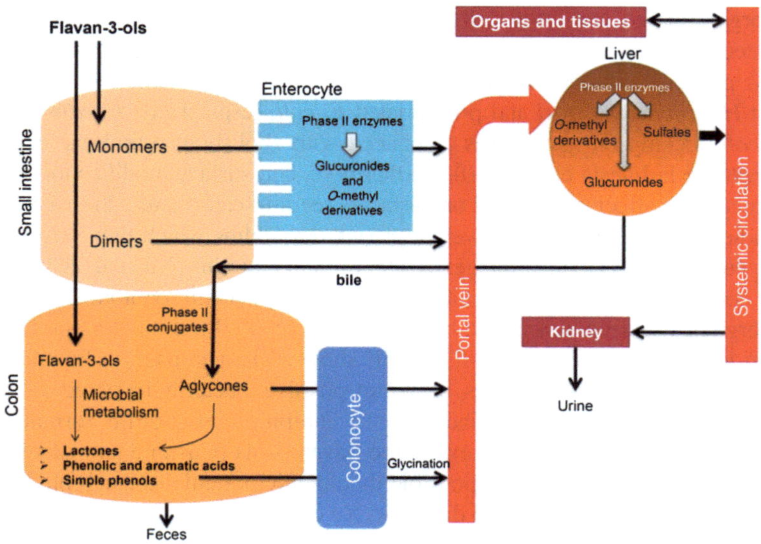

Fig. 8 Schematic diagram of bioavailability of flavan-3-ols (Monagas et al. 2010)

4 Beneficial Health Effects

The antioxidant, free radical scavenging, and anti-inflammatory role of proanthocyanidins and their monomers (besides other biochemical properties these phytochemicals) turns them into interesting natural compounds, which ultimately may also enhance human health.

4.1 Antimicrobial Activity

In vitro inhibition of several Gram-positive and Gram-negative bacteria (like *Clostridium perfringens*, *Streptococcusmutans*, *Vibrio cholerae*, *Campylobacter jejuni* and *Escherichia coli*) has revealed an antibacterial activity of flavan-3-ols (Daglia 2012). These effects of the flavanols are mostly caused by affecting the bacterial growth rate or reducing the pathogenicity of such organisms by decreasing the virulence factors.

The antibacterial mode of these phytochemicals is still not fully understood though some aspects are more likely than others. Positive adhesion of pathogenic microorganisms to the host cell is a prerequisite for virulence. Research has illustrated the *anti-adhesive mechanism* of *Pelargonium sidoides* extract (contains mainly polymeric proanthocyanidins) and a cranberry proanthocyanidin extract against *Helicobacter pylori* and p-fimbriated *E. coli*, respectively. *In vitro* inhibition of *E. coli* docking on tissues was associated with abolishing p-fimbriae expression of *E. coli* (de la Iglesia et al. 2010). Furthermore, cell surface modifications—which might occur by affecting proteins on surfaces, seem to cause anti-adhesive effects in the studies with high molecular weight proanthocyanidins which themselves are not able to penetrate cells to alter the gene expression of adhesins (Xu et al. 2012).

On certain occasions, wet surfaces are likely to bear microcolonies of different microorganisms, structures called biofilms. Despite the different strategies for controlling biofilms available, such films still show a great impact on human health. Inhibition of sortases—enzymes of Gram-positive bacteria involved in surface protein assembly—might contribute to the inhibition of biofilm formation and have been mentioned in the context of the flavan-3-ols (Cushnie and Lamb 2011). These properties point toward a possible therapeutic potential of such phytochemicals in the context of periodontal diseases (Xu et al. 2012).

Inhibition of cell membrane synthesis is another possible action explaining antimicrobial activity of these flavan-3-ols.

Furthermore, inhibition of ATP synthase is a mechanism associated with flavanols which interferes with energy metabolism of bacteria. Finally, it is noteworthy to mention that disrupting liposomes—due to the lipophilicity, metal depletion, and inhibition of cell associated proteolysis of flavanols may play an important role in the context of their antibacterial activity (Xu et al. 2012; Cushnie and Lamb 2011).

Recently, some tea flavan-3-ols at comparable concentrations were found to be more active than antibiotics, i.e., vancomycin or tetracycline. Furthermore, galloylated flavan-3-ols potentiate the antimicrobial activity of β-lactam antibiotics against methicillin-resistant *Staphylococcus aureus* (MRSA) through decreasing the MICs of these antibacterial agents (Cushnie and Lamb 2011).

Concerning the antibacterial activity of flavanols it should also be mentioned that these compounds are able to shift the colonic microbial flora to a more beneficial composition without changing the total count or anaerobe/aerobe ratio (Xu et al. 2012).

The antiviral activity of EGCG against the flu virus was discovered in the 1990s. It has been established that inhibition of the viral attachment to the target cell plays an important role in the anti-viral activity of flavanols. Binding of EGCG to hemagglutinin of the flu virus and also prevention of herpes simplex virus (HSV-1) attachment to Vero cells by oligomeric proanthocyanidin extracts have been implicated as other possible modes of action (Xu et al. 2012; Daglia 2012). Therein, modifications on the surface of the viral particles are the most probable explanation for the inhibition of attachment (Xu et al. 2012). Notably, galloylation of flavan-3-ols seems to enhance their antiviral activity (Daglia 2012). The inhibition of viral polymerase is another promising interaction that may explain the antiviral activity, together with potential redox modulation efficacy (Gallina et al. 2011). Adenovirus and enterovirus are among other viruses showing vulnerability to flavanols (Daglia 2012; Gallina et al. 2011).

Ultimately, prevalent fungal superinfection of the vagina, intestine, and oral cavity caused by *Candida albicans*—trigged, for instance, by the use of immunosuppressive agents, glucocorticoids, or long-term treatments by antibiotics—might be curable with flavan-3-ols. EGCG, in particular, demonstrated fungicidal activities against various fungi including *Candida albicans* (Daglia 2012).

4.2 Anticancer and Cardiovascular Effects

Proanthocyanidins, especially EGCG, are suspected of various anticancer activities. Such compounds have, for instance, the ability to induce a chemopreventive, antioxidant response via MAPK signaling, inhibit COX-2 overexpression, block native metalloprotein expression (the latter play an important role in tumor invasion and angiogenesis), and affect the microbial flora of the colon so to minimize the excretion of toxic compounds by various gut bacteria, hence helping to prevent the formation of gastrointestinal cancers.

Cardiovascular disease (CVD), is a leading cause of death in the developed world, and includes heart attacks, stroke, hypertension, congenital or rheumatic heart disease, heart failure and peripheral artery disease (De Pascual-Teresa et al. 2010; de la Iglesia et al. 2010).

Among these actions and mechanisms, one should mention the activation of nitric oxide synthesis (properly via eNOS) and hence vasodilatory and vasorelaxant effects, a pronounced decrease in atheriosclerotic factors [e.g., tumor necrosis factor-alpha (TNF-α), monocyte chemoattractant protein 1 (MCP-1), C reactive protein (CRP), and soluble intercellular adhesion molecule 1 (sICAM-1)], inhibition of the platelet-dependent inflammatory response and anti-thrombotic activity, as well as inhibition of LDL oxidation due to antioxidant properties. Notably, it is necessary to take 2–5 mugs of green tea per day to attain this reduction in LDL oxidation.

Other beneficial effects of proanthocyanidins and/or EGCG include a down-regulation of MCP-1 and hence of insulin resistance, an elevation of plasma insulin levels, and delayed onset of the disease (as observed in non-obese Type I diabetic mice), an insulin mimetic activity of certain proanthocyanidins, and an enhanced expression of proteins linked to the insulin pathway. Certain beneficial effects may be more indirect, such as an amelioration of diabetes-associated complications, as observed in the case of proanthocyanidin-rich grape seed extracts.

5 Conclusions

Proanthocyanidins as polyphenolic compounds represent a vast and diverse group of phytochemicals with substantial health benefits, however, few studies so far have provided convincing evidence regarding the 'therapeutic' applications of proanthocyanidins as nutraceuticals.

The bioavailability of proanthocyanidins is another controversial issue and may limit their interactions to a direct contact with the targets either on the surface of the body or inside the gastrointestinal tract. Moreover, monomeric proanthocyanidins can readily be absorbed intact from the gut. The application of proanthocyanidins as functional foods or green pesticides may also be practicable aside from their various therapeutic potentials.

Designing efficacious and facile methods for the synthesis of pure proanthocyanidins with a sufficient degree of polymerization is essential for future clinical studies and so far, an easy-to-conduct method for such a synthesis is mostly lacking.

Ultimately, the biochemical and medical properties of proanthocyanidins from different sources need to be considered together with the exact effects of food processing and storage on these phytochemicals and their roles as a part of a healthy human diet.

References

Aron PM, Kennedy JA (2008) Flavan-3-ols: nature, occurrence and biological activity. Mol Nutr Food Res 52(1):79–104

Beart JE, Lilley TH, Haslam E (1985) Polyphenol interactions. Part 2. Covalent binding of procyanidins to proteins during acid-catalysed decomposition; observations on some polymeric proanthocyanidins. J Chem Soc, Perkin Trans 2 (9):1439–1443

Beecher GR (2003) Overview of dietary flavonoids: nomenclature, occurrence and intake. J Nutr 133(10):3248S–3254S

Chen W, Fu C, Qin Y, Huang D (2009) One-pot depolymerizative extraction of proanthocyanidins from mangosteen pericarps. Food Chem 114(3):874–880

Crozier A, Jaganath IB, Clifford MN (2009) Dietary phenolics: chemistry, bioavailability and effects on health. Nat Prod Rep 26(8):1001–1043

Cushnie TP, Lamb AJ (2011) Recent advances in understanding the antibacterial properties of flavonoids. Int J Antimicrob Agents 38(2):99–107

Daglia M (2012) Polyphenols as antimicrobial agents. Curr Opin Biotechnol 23(2):174–181

de la Iglesia R, Milagro FI, Campion J, Boque N, Martinez JA (2010) Healthy properties of proanthocyanidins. BioFactors 36(3):159–168

De Pascual-Teresa S, Moreno DA, García-Viguera C (2010) Flavanols and anthocyanins in cardiovascular health: a review of current evidence. Int J Mol Sci 11(4):1679–1703

Dixon RA, Xie DY, Sharma SB (2005) Proanthocyanidins–a final frontier in flavonoid research? New Phytol 165(1):9–28

Erlejman AG, Fraga CG, Oteiza PI (2006) Procyanidins protect Caco-2 cells from bile acid- and oxidant-induced damage. Free Radical Biol Med 41(8):1247–1256

Ferreira D, Coleman CM (2011) Towards the synthesis of proanthocyanidins: half a century of innovation. Planta Med 77(11):1071–1085

Fletcher AC, Porter LJ, Haslam E, Gupta RK (1977) Plant proanthocyanidins. Part 3. Conformational and configurational studies of natural procyanidins. J Chem Soc, Perkin Trans 1 (14):1628–1637

Fraga CG, Galleano M, Verstraeten SV, Oteiza PI (2010) Basic biochemical mechanisms behind the health benefits of polyphenols. Mol Aspects Med 31(6):435–445

Gallina L, Dal Pozzo F, Galligioni V, Bombardelli E, Scagliarini A (2011) Inhibition of viral RNA synthesis in canine distemper virus infection by proanthocyanidin A2. Antiviral Res 92(3):447–452

Geissman TA, Yoshimura NN (1966) Synthetic proanthocyanidin. Tetrahedron Lett 7(24):2669–2673

Haslam E (1974) Biogenetically patterned synthesis of procyanidins. J Chem Soc, Chem Commun 0(15):594–595

Iglesias J, Pazos M, Torres JL, Medina I (2012) Antioxidant mechanism of grape procyanidins in muscle tissues: redox interactions with endogenous ascorbic acid and alpha-tocopherol. Food Chem 134(4):1767–1774

Kajiya K, Hojo H, Suzuki M, Nanjo F, Kumazawa S, Nakayama T (2004) Relationship between antibacterial activity of (+)-catechin derivatives and their interaction with a model membrane. J Agric Food Chem 52(6):1514–1519

Kajiya K, Kumazawa S, Nakayama T (2002) Effects of external factors on the interaction of tea catechins with lipid bilayers. Biosci Biotechnol Biochem 66(11):2330–2335

Kalili KM, de Villiers A (2009) Off-line comprehensive 2-dimensional hydrophilic interaction× reversed phase liquid chromatography analysis of procyanidins. J Chromatogr A 1216(35):6274–6284

Kashiwada Y, Nonaka G-I, Nishioka I (1986) Tannins and related compounds. XLV. Rhubarb. (5). Isolation and characterization of Flavan-3-ol and procyanidin glucosides. Chem Pharm Bull 34(8):3208–3222

Kelm MA, Hammerstone JF, Schmitz HH (2005) Identification and quantitation of flavanols and proanthocyanidins in foods: how good are the datas? Clin Dev Immunol 12(1):35–41

Kim S, Nimni ME, Yang Z, Han B (2005) Chitosan/gelatin-based films crosslinked by proanthocyanidin. J Biomed Mater Res B Appl Biomater 75(2):442–450

Kuzuhara T, Sei Y, Yamaguchi K, Suganuma M, Fujiki H (2006) DNA and RNA as new binding targets of green tea catechins. J Biol Chem 281(25):17446–17456

Kuzuhara T, Tanabe A, Sei Y, Yamaguchi K, Suganuma M, Fujiki H (2007) Synergistic effects of multiple treatments, and both DNA and RNA direct bindings on, green tea catechins. Mol Carcinog 46(8):640–645

Lazaro E, Castillo JA, Rafols C, Roses M, Clapes P, Torres JL (2007) Interaction of antioxidant biobased epicatechin conjugates with biomembrane models. J Agric Food Chem 55(8):2901–2905

Manach C, Williamson G, Morand C, Scalbert A, Remesy C (2005) Bioavailability and bioefficacy of polyphenols in humans. I. Review of 97 bioavailability studies. Am J Clin Nutr 81(1 Suppl):230S–242S

Mehansho H, Butler LG, Carlson DM (1987) Dietary tannins and salivary proline-rich proteins: interactions, induction, and defense mechanisms. Annu Rev Nutr 7:423–440

Monagas M, Urpi-Sarda M, Sanchez-Patan F, Llorach R, Garrido I, Gomez-Cordoves C, Andres-Lacueva C, Bartolome B (2010) Insights into the metabolism and microbial biotransformation of dietary flavan-3-ols and the bioactivity of their metabolites. Food Funct 1(3):233–253

Morimoto S, Nonaka G-I, Nishioka I (1986a) Tannins and related compounds. XXXIX. Procyanidin C-glucosides and an acylated Flavan-3-ol glucoside from the Barks of *Cinnamomum cassia* Blume and *C. obtusifolium* NEES. Chem Pharm Bull 34(2):643–649

Morimoto S, Nonaka G-I, Nishioka I (1986b) Tannins and related compounds. XXXVIII. Isolation and characterization of Flavan-3-ol glucosides and procyanidin oligomers from *Cassia bark*: *Cinnamomum cassia* Blume. Chem Pharm Bull 34(2):633–642

Oh HI, Hoff JE, Armstrong GS, Haff LA (1980) Hydrophobic interaction in tannin-protein complexes. J Agric Food Chem 28(2):394–398

Pan M-H, Lai C-S, Ho C-T (2010) Anti-inflammatory activity of natural dietary flavonoids. Food Funct 1(1):15–31

Paolocci F, Robbins MP, Madeo L, Arcioni S, Martens S, Damiani F (2007) Ectopic expression of a basic helix-loop-helix gene transactivates parallel pathways of proanthocyanidin biosynthesis. Structure, expression analysis, and genetic control of Leucoanthocyanidin 4-Reductase and Anthocyanidin Reductase genes in *Lotus corniculatus*. Plant Physiol 143(1):504–516

Plumb GW, De Pascual-Teresa S, Santos-Buelga C, Cheynier V, Williamson G (1998) Antioxidant properties of catechins and proanthocyanidins: effect of polymerisation, galloylation and glycosylation. Free Radical Res 29(4):351–358

Revesz K, Tutto A, Szelenyi P, Konta L (2011) Tea flavan-3-ols as modulating factors in endoplasmic reticulum function. Nutr Res 31(10):731–740

Ricardo da Silva JM, Rigaud J, Cheynier V, Cheminat A, Moutounet M (1991) Procyanidin dimers and trimers from grape seeds. Phytochemistry 30(4):1259–1264

Santos-Buelga C, Scalbert A (2000) Proanthocyanidins and tannin-like compounds–nature, occurrence, dietary intake and effects on nutrition and health. J Sci Food Agric 80(7):1094–1117

Sazuka M, Itoi T, Suzuki Y, Odani S, Koide T, Isemura M (1996) Evidence for the interaction between (−)-epigallocatechin gallate and human plasma proteins fibronectin, fibrinogen, and histidine-rich glycoprotein. Biosci Biotechnol Biochem 60(8):1317–1319

Schaart JG, Dubos C, De La Fuente IR, van Houwelingen AM, de Vos RC, Jonker HH, Xu W, Routaboul JM, Lepiniec L, Bovy AG (2013) Identification and characterization of MYB-bHLH-WD40 regulatory complexes controlling proanthocyanidin biosynthesis in strawberry (*Fragaria x ananassa*) fruits. New Phytol 197(2):454–467

Serafini M, Peluso I, Raguzzini A (2010) Flavonoids as anti-inflammatory agents. Proc Nutr Soc 69(03):273–278

Sharma SD, Katiyar SK (2010) Dietary grape seed proanthocyanidins inhibit UVB-induced cyclooxygenase-2 expression and other inflammatory mediators in UVB-exposed skin and skin tumors of SKH-1 hairless mice. Pharm Res 27(6):1092–1102

Smith DF (2013) Benefits of flavanol-rich cocoa-derived products for mental well-being: a review. J Funct Foods 5(1):10–15

Todd JJ, Vodkin LO (1993) Pigmented soybean (*Glycine max*) seed coats accumulate proanthocyanidins during development. Plant Physiol 102(2):663–670

Wang C-S, Vodkin L (1994) Extraction of RNA from tissues containing high levels of procyanidins that bind RNA. Plant Mol Biol Rep 12(2):132–145

Xu Z, Du P, Meiser P, Jacob C (2012) Proanthocyanidins: oligomeric structures with unique biochemical properties and great therapeutic promise. Nat Prod Commun 7(3):381–388

Yu X, Chu S, Hagerman AE, Lorigan GA (2011) Probing the interaction of polyphenols with lipid bilayers by solid-state NMR spectroscopy. J Agric Food Chem 59(12):6783–6789

Authors Biography

Hadi Ebrahimnejad (born 1983) graduated with his DVM from the School of Veterinary Medicine, Shiraz University, Iran in 2008. He received his PhD in Food Science and Public Health from the same university in 2013. During that year, he also spent several months with Prof. Claus Jacob at the Division of Bioorganic Chemistry at Saarland University, Germany, to fulfill his postgraduate expectations.

Currently, Hadi is Assistant Professor at the Department of Food Hygiene, Faculty of Veterinary Medicine, Shahid Bahonar University of Kerman, Iran. His major research interests include functional foods and nutraceuticals with a particular focus on enzymes.

Torsten Burkholz (born 1979) has been trained as an inorganic and medicinal chemist at the University of Saarland in Saarbruecken, Germany, graduating with a German 'Diplom' in Chemistry. After completing his PhD studies at the University of Saarland in the fields of Chemistry and Pharmacy under the supervision of Prof. Claus Jacob in 2010, he joined the European Marie Curie Initial Training Network "RedCat" as postdoctoral Experienced Researcher, conducting research in the field of Cell Biology in the group of Prof. Paul G. Winyard at the Peninsula College of Medicine and Dentistry, Exeter, UK. In 2012 Torsten moved back to the University of Saarland where he currently holds the position of an "Akademischer Rat" in Bioorganic Chemistry.

As part of this senior position, Torsten is managing the research laboratory of Bioorganic Chemistry, as well as the relevant teaching and the examinations of undergraduate students. Together with Prof. Claus Jacob, he has to date published over 30 publications in the field of oxidative stress, chalcogen containing natural compounds and their biological activity and, more recently, on the cellular thiolstat. Torsten's ongoing research includes synthetic and analytical chemistry, biological activity studies and 'intracellular diagnostics'.

In 2013, Torsten was appointed as Visiting Professor at the University of Applied Sciences Kaiserslautern, where he is lecturing Pharmacology. In the same year, he established his own small company named "Dr. Burkholz Life Science Consulting UG" which provides scientific consultations for small and medium sized companies in the field of nutrition, food supplements, and natural compounds. His company also offers training and consulting in Inorganic and Analytical Chemistry, as well as in Physics and Pharmacology.

Claus Jacob (born 1969) has been trained as a synthetic (in)organic and biological chemist at the Universities of Kaiserslautern, Leicester, Oxford, and Harvard. He graduated with a 1st class B.Sc. (Hons.) degree from the University of Leicester in 1993, and with a D.Phil. from the University of Oxford in 1997 ("Genetic engineering of redox active enzymes", supervisor Prof. Allen Hill FRS). He subsequently joined the institute of Prof. Bert Vallee at Harvard Medical School as a Feodor Lynen Fellow (Alexander von Humboldt-Foundation) to study processes controlling intracellular zinc homeostasis. During this time, he also obtained a Magister Artium degree in Philosophy, History, and Psychology from the University of Hagen in Germany (M.A. dissertation on

Protochemistry as constructivist foundation of chemistry). He left the US in 1999 to spend some time with Prof. Helmut Sies at the Heinrich-Heine-University in Duesseldorf, Germany, as part of a BASF Research Fellowship from the German Merit Foundation.

Claus started his independent scientific career as lecturer at the University of Exeter in the UK in 1999 and in 2005 moved to the University of Saarland where he currently holds the position of Professor of Bioorganic Chemistry. Claus is an expert in redox active compounds and their impact on biological systems and to date has published over 100 publications in this field. Over the years, his research has focused on Reactive Sulfur Species (RSS) and the cellular thiolstat, terms his team has introduced in 2001 and 2010, respectively. Besides his strong interest in redox active sulfur, Claus has also developed an active research program on synthetic 'sensor/effector' redox modulators based on selenium and tellurium, on redox active plant metabolites and on nanoscopic redox particles. His research includes synthetic and analytical chemistry, biological activity studies and 'intracellular diagnostics' to decipher and map out intracellular events and mode(s) of actions. Claus has coordinated the EU Marie Curie Initial Training Network "RedCat" (2008–2012), has been a partner in the technology transfer project "Corena" (2009–2012) and is currently partially in charge of the natural products project "NutriOx".

Throughout the years, Claus has undertaken many projects to become a highly skilled undertaker, but never a true philosopher, yet his more philosophical and cunning linguistic outpourings are famous and he still maintains a keen interest in various aspects related to the philosophy of chemistry.

Part VIII
Connecting Section Between Chapters 8 and 9

When discussing redox active substances from plants or fungi, the first compounds that come to mind are polyphenols. This may not be particularly surprising, as polyphenols exhibit a classical electron-transfer redox chemistry and are also ubiquitous in nature. From a chemical perspective, however, such compounds are somewhat problematic. First of all, these compounds are not really ideal redox modulators in a more biological context. Their ability to donate electrons is not particularly useful in biological systems, where most redox events proceed via exchange reactions. Indeed, the reaction kinetics associated with such antioxidants is often rather disappointing. Secondly, and perhaps even worse, polyphenols are not very selective as far as their reaction partners are concerned. It is even possible that some of these compounds—either directly or indirectly—reduce molecular oxygen to the highly damaging superoxide radical anion. Thirdly, polyphenols are redox cyclers. While they may initially act as reducing antioxidants, their oxidized forms, i.e., the quinones, may then oxidize cellular components. These oxidized forms are also highly reactive electrophiles that may cause serious cellular damage via uncontrolled Michael addition reactions. And fourthly, the bioavailability of these often poorly soluble compounds is rather limited, hence questioning whether it is possible to take up therapeutically relevant doses of these compounds in the form of food or food supplements at all (intravenous applications can obviously be ruled out as a realistic form of application). We have already discussed some of these issues in Chaps. 2 and 5 and will return to them later on.

In the next four chapters we will shift our focus to another class of redox active secondary metabolites, which avoids many of the drawbacks associated with flavonoids and related polyphenols. These chapters will open up the debate on organic sulfur compounds. The latter form a unique group of secondary metabolites found in many plants, fungi, and bacteria. These compounds are chemically diverse, yet resemble each other with regard to their high, yet selective reactivity with cysteine and selenocysteine residues in proteins and enzymes. While the traditional view on these compounds has been that any modifications induced by them in the cell are only temporarily and minor, and can be 'repaired' swiftly by glutathione (GSH) or glutathione-dependent enzymes, recent research has demonstrated that such compounds are not simply neutralized by GSH, that most intracellular reductases are unable to reverse cysteine modifications induced by these compounds, and that such modifications are indeed well suited to trigger significant cellular responses. Such responses may include the upregulation of antioxidant defenses as well as cell cycle arrest and the induction of apoptosis.

In order to understand better how such extrinsic, redox modulating agents trigger different, often temporary and reversible cellular responses, the concept of the 'cellular thiolstat' has been developed and will be introduced in the following chapter. The thiolstat is a cellular rheostat composed of various—redox sensitive—proteins and enzymes (such as tubulin), which together sense redox changes, provide a cellular redox feedback, and trigger an appropriate, yet often complex cellular response or responses. This medley of cysteine proteins and enzymes also seems to form the preferred target for thiol-specific redox modulating compounds, including many natural organic sulfur compounds.

Chapter 9
The Cellular 'Thiolstat' as an Emerging Potential Target of Some Plant Secondary Metabolites

Martin C. Gruhlke and Alan J. Slusarenko

Keywords Electron transfer • Cellular thiolstat • Glutathione • Reactive sulfur species • *S*-thiolation regulation

1 Introduction: Redox Maintenance and Regulation in Biological Systems

Several biological macromolecules can be reversibly oxidized or reduced and this can affect their properties and thus influence their function either positively or negatively. Indeed, this simple fact is the basis of some complex regulatory machinery in the cell. The redox environment in the cell needs to be closely buffered and monitored so that the multiplicity of the cell's biochemistry runs smoothly in an integrated fashion. Over-reduction, leading for example to misfolding of proteins in the endoplasmic reticulum (ER), and over-oxidation are both harmful to the cell's physiology (Delic et al. 2012; Higa and Chevet 2012). The redox state of particular cysteine thiols in the cell depends upon a number of factors such as their accessibility, specific pK_a, nature of surrounding amino acids and not just the thermodynamics but also the kinetics of possible oxidation/reduction reactions (Dalle-Donne et al. 2009; Nagy 2013; Winterbourn and Hampton 2008). Two paradigms, which are not mutually exclusive but perhaps also are not equally represented in cells, are relevant to the control of thiol-based micro-switches. In the first of these scenarios a particular thiol is in thermodynamic equilibrium within its subcellular environment and the ratio of oxidized to reduced forms is determined by the local redox potential. In the second scenario thermodynamic equilibrium is not assumed and the kinetics of oxidation of a relatively few target protein thiols 'sense' oxidative changes

M. C. Gruhlke · A. J. Slusarenko (✉)
Department of Plant Physiology (Bio III), RWTH Aachen University,
52056 Aachen, Germany
e-mail: alan.slusarenko@bio3.rwth-aachen.de

C. Jacob et al. (eds.), *Recent Advances in Redox Active Plant and Microbial Products*,
DOI: 10.1007/978-94-017-8953-0_9,

in the cell and function as signaling intermediates by relaying information, before being enzymatically reduced back to their basal degree of oxidation (Winterbourn and Hampton 2008). Many experimental data tend to support the second scenario but do not rule out the first for specific instances. It is often stated that the glutathione:glutathione disulfide redox couple (GSH:GSSG) buffers changes in cell redox. However, the situation is complex and although oxidative treatments often lead to a shift in the degree of total cellular glutathione oxidation, GSSG may be largely rapidly removed from the cytosol, for example into the vacuole, thus leaving the local electrochemical cell potential unaltered (Morgan et al. 2013). Thus, caution must be exercised in interpreting whole cell GSH:GSSG ratios in terms of electrochemical potentials in specific cellular compartments. Nevertheless, having stated this, it is a fact of physical chemistry that the redox environment, given the above provisos, will influence the oxidation state of accessible thiols. Therefore, it is important to consider how the redox environment in cells is maintained.

The redox status of the cell is constituted and buffered by a series of redox couples, which are pairs of molecules existing in reduced or oxidized states. For example: $NADH/NAD^+ + H^+$, $NADPH/NADP^+ + H^+$, reduced, and oxidized glutathione (GSH/GSSG) and reduced and oxidized ascorbate (ascorbate/dehydroascorbate) buffer redox systems in the water-soluble cell compartments and reduced and oxidized vitamin E molecules buffer redox changes in the lipophilic cell fractions (Foyer and Noctor 2005). These redox buffers can either directly react with redox active compounds or act as electron donors or reducing equivalents for enzymatic reactions. GSH, for instance, acts as electron donor for glutathione-dependent oxidoreductases ('glutaredoxins') and thioredoxins are reduced via Thioredoxin Reductases (TrxR) with electrons from NADPH. The redox potential (also called the oxidation-reduction potential or midpoint potential) of a redox pair represents the tendency of the oxidized form to acquire electrons and be reduced (and *vice versa*) and is defined by the half-cell electrochemical potential (standard redox potential) of the couple ($E^{0'}$ in mV). The redox buffering capacity of a redox couple is determined by the pool size. The standard-redox potential is empirically defined as the 'dimension' for the relative affinity of an electron-acceptor for electrons and is normalized against the half-cell potential of the standard reaction: $H^+ + e^- \rightarrow ½ H_2$ (oxidized and reduced forms both at 1 M, 298 K) which is arbitrarily given the value 0 V under standard conditions. In Biochemistry, most of the molecular species met with are not stable under the 'standard' conditions, so the reference potential ($E^{0'}$) for a redox couple is usually quoted at pH 7.0 (see also Chap. 4).

In most cells the $NADH/NAD^+$ and $NADPH/NADP^+ + H^+$ couples have the lowest redox potential ($E^{0'} = -315$ mV). The GSH/GSSG couple exhibits an $E^{0'} = -240$ mV, and the ascorbate/dehydroascorbate couple has a higher half cell potential ($E^{0'} = +54$ mV) (Schafer and Buettner 2001). The direction of electron flow is from lower (more reduced) to higher (more oxidized) redox potential [assuming that such a flow is possible mechanistically and not prohibited for kinetic reasons (see Chap. 4)].

The subcellular compartmentalization of the various redox couples and their *in situ* concentrations are important factors which determine local redox environments (redox state and buffering capacity) within cells. As defined by Schafer and Buettner "The redox environment of a linked set of redox couples as found in a biological fluid, organelle, cell, or tissue is the summation of the products of the reduction potential and reducing capacity of the linked redox couples present" (Schafer and Buettner 2001). When the oxidized and reduced forms of a redox couple are not present in a 1:1 ratio, as is usually the case in a cellular compartment, the redox potential can be calculated using the Nernst equation. The relative proportions of oxidized/reduced partners in the NAD(P)H/NAD(P)$^+$ and ascorbate/dehydroascorbate couples both determine and reflect the local redox potential (E_{hc}) independently from their overall absolute concentrations, whereas for the GSH/GSSG couple, not only the relative proportions of oxidized and reduced forms, but also their absolute concentrations must be taken into account. A consideration of the Nernst equation makes it clear why not only the proportions but also the concentrations of GSH/GSSG in the pool determine the redox potential (E_{hc}):

The Nernst equation:

$$E_{hc} = E^{0'} - \left[\left(\frac{RT}{nF} \right) \ln Q \right]$$

where E_{hc} = the electrochemical half-cell potential under the prevailing conditions; $E^{0'}$ = the reference half cell potential (pH = 7); R = the universal Gas Constant; T = 298 K or 25 °C; n = the number of electrons exchanged; F = the Faraday Constant; and Q is the mass action term.

Simplifying for the constants and converting from ln to $\log_{10}$ the expression becomes

$$E_{hc} = E^{0'} - \left[\left(\frac{59.1 \text{ mV}}{n} \right) \log Q \right]$$

For example for the NADP$^+$/NADPH half-reaction couple

$$\text{NADP}^+ + 2\text{e}^- + \text{H}^+ \rightarrow \text{NADPH}$$

$$E_{hc} = -320 \text{ mV} - \left[\left(\frac{59.1 \text{ mV}}{2} \right) \log \frac{[NADPH]}{[NADP^+]} \right]$$

at pH 7 and 25 °C.

Thus, irrespective of the absolute concentrations it is sufficient to know the relative proportions of NADP$^+$ and NADPH present in order to calculate the redox potential; whereas for GSH/GSSG:

$$\text{GSSG} + 2\text{e}^- + 2\text{H}^+ \rightarrow 2\text{GSH}$$

$$E_{hc} = -240 \text{ mV} - \left[\left(\frac{59.1 \text{ mV}}{2} \right) \log \frac{[GSH]^2}{[GSSG]} \right]$$

at pH 7 and 25 °C.

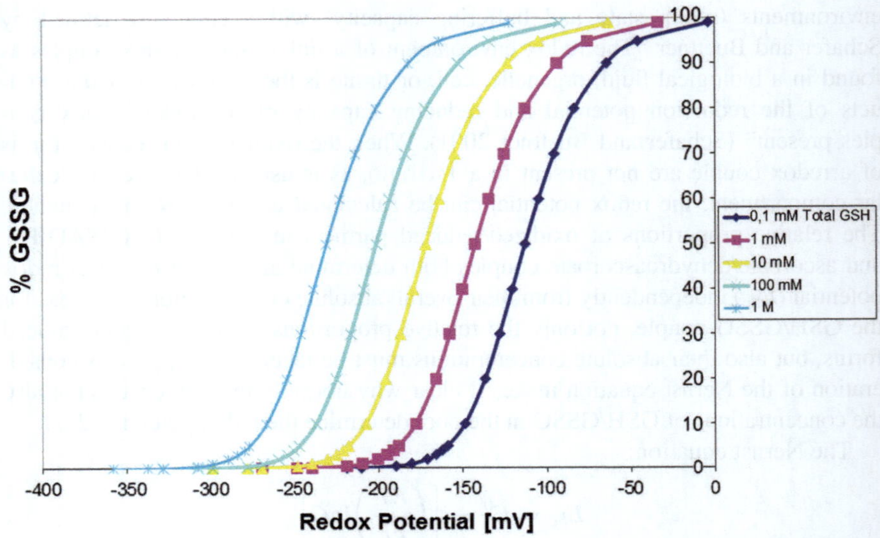

Fig. 1 The redox potential (E_{hc}) for the GSH:GSSG redox couple at varying degrees of oxidation for five GSH concentrations

Note: This form of the equation has the log concentration term for the reductant divided by the log concentration term for the oxidant. Another form of the equation has the log concentration term for the oxidant divided by the log concentration term for the reductant and avoids the use of the 'minus' before the term

$$E_{hc} = -240\text{ mV} + \left[\left(\frac{59.1\text{ mV}}{2}\right)\log\frac{[GSSG]}{[GSH]^2}\right]$$

at pH 7 and 25 °C.

Because of the $[GSH]^2$ 'squared term' introduced through the law of mass action, the relative proportion of GSH:GSSG in the pool is not sufficient to calculate the redox potential (E_{hc}); the absolute concentrations must be known and substituted in the equation. This effect is illustrated in Fig. 1. At this point it is pertinent to mention that much work published on GSH reports only changes in the relative amounts of GSH:GSSG. A change in the GSH:GSSG ratio documents a qualitative shift in the redox status of the system but it does not give information as to the degree of redox change or the absolute redox status (Fig. 1).

The GSH concentration in a plant cell is approximately ten times that of NADH and NADPH, respectively, and values in the range from 1 to 10 mM are commonly measured (Noctor 2006). Because of this high intracellular concentration it is generally held that GSH plays an important role as a cellular buffer against redox changes. Interestingly, the size of the ascorbate pool in plant cells can also be relatively large (~10–100 mM), endowing it with a high buffering capacity, but at $E^{0'} = +54$ mV it has a much higher standard half-cell potential than the $NAD(P)^+$/NAD(P)H and GSH/GSSG redox couples (Schafer and Buettner 2001).

Mutants completely unable to synthesize GSH are usually lethal but the traditional view of the GSH pool as a buffer against redox change has recently been challenged, at least for yeast (*Saccharomyces cerevisiae*). Here the major redox buffer role in the cytosol was attributed to Trx with only a back-up role proposed for GSH, which was, however, essential for Fe-S cluster synthesis, making the *gsh*1 deletion lethal. High GSH levels were postulated to be necessary because the pool size reduces under oxidative stress—i.e., the excess GSH was postulated to be simply necessary to ensure adequate supply for iron metabolism whereas Trx protected the redox environment (Kumar et al. 2011). In a further publication from this group the redox control in other yeast cell compartments was also elaborated upon. It was reported that while the cytosol possessed both Trx and GSH pathways in full, of which the Trx pathway was dominant, the mitochondrial matrix also possessed both pathways but here the GSH pathway had the major role in redox control. In both compartments GSH was essential for non-redox functions in Fe-S cluster synthesis. Furthermore, it was reported that the endoplasmic reticulum (ER) and mitochondrial intermembrane space (IMS) were sites of intense thiol oxidation but lacked thiol-reductase pathways except for GSH (Toledano et al. 2013). Furthermore, real-time measurements of the cytosolic redox potential in yeast using a Grx1-roGFP reporter, which is in thermodynamic equilibrium with the GSH/GSSG couple (Meyer and Dick 2010), showed that although under oxidative stress conditions the overall cellular glutathione pool became more oxidized, the cytosolic redox state was little affected (Morgan et al. 2013). The authors suggested that GSSG in the cytosol which was not immediately reduced was transported into the yeast vacuole and that the overall cellular GSH:GSSG ratio was a poor indicator of the actual cytosolic redox potential which tended to be approximately 100 mV lower than would be predicted (*ibid.*).

In cells a kind of 'redox flow' can be envisaged where electrons pass through an open ended system along a gradient from lower (more negative) to higher (more positive) redox potential, and 'new' redox potential is 'created' at the bottom end in plant cells by converting solar energy into electron transport along a series of electron carriers in the thylakoid membranes in chloroplasts to ferredoxin (Fd, $E^{0'} = -430$ mV). The enzyme ferredoxin-NADP$^+$ reductase uses reduced Fd to reduce NADP$^+$ to NADPH and these molecules are ultimately the sources of reducing potential at the beginning of the chain to re-reduce oxidized members of downstream redox couples. In animal cells and in non-photosynthetic plant tissues, highly reduced substrates, such as carbohydrates, are oxidized by NAD$^+$ to release reducing equivalents in the form of NADH. New NAD(P)H must be generated as required to keep the central cellular GSH and ascorbate redox buffers replenished. Ultimately, the majority of electrons flow to oxygen, reducing it to water. On their journey, however, partial one-electron and two-electron reductions of O_2 can occur and give rise to reactive oxygen species (ROS), such as the superoxide radical anion $O_2^{\bullet-}$ and hydrogen peroxide, respectively. Protecting the cell against these potentially damaging products of metabolism is very important. Intermediate on this 'electron highway' are thiol groups and here it must be emphasized that, quantitatively, oxidizable thiols in proteins may exceed the contribution of low molecular weight thiols (e.g., GSH) by up to five-fold (Dietz 2005; Konig et al. 2012). Furthermore, the thiol-disulfide exchange reactions entered into by thiols under oxidative conditions are perhaps better viewed as nucleophilic substitutions rather than electron transfer reactions, which is why NADPH as a reductant with a lower redox potential than thiols does not protect these from oxidation directly, but only indirectly by serving as a source of reducing power for enzymes such as glutathione reductase (Giles and Jacob 2002; Gruhlke and Slusarenko 2012). Conversely, excess GSH can directly protect protein thiols from oxidation (Foyer and Noctor 2005; Gruhlke and Slusarenko 2012; Noctor et al. 2011).

In the cell, it is often assumed, although this may not be the case, that redox couple partners tend to be in equilibrium between their reduced and oxidized forms and the global redox environment. Perhaps paradoxically, the position of the equilibrium for a given redox couple both depends upon and helps to determine the global cellular redox environment, which includes all thiols of a particular cellular compartment, including protein cysteine residues, free cysteine/cystine, and the tripeptide glutathione. The reduction of a disulfide by a thiol resulting in a mixed disulfide is known as thiol/disulfide exchange reaction and establishes the redox equilibrium of all thiol and disulfide groups (Jocelyn 1972). The nucleophilic exchange occurs best with the thiolate ion ($R{-}S^-$) rather than the thiol group and due to the high pK_a of most –SH groups (pK_a for free cysteine $= 8.3$) this means that below pH 7.0 the reaction will not proceed 'unaided' (Hofmann et al. 2002). As pointed out, however, earlier the redox state of a particular cysteine thiol depends on its accessibility, the specific pK_a, whether there are nearby basic amino acids and not just on the thermodynamics but also, very importantly, the kinetics of possible oxidation/reduction reactions (Dalle-Donne et al. 2009; Nagy 2013; Winterbourn and Hampton 2008). The prevailing opinion is perhaps that many protein thiols in the cell are not in thermodynamic equilibrium with the global redox environment and that oxidative stress signaling occurs predominantly upon the triggering of specific sensor trigger proteins which are direct targets for oxidants (Winterbourn and Hampton 2008). Many experimental data tend to support this scenario but do not rule out the influence of the thermodynamic equilibrium mechanism for specific instances.

The alternative view assumes a greater role for the global redox environment of the subcellular compartments and presumes that this is largely governed by the GSH/GSSG redox couple. The cysteine-containing tripeptide glutathione, the thiol species with the highest cellular concentration, can act as electron-donor for different redox reactions either directly (by reduction of different oxidants) or in an enzyme-catalyzed manner (e.g., enzymes that reduce protein-disulfides in a glutathione-dependent manner [glutaredoxins] or reduce peroxides using glutathione as electron source [glutathione-peroxidases]) and for the regeneration of other redox buffering systems like ascorbate/dehydroascorbate). Because of the central redox buffer position that GSH holds, and because of the likely equilibrium between GSH and at least some other cellular thiols, one can conceive of a cellular 'thiolstat' (Gruhlke and Slusarenko 2012; Jacob 2011) in analogy to a 'rheostat' as a 'variable resistor' or an 'instrument

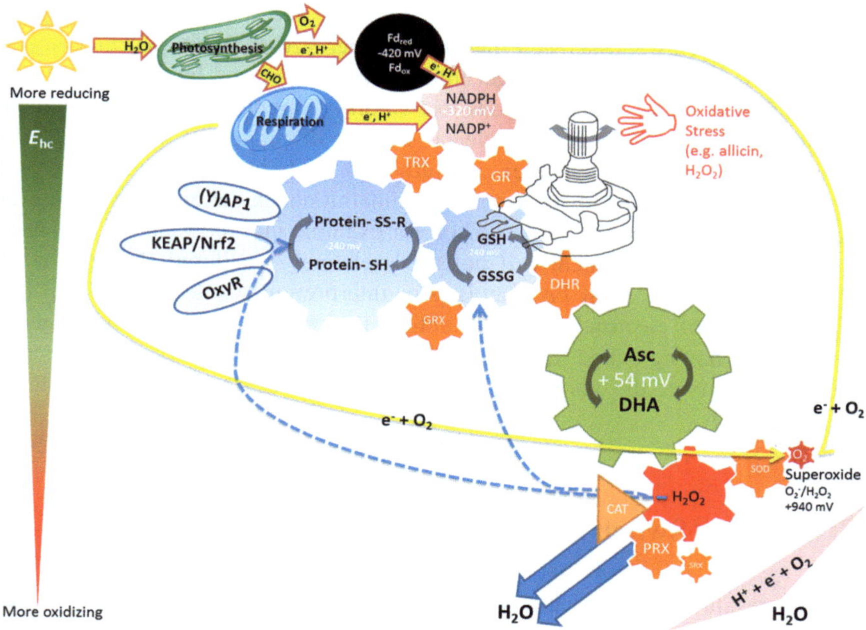

Fig. 2 Reductant–antioxidant–oxidant interactions in redox homeostasis and signaling in a typical plant cell. The flow of electrons through the thermodynamic 'open system' from photosynthetically produced reduction potential (reduced ferredoxin, NADPH) to oxygen along the electrochemical potential gradient is illustrated. *Asc* ascorbate, *CAT* catalase, *CHO* carbohydrate, *DHA* dehydroascorbate, *DHR* dehydroascorbate reductase, e^- electron, *Fdred/Fdox* reduced/oxidized ferredoxin, *GR* glutathione reductase, *Grx* glutaredoxin, *Prx* peroxiredoxin, *SOD* superoxide dismutase, *Srx* sulfiredoxin, *Trx* thioredoxin; KEAP/Nrf2, OxyR, and Y(AP1) are transcription factors involved in oxidative stress responses (see text)

for controlling and varying within limits the value of resistance in an (electrical) circuit' (Anon 1968). Thus, on the supposition that there is potentially an equilibrium between cellular thiols (glutathione and protein thiols) and other redox systems with either higher (e.g., ascorbate or tocopherol) or lower (e.g., $NAD(P)^+/NAD(P)H$) redox potentials, glutathione because of its high cellular abundance (around 5 mM in animal cells, which is around 500-fold higher than the NADPH/thioredoxin system) (Filomeni et al. 2002; Schafer and Buettner 2001) is often considered to be a central regulon of the redox state in the cell. Hence, the 'thiolstat' can be defined as the overall redox status, which is in equilibrium with the glutathione status in particular and so allows using the GSH/GSSG-redox couple as an indicator for all cellular thiols and thus cellular redox in general. Viewed simply, the cellular 'thiolstat' reflects the proportion of oxidized:reduced thiol groups in the cell (Fig. 2).

An excellent example of the thiolstat in action is the observation that seed viability can be predicted from knowing the redox potential of the GSH/GSSG redox couple (Kranner et al. 2006). The authors studied seeds from several plant families and species and showed that when under stress conditions, $E_{GSSG/2GSH}$ increased to −180 mV that seed germination rate decreased by ~50 % and at a potential higher than −160 mV seeds lost viability completely. The authors concluded that $E_{GSSG/2GSH}$ was a universal marker of plant cell viability (*ibid.*). However, the situation is complex and although oxidative treatments often lead to a shift in the degree of total cellular glutathione oxidation, GSSG may be largely rapidly removed from the cytosol, for example into the vacuole, thus leaving the local electrochemical cell potential relatively unaltered (Morgan et al. 2013). Thus, caution must be exercised in interpreting whole cell GSH:GSSG ratios in terms of electrochemical potentials in specific cellular compartments. Nevertheless, having stated this, it is a fact of physical chemistry that the redox environment, given the above provisos, will influence the oxidation state of accessible thiols.

2 The Redox Switch Concept

Cysteine residues are particularly sensitive to redox changes. The redox couple cysteine/cystine has a standard half-cell potential of around −220 mV (Jocelyn 1967). Since the half-cell potential describes the equilibrium conditions of the redox system, this means that at higher reduction potentials the equilibrium is shifted to a more oxidized, at lower potentials to a more reduced state. Nonetheless, especially for protein-thiols the equilibrium does not exclusively depend upon the half-cell potential, but also depends on the pK_a of the surrounding functional groups, which affect the degree of dissociation of the protein-thiol (Nagy 2013).

The translation of changes in the cellular thiol status to changes in the oxidation state of a single protein-thiol has been called a 'nano-switch' (Paget and Buttner 2003; Schafer and Buettner 2001). Since the catalytic or regulatory function of proteins is often dependent on the oxidation state of particular cysteine residues, a shift of the overall cellular thiolstat can change the position of many switches in a single sweep, and not necessarily only regulate a few specific targets in a conventionally 'linear' signaling cascade (Jacob et al. 2006; Jensen et al. 2009; Jones 2008; Kamata and Hirata 1999; Winterbourn and Hampton 2008).

It was shown empirically in cell cultures that physiological states like proliferation, differentiation, apoptosis, or necrosis correlated with the calculated cellular redox potential using the Nernst equation and the absolute glutathione concentrations (Cai and Jones 1998; Cai et al. 2000; Gruhlke et al. 2010; Hutter et al. 1997; Hwang et al. 1992; Jones et al. 1995; Kirlin et al. 1999).

Although these concrete correlations between redox and physiological status were shown in animal and human cell lines, the inference that the findings are applicable to other cell types, such as fungal, bacterial, or plant cells, seems likely, since the mechanism of influencing protein function by alteration of the thiol status is transferable. Indeed, this assumption has been shown to be valid at least for seed germination potential by Kranner et al. (2006). Hence, attempts to intentionally influence the thiolstat, based on the redox switch concept, need not be limited to animal systems (e.g., in cancer treatment), but would also represent a valid strategy in agricultural plant protection.

The change in a protein's redox status needs to be transduced into a physiological response, i.e., by affecting the cell's metabolism or altering gene transcription. An important point in this respect is how global redox change is translated into specific signaling events regulating specific responses. Well-known examples of transcription factors which transduce the redox switch into a physiological response are the OxyR and SoxS transcription factors from *Escherichia coli* and members of the mammalian AP1 family (see also Chap. 5). The latter have been well studied for the yeast AP-1 homolog (YAP1). A comparison of these two redox switch models is given in Choi et al. (2001).

Bacteria have developed a broad range of different redox sensors and resistance mechanisms to adapt to and to defend against oxidative stress conditions (Green and Paget 2004). OxyR regulates the response to H_2O_2 and the SoxRS system is important for resistance against redox active natural products with antibiotic activity (Chater 2006; Dietrich et al. 2008; Mavrodi et al. 2006). The SoxS transcription factor regulates the superoxide response (Demple 1991; Green and Paget 2004). Mechanistically the change in redox status is transduced to changes in the cysteine residues of the transcription factor that becomes activated and subsequently facilitates the transcription of stress-related and antioxidative enzymes in a coordinated way.

How the thiol microswitches in Yap1 are differently regulated by oxidants such as H_2O_2 and diamide is relatively well understood (Azevedo et al. 2003, 2007; Delaunay et al. 2000) and specific switching by H_2O_2 or diamide leads to largely oxidant-specific protection responses via the activation of characteristic sets of defense genes (Morano et al. 2012; Ouyang et al. 2011). The Yap1 transcription factor has N- and C-terminal cysteine rich domains (n-CRD and c-CRD) and a nuclear export protein (Crm1) binding domain. In the absence of oxidative stress Yap1 in the nucleus is bound by Crm1 and rapidly exported into the oxidative stress cytosol. In conjunction with a glutathione peroxidase (Gpx3) and the Yap1-binding protein, H_2O_2 leads to an oxidative intramolecular folding of Yap1 involving both n-CRD and c-CRD domains. The intramolecular folding of Yap1 masks the Crm1 binding site and allows Yap1 to accumulate in the nucleus. Thiol reagents such as NEM (*N*-ethylmaleimide), diamide and others, however, have been shown to form adducts to cysteines in the c-CRD and thus block access of Crm1 to the Crm1-binding site (Morano et al. 2012). Accumulation of Yap1 in the nucleus leads to the transcription of sets of oxidative stress induced genes which can be different, depending on whether H_2O_2 or NEM/diamide activated Yap1. Oxidized Yap1 is reduced back to its initial state by the thioredoxin Trx2 (Delaunay et al. 2000, 2002; Meyer et al. 2009; Wood et al. 2004).

3 Examples of How Physiologically Active Secondary Metabolites Can Affect the Overall Thiol Status of the Cell: Plant Defense Compounds

Many metabolites, e.g., glucose or alanine, are present in all living organisms and are thus called 'primary metabolites'. Some metabolites in contrast have a limited taxonomic distribution and are often produced in specific organs or at particular developmental stages; these are called 'secondary metabolites'. Secondary metabolites often accumulate to very large amounts. A variety of natural products of so-called 'secondary metabolism', especially in plants, play important roles in defense against pathogens, predators, and competitors (see Chap. 1). It might be mentioned here that organisms other than plants produce secondary metabolites, for example fungi and bacteria, and here crucial roles are often to be found in ecological relationships such as competitor suppression. Plants produce a large variety of chemically diverse bioactive molecules. Plant defense substances are classified into two broad groups as either 'phytoanticipins' or 'phytoalexins'. Preformed substances which are present before the plant is attacked, or which are produced rapidly and spontaneously from a preformed substrate by simple chemical or enzymatic modification via a pre-existing enzyme, are called 'phytoanticipins' (Van Etten et al. 1994). They build a first line of chemical defense which can be likened to a booby trap which, when triggered, has immediate unpleasant consequences for any attacker. Allicin from garlic, synthesized when cell damage results in the mixing of the preformed alliin substrate with the preformed alliinase enzyme, is a good example of a phytoanticipin. The second group of chemical defense substances, phytoalexins, are synthesized from distant precursors after pathogen attack and require *de novo* gene expression and the production of enzymes leading to the installation of new biosynthetic pathways not usually present in the unchallenged plant. Thus, in the biological sense, phytoanticipin and phytoalexin production can be viewed as components of 'passive' and 'active' defense strategies, respectively (see also Chap. 2) (Mansfield 2000).

The spectrum of organisms that attack plants is broad and from an evolutionary point of view it is advantageous to produce defense molecules with a global mode of action that is not restricted to a narrow class of organisms attacking the plant. Therefore, plant defense substances, far from being 'magic bullets', often have rather general mechanisms of action. Thus, many destroy the selective permeability of membranes by creating channels, e.g., the steroidal glycoside and phytoanticipin α-tomatine from tomato (*Lycopersicon esculentum*), and the isoflavonoid phytoalexins from the *Leguminosae*. Some defense compounds work by inducing oxidative stress, for example α-tomatine and allicin (Ito et al. 2007). Thus, α-tomatine is an example of a defense compound with a dual mode of action. In this context cellular redox homoeostasis can be viewed similarly to cell membrane integrity, in that both are essential for normal cell function and viability (homeostasis) and make good general targets for broadly acting defense compounds (Dubreuil-Maurizi and Poinssot 2012; Kerchev et al. 2012).

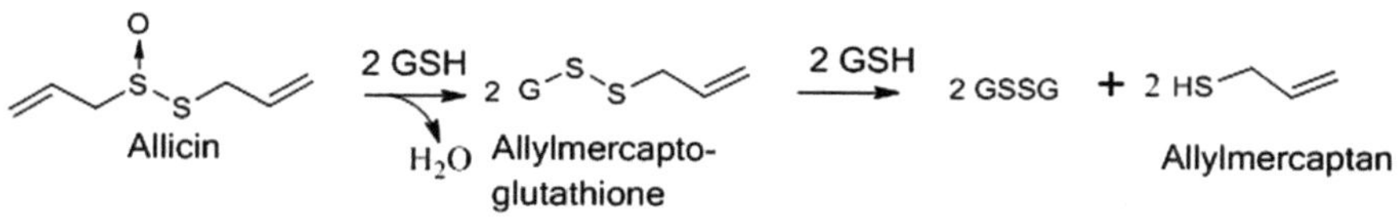

Fig. 3 The reaction of one mol of allicin with in total four mol GSH to yield two mol GSSG

Theoretical considerations suggest at least four ways in which redox active secondary metabolites might affect the cellular redox homoeostasis, none of which are mutually exclusive. Indeed quite the opposite is true, and effects might be expected to occur concomitantly:

1. *Function* via *direct interaction with redox sensitive proteins*

Secondary metabolites might affect cellular redox homeostasis by reacting directly with redox sensitive proteins. Since protein thiols make up the largest group of thiol compounds in the cell, the direct quantitative effect of 'titrating out' protein thiol groups on redox homoeostasis cannot be neglected, i.e., protein thiols themselves constitute part of the cellular thiolstat machinery. Furthermore, oxidation of sensitive regulatory proteins which coordinate cellular oxidative stress adaptations will lead to a cascade effect on the redox environment (see Sect. 3.4).

2. *Function* via *GSH:GSSG causing direct redox shift*

Secondary metabolites can affect the cellular redox potential by direct oxidation of GSH to GSSG. This can occur in a direct chemical manner as is the case with some natural compounds containing oxidized sulfur, like allicin (Gruhlke and Slusarenko 2012) resulting in intermediate mixed-disulfide adducts but ultimately in the formation of GSSG (Fig. 3). Similarly, the reaction of a substance like allicin with other cellular thiol groups, e.g., protein-thiols, affects the glutathione pool since glutathione is potentially in equilibrium with other cellular thiol groups via the thiol disulfide exchange reaction (TDER) and thus such compounds can oxidize GSH indirectly.

3. *Function* via *GSH-depletion*

Besides a direct reaction mechanism, a glutathione *S*-transferase (GST)-mediated reaction with a secondary cassette metabolite, e.g., in the course of detoxification, can lead to a depletion of the glutathione pool because glutathione-adducts are often exported out of the cell or, in the case of plants and fungi, into the vacuole. Multidrug resistant pumps (ATP-Binding Cassette, ABC-Transporters) often play an important role in this form of detoxification (Franco and Cidlowski 2012). Depletion of the GSH pool reduces the redox buffering capacity of the cell. As a consequence, even minor oxidative stress could then lead to strong influences on the thiolstat.

4. *Function* via *ROS* (*or RNS*) *induction*

Secondary metabolites can be sources of ROS or RNS which can act as strong oxidants for thiol groups thus shifting the thiolstat to a more oxidized state.

3.1 Throwing a Nano-Switch: Might Allicin Lead to Apoptosis by Affecting Tubulin Polymerization?

It has been shown that cytoskeletal disruption can lead into apoptosis in yeast (Leadsham and Gourlay 2008). Furthermore, it was shown that allicin destroys the tubulin cytoskeleton and this effect is abolished by treatment with reducing agents such as DTT or β-mercaptoethanol in mouse fibroblasts (Prager-Khoutorsky et al. 2007). Allicin induces apoptosis in yeast cells by shifting the cellular redox state and the concentration of GSH and GSSG. Allicin treatment showed both a loss in total glutathione and an increase in GSSG concentration (Gruhlke et al. 2010). Induction of apoptosis by allicin was also demonstrated in cancer cell lines (Miron et al. 2008). These observations suggest an influence of allicin on thiol groups of the tubulin protein. It is not yet clear, however, whether allicin acts directly by forming mixed disulfides with cysteine thiols in tubulin, which it can do *in vitro*, or indirectly via the oxidation of glutathione, hence affecting the thiolstat and leading to cysteine-based mixed disulfide formation in tubulin (e.g., *S*-glutathiolation). Furthermore, tubulin has been shown to be an important target of protein-glutathiolation and the formation of mixed disulfides affects the polymerization of the tubulin monomers (Landino et al. 2010).

3.2 Secondary Metabolites Depleting Cellular GSH

Isothiocyanates are well-studied bioactive reactive sulfur species (RSS) widely distributed in members of the *Brassicaceae*. Isothiocyanates are produced by myrosinase enzymes from glucosinolates (see also Chap. 10 in this book) (Nwachukwu et al. 2012).

Isothiocyanates can react directly with thiol groups to produce a dithiocarbamate that cannot be reduced by regenerating enzymes like thioredoxins, glutaredoxins, or glutathione reductases; thus, these compounds take glutathione out of the pool and consequently affect the cellular thiolstat. The benzoxazinoid DIMBOA (2,4-dihydroxy-7-methoxy-2H-1,4-benzoxazin-3-one), a phytoanticipin from maize and some other cereals also removes glutathione from the thiolstat pool by forming a spirocyclic adduct (Dixon et al. 2012). Furthermore, the authors showed that DIMBOA reacted with exposed cysteine thiols in proteins, producing adducts that could not be re-reduced with DTT (*ibid.*).

3.3 Secondary Metabolites Inducing ROS

Some redox active natural compounds affect the cellular thiolstat indirectly by induction of ROS that in turn leads to oxidation of components in the cell.

There are many instances of ROS induction in cells by natural compounds; some organic polysulfanes, for instance, are known to induce ROS, because they undergo complex redox reactions with thiol groups, forming products that can generate ROS in the presence of glutathione and Fe^{2+} (see also Chap. 10) (Munday et al. 2003; Nwachukwu et al. 2012; Schneider et al. 2011).

Organic polysulfane treatment can lead to the generation of O_2^- which can react with thiol groups and produce GSSG and glutathione sulfonic acid (Winterbourn and Metodiewa 1994). Furthermore, H_2O_2 can be generated from O_2^- by the activity of superoxide-dismutase and can directly oxidize GSH to GSSG.

Some natural products produce ROS on exposure to light via a photodynamic effect, e.g., the accumulation of singlet oxygen ($^1O_2$) on light exposure of some furanocoumarins in the *Apiaceae* (Bode and Hansel 2005) or by the sulfur-containing thiophenes (Champagne et al. 1986; Hudson et al. 1993; Nwachukwu et al. 2012). Although lipophilic redox buffer systems are efficient at scavenging singlet oxygen, the latter can react directly with glutathione or, more importantly, can be quenched by ascorbate (Devasagayam et al. 1991; Triantaphylides and Havaux 2009). Nevertheless, dehydroascorbate is recycled by GSH in the Halliwell-Asada-Cycle, resulting in glutathione oxidation, which could also affect the thiolstat. Furthermore, methionine seems to be able to form methionine sulfoxide via a photodynamic effect involving $^1O_2$ (Devasagayam et al. 1991; Triantaphylides and Havaux 2009).

Thiophenes have been shown to inhibit superoxide dismutases, which lowers the effectivity of ROS detoxification mechanisms and thus results in greater accumulation of these reactive species (Nivsarkar et al. 1991). Thus, it is not always the natural compound itself which alters the thiol status but sometimes the ROS which are produced as result of the natural compound.

3.4 Increasing the Cellular GSH Pool by Activating *γ*-Glutamylcysteine Ligase, the First Enzyme in the Biosynthetic Pathway

The examples discussed so far are of natural compounds that shift the thiol status to a more oxidized state. The converse, however, is also possible. Thus, some natural compounds can act to increase the glutathione pool and reinforce the buffering capacity of the thiolstat, and could therefore be expected to increase resistance against oxidative stress and its consequences.

The rate-limiting step in glutathione biosynthesis is the ligation of glutamate and cysteine to γ-glutamylcysteine, a reaction catalyzed by the enzyme γ-glutamylcysteine ligase (GSH1, GCL, and GSHA in plants and fungi, animals and bacteria,

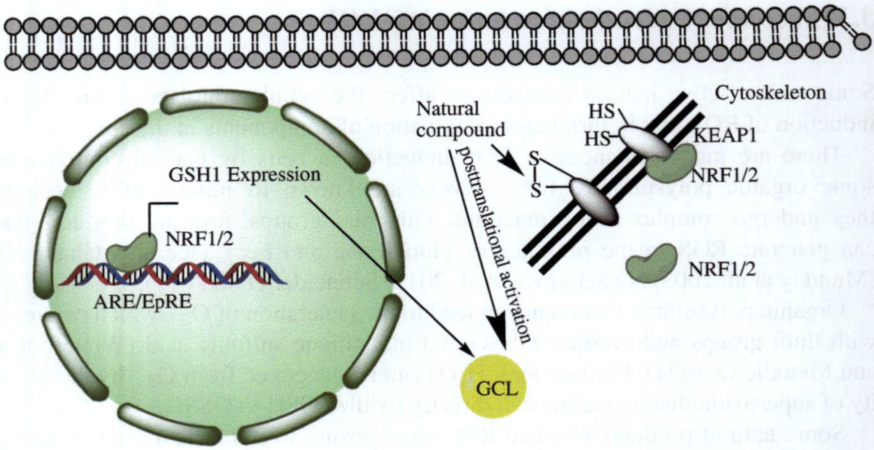

Fig. 4 Natural products can stimulate glutathione biosynthesis via two avenues. First, by post-translational activation of regulatory cysteines in GCL, and second, by releasing the Nrf1/2 transcription factor from its actin-bound complex with KEAP1 and its subsequent migration to the nucleus where GCL gene expression is stimulated by Nrf1/2 binding to ARE elements in the GCL promoter (Moskaug et al. 2005; Surh et al. 2008)

respectively), and both transcriptional and posttranscriptional regulation has been documented. Activation of γ-glutamyl-cysteine ligase by oxidation of specific cysteine thiols has been shown in the model plant *Arabidopsis* and in mammals such as rat and human (Hicks et al. 2007; Huang et al. 1993; Krejsa et al. 2010).

It was shown in oxidatively stressed yeast that *GSH*1 expression was regulated by the oxidation-sensitive transcription factor Yap1 (see Sect. 2) (Dormer et al. 2002). In humans GCL has regulatory (GCSl) and catalytic (GCSh) subunits. The expression of both subunits is controlled by the Antioxidant Response Element (ARE) motifs in the promoter sequence interacting with the Nrf1/2 transcription factor. The localization of the Nrf1/2-proteins depends on the Kelch like-ECH-associated protein 1 (KEAP1). KEAP1 binds to Nrf1/2 and this complex is normally associated with the actin cytoskeleton (Kang et al. 2004). Under the influence of oxidative stress, the Nrf1/2-KEAP1 complex dissociates and Nrf1/2 migrates into the nucleus where it can bind to ARE elements in the promoters of genes coding for antioxidative enzymes (Fig. 4).

It has been shown that certain plant phenolics, for example naringin (a flavone in grapefruit), can stimulate the expression of the genes coding for both the regulatory and catalytic subunits of GCL in humans leading to increased glutathione synthesis (Gopinath and Sudhandiran 2012; Moskaug et al. 2005; Surh et al. 2008).

Although allicin has been shown to rapidly deplete cells and blood of glutathione (Gruhlke et al. 2010; Rabinkov et al. 2000) at low concentrations and in the long term allicin up-regulates the intracellular glutathione concentration by affecting the GCL enzyme, as shown in vascular endothelial cells (Horev-Azaria et al. 2009). This effect is tied together with the observation that although allicin clearly has oxidative properties, physiologically at low concentrations it works as an 'antioxidant' by inducing protection mechanisms (Munday et al. 2003). Because this process invigorates the 'antioxidative shield' of a cell, induction of glutathione biosynthesis by natural compounds, e.g., from nutrition, might be of great interest for health care (Masella et al. 2005). In this way *oxidatively* active substances can, paradoxically, be useful indirectly as 'antioxidants' in a physiological sense (see also Chap. 5) (Jung and Kwak 2010).

4 The Biological Consequences of Altering the Thiolstat: Targets and Signaling Pathways

4.1 Endogenous Disulfide Formation

Upon shifting the cellular thiolstat and thus the cellular redox environment to a more oxidized position, the formation of disulfides will be triggered. Two principle options exist. If two thiol groups in the same molecule are positioned such that they can react with each other, an internal disulfide bridge can form, whereas in the case that no other thiol group in the same molecule is available, the reaction might occur with thiol groups in other molecules. Glutathiolation (discussed in Sect. 4.2) is a special case of this more general scheme. Overproduction of disulfides in cells leads to the condition known as 'disulfide stress' (Aslund and Beckwith 1999). A consequence of disulfide stress can be further oxidation of protein thiols with serious implications on the catalytic function of enzymes affected. Furthermore, protein-disulfide isomerases (PDIs), which catalyze disulfide bond formation between protein cysteines, can be important targets of natural compounds. Inhibition of the ER-localized PDIs leads to denaturation and misfolding of proteins in the ER. In turn this leads to the ubiquitous 'unfolded protein response' (UPR) which aims to restore normal cell function by stopping mRNA translation and activating pathways leading to the production of molecular chaperones to restore proper protein folding. If normalization is not achieved rapidly enough, the UPR pushes cells into apoptosis (Walter and Ron 2011). It has been shown that the natural sesquilactone juniferidin from *Ferula malacophylla* (Sagitdinova et al. 1978) inhibits PDI and induces apoptosis via the UPR (Khan et al. 2011).

4.2 Protein-Glutathiolation as a Consequence of a Thiolstat Shift

Under oxidizing conditions, glutathione readily form mixed disulfides with protein thiols; this process is called *protein-glutathiolation* (more precisely: protein-*S*-gluta-thiolation) (Dalle-Donne et al. 2009). A list of proteins found to be glutathiolated under oxidative stress conditions is documented in Table 1 (modified from Michelet et al. 2006). The extent to which this reaction occurs

Table 1 A selection of proteins that have been shown to become glutathiolated under oxidative conditions, together with the effect of glutathiolation on protein function (modified from Michelet et al. 2006)

Protein	Impact of glutathiolation on protein activity	References
Actin	Glutathiolation of actin leads to reduced polymerization of G-actin and thus regulates turnover between F- and G-actin	Dalle-Donne et al. (2003, 2005), Eaton and Shattock (2002), Fratelli et al. (2003, 2004), Lind et al. (2002), Wang et al. (2001, 2003)
AP-1 (C-subunit)	Reduces DNA-binding affinity of AP-1 protein	Klatt et al. (1999a, b, 2000)
Caspase-3	Glutathiolation inhibits proteolytic activation and enzyme activity of caspase-3 in HL-60 cells	Huang et al. (2008), Klatt et al. (2000)
Cofilin	? (Most likely reduced actin-depolymerization)	Fratelli et al. (2002, 2004)
HSP90	Possibly, in analogy to HSP90 nitrosation, inhibition of its ATPase activity	Lind et al. (2002), Martinez-Ruiz et al. (2005)
JNK	Inhibition of JNK-signaling	Adler et al. (1999)
Myosin	Lower ATPase activity and higher sensitivity to proteolytic cleavage	Fratelli et al. (2002)
NFκB (p50 subunit)	Glutathiolation of p50 inhibits NFκB-ediated gene expression	Ji et al. (1999), Klatt et al. (2000), Pineda-Molina et al. (2001)
Protein phosphatase (1B and 2A)	Glutathiolation inhibits protein phosphatases	Rao and Clayton (2002)
Protein kinase A	The catalytic subunit of protein-kinase A in humans is inhibited by glutathiolation	Humphries et al. (2002)
Protein kinase C	Inactivation	Chu et al. (2003), Eaton and Shattock (2002), Ward et al. (2000, 2002)
h-Ras	Glutathiolation of Cys^{118} in Ras2 enhances the activity of RAS2	Adachi et al. (2004), Eaton and Shattock (2002), Ji et al. (1999), Mallis et al. (2001)
Tubulin $\beta 1$	?	Landino et al. (2004), Lind et al. (2002)

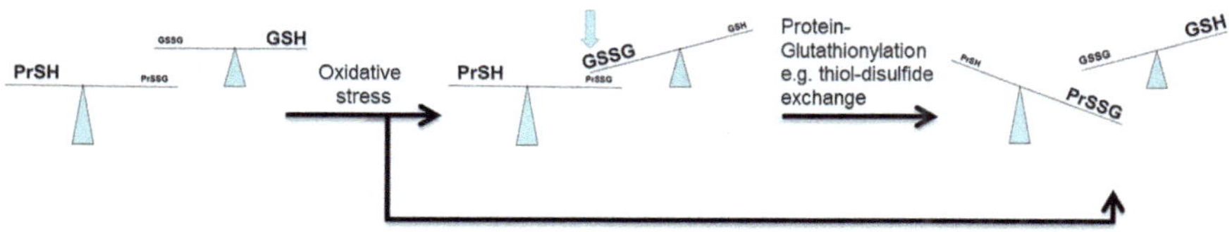

Fig. 5 The redox equilibrium is comparable to an asymmetric balance. In an unstressed cell the majority of the glutathione pool is in the reduced GSH form and protein 'switch' thiols (*PrSH*) are largely reduced. Upon oxidative stress there is not only direct glutathiolation of protein thiols, but the increased GSSG pool leads indirectly also to enhanced glutathiolation of protein thiols via a spontaneous enzyme catalyzed TDER. Please note that PrSSPr may be found in addition to PrSSG

as a result of oxidative stress depends on a number of factors, e.g., the local redox potential, the absolute GSH concentration (law of mass action) and the pK_a of the protein thiol in question (Dalle-Donne et al. 2009). The lower the pK_a of a particular cysteine, and thus the greater its tendency to dissociate at physiological pH, the higher is usually its reactivity with GSSG to build a disulfide. Depending on the redox environment it is also possible for a protein-cystine to become glutathiolated by reacting with a molecule of GSH in thiol-disulfide exchange reaction (Fig. 5).

Protein glutathiolation is central to the concept of the thiolstat and redox switch-regulation of cell metabolism. The pi class of human glutathione-*S*-transferases (GST-P) can catalyze this reaction (Tew et al. 2011) and several secondary metabolites stimulate (e.g., isothiocyanates) GST activity via the already mentioned Nrf2/ARE pathway (Andorfer et al. 2004; Xu et al. 2006). Furthermore, for several sulfur-containing garlic metabolites a direct, non-enzyme-catalyzed stimulation of protein *S*-glutathiolation has been documented (Pinto et al. 2006).

Elements of the cytoskeleton are targets for glutathiolation, regulating actin polymerization via the proportion of filamentous (F-) to monomeric (G-) actin (Sakai et al. 2012; Wang et al. 2001). It has been shown in yeast that the degree of actin polymerization correlates with the mitochondrial membrane potential ($\Delta\psi_m$). A similar situation can be expected in mammalian cells (for an appropriate 'intracellular diagnostics' of such phenomena see Explanatory Box 1) (Dalle-Donne et al. 2003).

Explanatory Box 1: Intracellular Diagnostics

Several chapters of this book discuss intracellular processes, for instance in the context of redox modifications and signaling cascades. Interestingly, many of these processes are rather subtle and their investigation requires sophisticated tools. Some of these cell bioanalytical methods have already been discussed in Chap. 3. Here, we will briefly consider some aspects of 'intracellular diagnostics'. During this discussion, it is important to bear in mind that most cellular events can only be measured reliably in living cells, and that fixed or even lysed cells only provide a static picture of real events. This picture is often also marred by artifacts due to the fixing, staining, and lysis techniques, and due to the fact that the normal processes have come to a standstill (and decay may have set in) and that cells (or parts thereof) become exposed to dioxygen. It is therefore not surprising that techniques involving living—or at least intact—cells have recently gained considerable prominence. Here, we find some of the most cutting-edge techniques of "Live Cell Imaging". Western blots cannot be performed on living or intact cells and I have never heard of chromogenomics. As redox active agents often cause disturbance in the intracellular redox balance, several fluorescent dyes have been developed which enable researchers to stain cells in order to subsequently quantify oxidative stress (OS) in general or certain Reactive Oxygen Species in particular. Here, one may consider the OS-sensitive 2′,7′–dichlorofluorescein diacetate (DCFDA), the superoxide radical anion-sensitive hydroethidium bromide or the singlet oxygen-sensitive *meso*-tetraphenylporphyrin (H_2TTPS). At the same time, stains are available to quantify intracellular thiols. A combination of fluorescent staining and microscopy and/or a plate reader can also be employed to analyze for disruption of cellular organelles and their function. Dyes such as MitoTracker® and ER-Tracker® can be used to stain mitochondria and the Endoplasmatic Reticulum, respectively. In contrast, the MitoSOX™ dye (JA-1), does not simply stain the mitochondria but itself is redox sensitive and changes fluorescence according to the mitochondrial membrane potential $\Delta\Psi_M$. Fluorescent dyes can also be used to visualize specific intracellular proteins under the microscope. Here, larger protein aggregates, such as tubulin and actin networks, which form the cytoskeleton, can be stained by fluorescent dyes attached to specific antibodies which recognize and hence attach to these proteins. Smaller aggregates or single proteins stained with fluorescent antibodies cannot be spotted individually under a simple inverted light microscope as they are too small to be seen at the magnification available. However, their presence in the cell can be quantified by using such dyes and an appropriate plate reader. At the same time, the activity of such enzymes can also be measured using specific substrates which become fluorescent or change their emission wavelength or intensity once they are turned over.

These whole-cell-based techniques can be supplemented by a wide spectrum of more invasive techniques. For instance, the concentration of a specific protein in the cell can also be determined fairly adequately using Western Blotting techniques. More recently, methods to measure mRNA levels as indicators of gene expression in the cell, such as real-time PCR, have also become available. Indeed, there are many proteomic and genomic methods now available to map out specific gene expression patterns, protein levels (and changes thereof), posttranslational modifications in proteins (such as sulfenic acid formation), and the activity of certain transcription factors. Such techniques are often associated with genomics, proteomics, transcriptomics, or redoxomics, and share the desire to analyze the entire cellular network of entities or process in one go, and to compare its appearance to similar maps of sick or healthy cells, or cells responding to a specific external stimulus, such as an administered drug.

In natural products research, these techniques are complemented by a range of methods used to validate the intracellular targets of such compounds. Here, we find sophisticated mass spectrometric methods to identify small modifications on proteins and enzymes, such as the oxidation of cysteine residues to sulfenic acids (i.e., to map out the cellular 'sulfenome'). Chemogenomics based on an extensive depository of yeast mutants enables researchers to identify mutants, and hence proteins involved in the action of a specific substance. Here, mutants particularly sensitive or resistant to a specific compound direct the researcher to the cellular pathway(s) most likely to be involved or affected. Another comparably new method worth mentioning in this context is "drug affinity responsive target stability" (DARTS), which again is particularly suited to identify intracellular targets of redox active substances. Here, the compound in question is applied to the cell, which is subsequently lysed and its protein content digested by a mixture of powerful proteases. As such proteases cannot digest chemically modified proteins, however, the latter survive and can be identified as potential targets using a combination of Western blots and mass spectrometry.

This list of modern and emerging techniques, of course, is far from complete and still expanding rapidly. It is therefore worthwhile to keep a close eye on the progress in this field of intracellular diagnostics. During the next decade or two, such emerging methods will ultimately be able to address many rather difficult questions in the field and also resolve a number of riddles, puzzles, and apparent contradictions currently challenging the research community.

4.3 Effects on Metalloproteins or How to Affect Electron Transport and DNA Transcription

Thiol groups are able to bind to Lewis acids, for example to Zn^{2+} ions. Thus, cysteine-coordinated zinc-centers can also act as redox switches (Ilbert et al. 2006). Zinc coordination is of great importance for catalytic activity and protein structure in a variety of proteins and perturbation of zinc coordination can lead to inactivation of proteins. Hence, changes in the cellular thiolstat can be transduced into physiological responses via the effect of redox changes on cysteine-coordinated zinc-clusters.

So-called 'zinc-finger' motifs are well-known zinc-chelating elements and the binding of a Zn^{2+} ion leads to the formation of a ribbon, which allows the peptide chain to interact with nucleic acids. Thus, zinc-fingers are common motifs in transcription factors and perturbation of the structure, which depends on the zinc-center, results in loss of DNA-binding activity. This highlights a further possible way in which changes in the cellular thiolstat might be transduced to changes in gene expression.

In murine macrophages, for example, it was shown that allicin is able to trigger the release of zinc from proteins in a concentration-dependent manner, leading to an inhibition of phosphatase activity and subsequently to enhanced ERK1/2 phosphorylation (Haase et al. 2012). This example illustrates that metalloproteins can also be important 'transducers' of changes in the cellular thiolstat.

Metalloproteins with iron-sulfur clusters play important roles in numerous electron transfer redox reactions in cells. These proteins complex two to four iron atoms via cysteine-sulfur atoms or through complex formation with histidine residues. Changes in such Fe-S clusters resulting from changes in the thiolstat can have significant consequences for cell metabolism and this turns Fe-S clusters into targets for oxidative stress (Gruhlke and Slusarenko 2012).

5 Redox Activity as the Heart of Antibiotic Activity

Up to now this chapter has dealt with the influence of the thiolstat on eukaryotic systems and in particular on mammalian cells. Nevertheless, shifting the cellular thiolstat can cause cell death in prokaryotes and in non-mammalian eukaroytes (e.g., in fungi). The induction of ROS was shown to contribute to the antibiotic activity of compounds toward bacteria. Thus, three major classes of bactericidal antibiotics, regardless of drug–target interaction, stimulate the production of highly deleterious hydroxyl radicals in Gram-negative and Gram-positive bacteria, which ultimately contribute to cell death (Kohanski et al. 2007). Elucidating how changing the cellular thiolstat correlates with the antibiotic properties of compounds is of great interest for developing antibiotic therapies and in plant protection. Targets of antibiotic compounds are diverse. Thus, commercial antibiotics like ampicillin (which targets cell wall biosynthesis of bacteria) or kanamycin (which inhibits protein biosynthesis at the 30S ribosomal subunit) have specific targets that are necessary for the cells to survive or to proliferate. As discussed before, targeting the thiolstat of a cell is a potential mechanism for antibiotic action which may affect *several* cellular functions simultaneously (and hence may also avoid the development of resistance). Thus, substances that specifically change the cellular redox environment could be promising antibiotics.

Interesting examples of bacterially produced antibiotics which are redox active are pyocyanin from *Pseudomonas aeruginosa* and actinorhodin from *Streptomyces coelicolor*. Both are able to cause oxidative stress in bacteria, presumably via O_2^- production and cause activation of the O_2^--regulated SOXR-regulon (Dietrich et al. 2008). Another interesting example is the phytoanticipin 'tomatine' from the *Solanaceae* that was thought to target exclusively the fungal membrane (Roddick and Drysdale 1984). However, recent studies imply that tomatine causes oxidative stress, as shown in the fungus *Fusarium oxysporum* and induces apoptosis via changing the cellular redox state in the fungus (Ito et al. 2007).

6 Conclusions and Outlook

As part of this chapter we have described and discussed some examples as to how a change in the cellular redox environment, or thiolstat, can be transduced to physiological effects. In mammals and fungi the induction of apoptosis is an important consequence of redox perturbation. Some studies with sulfur-containing molecules like polysulfanes and allicin in yeast (*Candida albicans* or *Saccharomyces cerevisiae*) demonstrated that a redox shift is responsible for the fungicidal activity of these compounds (Gruhlke et al. 2010; Lemar et al. 2005, 2007).

While the biochemical consequences of shifting the thiolstat (e.g., protein *S*-glutathiolation, effects on Fe-S clusters) are conserved between prokaryotes and eukaryotes, the consequences are thought to be different. For example, programmed cell death (PCD), one form of which is apoptosis, is held to be a solely eukaryotic phenomenon. Indeed, until a few years ago it was widely believed that apoptosis occurred only in multicellular eukaryotes. The overwhelming body of evidence for apoptosis in the unicellular model microorganism *Saccharomyces cereviseae* (Baker's yeast) led to a shift in the paradigm. In this sense it is important to take notice of recent reports of PCD in prokaryotes, and that this is a possible mechanism for the action of antibiotics (Engelberg-Kulka et al. 2004). Thus, some typical markers of apoptosis have been observed in bacteria (Dwyer et al. 2012; Kohanski et al. 2007, 2010). Nevertheless, how far the principle concept of redox dependence of PCD in the sense of Schafer and Buettner's model (2001) can be transferred to bacteria is, to our knowledge, not yet known. Since a shift in redox state in general affects a plethora of different cellular targets (in contrast to a concept of 'one compound-one target'), it is likely that the bacterial PCD might also be redox dependent and by this provides an avenue for antibiotics to act via redox modification(s) (Schafer and Buettner 2001).

Ultimately, a deeper understanding of how natural products can influence the thiolstat may allow us a targeted approach for designing new uses as nutriceuticals and in plant protection.

Acknowledgments Financial support from the RWTH Aachen and the EU "RedCat" 21005 Marie Curie ITN is gratefully acknowledged. Dr. Nikolaus Schlaich (RWTH Aachen University) is thanked for critical reading of the manuscript.

References

Adachi T, Pimentel DR, Heibeck T, Hou XY, Lee YJ, Jiang BB, Ido Y, Cohen RA (2004) S-glutathiolation of Ras mediates redox-sensitive signaling by angiotensin II in vascular smooth muscle cells. J Biol Chem 279(28):29857–29862

Adler V, Yin ZM, Fuchs SY, Benezra M, Rosario L, Tew KD, Pincus MR, Sardana M, Henderson CJ, Wolf CR, Davis RJ, Ronai Z (1999) Regulation of JNK signaling by GSTp. EMBO J 18(5):1321–1334

Andorfer JH, Tchaikovskaya T, Listowsky I (2004) Selective expression of glutathione S-transferase genes in the murine gastrointestinal tract in response to dietary organo sulfur compounds. Carcinogenesis 25(3):359–367

Anon (1968) Collins dictionary and encyclopedia. Collins, London Glasgow

Aslund F, Beckwith J (1999) Bridge over troubled waters: sensing stress by disulfide bond formation. Cell 96(6):751–753

Azevedo D, Tacnet F, Delaunay A, Rodrigues-Pousada C, Toledano MB (2003) Two redox centers within Yap1 for H2O2 and thiol-reactive chemicals signaling. Free Radical Biol Med 35(8):889–900

Azevedo D, Nascimento L, Labarre J, Toledano MB, Rodrigues-Pousada C (2007) The S-cerevisiae Yap1 and Yap2 transcription factors share a common cadmium-sensing domain. FEBS Lett 581(2):187–195

Bode CW, Hansel W (2005) 5-(3-Phenylpropoxy)psoralen and 5-(4-phenylbutoxy)psoralen: mechanistic studies on phototoxicity. Pharmazie 60(3):225–228

Cai JY, Jones DP (1998) Communication—superoxide in apoptosis—mitochondrial generation triggered by cytochrome c loss. J Biol Chem 273(19):11401–11404

Cai JY, Wallace DC, Zhivotovsky B, Jones DP (2000) Separation of cytochrome c-dependent caspase activation from thiol-disulfide redox change in cells lacking mitochondrial DNA. Free Radical Biol Med 29(3–4):334–342

Champagne DE, Arnason JT, Philogene BJR, Morand P, Lam J (1986) Light-mediated allelochemical effects of naturally-occurring polyacetylenes and thiophenes from asteraceae on herbivorous insects. J Chem Ecol 12(4):835–858

Chater KF (2006) Streptomyces inside-out: a new perspective on the bacteria that provide us with antibiotics. Philos Trans Royal Soc B-Biol Sci 361(1469):761–768

Choi HJ, Kim SJ, Mukhopadhyay P, Cho S, Woo JR, Storz G, Ryu SE (2001) Structural basis of the redox switch in the OxyR transcription factor. Cell 105(1):103–113

Chu F, Ward NE, O'Brian CA (2003) PKC isozyme S-cysteinylation by cystine stimulates the pro-apoptotic isozyme PKC delta and inactivates the oncogenic isozyme PKC epsilon. Carcinogenesis 24(2):317–325

Dalle-Donne I, Rossi R, Giustarini D, Colombo R, Milzani A (2003) Actin S-glutathionylation: eidence against a thiol-disulphide exchange mechanism. Free Radical Biol Med 35(10):1185–1193

Dalle-Donne I, Giustarini D, Colombo R, Milzani A, Rossi R (2005) S-glutathionylation in human platelets by a thiol-disulfide exchange-independent mechanism. Free Radic Biol Med 38(11):1501–1510

Dalle-Donne I, Rossi R, Colombo G, Giustarini D, Milzani A (2009) Protein S-glutathionylation: a regulatory device from bacteria to humans. Trends Biochem Sci 34(2):85–96

Delaunay A, Isnard AD, Toledano MB (2000) H2O2 sensing through oxidation of the Yap1 transcription factor. EMBO J 19(19):5157–5166

Delaunay A, Pflieger D, Barrault MB, Vinh J, Toledano MB (2002) A thiol peroxidase is an H2O2 receptor and redox-transducer in gene activation. Cell 111(4):471–481

Delic M, Rebnegger C, Wanka F, Puxbaum V, Haberhauer-Troyer C, Hann S, Kollensperger G, Mattanovich D, Gasser B (2012) Oxidative protein folding and unfolded protein response elicit differing redox regulation in endoplasmic reticulum and cytosol of yeast. Free Radical Biol Med 52(9):2000–2012

Demple B (1991) Regulation of bacterial oxidative stress genes. Annu Rev Genet 25:315–337
Devasagayam TPA, Sundquist AR, Di Mascio P, Kaiser S, Sies H (1991) Activity of thiols as singlet molecular-oxygen quenchers. J Photochem Photobiol B-Biol 9(1):105–116
Dietrich LEP, Teal TK, Price-Whelan A, Newman DK (2008) Redox-active antibiotics control gene expression and community behavior in divergent bacteria. Science 321(5893):1203–1206
Dietz K-J (2005) Plant thiol enzymes and thiol homeostasis in relation to thiol-dependent redox-regulation and oxidative stress. In: NS (ed) Antioxidants and reactive oxygen species in plants. Blackwell Publishing Ltd., New Jersey, p 302
Dixon DP, Sellars JD, Kenwright AM, Steel PG (2012) The maize benzoxazinone DIMBOA reacts with glutathione and other thiols to form spirocyclic adducts. Phytochemistry 77:171–178
Dormer UH, Westwater J, Stephen DWS, Jamieson DJ (2002) Oxidant regulation of the Saccharomyces cerevisiae GSH1 gene. Biochim Biophys Acta-Gene Struct Expr 1576(1–2): 23–29
Dubreuil-Maurizi C, Poinssot B (2012) Role of glutathione in plant signaling under biotic stress. Plant Signal Behav 7(2):210–212
Dwyer DJ, Camacho DM, Kohanski MA, Callura JM, Collins JJ (2012) Antibiotic-induced bacterial cell death exhibits physiological and biochemical hallmarks of apoptosis. Mol Cell 46(5):561–572
Eaton P, Shattock MJ (2002) Purification of proteins susceptible to oxidation at cysteine residues: Identification of malate dehydrogenase as a target for S-glutathiolation. Cell Signaling Transcription Trans Ther Targets 973:529–532
Engelberg-Kulka H, Sat B, Reches M, Amitai S, Hazan R (2004) Bacterial programmed cell death systems as targets for antibiotics. Trends Microbiol 12(2):66–71
Filomeni G, Rotilio G, Ciriolo MR (2002) Cell signalling and the glutathione redox system. Biochem Pharmacol 64(5–6):1057–1064
Foyer CH, Noctor G (2005) Redox homeostasis and antioxidant signaling: a metabolic interface between stress perception and physiological responses. Plant Cell 17(7):1866–1875
Franco R, Cidlowski JA (2012) Glutathione efflux and cell death. Antioxid Redox Signal 17(12):1694–1713
Fratelli M, Demol H, Puype M, Casagrande S, Eberini I, Salmona M, Bonetto V, Mengozzi M, Duffieux F, Miclet E, Bachi A, Vandekerckhove J, Gianazza E, Ghezzi P (2002) Identification by redox proteomics of glutathionylated proteins in oxidatively stressed human T lymphocytes. Proc Natl Acad Sci USA 99(6):3505–3510
Fratelli M, Demol H, Puype M, Casagrande S, Villa P, Eberini I, Vandekerckhove J, Gianazza E, Ghezzi P (2003) Identification of proteins undergoing glutathionylation in oxidatively stressed hepatocytes and hepatoma cells. Proteomics 3(7):1154–1161
Fratelli M, Gianazza E, Ghezzi P (2004) Redox proteomics: identification and functional role of glutathionylated proteins. Expert Rev Proteomics 1(3):365–376
Giles GI, Jacob C (2002) Reactive sulfur species: an emerging concept in oxidative stress. Biol Chem 383(3–4):375–388
Gopinath K, Sudhandiran G (2012) Naringin modulates oxidative stress and inflammation in 3-nitropropionic acid-induced neurodegeneration through the activation of nuclear factor-erythroid 2-related factor-2 signalling pathway. Neuroscience 227:134–143
Green J, Paget MS (2004) Bacterial redox sensors. Nat Rev Microbiol 2(12):954–966
Gruhlke MCH, Slusarenko AJ (2012) The biology of reactive sulfur species (RSS). Plant Physiol Biochem 59:98–107
Gruhlke MCH, Portz D, Stitz M, Anwar A, Schneider T, Jacob C, Schlaich NL, Slusarenko AJ (2010) Allicin disrupts the cell's electrochemical potential and induces apoptosis in yeast. Free Radical Biol Med 49(12):1916–1924
Haase H, Hieke N, Plum LM, Gruhlke MCH, Slusarenko AJ, Rink L (2012) Impact of allicin on macrophage activity. Food Chem 134(1):141–148

Hicks LM, Cahoon RE, Bonner ER, Rivard RS, Sheffield J, Jez JM (2007) Thiol-based regulation of redox-active glutamate-cysteine ligase from arabidopsis thaliana. Plant Cell 19(8):2653–2661
Higa A, Chevet E (2012) Redox signaling loops in the unfolded protein response. Cell Signal 24(8):1548–1555
Hofmann B, Hecht HJ, Flohe L (2002) Peroxiredoxins. Biol Chem 383(3–4):347–364
Horev-Azaria L, Eliav S, Izigov N, Pri-Chen S, Mirelman D, Miron T, Rabinkov A, Wilchek M, Jacob-Hirsch J, Amariglio N, Savion N (2009) Allicin up-regulates cellular glutathione level in vascular endothelial cells. Eur J Nutr 48(2):67–74
Huang CS, Chang LS, Anderson ME, Meister A (1993) Catalytic and regulatory properties of the heavy subunit of rat-kidney gamma-glutamylcysteine synthetase. J Biol Chem 268(26):19675–19680
Huang Z, Pinto JT, Deng H, Richie JP (2008) Inhibition of caspase-3 activity and activation by protein glutathionylation. Biochem Pharmacol 75(11):2234–2244
Hudson JB, Graham EA, Rossi R, Carpita A, Neri D, Towers GHN (1993) Biological-activities of terthiophenes and polyynes from the asteraceae. Planta Med 59(5):447–450
Humphries KM, Juliano C, Taylor SS (2002) Regulation of cAMP-dependent protein kinase activity by glutathionylation. J Biol Chem 277(45):43505–43511
Hutter DE, Till BG, Greene JJ (1997) Redox state changes in density-dependent regulation of proliferation. Exp Cell Res 232(2):435–438
Hwang C, Sinskey AJ, Lodish HF (1992) Oxidized redox state of glutathione in the endoplasmic-reticulum. Science 257(5076):1496–1502
Ilbert M, Graf PCF, Jakob U (2006) Zinc center as redox switch—new function for an old motif. Antioxid Redox Signal 8(5–6):835–846
Ito S, Ihara T, Tamura H, Tanaka S, Ikeda T, Kajihara H, Dissanayake C, Abdel-Motaal FF, El-Sayed MA (2007) Alpha-tomatine, the major saponin in tomato, induces programmed cell death mediated by reactive oxygen species in the fungal pathogen fusarium oxysporum. FEBS Lett 581(17):3217–3222
Jacob C (2011) Redox signalling via the cellular thiolstat. Biochem Soc Trans 39:1247–1253
Jacob C, Knight I, Winyard PG (2006) Aspects of the biological redox chemistry of cysteine: from simple redox responses to sophisticated signalling pathways. Biol Chem 387(10–11):1385–1397
Jensen KS, Hansen RE, Winther JR (2009) Kinetic and thermodynamic aspects of cellular thiol-disulfide redox regulation. Antioxid Redox Signal 11(5):1047–1058
Ji Y, Akerboom TP, Sies H, Thomas JA (1999) S-nitrosylation and S-glutathiolation of protein sulfhydryls by S-nitroso glutathione. Arch Biochem Biophys 362(1):67–78
Jocelyn PC (1967) The standard redox potential of cysteine-cystine from the thiol-disulphide exchange reaction with glutathione and lipoic acid. Eur J Biochem 2(3):327–331
Jocelyn PC (1972) Biochemistry of the SH-group. The occurrence, chemical properties, metabolism and biological functions of thiols and disulfides. Academic Press, London
Jones DP (2008) Radical-free biology of oxidative stress. Am J Physiol-Cell Physiol 295(4):849–868
Jones DP, Maellaro E, Jiang SN, Slater AFG, Orrenius S (1995) Effects of N-acetyl-L-cysteine on T-cell apoptosis are not mediated by increased cellular glutathione. Immunol Lett 45(3):205–209
Jung KA, Kwak MK (2010) The Nrf2 system as a potential target for the development of indirect antioxidants. Molecules 15(10):7266–7291
Kamata H, Hirata H (1999) Redox regulation of cellular signalling. Cell Signal 11(1):1–14
Kang MI, Kobayashi A, Wakabayashi N, Kim SG, Yamamoto M (2004) Scaffolding of Keap1 to the actin cytoskeleton controls the function of Nrf2 as key regulator of cytoprotective phase 2 genes. Proc Natl Acad Sci USA 101(7):2046–2051
Kerchev PI, Fenton B, Foyer CH, Hancock RD (2012) Plant responses to insect herbivory: interactions between photosynthesis, reactive oxygen species and hormonal signalling pathways. Plant Cell Environ 35(2):441–453
Khan MMG, Simizu S, Kawatani M, Osada H (2011) The potential of protein disulfide isomerase as a therapeutic drug target. Oncol Res 19(10–11):445–453

Kirlin WG, Cai JY, Thompson SA, Diaz D, Kavanagh TJ, Jones DP (1999) Glutathione redox potential in response to differentiation and enzyme inducers. Free Radical Biol Med 27(11–12):1208–1218

Klatt P, Molina EP, De Lacoba MG, Padilla CA, Martinez-Galisteo E, Barcena JA, Lamas S (1999a) Redox regulation of c-Jun DNA binding by reversible S-glutathiolation. Faseb J 13(12):1481–1490

Klatt P, Molina EP, Lamas S (1999b) Nitric oxide inhibits c-Jun DNA binding by specifically targeted S-glutathionylation. J Biol Chem 274(22):15857–15864

Klatt P, Molina EP, Perez-Sala D, Lamas S (2000) Novel application of S-nitrosoglutathione-sepharose to identify proteins that are potential targets for S-nitrosoglutathione-induced mixed-disulphide formation. Biochem J 349:567–578

Kohanski MA, Dwyer DJ, Hayete B, Lawrence CA, Collins JJ (2007) A common mechanism of cellular death induced by bactericidal antibiotics. Cell 130(5):797–810

Kohanski MA, Dwyer DJ, Collins JJ (2010) How antibiotics kill bacteria: from targets to networks. Nat Rev Microbiol 8(6):423–435

Konig J, Muthuramalingam M, Dietz KJ (2012) Mechanisms and dynamics in the thiol/disulfide redox regulatory network: transmitters, sensors and targets. Curr Opin Plant Biol 15(3):261–268

Kranner I, Birtic S, Anderson KM, Pritchard HW (2006) Glutathione half-cell reduction potential: a universal stress marker and modulator of programmed cell death? Free Radical Biol Med 40(12):2155–2165

Krejsa CM, Franklin CC, White CC, Ledbetter JA, Schieven GL, Kavanagh TJ (2010) Rapid activation of glutamate cysteine ligase following oxidative stress. J Biol Chem 285(21):16116–16124

Kumar C, Igbaria A, D'autreaux B, Planson AG, Junot C, Godat E, Bachhawat AK, Delaunay-Moisan A, Toledano MB (2011) Glutathione revisited: a vital function in iron metabolism and ancillary role in thiol-redox control. EMBO J 30(10):2044–2056

Landino LM, Moynihan KL, Todd JV, Kennett KL (2004) Modulation of the redox state of tubulin by the glutathione/glutaredoxin reductase system. Biochem Biophys Res Commun 314(2):555–560

Landino LM, Brown CM, Edson CA, Gilbert LJ, Grega-Larson N, Wirth AJ, Lane KC (2010) Fluorescein-labeled glutathione to study protein S-glutathionylation. Anal Biochem 402(1):102–104

Leadsham JE, Gourlay CW (2008) Cytoskeletal induced apoptosis in yeast. Biochim Biophys Acta-Mol Cell Res 1783(7):1406–1412

Lemar KM, Passa O, Aon MA, Cortassa S, Muller CT, Plummer S, O'Rourke B, Lloyd D (2005) Allyl alcohol and garlic (allium sativum) extract produce oxidative stress in Candida albicans. Microbiology-Sgm 151:3257–3265

Lemar KM, Aon MA, Cortassa S, O'Rourke B, Muller CT, Lloyd D (2007) Diallyl disulphide depletes glutathione in Candida albicans: oxidative stress-mediated cell death studied by two-photon microscopy. Yeast 24(8):695–706

Lind C, Gerdes R, Hamnell Y, Schuppe-Koistinen I, von Lowenhielm HB, Holmgren A, Cotgreave IA (2002) Identification of S-glutathionylated cellular proteins during oxidative stress and constitutive metabolism by affinity purification and proteomic analysis. Arch Biochem Biophys 406(2):229–240

Mallis RJ, Buss JE, Thomas JA (2001) Oxidative modification of H-ras: S-thiolation and S-nitrosylation of reactive cysteines. Biochem J 355:145–153

Mansfield JW (2000) Antimicrobial compounds and resistance—the role of phytoalexins and phytoanticipins. Mechanisms of resistance to plant diseases. Kluwer Academic Publishers, Dordrecht

Martinez-Ruiz A, Villanueva L, de Orduna CG, Lopez-Ferrer D, Higueras MA, Tarin C, Rodriguez-Crespo I, Vazquez J, Lamas S (2005) S-nitrosylation of Hsp90 promotes the inhibition of its ATPase and endothelial nitric oxide synthase regulatory activities. Proc Natl Acad Sci USA 102(24):8525–8530

Masella R, Di Benedetto R, Vari R, Filesi C, Giovannini C (2005) Novel mechanisms of natural antioxidant compounds in biological systems: involvement of glutathione and glutathione-related enzymes. J Nutr Biochem 16(10):577–586

Mavrodi DV, Blankenfeldt W, Thomashow LS (2006) Phenazine compounds in fluorescent *Pseudomonas* spp. biosynthesis and regulation. Annu Rev Phytopathol 44:417–445

Meyer AJ, Dick TP (2010) Fluorescent protein-based redox probes. Antioxid Redox Signal 13(5):621–650

Meyer Y, Buchanan BB, Vignols F, Reichheld JP (2009) Thioredoxins and glutaredoxins: unifying elements in redox biology. Annu Rev Genet 43:335–367

Michelet L, Zaffagnini M, Massot V, Keryer E, Vanacker Hn, Miginiac-Maslow M, Issakidis-Bourguet E, Lemaire SD (2006) Thioredoxins, glutaredoxins, and glutathionylation: new crosstalks to explore. Photosynth Res 89:225–245

Miron T, Wilehek M, Sharp A, Nakagawa Y, Naoi M, Nozawa Y, Akao Y (2008) Allicin inhibits cell growth and induces apoptosis through the mitochondrial pathway in HL60 and U937 cells. J Nutr Biochem 19(8):524–535

Morano KA, Grant CM, Moye-Rowley WS (2012) The response to heat shock and oxidative stress in saccharomyces cerevisiae. Genetics 190(4):1157–1195

Morgan B, Ezerina D, Amoako TNE, Riemer J, Seedorf M, Dick TP (2013) Multiple glutathione disulfide removal pathways mediate cytosolic redox homeostasis. Nat Chem Biol 9(2):119–125

Moskaug JO, Carlsen H, Myhrstad MCW, Blomhoff R (2005) Polyphenols and glutathione synthesis regulation. Am J Clin Nutr 81(1):277s–283s

Munday R, Munday JS, Munday CM (2003) Comparative effects of mono-, di-, tri-, and tetrasulfides derived from plants of the allium family: redox cycling in vitro and hemolytic activity and phase 2 enzyme induction in vivo. Free Radical Biol Med 34(9):1200–1211

Nagy P (2013) Kinetics and mechanisms of thiol-disulfide exchange covering direct substitution and thiol oxidation-mediated pathways. Antioxid Redox Signal 18(13):1623–1641

Nivsarkar M, Kumar GP, Laloraya M, Laloraya MM (1991) Superoxide-dismutase inactivation by alpha-terthienyl—a novel observation in thiophene photochemistry. Pestic Biochem Physiol 41(1):53–59

Noctor G (2006) Metabolic signalling in defence and stress: the central roles of soluble redox couples. Plant Cell Environ 29(3):409–425

Noctor G, Queval G, Mhamdi A, Chaouch S, Foyer CH (2011) Glutathione—the arabidopsis book. The American Society of Plant Biologists

Nwachukwu ID, Slusarenko AJ, Gruhlke MCH (2012) Sulfur and sulfur compounds in plant defence. Nat Prod Commun 7(3):395–400

Ouyang XG, Tran QT, Goodwin S, Wible RS, Sutter CH, Sutter TR (2011) Yap1 activation by H2O2 or thiol-reactive chemicals elicits distinct adaptive gene responses. Free Radical Biol Med 50(1):1–13

Paget MS, Buttner MJ (2003) Thiol-based regulatory switches. Annu Rev Genet 37:91–121

Pineda-Molina E, Klatt P, Vazquez J, Marina A, de Lacoba MG, Perez-Sala D, Lamas S (2001) Glutathionylation of the p50 subunit of NF-kappa B: a mechanism for redox-induced inhibition of DNA binding. Biochemistry 40(47):14134–14142

Pinto JT, Krasnikov BF, Coopert AJL (2006) Redox-sensitive proteins are potential targets of garlic-derived mercaptocysteine derivatives. J Nutr 136(3):835s–841s

Prager-Khoutorsky M, Goncharov I, Rabinkov A, Mirelman D, Geiger B, Bershadsky AD (2007) Allicin inhibits cell polarization, migration and division via its direct effect on microtubules. Cell Motil Cytoskelet 64(5):321–337

Rabinkov A, Miron T, Mirelman D, Wilchek M, Glozman S, Yavin E, Weiner L (2000) S-Allylmercaptoglutathione: the reaction product of allicin with glutathione possesses SH-modifying and antioxidant properties. Biochim Biophys Acta-Mol Cell Res 1499(1–2):144–153

Rao RK, Clayton LW (2002) Regulation of protein phosphatase 2A by hydrogen peroxide and glutathionylation. Biochem Biophys Res Commun 293(1):610–616

Roddick JG, Drysdale RB (1984) Destabilization of liposome membranes by the steroidal glycoalkaloid alpha-tomatine. Phytochemistry 23(3):543–547

Sagitdinova GV, Saidkhodzhaev AI, Malikov VM (1978) The structures of juferin and juniferidin. Chem Nat Compd 14:693–694

Sakai J, Li JY, Subramanian KK, Mondal S, Bajrami B, Hattori H, Jia YH, Dickinson BC, Zhong J, Ye KQ, Chang CJ, Ho YS, Zhou J, Luo HBR (2012) Reactive oxygen species-induced actin glutathionylation controls actin dynamics in neutrophils. Immunity 37(6):1037–1049

Schafer FQ, Buettner GR (2001) Redox environment of the cell as viewed through the redox state of the glutathione disulfide/glutathione couple. Free Radical Biol Med 30(11):1191–1212

Schneider T, Ba LA, Khairan K, Zwergel C, Bach ND, Bernhardt I, Brandt W, Wessjohann L, Diederich M, Jacob C (2011) Interactions of polysulfanes with components of red blood cells. Med Chem Comm 2(3):196–200

Surh YJ, Kundu JK, Na HK (2008) Nrf2 as a master redox switch in turning on the cellular signaling involved in the induction of cytoprotective genes by some chemopreventive phytochemicals. Planta Med 74(13):1526–1539

Tew KD, Manevich Y, Grek C, Xiong Y, Uys J, Townsend DM (2011) The role of glutathione S-transferase P in signaling pathways and S-glutathionylation in cancer. Free Radical Biol Med 51(2):299–313

Toledano MB, Delaunay-Moisan A, Outten CE, Igbaria A (2013) Functions and Cellular Compartmentation of the Thioredoxin and Glutathione Pathways in Yeast. Antioxid Redox Signal 18(13):1699–1711

Triantaphylides C, Havaux M (2009) Singlet oxygen in plants: production, detoxification and signaling. Trends Plant Sci 14(4):219–228

Van Etten HD, Mansfield JW, Bailey JA, Farmer EE (1994) Two classes of plant antibiotics: phytoalexins versus "phytoanticipins". Plant Cell 6(9):1191–1192

Walter P, Ron D (2011) The unfolded protein response: from stress pathway to homeostatic regulation. Science 334(6059):1081–1086

Wang J, Boja ES, Tan WH, Tekle E, Fales HM, English S, Mieyal JJ, Chock PB (2001) Reversible glutathionylation regulates actin polymerization in A431 cells. J Biol Chem 276(51):47763–47766

Wang J, Tekle E, Oubrahim H, Mieyal JJ, Stadtman ER, Chock PB (2003) Stable and controllable RNA interference: investigating the physiological function of glutathionylated actin. Proc Natl Acad Sci USA 100(9):5103–5106

Ward NE, Stewart JR, Ioannides CG, O'Brian CA (2000) Oxidant-induced S-gutathiolation inactivates protein kinase C-alpha (PKC-alpha): a potential mechanism of PKC isozyme regulation. Biochemistry 39(33):10319–10329

Ward NE, Chu F, O'Brian CA (2002) Regulation of protein kinase c isozyme activity by S-glutathiolation Methods in Enzymology. Academic Press, London

Winterbourn CC, Hampton MB (2008) Thiol chemistry and specificity in redox signaling. Free Radical Biol Med 45(5):549–561

Winterbourn CC, Metodiewa D (1994) The reaction of superoxide with reduced glutathione. Arch Biochem Biophys 314(2):284–290

Wood MJ, Storz G, Tjandra N (2004) Structural basis for redox regulation of Yap1 transcription factor localization. Nature 430(7002):917–921

Xu CJ, Yuan XL, Pan Z, Shen GX, Kim JH, Yu SW, Khor TO, Li WG, Ma JJ, Kong ANT (2006) Mechanism of action of isothiocyanates: the induction of ARE-regulated genes is associated with activation of ERK and JNK and the phosphorylation and nuclear translocation of Nrf2. Mol Cancer Ther 5(8):1918–1926

Authors Biography

Martin C. Gruhlke (born 1983) studied Biology at the RWTH Aachen University in Germany. He subsequently joined the laboratory of Prof. Alan Slusarenko in 2008, where he currently conducts research on molecular mechanisms of sulfur-containing natural compounds in fungi, plants, and animal cell lines with a certain focus on allicin as a potential 'redox toxin'.

Alan Slusarenko is Head of the Plant Physiology Department at RWTH Aachen University. His research has centered on resistance mechanisms of *Arabidopsis* to infection and more recently on Natural Products in Plant Protection. Alan obtained a PhD in Plant Pathology from Imperial College in 1981 and was a lecturer in the Department of Plant Biology at Hull University in the UK from 1983 until moving in 1988 to an Assistant Professorship in Molecular Plant Pathology at the University of Zuerich in Switzerland and subsequently in 1995 to the Chair of Plant Physiology at RWTH Aachen in Germany.

Part IX
Connecting Section Between Chapters 9 and 10

The previous chapter provided a brief description of the cellular thiolstat, its major components and implications for intracellular redox signaling. It has paved the way for a further discussion of redox signaling and control, triggered, for instance, by exogenous redox modulating agents. The next chapter will use this emerging concept of the thiolstat to consider cellular responses toward highly reactive and effective, yet at the same time also considerably selective redox modulating secondary metabolites. In doing so, it will show how chemically rather simple organic sulfur compounds are able to trigger extensive cellular responses, which often differ significantly from cell type to cell type.

Here, the reader will be confronted with a number of chemically rather unusual compounds, such as thiosulfinates and polysulfanes, which occur in many plants and fungi, and exert a facet-rich spectrum of biological activities. Importantly, these compounds possess considerable chemical reactivity toward thiol groups, and hence react efficiently with cysteine residues in many proteins and enzymes (e.g., tubulin), while other amino acid residues hardly become affected (methionine may sometimes also be modified, although its chemistry differs considerably from the one of cysteine and selenocysteine). As the previous chapter has already emphasized, the resulting *S*-thiolation of cysteine residues (i.e., disulfide bond formation with the cysteine thiol) often impacts on the protein or enzyme affected. Depending on the severity, extent, and duration of such modifications, widespread cellular signaling is likely, which frequently results in an increased antioxidant response and/or in cell cycle arrest and apoptosis. Indeed, it appears that various thiosulfinates and polysulfanes employ this kind of thiol-reactive chemistry to cause selective cell death, for instance, of oxidatively stressed cancer cells, of macrophages, and of out-of-control fibroblasts in scleroderma.

Importantly, the modification of the cysteine residues in proteins and enzymes is not limited to mammalian cells. Recent studies have provided convincing evidence that the concept of the thiolstat can also be applied to yeast, bacteria, and fungi. Some of these microorganisms are particularly sensitive toward redox modulation, and hence may provide a prime target for thiol-reactive sulfur compounds, such as allicin and diallylsulfanes from garlic or dipropylsulfanes from onions.

The next chapter will discuss some of these reactive organic sulfur species and compounds, and how they interact with different target cells and organisms.

Chapter 10
Thiosulfinates, Organic Polysulfanes, and Related Compounds: From an Unusual Chemistry Toward a Wealth of Potential Applications

Ifeanyi D. Nwachukwu and Alan J. Slusarenko

Keywords Allicin • Ajoene • Antibiotic activity • Haarlem oil • Organic polysulfanes

1 Introduction

It is often said that plants are brilliant synthetic chemists. While that expression may have become trite, what has not ceased to evoke interest and wonder is the sheer myriad of uses, both old and novel, to which many compounds of plant (and generally natural) origin have continued to be put. From medicine to agriculture, and from chemistry to the culinary arts, these compounds continue to receive the attention of experts and laymen alike as a result of the enormous potential for varied applications which they hold. What is more, about 60 % of currently available drugs were either directly or indirectly derived from natural products (Paterson and Anderson 2005), and, not too surprisingly, nutriceuticals—"substances that may be considered a food or part of a food, and (which provide) medical or health benefits, including the prevention and treatment of disease" have been described as the fastest growing sector of today's food industry (see also Chap. 2 and Explanatory Box 1. on the 'synthetic natural' mixture of compounds known as 'Haarlem Oil') (Hardy 2000; Medicine 1992).

I. D. Nwachukwu · A. J. Slusarenko (✉)
Department of Plant Physiology (Bio III), RWTH Aachen University, 52056 Aachen, Germany
e-mail: alan.slusarenko@bio3.rwth-aachen.de

C. Jacob et al. (eds.), *Recent Advances in Redox Active Plant and Microbial Products*,
DOI: 10.1007/978-94-017-8953-0_10,

Explanatory Box 1: Haarlem Oil

Most chapters of this book, including the present one, consider redox active natural products isolated from plants, fungi, and other organisms. The logic behind this approach demands that these compounds are already present in the source and simply need to be 'extracted' in order to be used. In some instances, the extraction procedure may also result in a small change of the molecular structure of the natural metabolite, for instance by cleaving an ester bond (as is the case with many flavonoids). Yet these changes are minor in comparison and do not change the natural product dramatically.

There is, however, also another approach. Here, a given biological material, such as garlic oil or a specific sap, is heated up to temperatures which trigger chemical processes, such as pyrolysis or other forms of decomposition. In the end, entirely novel and 'not really entirely natural' compounds are generated which show their own, rather distinct chemical properties. Interestingly, this kind of 'natural synthetic chemistry' is not limited to simple pyrolysis and the collection of fractions in the headspace of a distillation apparatus. It is also possible to add various natural products together and hence to carry out a rather intriguing chemistry based on two or more natural products. The result of such a process is often a mixture of chemical compounds which are highly unusual and, at the same time, biologically rather active. Amazingly, as such processes are the result of a 'kitchen chemistry' using only the finest, organically grown ingredients, and standard cooking, frying, or baking procedures, the resulting chemical cocktail in the past has often escaped the kind of regulatory constraints usually associated with the use of 'artificial' chemical products.

Haarlem Oil reflects this particular approach. It is an old remedy used for centuries by sailors on their long journeys around the globe against infections and other ailments. Nowadays, it is sold fairly freely as a cure against a wide range of apparent disorders, which range from general fatigue to more serious diseases. While the effects of Haarlem Oil on the human body are still not understood fully, the oil is also used to improve the health and performance of horses, including equine and human stallions. Interestingly, Haarlem Oil itself is not really a natural product according to the normal meaning of this term. It is generated by heating up elementary sulfur in a mixture of pine turpentine and linseed oil. According to chemical analysis, the resulting product contains a fair amount of Organic Sulfur Species, including 1,2-dithiole-3-thiones, such as alpha- and beta-pinene 1,2-dithiole-3-thione. 1,2-Dithiole-3-thiones are well known in the pharmaceutical literature. Their complex redox chemistry results in a wide range of biological activities, which in the case of 4-methyl-5-(2-pyrazinyl)-1,2-dithiole-3-thione (Oltipraz®) include a pronounced activity against parasitic schistosoma (blood-flukes). Oltipraz and related 1,2-dithiole-3-thiones, such as anethole 1,2-dithiole-3-thione (Sulfarlem®) are currently also under investigation as chemopreventive antioxidants or potential cytotoxic anticancer agents. In all cases, the pronounced redox activity of these compounds is key to biological activity. Here, natural products, such as pine turpentine, linseed oil, and sulfur are not simply extracted but processed together to form a new product, which contains entirely new substances endowing this 'not quite natural' end-product with its distinct biological activities.

Prominent among plants with physiologically active compounds are Alliums, members of the onion and garlic family, which have enjoyed a generous dose of attention since ancient times because of their plethora of uses, ranging from antidote for snake bites (Wills 1956) to therapy for pneumonia (Castiglioni 1985). It is bioactive compounds like the thiosulfinates and their derivatives, the organic polysulfanes, with robust oxidative properties, formed in plants like those of the genus *Allium* or generated from them after cellular damage, that are mainly responsible for the pharmacological (medicinal), chemical, agricultural (antimicrobial), and thus economic importance of such plants (see also Chap. 12) (Block et al. 2010) (Fig. 1).

The first set of chemical syntheses of alkyl thiosulfinates was achieved by the oxidation of corresponding disulfides with an organic peracid, e.g., perbenzoic, percamphoric, perfuroic, or peracetic acid, in an appropriate solvent such as chloroform (Small et al. 1947). Conversely, thiosulfinates are generated in nature upon wounding in certain plants. For instance, diallylthiosulfinate (2-propene-1-sulfinothioic acid *S*-2-propenyl ester) or 'allicin', arguably the best known thiosulfinate, is released following tissue damage in garlic (see Sect. 2). Other thiosulfinates with varying organic substituent combinations, including *n*-propyl, 1-propenyl, and *n*-butyl, are also formed upon tissue injury in some plants of the *Allium* and *Brassica* genera (Block et al. 2010; Kubec et al. 2010).

Although there are three major pathways by which thiosulfinates could be formed *in vivo* (Jacob 2006), it is their enzymatic formation from virtually redox inactive sulfoxides in the Alliums which is of primary interest to us here (Block 1992).

While the aggressively oxidative allicin from garlic, which consumes up to four thiol equivalents per molecule (Giles et al. 2001, 2002) is probably the most bioactive *Allium*-derived thiosulfinate, related compounds like thiosulfonates (which possess one more oxygen than a thiosulfinate), e.g., methyl methanethiosulfonate and 2-propenyl 2-propenethiosulfonate (pseudoallicin) are also known to possess antimicrobial properties (Jacob 2006). The specific oxidizing action of thiosulfinates is

Fig. 1 General chemical structure of a thiosulfinate

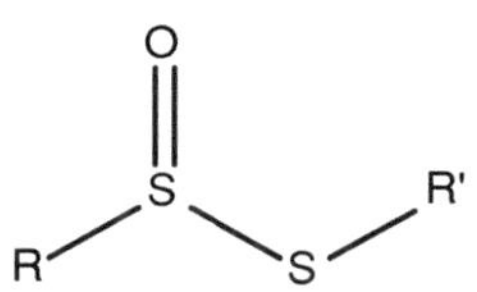

somewhat preferably targeted at thiols (Giles et al. 2002; Jacob 2006) and cysteine proteins in what is generally described as thiol-disulfide exchange reactions (Jacob and Anwar 2008), and has been found not to ordinarily affect other reducing equivalents like NADH or L-ascorbate *in vitro* (Giles et al. 2002). Here, thiosulfinates could have far-reaching effects on the cell's redox potential as a result of the possible cascade of events triggered by an initial reaction including the formation of mixed disulfides and sulfenic acids, while the oxidation of –SH groups in proteins and low molecular weight thiols such as glutathione, could trigger widespread changes to the cellular thiolstat (Giles et al. 2002; see Chap. 9).

Also known for intense bioactivity and similarly derived from diverse natural sources including the Alliums and trees like *Scorodophloeus zenkeri* Harms are organic polysulfanes like lenthionine, 1,2-vinyldithiin and 1,3-vinyldithiin,and related methyl-, ethyl-, and propyl-based tri- and tetrasulfanes (Agarwal 1996; Block 1992; Kouokam et al. 2002). In this work, we focus on organic polysulfanes of the type $R–S_n–R$ with $n \geq 3$, $R =$ an organic residue and $R \neq H$ (Schneider et al. 2011; Steudel 2002). These compounds may be cyclic or acyclic and the organic substituents R which either terminate the sulfur chain or close the chain to form a ring are connected to the sulfur chain via carbon atoms (Steudel 2002).

Organic polysulfanes not only possess demonstrated antibacterial, antifungal, and anticancer properties (Kouokam et al. 2002; Munday et al. 2003) but are also known to be involved in intricate cellular redox modulation events (Schneider et al. 2011; see Chap. 4). For instance, polysulfanes are known to undergo complex oxidation-reduction reactions *in vitro* and *in vivo* with –SH groups and in the process form perthiols which could generate 'reactive oxygen species' (ROS) in the presence of glutathione, Fe^{2+} from hemoglobin or other iron-containing proteins (Munday et al. 2003; Schneider et al. 2011). Additionally, some hydropolysulfanes produced by the reduction of the cyclic pentasulfane varacin (see later) are known to subsequently yield S_2 and possibly S_3 breakdown products which are thought to possess the capacity to attack the DNA with adverse cell survival consequences (Brzostowska and Greer 2003; Jacob 2006).

Detailed examination of these obviously important compounds from various natural sources is our objective in this chapter. We begin with allicin, the major antimicrobial substance found in garlic extracts.

2 Allicin from Garlic

The disruption of garlic (*Allium sativum* L., *Amaryllidaceae*) tissues by way of crushing, piercing, wounding, etc., releases the volatile organosulfur compound, allicin (diallylthiosulfinate), garlic's most studied and arguably its biologically most active compound. Allicin is a 'phytoanticipin'-a term used to refer to any "…low molecular weight, antimicrobial compound present in plants before challenge by microorganisms, or produced after infection solely from pre-existing constituents." (Van Etten et al. 1994). The synthesis of allicin follows the general

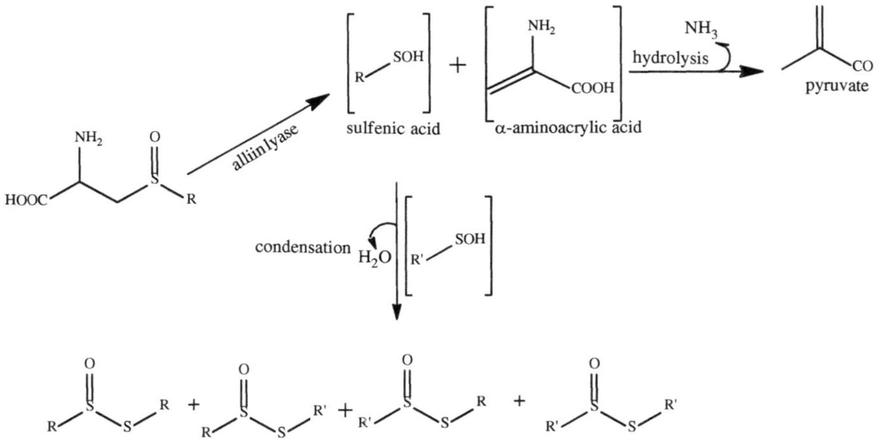

Scheme 1 General model for the alliinase-catalyzed synthesis of thiosulfinates from cysteine sulfoxide precursors (modified from Block 1992; Musah et al. 2009; Shimon et al. 2007). R $= CH_2 = CHCH_2^-$ in alliin (E)$CH_3CH = CH^-$ in isoalliin, CH_3^- in methiin and $CH_3CH_2CH_2^-$ in propiin

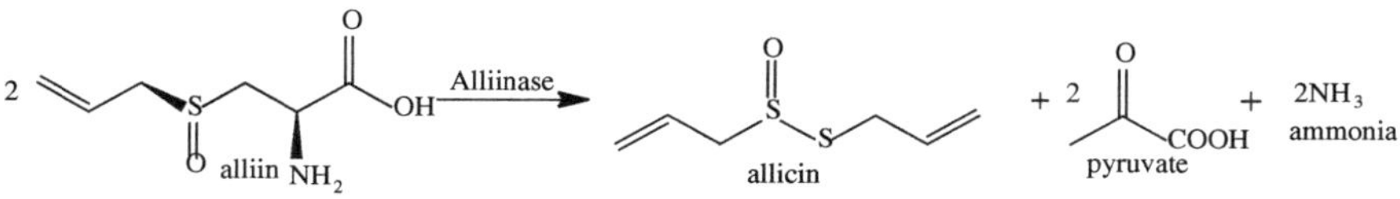

Scheme 2 Formation of allicin from S-allyl-L-cysteine sulfoxide, alliin

pattern found in some other *Allium* species including leek, onion, shallot, chive, scallion, and in the medicinal plant *Petiveria alliacea*, in which the degradation of cysteine sulfoxide derivatives is mediated by pyridoxal 5′-P (PLP)-dependent cysteine sulfoxide lyases (alliinases), releasing short-lived sulfenic acid intermediates and α-aminoacrylic acid (Scheme 1) (Ankri and Mirelman 1999; Block 1992; Musah et al. 2009). The transient sulfenic acids so formed generally undergo instant condensation to form thiosulfinates while the α-aminoacrylic acid is hydrolyzed to yield pyruvic acid with the concomitant loss of ammonia.

The formation of allicin occurs instantaneously when the inactive cysteine sulfoxide precursor, alliin (*S*-allyl-L-cysteine sulfoxide), usually found in the cytoplasmic matrix interacts with the previously vacuole-bound alliin lyase (EC 4.4.1.4) following a breach in tissue structural integrity (Scheme 2).

Allicin demonstrably oxidizes several –SH containing enzymes, such as alcohol dehydrogenase and succinic dehydrogenase *in vitro*, through a thiol-disulfide exchange reaction to form an *S*-allylthioderivative, and this reaction has been proposed as the basis of its antimicrobial properties *in vivo* (Ankri and Mirelman 1999; Davis 2005; Wills 1956). The typical reaction between allicin and a biologically relevant thiol was first reported between the organosulfur compound and

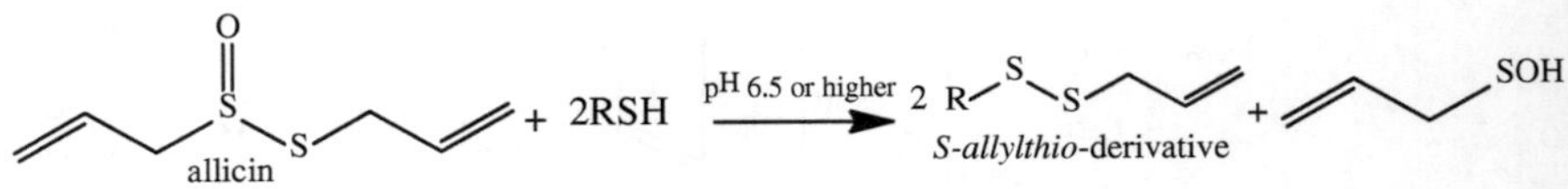

Scheme 3 An example of the kind of rapid thiol-disulfide exchange reaction allicin and other thiosulfinates undergo with thiol compounds like glutathione and cysteine (Cavallito et al. 1944). In this kind of reaction, cysteine becomes attached to allicin's allylthio group forming S-allylmercaptocysteine while for glutathione, the reaction produces S-allylmercaptoglutathione. Other related compounds like thiosulfonates interact similarly with low molecular weight thiols and other –SH containing groups (Small et al. 1949)

the –SH group of the amino acid cysteine (Cavallito and Bailey 1944), a reaction paralleled by that which occurs between the allicin follow-on product, ajoene and cysteine (see later). While allicin may kill microbial cells by oxidatively inactivating essential enzymes, it similarly reacts with the –SH group of acetyl CoASH, the building block for cholesterol, and triglyceride biosynthesis, and this has been suggested to play a crucial role in the cholesterol-lowering effect of garlic (Agarwal 1996; Gardner et al. 2007; Sendl et al. 1992) (Scheme 3).

Allicin exhibits proven antibacterial, antifungal, antiviral, and antiparasitic properties (Ankri and Mirelman 1999), including an activity against agriculturally important pathogens like *Fusarium spp., Phytophthora spp., Pseudoperonospora spp., Magnaporthe spp.* (Arya et al. 1995; Bianchi et al. 1997; Curtis et al. 2004; Portz et al. 2008; Russell and Mussa 1977), thus underlining its potential as a viable crop protectant (see also Chap. 12).

In 2010, Gruhlke et al. proposed an additional facet of allicin's cell-killing mechanism in yeast (Gruhlke et al. 2010). Given allicin's redox active properties and ability to readily penetrate the cellular plasma membrane, the authors suggested that in addition to its specific reaction with –SH groups in essential enzymes and proteins, it could also have a dramatic effect on the cell's global redox status. Using *S. cerevisiae* as a tool, evidence was presented showing that at threshold inhibitory concentrations, allicin has the capacity to alter the cell's entire redox potential, and thus shunt cells into apoptosis. Although allicin has been known to induce apoptosis in other cell types (Bat-Chen et al. 2010; Hirsch et al. 2000; Miron et al. 2008; Oommen et al. 2004; Park et al. 2005; Zhang et al. 2010), this study represents the first report of allicin-induced apoptosis in fungi, and thus deepens our knowledge of its multifaceted antimicrobial capacity.

Allicin has been described as a *pro-oxidant*, which by definition are "chemicals that induce oxidative stress, either through creating ROS or inhibiting antioxidant systems" (Block 2010; Gruhlke et al. 2010). While it is logical to think that allicin would deplete cellular levels of the antioxidant glutathione, given its interaction with –SH groups, results from one study suggest that such an effect is only transient (Hirsch et al. 2000) although this may be explained by the conversion of a proportion of intracellular reduced glutathione to *S*-allylmercaptoglutathione (Rabinkov et al. 2000). Interestingly, allicin has been reported to in fact increase cellular glutathione levels in vascular endothelial cells (Horev-Azaria et al. 2009) and to suppress ROS/ROS-signaling pathways in certain cell types (Liu et al. 2010). To explain such seeming contradictions of allicin acting as both an antioxidant and a pro-oxidant, it is important to note that at low concentrations, allicin induces the intracellular synthesis of Phase II protective enzymes in mammals; these enzymes function to palliate the effect of subsequent oxidative attacks (see also Chap. 5) (Munday et al. 2003). Secondly, allicin readily decomposes to 2-propenyl sulfenic acid, a well-known reducing agent (Vaidya et al. 2009). Thus allicin, while acting as an oxidant, simultaneously yields a strong reducing agent.

3 Other Sulfur-Containing Natural Products

3.1 Vinyldithiins

The breakdown products of allicin include vinyldithiins, which are the most abundant sulfur compounds in garlic oil from macerates (Scheme 4) (Voigt and Wolf 1986). The pure dithiins are yellowish, oily liquids with a flavor distinct from allicin and a garlic-like aroma (Iberl et al. 1990). Although exact proportions tend to vary depending on the extraction solvent, various studies indicate that in commercial garlic oil preparations, the amount of 1,3-vinyldithiin (2-vinyl-4*H*-1,3-dithiin) is usually higher than that of 1,2-vinyldithiin (3-vinyl-4*H*-1,2-dithiin) (Iberl et al. 1990; Lawson et al. 1991; Voigt and Wolf 1986). Compared to allicin, the dithiins are less rapidly metabolized in the liver, with the more hydrophilic 1,3-vinyldithiin taking a longer time to be metabolized than 1,2-vinyldithiin both *in vitro* in liver homogenates and in *in vivo* studies (Egen-Schwind et al. 1992).

1,3-vinyldithiin has been found to inhibit cholesterol synthesis *in vitro* at a concentration of 1.0 mM thus leading to the suggestion that it could possibly play a role in lowering serum cholesterol levels (Sendl et al. 1992), an outcome that would be at variance with the results obtained by Gardner et al. following their investigation of the effects of raw garlic, powdered garlic supplement, and aged garlic extract supplement on *in vivo* cholesterol levels (Gardner et al. 2007). Other reported biological activities of the dithiins include increasing the activities of glutathione reductase and glutathione *S*-transferase as well as reducing aflatoxin B1-induced DNA damage in primary rat hepatocytes (Kensler et al. 2000).

3.2 Varacin

Varacin, 2-(6,7-dimethoxy-1,2,3,4,5-benzopentathiepin-9-yl) ethanamine, is the bicyclic, organic polysulfane isolated from the Far Eastern marine ascidian, *Polycitor sp*. Varacin possesses considerable antifungal and cytotoxic properties with an activity 100-fold more than that of the anticancer drug 5-fluorouracil in the human colon cancer cell line HCT 116 (Ford and Davidson 1993). Its cytotoxic

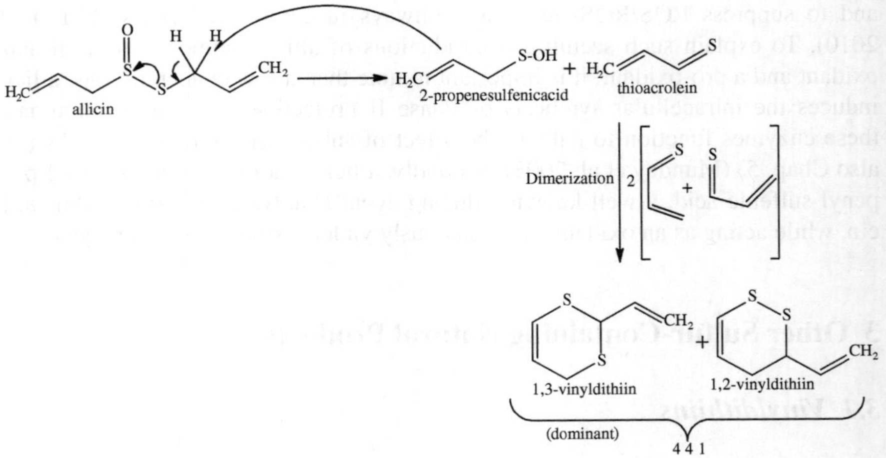

Scheme 4 1,3-dithiin and 1,2-dithiin, the major and minor unsaturated heterocyclic Diels-Alder dimers of thioacrolein, respectively, are formed during the decomposition of allicin. The amount of 1,3-dithiin formed from this reaction is more than four times that of 1,2-dithiin (modified from Block 2010)

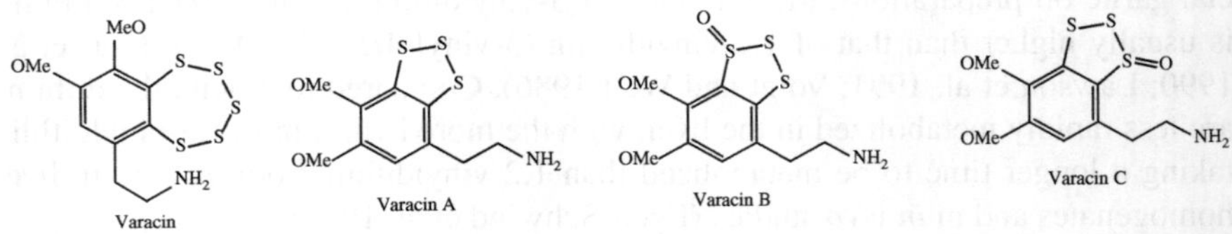

Fig. 2 Structures of varacin and related organic polysulfanes

property has been attributed to its capacity to cause DNA damage (Davidson et al. 1991). Three other biologically active marine polysulfanes with similar structures to varacin's pentathiepin ring and accordingly named varacin A, B, C have also been isolated from the same marine source (Makarieva et al. 1995). Varacin, varacin acetate, and varacin A-C acetates have all shown potent antimicrobial activities against *Candida albicans* and *Bacillus subtilis* at low concentrations, an indication that the presence of a free amino group in their side chains has virtually no influence on the activities of the pentathiepins (Makarieva et al. 1995) (Fig. 2).

3.3 Organic Polysulfanes

Organic polysulfanes are found in varying amounts in extracts of garlic, onion, Shiitake mushroom, Durian fruit, *Petiveria alliacea* (Steudel 2002), and other plant species. For instance, considerable amounts of diallyldisulfide DADS, diallyltrisulfide DATS, and diallyltetrasulfide DATTS, are commonly present in garlic oil preparations and are well studied, while higher polysulfanes like diallylpentasulfide, diallylhexasulfide, and diallylheptasulfide are not as frequently and easily extracted (Munchberg et al. 2007), thus limiting the extent to which studies employing them have been carried out (Block 2010). Polysulfanes have also been synthesized enzymatically from mixtures of alliinase, alliin, and cystine-*S*-oxide (Keusgen 2008; Lancaster et al. 2000).

Not surprisingly, the uses to which polysulfanes could be potentially put are as varied as their chemical structures, and span fields as wide as agriculture and medicine (see also Chap. 12). For instance, disulfanes and higher polysulfanes terminated by sulfinate functions [–S(O)O–] have been studied for potential use as antiradiation agents (Bowman et al. 1986); DADS is implicated in the neutralization of their toxic effects of cyanide in mouse tissues (Iciek et al. 2005); DATTS shields cells from the poisonous effects of cadmium (Pari et al. 2007); DATS has been reported to suppress egg hatching, as well as larval and adult emergence in both maize weevils, *Sitophilus zeamais,* and red flour beetle, *Tribolium castaneum* (Huang et al. 2000); while DATS and DATTS are active against the blue mold, *Penicillium expansum*, responsible for the decay of harvested fruits (Liu et al. 2009), and which synthesizes the carcinogenic toxin patulin.

Of particular importance is the potential for using garlic-derived polysulfanes in cancer chemotherapy. Cerella et al. (2009) found that DATTS selectively induces early mitotic arrest, sequentially activates proapoptotic Bak and Bax, and triggers caspase-dependent apoptosis in U937 cells, a model for human histiocytic lymphoma (a rare type of non-Hodgkin's lymphoma marked by large tumor cells that are similar in shape to histiocytes), without affecting peripheral blood mononuclear cells (PBMCs) from healthy donors. It is now believed that DATTS primarily targets tubulin in U937 cells, and thus disrupts the formation of normal spindle microtubules and leads to G2/M arrest, while JNK activated early upon DATTS challenge, mediates multi-site phosphorylation of the anti-apoptotic protein Bcl-2, and subsequently its proteolysis (Kelkel et al. 2012). This result is in line with similar ones highlighting the selective toxicity of polysulfanes including DADS and DATS against various cancer cell lines (Antosiewicz et al. 2006; Busch et al. 2010; Seki et al. 2008; Xiao et al. 2009; Yang et al. 2009). Generally, it appears that the bioactivity of these polysulfanes increases with the increasing number of sulfur atoms (O'Gara et al. 2000; Tsao and Yin 2001) but as Muenchberg et al. (2007) pointed out, this relationship is not always linear.

It is important, in discussing polysulfanes, to highlight that it is the sulfur-centered functional groups, for example S-S in disulfanes and S-S-S in trisulfanes, that are taken into account and not the total number of sulfur atoms in the compound (Jacob 2006). Consequently, a compound like trithiophene, the aromatic sulfur-containing antibacterial (Hudson et al. 1993), antifungal (Romagnoli et al. 1994), and nematicidal (Bakker et al. 1979) heterocyclic secondary metabolite from African marigold (*Tagetes spp.*) would not be described as a trisulfane, its bioactivity notwithstanding. Similarly, while it is common to refer to polysulfanes as 'polysulfides', this is somewhat incorrect as only (an-)ionic compounds like $S_x^{2-}, x \geq 3$, for instance, should be so called (Doering et al. 2010; Steudel 2002).

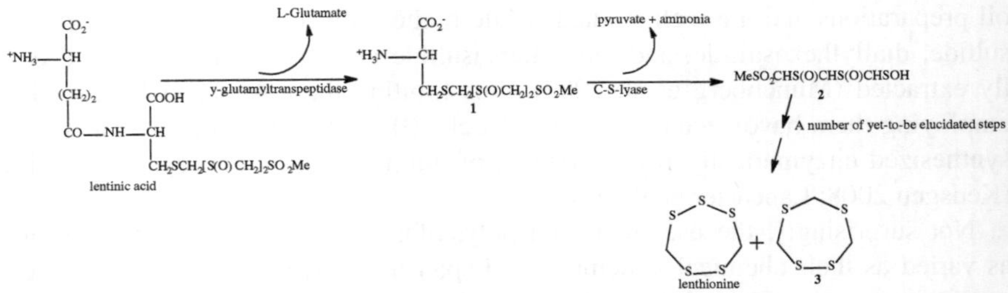

Scheme 5 The synthesis of lenthionine is believed to involve the transformation by a γ-glutamyltranspeptidase of lentinic acid to the substituted cysteine, *1* (Kjaer and Larsen 1977) which is subsequently cleaved by a C-S-lyase to yield *2*. A series of reactions still unclear which possibly include polymerization of dithiiranes (Höfle et al. 1976) results in lenthionine and related organic polysulfanes like *3* (Block and Deorazio 1994)

3.4 Lenthionine

Lenthionine (1,2,3,5,6-pentathiepane) is the organic polysulfane chiefly responsible for the characteristic aroma of the edible *Shiitake* mushroom, a widely popular component of the diet in parts of Southeast Asia including Japan and China (Morita and Kobayash 1966). Since the time of its isolation from *Shiitake*, *Lentinus edodes* (*Berk*) *Sing*., was first reported nearly 50 years ago (Morita and Kobayash 1966), subsequent studies have shown that the cyclic organosulfur compound is also produced in the red alga *Chondria californica* (Wratten and Faulkner 1976) and in bitter beans (*Parkia speciosa*) (Gmelin et al. 1981). Although not all the details of its production are yet known, the synthesis of lenthionine in Shiitake mushrooms is known to proceed by the enzymatic cleavage of lentinic acid by a gamma-glutamyl transpeptidase to yield a substituted cysteine (Scheme 5) (Kjaer and Larsen 1977). The next step involves the cleavage of the substituted cysteine by a pyridoxal phosphate-dependent *C*-*S*-lyase to give the complex sulfone, sulfoxide, and sulfenic acid-sulfone containing compound $MeSO_2CHS(O)CHS(O)CHSOH$ in a reaction generally similar to the pre-odor/flavor production steps in *Brassica* and *Allium* plants (Block and Deorazio 1994). Lenthionine is eventually produced in a series of subsequent chemical reactions that are still not fully understood.

In addition to its antibacterial and antifungal properties (Morita and Kobayash 1967; Yasumoto et al. 1971) lenthionine has also been implicated in the prevention of platelet aggregation in experimental rats thus suggesting possible clinical uses as a remedy for thrombosis (Shibuya et al. 2005; Shimada et al. 2004).

3.5 Lachrymatory Factors and Attendant Thiosulfinates

The disruption of onion (*Allium cepa*) tissues by chopping, crushing, wounding, or piercing releases the tear-inducing volatile sulfine (*Z*)-propanethial *S*-oxide, PTSO (Kupiecki and Virtanen 1960) trivially known as the onion lachrymatory

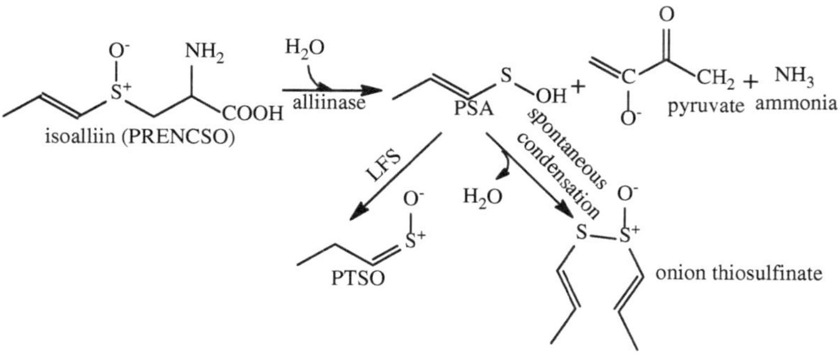

Scheme 6 This two-step reaction shows the synthesis of the onion lachrymator and the mandatory participation of LF synthase (LFS) in the production of LF as first put forward by Imai et al. (2002). In the first step, PRENCSO is enzymatically cleaved to 1-propenylsulfenic acid (*PSA*) by alliinase releasing pyruvate and ammonia in the process. PSA then reacts *via* LFS in the second step to generate LF while the instantaneous condensation of some PSA molecules leads to the production of the onion thiosulfinate, responsible for onion's spicy flavor and known to be converted to compounds with hypolipodemic (Adamu et al. 1982) and anticoagulant properties (Ariga et al. 1981; Makheja and Bailey 1990)

factor (LF) in addition to the onion thiosulfinate to which the spicy flavor of the edible *Allium* is attributed (Imai et al. 2002). The onion LF was previously thought to be formed as a spontaneous by-product of the action of alliinase on isoalliin (*E*)-*S*-(1-propenyl)-L-cysteine sulfoxide (PRENCSO) as a result of the assumption that PRENCSO's unique structural properties in relation to other known naturally occurring *S*-substituted cysteine sulfoxides facilitate the uncatalyzed intramolecular rearrangement of the intermediate 1-propenylsulfenic acid to yield PTSO (Bayer et al. 1989; Block et al. 1996; Musah et al. 2009). Seminal work by Imai et al. (2002), however, has since demonstrated that the formation of the onion LF actually involves an *essential catalyzed* step by the onion lachrymatory factor synthase (LFS) (Scheme 6). PTSO is a popular example of a herbivore-induced plant volatile (HIPV) (Dicke 1999) and is thought to deter (grazing) herbivores from attacking onion plants in the same manner that the herbivorous green peach aphid *Myzus persicae* is repelled by low concentrations of the sulfur volatiles found in garlic or onion oils (Amarawardana et al. 2007; Assis et al. 2007; Jones 2010).

Although the formation of LF to a degree mirrors that of allicin from garlic in that it involves the enzyme alliinase, the disruption of *Allium* tissues, and the release of $CH_3C(O)COOH$ and NH_3 as by-products, LF is not produced in garlic (Imai et al. 2002). Furthermore, LF has not been credited with a similarly potent antimicrobial activity as allicin. Interestingly, potent antimicrobial properties have been recorded with the thiosulfinate compounds of another plant, the Amazonian non-alliaceous *Petiveria alliacea* L. (*Phytolaccaceae*), recently discovered to contain a LFS/alliinase type system similar to the one found in onions (Scheme 7) (Kim et al. 2006; Musah et al. 2009).

An important distinction between the onion and *P. alliacea* LF-producing systems is that while the onion LFS 'simply' catalyzes a *rearrangement* of

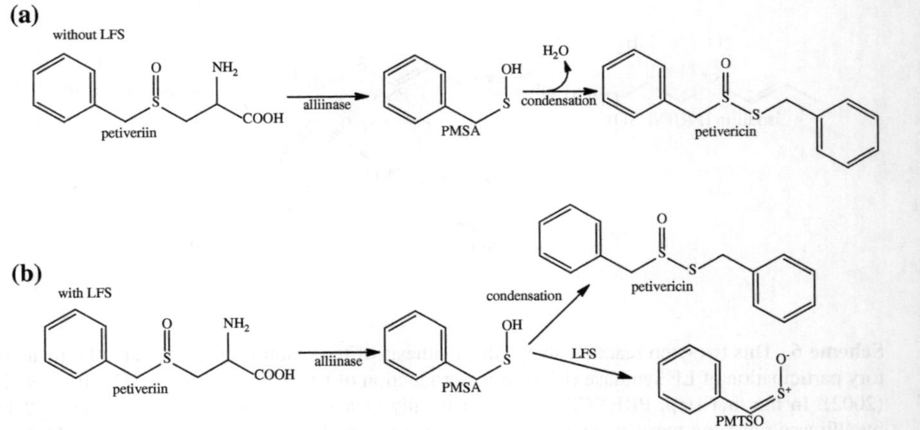

Scheme 7 In *Petiveria alliacea* (also commonly called Anamu), alliinase catalyzes the conversion of the nonprotein aromatic amino acid, petiveriin, to phenylmethanesulfenic acid (PMSA) which then **a** in the absence of the *P. alliacea* LF, as experimentally determined by Musah et al. (2009), spontaneously condenses to the antibacterial and antifungal thiosulfinate, petivericin only, or **b** in the presence of the LF yields both petivericin and the lachrymatory sulfine (*Z*)-phenylmethanethial S-oxide (PMTSO)

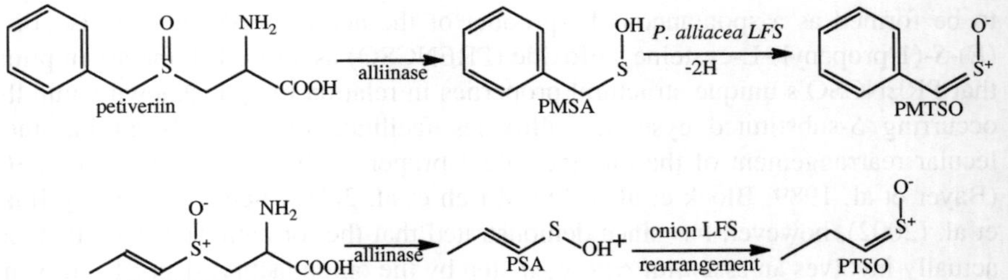

Scheme 8 Although the onion LFS and the LFS from *Petiveria alliacea* both catalyze the synthesis of lachrymatory sulfines from onion and *P. alliacea* respectively, they fundamentally differ in the specific mechanism by which they accomplish this. While the onion LFS acts like a typical isomerase by inducing the chemical rearrangement of PSA to PTSO (the double bond is 'shifted'), the LFS from Anamu behaves like a dehydrogenase in the enzymatic synthesis of PMTSO from PMSA (the double bond is 'created')

1-propenylsulfenic acid to PTSO, the *P. alliacea* LFS drives the enzymatic *dehydrogenation* of PMSA to yield PMTSO (Scheme 8) (Musah et al. 2009).

Finally, a volatile lachrymator was recently reported to be produced following the wounding of the decorative plant *Allium siculum*, sometimes commonly called the Sicilian honey garlic (Block 2010; Kubec et al. 2010). Interestingly, this tear-inducing compound (*Z*)-butanethial *S*-oxide, with a molecular formula of C_4H_8SO is produced along with several 1-butenyl/butyl and dibutyl thiosulfinates, and together these compounds are believed to constitute a powerful arsenal of irritant defensive compounds employed by the plant to ward off pests (Kubec et al. 2010). It is instructive to note that a compound identical in structure to this recently isolated lachrymator had been chemically synthesized in the laboratory nearly 50 years ago from the isoalliin synthetic analog, *S*-(buten-1-yl)-*L*-cysteine sulfoxide (Block 2010; Mueller and Virtanen 1966).

3.6 Ajoene

The allicin breakdown product, ajoene, exists as a colorless and odorless (Agarwal 1996) mixture of two distinct isomers: *E*- and *Z*-4,5,9-trithiadodeca-1,6,11-triene 9-oxide (Block et al. 1984). The amount of ajoene formed from allicin/garlic depends on a number of factors. For instance, ajoene constitutes approximately 12 % of the breakdown products formed from allicin incubated in low polar solvents like hexane, 8 % of garlic homogenate incubated in a polar solvent like ethanol, and 55 % of pure allicin incubated in ethanol (Koch and Lawson 1996). Ajoene, like allicin, is more soluble in organic solvents than in water but, in the absence of solvent, shows a far greater stability than allicin with no decomposition observed at room temperature after 6 days (Koch and Lawson 1996). The nature of the solvent in which synthetic allicin is solubilized as well as reaction/storage temperature also influences the fractional distribution of ajoene isomers with increasing solvent polarity and heat was found to favor an increase in the ratio of *E* over *Z* isomers (Iberl et al. 1990).

Ajoene is widely recognized as an excellent antithrombotic agent and is the major antiplatelet compound derived from garlic (Apitz-Castro et al. 1994), with a higher inhibitory effect (IC_{50}, 60 μM) on platelet aggregation in human whole blood than allicin (IC_{50}, 90 μM) (Lawson et al. 1992). Ajoene, unlike some other platelet aggregation inhibitors, inhibits *in vivo* platelet aggregation induced by all known platelet stimulators including collagen, arachidonic acid, epinephrine, ADP, calcium ionophores, platelet activating factor (PAF), and thrombin (Apitz-Castro et al. 1983, 1986; Jain and Apitz-Castro 1987). The capacity of ajoene to inhibit *in vivo* platelet aggregation by virtually all known agonists and its ability to exert its antithrombotic effect in species as varied as cow, dog, guinea-pig, horse, monkey, pig, rabbit, and rat (Apitz-Castro et al. 1986; Jain and Apitz-Castro 1987) suggest a promising potential for a future application as an antithrombotic agent, and in fact, ajoene has been reported to be undergoing development for use in the treatment of thromboembolic disorders (Agarwal 1996). The matter of the mechanism by which ajoene executes its anti-coagulatory action is not completely settled. While Apitz-Castro et al. (1986) suggest that the organosulfur compound's specific direct interaction with the fibrinogen receptor of platelets is responsible for its inhibition of platelet aggregation, it has been argued that an additional, or even alternative mode of action, may be represented by ajoene's alteration of the ligand-binding properties of a hemoprotein involved in platelet activation (Jamaluddin et al. 1988).

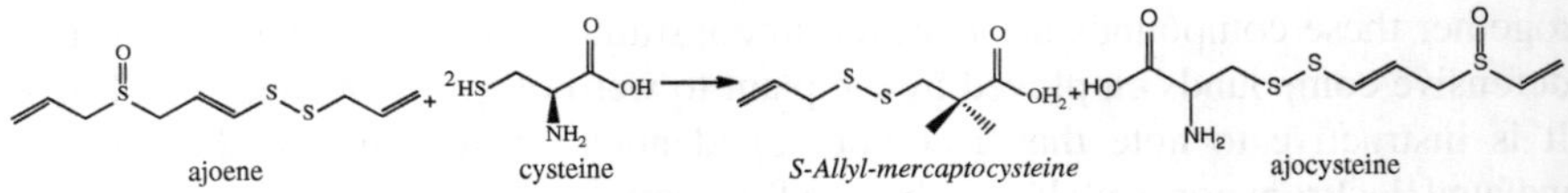

Scheme 9 The reaction of ajoene with cysteine yields **a** S-allyl-mercaptocysteine which has been extensively studied as an inhibitor of proliferating cancer cells, and **b** ajocysteine

In recent years, ajoene has become increasingly important and popular as a result of various studies (Dirsch et al. 1998; Li et al. 2002; Nishikawa et al. 2002) indicating its possible use as an anti-tumor agent, highlighting its induction of apoptosis in various mammalian tumor cell lines (Dirsch et al. 1998; Zheng et al. 1998), and demonstrating its relatively higher activity against different cancer cell lines in comparison to allicin (Scharfenberg et al. 1990). *Z*-ajoene has been reported to inhibit 12-*O*-tetradecanoylphorbol-13-acetate (TPA)-promoted carcinogenesis in a dose-dependent fashion, with 250 μg resulting in 95.1 % inhibition of mice skin tumor formation (Nishikawa et al. 2002). Similarly, ajoene purified from alcoholic extracts of garlic was found to be active against basal cell carcinoma (BCC), the most common type of cancer among Caucasians, reducing the size of 84 % of treated tumors after 6 months. Ajoene is believed to do so by induction of apoptosis as evidenced by the reduced expression of the antiapoptotic Bcl-2 protein in treated tumor cells (Tilli et al. 2003). Results demonstrating *Z*-ajoenes reduced cytotoxic effect on normal in comparison to neoplastic cells, as well as its capacity to disrupt tubulin polymerization, hence blocking the progression from G2 to M phase in various cancer cell lines, and essentially limiting cell proliferation (Li et al. 2002), come together to turn ajoene into a strong candidate for the development of novel anticancer agents.

First isolated and characterized as an antithrombotic agent by Apitz-Castro et al. (1983) ajoene has also shown promising antimicrobial activities. Early studies indicated that while ajoene inhibited the growth of *A. niger* and *C. albicans* (more than allicin) by a poorly understood interaction with the fungal cell wall, the allicin breakdown product had no activity against screened bacteria with the exception of *S. aureus* (Yoshida et al. 1987). Ajoene, however, has subsequently been reported to possess both biostatic and biocidal broadspectrum activities against a number of bacteria including (a) **bacteriostatic** activities against *Bacillus cereus, B. subtilis*, *Mycobacterium smegmatis* and *Streptomyces griseus* at 5 μg/ml ajoene; *Staphylococcus aureus* and *Lactobacillus plantarum* at <20 μg/ml; *Escherichia coli*, *Klebsiella pneumoniae*, and *Xanthomonas maltophilia* at 100–160 μg/ml; and (b) **bacteriocidal** activities against *B. cereus* and *Saccharomyces cerevisiae* at 30 μg/ml (Naganawa et al. 1996). Apart from possibly acting on the cell wall (Yoshida et al. 1987), other suggested mechanisms of ajoenes antimicrobial activities include inhibiting phosphatidylcholine biosynthesis (Ledezma and Apitz-Castro 2006) and oxidizing –SH groups (Naganawa et al. 1996). In fact, one mol of ajoene reacts rapidly *in vitro* with two mols of –SH group-containing cysteine to yield one mol each of *S*-allylmercaptocysteine and ajocysteine (Scheme 9) (Koch and Lawson 1996), a reaction responsible for the observed depletion of ajoene in growth media and blunting of the compound's antimicrobial activity (Naganawa et al. 1996). Ajoene's antiviral (Tatarintsev et al. 1992) and anti-protozoan (Perez et al. 1994; Urbina et al. 1993) activities are also well documented.

4 Conclusion

There has been a growing interest in recent years in relatively little known thiosulfinates and organic polysulfanes derived from natural sources, such as the vinyldithiins and varacin, and the recently discovered onion lachrymatory factor LF. Compared to the more extensively studied allicin, for instance, there is scant information in the literature on most of these compounds. Available data, however, indicate a massive potential in the potential uses to which these compounds may ultimately be put to.

Notwithstanding the great promise these compounds may hold, the relative dearth of published information regarding these substances is not the only hurdle between them and full commercialization and maximum utilization. Like allicin, which has several potential and actual applications, but so far has not successfully made the transition to large-scale commercial use (Slusarenko et al. 2008), most of these compounds are yet to be put into industrial-scale efficient use. Considering the myriad of uses to which chemically synthesized organic polysulfanes have been put (Steudel 2002) it is safe to state that organic sulfur compounds of biogenic origin are a long way from being put to their maximum possible application (see also Explanatory Box 2 for other unusual sulfur compounds already used in Medicine).

Explanatory Box 2: Reactive Sulfur Species as Medical Drugs

As we have seen, organic sulfur compounds (OSCs) derived from natural sources are often highly reactive and hence also frequently possess a pronounced biological activity. Several chapters of this book present OSCs which react fast and efficiently with cysteine residues in proteins and enzymes, and hence interfere with or even regulate the function or activity of these important biomolecules. At the forefront of this discussion, we find Reactive Sulfur Species such as thiosulfinates, disulfides, polysulfanes, and isothiocyanates. Interestingly, these sulfur compounds only represent a small—but important—part of a much wider spectrum of reactive and biologically active sulfur species. Some of these sulfur-based agents are also of pharmaceutical interest and already used in therapy, and will be discussed here briefly.

We have already seen that some 1,2-dithiole-3-thiones, such as Oltipraz (4-methyl-5-(2-pyrazinyl)-1,2-dithiole-3-thione) are used as an effective schistosomicide. These 1,2-dithiole-3-thiones represent an important class of (mostly artificial) sulfur compounds which exhibit a unique reactivity and pronounced biological activity. Recent studies associate these compounds also with antioxidant, chemopreventive and anticancer activity. Indeed, Oltipraz activates the transcription factor Nrf2 which binds to antioxidant response elements (AREs).

A similar reactivity is found in natural products containing the isothiocyanate group. Various organic isothiocyanates (ITCs), including sulforaphane and allyl isothiocyanate, occur as secondary metabolites in plants, most notably in species of the Brassicaceae family. These ITCs form part of our daily diet and are synthesized in the plant from chemically rather unreactive glucosinolates in an enzymatic conversion involving myrosinase enzymes. Such ITCs react readily with thiol and amine groups, are able to modify and hence regulate key proteins and enzymes and therefore may trigger widespread cellular responses, including the expression of Phase 2 enzymes. Whilst usually being considered as antioxidants, ITCs are also able to increase intracellular levels of ROS in cancer cells, ultimately resulting in apoptosis of these cells while healthy cells are often less affected.

Sulfenic acids occupy a special place among the most reactive sulfur species. As sulfenic acids are highly reactive, they are also easily oxidized or sequestered by other compounds. Hence there are only a few examples of 'stabilized' sulfenic acids, which escape capture either by electronic effects or steric hindrance. Nonetheless, sulfenic acids can be formed *in situ* in biological systems, where they show presence and also take part in several important processes. They are found, for instance, as part of the catalytic cycle of the peroxiredoxins and also represent a precursor of allicin. There are also certain drugs which rely on sulfenic acid chemistry. Prominent examples include the prazole proton-pump inhibitors, such as Omeprazole®, which are used in the treatment of stomach ulcers. Prazoles are pro-drugs which contain a sulfoxide moiety. Under acidic conditions (e.g., in the stomach), a carbon–sulfur bond of the prazole is cleaved and the resulting molecule rearranges to form a sulfenic acid. The latter reacts with a critical cysteine thiol of H^+/K^+—ATPase, subsequently inhibiting the enzyme and raising the pH value in the stomach.

A similar sulfur-centered activation of a pro-drug in response to a particular physiological process is observed in the case of the tetrahydrothienopyridine-based antithrombolytic drug Clopidogrel. In this case, hydroxylation of the tetrahydrothienopyridine moiety by cellular oxidases, such as cytochrome P450 2C19, results in a thiolactone, which is subsequently hydrolyzed to an acid and a thiol. The latter, possibly via a transiently formed sulfenic acid, reacts with a particular cysteine residue of the

ADP-receptor P2Y12 to form a disulfide, a modification of P2Y12 which ultimately inhibits the aggregation of blood platelets.

Other reactive sulfur species of interest include α,β-unsaturated disulfides, such as ajoene and 3-vinyl-3,4-dihydro-1,2-dithiin (1,2-DT), and the thiuram disulfides. These sulfur species exhibit high reactivity toward thiol groups in proteins and enzymes and, once 'activated', lead to an extensive follow-on chemistry. Ajoene and 1,2-DT, for instance, have formed part of the human diet for several millennia, and provide a lead for therapeutically active agents, which appear to be largely nontoxic for humans. These unsaturated compounds are considerably more reactive than a 'normal' disulfide and the initial thiol/disulfide exchange reaction results in a whole cascade of biologically relevant, reactive species, which also include a highly reactive thioaldehyde. A similar behavior has been found in the case of thiuram disulfides. Disulfiram (1,1′,1″,1‴-[disulfanediylbis (carbonothioylnitrilo)]tetraethane, tetraethylthiuramdisulfide) is a highly reactive, thiol-specific reagent, which in the past has been used under the name 'Antabuse®' to treat alcoholism. It reacts with a critical cysteine residue of the enzyme aldehyde dehydrogenase, which is inhibited and hence can no longer metabolize acetaldehyde (itself a metabolite of ethanol metabolism in humans). Upon ingestion of ethanol, this particular inhibition leads to an accumulation of acetaldehyde in the body and to a range of unpleasant physiological effects associated with this aldehyde (the latter are supposed to discourage the further consumption of alcohol). The reduced form of the thiuram disulfide, i.e., the thiocarbamate, is still reactive and may also interact with metal ions and metalloproteins, or even release the signaling molecule hydrogen sulfide (H_2S). Indeed, disulfiram derivatives are currently being considered as potential antimicrobial and anticancer agents.

While there are numerous other examples of pharmaceutically relevant and chemically 'exotic' RSS, these examples should suffice to underline the importance of these molecules in modern drug development.

Nonetheless, and regardless of the aforementioned challenges, it is safe to expect that solutions to these challenges will ultimately be found, and that in the future, these compounds could be put to more practical uses than is presently the case.

Acknowledgment Ifeanyi D. Nwachukwu was supported by funding from the European Community's 7th Framework Programme [FP7/2007-2013] under grant agreement No: 215009.

References

Adamu I, Joseph PK, Augusti KT (1982) Hypolipidemic action of onion and garlic unsaturated oils in sucrose fed rats over a two-month period. Experientia 38(8):899–901

Agarwal KC (1996) Therapeutic actions of garlic constituents. Med Res Rev 16(1):111–124

Amarawardana L, Bandara P, Kumar V, Pettersson J, Ninkovic V, Glinwood R (2007) Olfactory response of *Myzus persicae* (Homoptera: Aphididae) to volatiles from leek and chive: potential for intercropping with sweet pepper. Acta Agr Scand B-S P 57(1):87–91

Ankri S, Mirelman D (1999) Antimicrobial properties of allicin from garlic. Microbes Infect 1(2):125–129

Antosiewicz J, Herman-Antosiewicz A, Marynowski SW, Singh SV (2006) C-Jun NH2-terminal kinase signaling axis regulates diallyl trisulfide-induced generation of reactive oxygen species and cell cycle arrest in human prostate cancer cells. Cancer Res 66(10):5379–5386

Apitz-Castro R, Cabrera S, Cruz MR, Ledezma E, Jain MK (1983) Effects of garlic extract and of three pure components isolated from it on human platelet aggregation, arachidonate metabolism, release reaction and platelet ultrastructure. Thromb Res 32(2):155–169

Apitz-Castro R, Ledezma E, Escalante J, Jain MK (1986) The molecular basis of the antiplatelet action of ajoene: direct interaction with the fibrinogen receptor. Biochem Biophys Res Commun 141(1):145–150

Apitz-Castro R, Badimon JJ, Badimon L (1994) A garlic derivative, ajoene, inhibits platelet deposition on severely damaged vessel wall in an in vivo porcine experimental-model. Thromb Res 75(3):243–249

Ariga T, Oshiba S, Tamada T (1981) Platelet aggregation inhibitor in garlic. Lancet 1(8212):150–151

Arya A, Chauhan RCA (1995) Effect of allicin and extracts of garlic and bignonia on two fungi. Indian J Mycol Plant Pathol 25:316–318

Assis FA, Moraes JC, Assis GA (2007) Effect of the aqueous extract of garlic bulbs on the aphid *Myzus persicae* (Sulzer) (Hemiptera: Aphididae) on potato. Ecossistema 32(1–2):63–66

Bakker J, Gommers FJ, Nieuwenhuis I, Wynberg H (1979) Photoactivation of the nematicidal compound alpha-terthienyl from roots of marigolds (*Tagetes* species). A possible singlet oxygen role. J Biol Chem 254(6):1841–1844

Bat-Chen W, Golan T, Peri I, Ludmer Z, Schwartz B (2010) Allicin purified from fresh garlic cloves induces apoptosis in colon cancer cells via Nrf2. Nutr Cancer 62(7):947–957

Bayer T, Wagner H, Block E, Grisoni S, Zhao SH, Neszmelyi A (1989) Zwiebelanes—novel biologically-active 2,3-dimethyl-5,6-dithiabicyclo[2.1.1]hexane 5-oxides from onion. J Am Chem Soc 111(8):3085–3086

Bianchi A, Zambonelli A, DAulerio AZ, Bellesia F (1997) Ultrastructural studies of the effects of *Allium sativum* on phytopathogenic fungi in vitro. Plant Dis 81(11):1241–1246

Block E (1992) The organosulfur chemistry of the genus *Allium*—implications for the organic-chemistry of sulfur. Angew Chem Int Edit 31(9):1135–1178

Block E (2010) Garlic and other *Alliums*: the lore and the science

Block E, Deorazio R (1994) Chemistry in a salad bowl—comparative organosulfur chemistry of garlic, onion and shiitake mushrooms. Pure Appl Chem 66(10–11):2205–2206

Block E, Ahmad S, Jain MK, Crecely RW, Apitzcastro R, Cruz MR (1984) (E, Z)-ajoene—a potent antithrombotic agent from garlic. J Am Chem Soc 106(26):8295–8296

Block E, Bayer T, Naganathan S, Zhao SH (1996) Allium chemistry: Synthesis and sigmatropic rearrangements of alk(en)yl 1-propenyl disulfide S-oxides from cut onion and garlic. J Am Chem Soc 118(12):2799–2810

Block E, Dane AJ, Thomas S, Cody RB (2010) Applications of direct analysis in real time mass spectrometry (DART-MS) in allium chemistry. 2-propenesulfenic and 2-propenesulfinic acids, diallyl trisulfane S-oxide, and other reactive sulfur compounds from crushed garlic and other *Alliums*. J Agr Food Chem 58(8):4617–4625

Bowman GT, Clement JJ, Davidson DE Jr, Eswarakrishnan V, Field L, Hoch JM, Musallam HA, Pick RO, Ravichandran R, Srivastava PK (1986) Potential antiradiation drugs containing no nitrogen, and related compounds. Chem Biol Interact 57(2):161–174

Brzostowska EM, Greer A (2003) The role of amine in the mechanism of pentathiepin (polysulfur) antitumor agents. J Am Chem Soc 125(2):396–404

Busch C, Jacob C, Anwar A, Burkholz T, Ba LA, Cerella C, Diederich M, Brandt W, Wessjohann L, Montenarh M (2010) Diallylpolysulfides induce growth arrest and apoptosis. Int J Oncol 36(3):743–749

Castiglioni A (1985) A history of medicine. Jason Aronson, New York

Cavallito CJ, Bailey JH (1944) Allicin, the antibacterial principle of *Allium sativum*. I. Isolation, physical properties and antibacterial action. J Am Chem Soc 66(11):1950–1951

Cavallito CJ, Buck JS, Suter CM (1944) Allicin, the antibacterial principle of *Allium sativum*. II. Determination of the chemical structure. J Am Chem Soc 66(11):1952–1954

Cerella C, Scherer C, Cristofanon S, Henry E, Anwar A, Busch C, Montenarh M, Dicato M, Jacob C, Diederich M (2009) Cell cycle arrest in early mitosis and induction of caspase-dependent apoptosis in U937 cells by diallyltetrasulfide (Al2S4). Apoptosis 14(5):641–654

Curtis H, Noll U, Stoermann J, Slusarenko AJ (2004) Broad-spectrum activity of the volatile phytoanticipin allicin in extracts of garlic (*Allium sativum L.*) against plant pathogenic bacteria, fungi and oomycetes. Physiol Mol Plant 65(2):79–89

Davidson BS, Molinski TF, Barrows LR, Ireland CM (1991) Varacin—a novel benzopentathiepin from *Lissoclinum*-Vareau that is cytotoxic toward a human colon-tumor. J Am Chem Soc 113(12):4709–4710

Davis SR (2005) An overview of the antifungal properties of allicin and its breakdown products—the possibility of a safe and effective antifungal prophylactic. Mycoses 48(2):95–100

Dicke M (1999) Are herbivore-induced plant volatiles reliable indicators of herbivore identity to foraging carnivorous arthropods? Entomol Exp Appl 91(1):131–142

Dirsch VM, Gerbes AL, Vollmar AM (1998) Ajoene, a compound of garlic, induces apoptosis in human promyeloleukemic cells, accompanied by generation of reactive oxygen species and activation of nuclear factor kappa B. Mol Pharmacol 53(3):402–407

Doering M, Ba LA, Lilienthal N, Nicco C, Scherer C, Abbas M, Zada AA, Coriat R, Burkholz T, Wessjohann L, Diederich M, Batteux F, Herling M, Jacob C (2010) Synthesis and selective anticancer activity of organochalcogen based redox catalysts. J Med Chem 53(19):6954–6963

Egen-Schwind C, Eckard R, Jekat FW, Winterhoff H (1992) Phamacokinetics of vinyldithiins, transformation products of allicin. Planta Med 58(1):8–13

Ford PW, Davidson BS (1993) Synthesis of varacin, a cytotoxic naturally-occurring benzopentathiepin isolated from a marine ascidian. J Org Chem 58(17):4522–4523

Gardner CD, Lawson LD, Chatterjee LM, Kiazand A, Balise RR, Kraemer HC (2007) Effect of raw garlic vs commercial garlic supplements on plasma lipid concentrations in adults with moderate hypercholesterolemia—a randomized clinical trial. Arch Intern Med 167(4):346–353

Giles GI, Tasker KM, Jacob C (2001) Hypothesis: the role of reactive sulfur species in oxidative stress. Free Radical Bio Med 31(10):1279–1283

Giles GI, Tasker KM, Collins C, Giles NM, O'Rourke E, Jacob C (2002) Reactive sulphur species: an in vitro investigation of the oxidation properties of disulphide S-oxides. Biochem J 364:579–585

Gmelin R, Susilo R, Fenwick GR (1981) Cyclic polysulfides from parkia-speciosa. Phytochemistry 20(11):2521–2523

Gruhlke MCH, Portz D, Stitz M, Anwar A, Schneider T, Jacob C, Schlaich NL, Slusarenko AJ (2010) Allicin disrupts the cell's electrochemical potential and induces apoptosis in yeast. Free Radical Bio Med 49(12):1916–1924

Hardy G (2000) Nutraceuticals and functional foods: introduction and meaning. Nutrition 16(7–8):688–689

Hirsch K, Danilenko M, Giat J, Miron T, Rabinkov A, Wilchek M, Mirelman D, Levy J, Sharoni Y (2000) Effect of purified allicin, the major ingredient of freshly crushed garlic, on cancer cell proliferation. Nutr Cancer 38(2):245–254

Höfle G, Gmelin R, Luxa HH, Galamulumetreves MN, Hatanaka SI (1976) Structure of lentinic acid—2-(Gamma-Glutamylamino)-4,6,8,10,10-pentaoxo-4,6,8,10-tetrathiaundecanoic acid. Tetrahedron Lett 36:3129–3132

Horev-Azaria L, Eliav S, Izigov N, Pri-Chen S, Mirelman D, Miron T, Rabinkov A, Wilchek M, Jacob-Hirsch J, Amariglio N, Savion N (2009) Allicin up-regulates cellular glutathione level in vascular endothelial cells. Eur J Nutr 48(2):67–74

Huang Y, Chen SX, Ho SH (2000) Bioactivities of methyl allyl disulfide and diallyl trisulfide from essential oil of garlic to two species of stored-product pests, *Sitophilus zeamais* (Coleoptera: Curculionidae) and *Tribolium castaneum* (Coleoptera: Tenebrionidae). J Econ Entomol 93(2):537–543

Hudson JB, Graham EA, Rossi R, Carpita A, Neri D, Towers GHN (1993) Biological-activities of terthiophenes and polyynes from the *Asteraceae*. Planta Med 59(5):447–450

Iberl B, Winkler G, Knobloch K (1990) Products of allicin transformation: ajoenes and dithiins, characterization and their determination by HPLC. Planta Med 56(2):202–211

Iciek M, Bilska A, Ksiazek L, Srebro Z, Wlodek L (2005) Allyl disulfide as donor and cyanide as acceptor of sulfane sulfur in the mouse tissues. Pharmacol Rep 57(2):212–218

Imai S, Tsuge N, Tomotake M, Nagatome Y, Sawada H, Nagata T, Kumagai H (2002) Plant biochemistry: an onion enzyme that makes the eyes water. Nature 419(6908):685

Jacob C (2006) A scent of therapy: pharmacological implications of natural products containing redox-active sulfur atoms. Nat Prod Rep 23(6):851–863

Jacob C, Anwar A (2008) The chemistry behind redox regulation with a focus on sulphur redox systems. Physiol Plant 133(3):469–480

Jain MK, Apitz-Castro R (1987) Garlic: molecular basis of the putative 'vampire-repellant' action and other matters related to heart and blood. Trends Biochem Sci 12:252–254

Jamaluddin MP, Krishnan LK, Thomas A (1988) Ajoene inhibition of platelet aggregation: possible mediation by a hemoprotein. Biochem Biophys Res Commun 153(1):479–486

Jones MG (2010) The biosynthesis of volatile sulfur flavour compounds. In: Herrmann A (ed) The chemistry and biology of volatiles. Wiley, West Sussex, United Kingdom

Kelkel M, Cerella C, Mack F, Schneider T, Jacob C, Schumacher M, Dicato M, Diederich M (2012) ROS-independent JNK activation and multisite phosphorylation of Bcl-2 link diallyl tetrasulfide-induced mitotic arrest to apoptosis. Carcinogenesis 33(11):2162–2171

Kensler TW, Curphey TJ, Maxiutenko Y, Roebuck BD (2000) Chemoprotection by organosulfur inducers of phase 2 enzymes: dithiolethiones and dithiins. Drug Metab Drug Interact 17(1–4):3–22

Keusgen M (2008) Unusual cystine lyase activity of the enzyme alliinase: direct formation of polysulphides. Planta Med 74(1):73–79

Kim S, Kubec R, Musah RA (2006) Antibacterial and antifungal activity of sulfur-containing compounds from *Petiveria alliacea L.* J Ethnopharmacol 104(1–2):188–192

Kjaer A, Larsen PO (1977) Non-protein amino acids, cyanogenic glycosides and glucosinolates. In: Bulock JD (ed) Biosynthesis, specialist periodical reports, vol 2., Royal Society of ChemistryCambridge, United Kingdom, pp 121–122

Koch HP, Lawson LD (1996) Garlic: the science and therapeutic application of Allium sativum L and related species. Williams and Wilkins, Baltimore, Md

Kouokam JC, Jahns T, Becker H (2002) Antimicrobial activity of the essential oil and some isolated sulfur-rich compounds from *Scorodophloeus zenkeri*. Planta Med 68(12):1082–1087

Kubec R, Cody RB, Dane AJ, Musah RA, Schraml J, Vattekkatte A, Block E (2010) Applications of direct analysis in real time-mass spectrometry (DART-MS) in allium chemistry. (Z)-Butanethial S-oxide and 1-butenyl thiosulfinates and their S-(E)-1-butenylcysteine S-oxide precursor from *Allium siculum*. J Agr Food Chem 58(2):1121–1128

Kupiecki FP, Virtanen AI (1960) Cleavage of alkyl cysteine sulphoxides by an enzyme in onion (*Allium-Cepa*). Acta Chem Scand 14(9):1913–1918

Lancaster JE, Shaw ML, Walton EF (2000) S-alk(en)yl-L-cysteine sulfoxides, alliinase and aroma in *Leucocoryne*. Phytochemistry 55(2):127–130

Lawson LD, Wood SG, Hughes BG (1991) Hplc analysis of allicin and other thiosulfinates in garlic clove homogenates. Planta Med 57(3):263–270

Lawson LD, Ransom DK, Hughes BG (1992) Inhibition of whole-blood platelet-aggregation by compounds in garlic clove extracts and commercial garlic products. Thromb Res 65(2):141–156

Ledezma E, Apitz-Castro R (2006) [ajoene the main active compound of garlic (*Allium sativum*): a new antifungal agent]. Revista iberoamericana de micologia 23(2):75–80

Li M, Ciu JR, Ye Y, Min JM, Zhang LH, Wang K, Gares M, Cros J, Wright M, Leung-Tack J (2002) Antitumor activity of Z-ajoene, a natural compound purified from garlic: antimitotic and microtubule-interaction properties. Carcinogenesis 23(4):573–579

Liu P, Guo JH, Liu HM, Cheng YJ, Wang BQ, Long CA, Deng BX (2009) Diallyl trisulfide (DATS) effectively induced apoptosis of postharvest disease *Penicillium expansum* of citrus. Ann Microbiol 59(4):675–679

Liu C, Cao F, Tang QZ, Yan L, Dong YG, Zhu LH, Wang L, Bian ZY, Li HL (2010) Allicin protects against cardiac hypertrophy and fibrosis via attenuating reactive oxygen species-dependent signaling pathways. J Nutr Biochem 21(12):1238–1250

Makarieva TN, Stonik VA, Dmitrenok AS, Grebnev BB, Isakov VV, Rebachyk NM, Rashkes YW (1995) Varacin and three new marine antimicrobial polysulfides from the far-eastern ascidian Polycitor sp. J Nat Prod 58(2):254–258

Makheja AN, Bailey JM (1990) Antiplatelet constituents of garlic and onion. Agents Actions 29(3–4):360–363

Medicine TFfI (1992) The nutraceutical initiative—a proposal for economic and regulatory reform. Food Technol Chicago 46(4):77–79

Miron T, Wilehek M, Sharp A, Nakagawa Y, Naoi M, Nozawa Y, Akao Y (2008) Allicin inhibits cell growth and induces apoptosis through the mitochondrial pathway in HL60 and U937 cells. J Nutr Biochem 19(8):524–535

Morita K, Kobayash S (1966) Isolation and synthesis of lenthionine an odorous substance of shiitake an edible mushroom. Tetrahedron Lett 6:573

Morita K, Kobayash S (1967) Isolation structure and synthesis of lenthionine and its analogs. Chem Pharm Bull 15(7):988

Mueller AL, Virtanen AI (1966) Synthesis of S-(buten-1-yl)-L-cysteine sulfoxide and its enzymic cleavage to 1-butenylsulfenic acid. Acta Chem Scand B 20(4):1163–1165

Munchberg U, Anwar A, Mecklenburg S, Jacob C (2007) Polysulfides as biologically active ingredients of garlic. Org Biomol Chem 5(10):1505–1518

Munday R, Munday JS, Munday CM (2003) Comparative effects of mono-, di-, tri-, and tetrasulfides derived from plants of the allium family: redox cycling in vitro and hemolytic activity and phase 2 enzyme induction in vivo. Free Radical Bio Med 34(9):1200–1211

Musah RA, He Q, Kubec R (2009) Discovery and characterization of a novel lachrymatory factor synthase in *Petiveria alliacea* and its influence on alliinase-mediated formation of biologically active Organosulfur compounds. Plant Physiol 151(3):1294–1303

Naganawa R, Iwata N, Ishikawa K, Fukuda H, Fujino T, Suzuki A (1996) Inhibition of microbial growth by ajoene, a sulfur-containing compound derived from garlic. Appl Environ Microb 62(11):4238–4242

Nishikawa T, Yamada N, Hattori A, Fukuda H, Fujino T (2002) Inhibition by ajoene of skin-tumor promotion in mice. Biosci Biotech Bioch 66(10):2221–2223

O'Gara EA, Hill DJ, Maslin DJ (2000) Activities of garlic oil, garlic powder, and their diallyl constituents against Helicobacter pylori. Appl Environ Microb 66(5):2269–2273

Oommen S, Anto RJ, Srinivas G, Karunagaran D (2004) Allicin (from garlic) induces caspase-mediated apoptosis in cancer cells. Eur J Pharmacol 485(1–3):97–103

Pari L, Murugavel P, Sitasawad SL, Kumar KS (2007) Cytoprotective and antioxidant role of diallyl tetrasulfide on cadmium induced renal injury: an in vivo and in vitro study. Life Sci 80(7):650–658

Park SY, Cho SJ, Kwon HC, Lee KR, Rhee DK, Pyo S (2005) Caspase-independent cell death by allicin in human epithelial carcinoma cells: involvement of PKA. Cancer Lett 224(1):123–132

Paterson I, Anderson EA (2005) The renaissance of natural products as drug candidates. Science 310(5747):451–453

Perez HA, De la Rosa M, Apitz R (1994) In vivo activity of ajoene against rodent malaria. Antimicrob Agents Chemother 38(2):337–339

Portz D, Koch E, Slusarenko AJ (2008) Effects of garlic (*Allium sativum*) juice containing allicin on *Phytophthora infestans* and downy mildew of cucumber caused by *Pseudoperonospora cubensis*. Eur J Plant Pathol 122(1):197–206

Rabinkov A, Miron T, Mirelman D, Wilchek M, Glozman S, Yavin E, Weiner L (2000) S-Allylmercaptoglutathione: the reaction product of allicin with glutathione possesses SH-modifying and antioxidant properties. Bba-Mol Cell Res 1499(1–2):144–153

Romagnoli C, Mares D, Fasulo MP, Bruni A (1994) Antifungal effects of alpha-terthienyl from tagetes-patula on 5 dermatophytes. Phytother Res 8(6):332–336

Russell PE, Mussa AEA (1977) Use of garlic (*Allium sativum*) extracts to control foot rot of phaseolus-vulgaris caused by fusarium-solani fsp phaseoli. Ann Appl Biol 86(3):369–372

Scharfenberg K, Wagner R, Wagner KG (1990) The cytotoxic effect of ajoene, a natural product from garlic investigated differ cell-lines. Cancer Lett 53(2–3):103–108

Schneider T, Ba LA, Khairan K, Zwergel C, Bach ND, Bernhardt I, Brandt W, Wessjohann L, Diederich M, Jacob C (2011) Interactions of polysulfanes with components of red blood cells. Medchemcomm 2(3):196–200

Seki T, Hosono T, Hosono-Fukao T, Inada K, Tanaka R, Ogihara J, Ariga T (2008) Anticancer effects of diallyl trisulfide derived from garlic. Asia Pac J Clin Nutr 17:249–252

Sendl A, Schliack M, Loser R, Stanislaus F, Wagner H (1992) Inhibition of cholesterol synthesis in vitro by extracts and isolated compounds prepared from garlic and wild garlic. Atherosclerosis 94(1):79–85

Shibuya T, Shimada S, Sakurai H, Kumagai H (2005) Mechanism of inhibition of platelet aggregation by lenthionine, a flavour component from shiitake mushroom. In: IFT annual meeting, New Orleans, LA, 15–20 July 2005

Shimada S, Komamura K, Kumagai H, Sakurai H (2004) Inhibitory activity of shiitake flavor against platelet aggregation. Bio Factors 22(1–4):177–179

Shimon LJ, Rabinkov A, Shin I, Miron T, Mirelman D, Wilchek M, Frolow F (2007) Two structures of alliinase from *Alliium sativum* L.: apo form and ternary complex with aminoacrylate reaction intermediate covalently bound to the PLP cofactor. J Mol Biol 366:611–625

Slusarenko AJ, Patel A, Portz D (2008) Control of plant diseases by natural products: allicin from garlic as a case study. Eur J Plant Pathol 121(3):313–322

Small LD, Bailey JH, Cavallito CJ (1947) Alkyl thiolsulfinates. J Am Chem Soc 69(7):1710–1713

Small LD, Bailey JH, Cavallito CJ (1949) Comparison of some properties of thiosulfinates and thiosulfonates. J Am Chem Soc 71(10)

Steudel R (2002) The chemistry of organic polysulfanes R-S-n-R (n > 2). Chem Rev 102(11):3905–3945

Tatarintsev AV, Vrzheshch PV, Schegolev AA, Yershov DE, Turgiev AS, Varfolomeyev SD, Kornilayeva GV, Makarova TV, Karamov EV (1992) Ajoene antagonizes integrin-dependent processes in HIV-infected T-lymphoblasts. Aids 6(10):1215–1217

Tilli CMLJ, Stavast-Kooy AJW, Vuerstaek JDD, Thissen MRTM, Krekels GAM, Ramaekers FCS, Neumann HAM (2003) The garlic-derived organosulfur component ajoene decreases basal cell carcinoma tumor size by inducing apoptosis. Arch Dermatol Res 295(3):117–123

Tsao SM, Yin MC (2001) In vitro antimicrobial activity of four diallyl sulphides occurring naturally in garlic and Chinese leek oils. J Med Microbiol 50(7):646–649

Urbina JA, Marchan E, Lazardi K, Visbal G, Apitz-Castro R, Gil F, Aguirre T, Piras MM, Piras R (1993) Inhibition of phosphatidylcholine biosynthesis and cell proliferation in *trypanosoma cruzi* by ajoene, an antiplatelet compound isolated from garlic. Biochem Pharmacol 45(12):2381–2387

Vaidya V, Ingold KU, Pratt DA (2009) Garlic: source of the ultimate antioxidants-sulfenic acids. Angew Chem-Int Ed 48(1):157–160
Van Etten HD, Mansfield JW, Bailey JA, Farmer EE (1994) Two classes of plant antibiotics: phytoalexins versus "phytoanticipins". Plant Cell 6(9):1191–1192
Voigt M, Wolf E (1986) Knoblauch: HPLC-Bestimmung von Knoblauchwirkstoffen in Extrakten, Pulver und Fertigarzneimitteln. Deutsche Apotheker Zeitung 126:591–593
Wills ED (1956) Enzyme inhibition by allicin, the active principle of garlic. Biochem J 63(3):514–520
Wratten SJ, Faulkner DJ (1976) Cyclic polysulfides from the red alga Chondria Californica. J Org Chem 41(14):2465–2467
Xiao D, Zeng Y, Hahm ER, Kim YA, Ramalingam S, Singh SV (2009) Diallyl trisulfide selectively causes Bax- and Bak-mediated apoptosis in human lung cancer cells. Environ Mol Mutagen 50(3):201–212
Yang JS, Chen GW, Hsia TC, Ho HC, Ho CC, Lin MW, Lin SS, Yeh RD, Ip SW, Lu HF, Chung JG (2009) Diallyl disulfide induces apoptosis in human colon cancer cell line (COLO 205) through the induction of reactive oxygen species, endoplasmic reticulum stress, caspases casade and mitochondrial-dependent pathways. Food Chem Toxicol 47(1):171–179
Yasumoto K, Iwami K, Mitsuda H (1971) Studies on shiitake flavor 1, new sulfur-containing peptide from lentinus-edodes acting as a precursor for lenthionine. Agr Biol Chem Tokyo 35(13):2059
Yoshida S, Kasuga S, Hayashi N, Ushiroguchi T, Matsuura H, Nakagawa S (1987) Antifungal activity of ajoene derived from garlic. Appl Environ Microb 53(3):615–617
Zhang WL, Ha MW, Gong YH, Xu Y, Dong NA, Yuan YA (2010) Allicin induces apoptosis in gastric cancer cells through activation of both extrinsic and intrinsic pathways. Oncol Rep 24(6):1585–1592
Zheng DX, Zhang SQ, Zhao S, Min JM, Wang K, Liu SL (1998) Z-ajoene induces tumor cells to die by apoptosis. Chinese Sci Bull 43(13):1135–1140

Authors Biography

Ifeanyi Nwachukwu completed his B.Sc. degree in Biochemistry and Microbiology (with First Class Honours) at the University of Nigeria before moving to England on an Andrew Stratton Scholarship Award and a Full Fees Master's Scholarship for the M.Sc. program in Biocatalysis at the University of Exeter. He subsequently spent three years as an Early Stage Researcher of the European Union/Marie Curie Initial Training Network under the "RedCat" consortium at the laboratory of Prof Alan Slusarenko at the RWTH Aachen University in Germany. During this time, he studied the resistance mechanisms of certain bacteria to the antimicrobial substance allicin derived from garlic.

Shortly after completion of his project in October 2012, he moved to the University of Manitoba in Canada where he is currently conducting research elucidating the bioavailability and membrane permeability potential of antihypertensive bioactive peptides from natural sources.

Alan Slusarenko is Head of the Plant Physiology Department at RWTH Aachen University. His research has centered on resistance mechanisms of Arabidopsis to infection and more recently on Natural Products in Plant Protection. Alan obtained a PhD in Plant Pathology from Imperial College in 1981 and was a lecturer in the Department of Plant Biology at Hull University in the UK from 1983 until moving in 1988 to an Assistant Professorship in Molecular Plant Pathology at the University of Zuerich in Switzerland and subsequently in 1995 to the Chair of Plant Physiology at RWTH Aachen in Germany.

Part X
Connecting Section Between Chapters 10 and 11

The previous two chapters have discussed the thiolstat and its role in redox signaling and response primarily from the perspective of mammalian cells. In these cells, the intracellular, thiol-based redox buffer glutathione (GSH/GSSG) is responsible for the maintenance of a fairly reducing cytosolic redox environment (the redox status of individual cellular organelles may be considerably more oxidizing). Traditionally, GSH has also been considered as the first line of defense against external redox assaults, such as the ones caused by the thiosulfinates and polysulfanes present in certain food constituents. Indeed, even if some of these oxidants initially were to escape sequestration by GSH and would manage to modify cysteine residues in proteins, the resulting modifications would be repaired rapidly and efficiently by glutathione in tandem with certain GSH-dependent reductases. In essence, GSH has long been considered as an extremely reliable defender of the intracellular redox state.

As we have seen in the previous chapters, this traditional view is too narrow and does not account for many observations, such as widespread *S*-thiolation in cells affected by thiol-specific reagents. It now appears that GSH is rather slow in reacting with thiol-specific reagents, and despite its high abundance in the cytosol is unable to sequester many oxidants efficiently. Furthermore, *S*-thiolation often results in disulfides that are chemically 'alien' to the mammalian cell, such as allyl- and propyl-based mixed disulfides, and hence cannot be reduced easily by cellular reductases. Indeed, some studies have even demonstrated that GSSG, i.e., the oxidized from of GSH, is not as benign as thought previously and may represent an oxidative stressor in its own right.

The demise of the 'mythos glutathione' in mammalian biochemistry is reflected by recent discoveries in the field of bacteria. Here, glutathione is not the only redox buffer in cells, and various bacterial thiols have been identified during the last decades (together with other major cellular thiols, such as ovothiols found in marine invertebrates and ergothioneine found in the ergot fungus).

Chapter 11
Beyond Glutathione: Different Low Molecular Weight Thiols as Mediators of Redox Regulation and Other Metabolic Functions in Lower Organisms

Chris J. Hamilton, Miriam Arbach and Murree Groom

Keywords Bacillithiol • Coenzyme A • Ergothioneine • Metabolic functions • Redox regulation

1 Introduction

Non-protein low molecular weight (LMW) thiols play an important role in the maintenance and regulation of a diverse array of cellular processes (Fig. 1). For example, glutathione (GSH) is the major LMW thiol in eukaryotes and many Gram-negative bacteria (Fig. 2). GSH plays a critical role in maintaining an intracellular reducing environment via the reduction of toxic oxidants (e.g., hydrogen peroxide and nitric oxide) (Fahey and Newton 1983). It also facilitates metal ion homeostasis and the detoxification of electrophilic xenobiotics/metabolites. In some bacteria it serves as a cofactor in the decomposition of aromatic compounds (e.g., via the gentisate pathway). Protein glutathiolation (formation of GS-S-Protein mixed disulfides) is also an important posttranslational modification for the redox regulation of many proteins, as well as for protecting exposed cysteine (Cys) residues from irreversible damage during oxidative stress (Dalle-Donne et al. 2007; Ghezzi 2005). Not all organisms, however, produce or utilize GSH. In particular, many Gram-positive bacteria produce other LMW thiols such as mycothiol (MSH), bacillithiol (BSH), trypanothione ($T(SH)_2$) and gamma-glutamylcysteine (γGC), which appear to serve as GSH surrogates. In addition

C. J. Hamilton (✉) · M. Arbach
School of Pharmacy, University of East Anglia, Norwich Research Park, Norwich NR4 7TJ, UK
e-mail: C.Hamilton@uea.ac.uk

M. Arbach · M. Groom
ECOspray Limited, Grange Farm Hilborough, Thetford, Norfolk IP26 5BT, UK

C. Jacob et al. (eds.), *Recent Advances in Redox Active Plant and Microbial Products*, 291
DOI: 10.1007/978-94-017-8953-0_11, © Springer Science+Business Media Dordrecht 2014

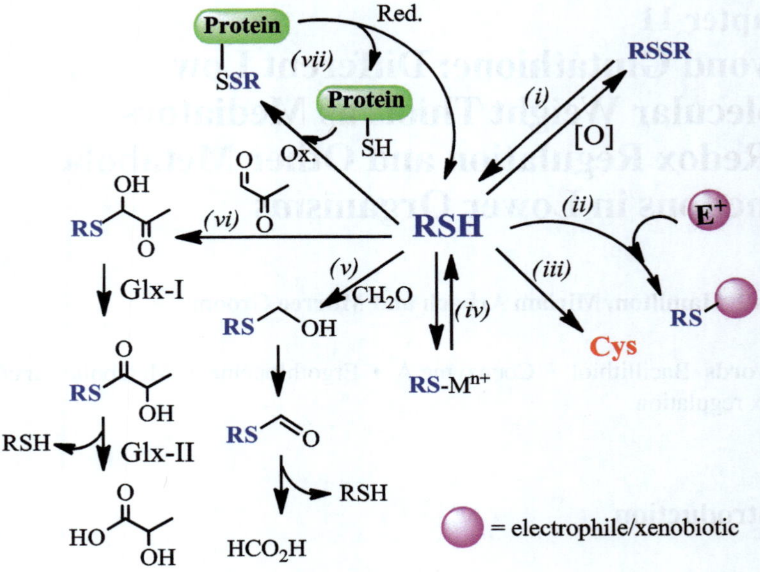

Fig. 1 The main cellular functions of low molecular weight thiols: (*i*) as a redox buffer; (*ii*) in xenobiotic detoxification; (*iii*) as an intracellular cysteine reservoir; (*iv*) in metal ion homeostasis; (*v*) in the detoxification of reactive carbonyl electrophiles such as formaldehyde and (*vi*) methylglyoxal; and (*vii*) in the redox regulation/protection of protein functions

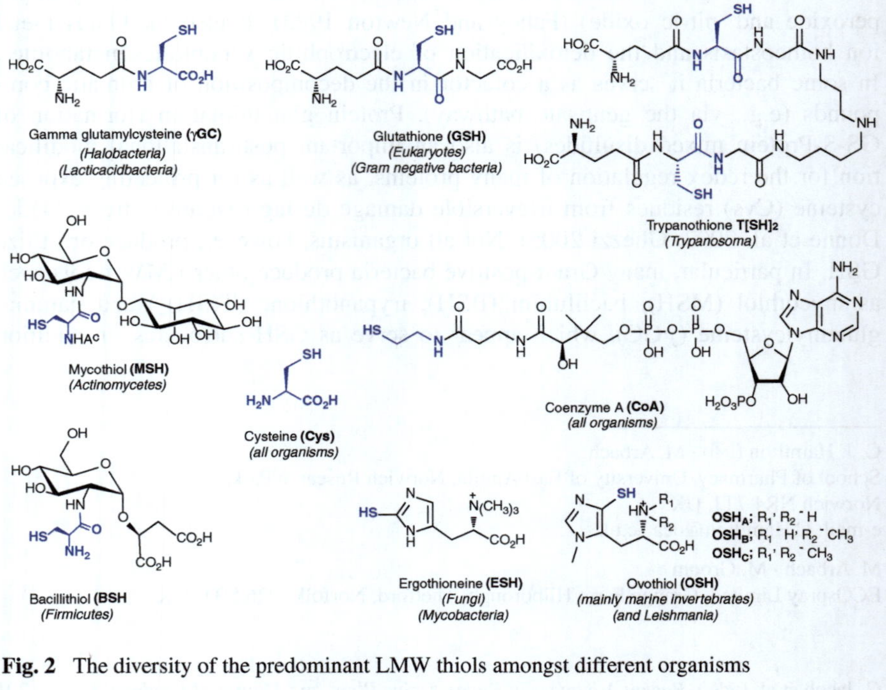

Fig. 2 The diversity of the predominant LMW thiols amongst different organisms

Thiol	$E^{0'}$(mV)[a]	$E^{0'}$	Thiol pK_a	pK_a
GSH	-240 (Åslund et al.)	1997	8.93 (Rabenstein)	1973
γGC	-300 (Rubino et al.)[b]	2008	9.9 (Benesch and Benesch)	1955
MSH	ND		ND	
BSH	-221 (Sharma et al.)	2013	pk_s7.63 (Sharma et al.) pk_{ns} 9.55 (Sharma et al.)	2013 2013
ESH	-60 (Jocelyn)	1972	10.75 (Stanovnik and Tišler) 11.5 (Motohashi et al.)	1964 1976
OSH-A	-94 (pH 3) (Weaver and Rabenstein) -67 (pH 7) (Weaver and Rabenstein) -141 (pH 10) (Weaver and Rabenstein)	1995 1995 1995	1.42 (Weaver and Rabenstein)	1995
Cys	-223 (Keire et al.)	1992b	pk_s8.45 (Clement and Hartz) pk_{ns} 10.00 (Clement and Hartz)	1971 1971
CoA	-234 (Keire et al.)	1992b	9.83 (Keire et al.)	1992a
$T(SH)_2$	-252 (Fairlamb and Cerami)	1992	7.4 (Moutiez et al.)	1994

Fig. 3 Thiol pKa and redox potentials of common LMW thiols. [a] GSH is commonly used as the reference standard when calculating thiol redox potentials, but different reports use different values of $E^{0'}{}_{(GSH/GSSG)}$. Herein all literature values have been adjusted with respect to the now more robustly determined value for GSH ($E^{0'}{}_{(GSH/GSSG)} = -240\,\text{mV}$) (Åslund et al. 1997) so that comparisons can be made between them on the same standardized scale. [b] This value was determined by a recently developed mass spectrometry procedure. It is worth noting, however, that the same procedure provided an unusually low value for cysteine itself (−283 mV)

to their major LMW thiol, many organisms also utilize additional thiol cofactors as redox buffers such as coenzymeA (CoA), ovothiol (OSH), and ergothioneine (ESH) (Fig. 2).

This chapter provides a broad overview of what is currently known about these other LMW thiols which occur in numerous organisms alongside, or instead of, GSH. Some of the LMW thiols discussed herein (e.g., BSH) have only very recently been discovered, whereas the existence of others has long been known. However, two things they all have in common are: (1) the full extent of their metabolic functions are far from completely known and (2) while some of their cellular functions parallel those of GSH, many of their biophysical properties (Fig. 3) and biochemical capabilities evidently differ. Therefore, we will also offer some perspective on future avenues of investigation that could provide a more comprehensive picture of their various metabolic functions.

2 Glutathione and Monoglutathionyl Spermidine

Glutathione (GSH), the most widely distributed of the LMW thiols is biosynthesized and utilized by all eukaryotes and most Gram-negative bacteria. For this reason GSH is the most extensively studied LMW thiol to date. Its metabolic functions include all of those summarized in Fig. 1, and the reader is directed to several excellent reviews of GSH metabolism for further information (Wang and Ballatori 1998;

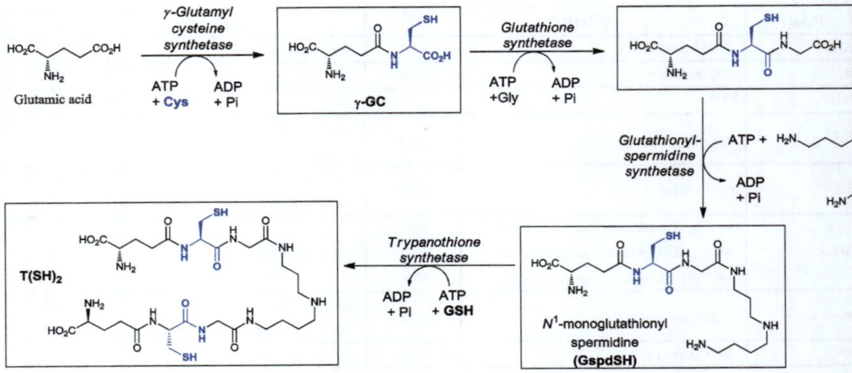

Fig. 4 Biosynthesis of GSH and structurally related LMW thiols

Meister 1988; Pompella et al. 2003; Duan et al. 2010; Masip et al. 2006). It is worth noting the commonalities between the GSH biosynthetic pathway and those for γGC and $T(SH)_2$ (Fig. 4). The first step in GSH biosynthesis involves the enzyme γ-glutamyl cysteine synthetase which catalyzes the ATP-dependent amide bond formation between cysteine and the carboxylate sidechain of glutamic acid to result in γGC. This is followed by the addition of glycine onto the cysteine carboxylate, by the enzyme glutathione synthetase, to yield GSH. Interestingly, in *Escherichia coli* it has been shown that while GSH is the predominant LMW thiol, under anaerobic conditions and during stationary phase ~80 % of this compound exists as its monoglutathionyl spermidine conjugate (GspdSH) (Smith et al. 1995). GspdSH is produced by a dual function synthetase that ligates GSH with the polyamine spermidine, and can also display amidase activity in hydrolyzing GspdSH to regenerate GSH (see Explanatory Box 1 for details on spermidine and spermine) (Bollinger et al. 1995). This suggests a probable regulatory role for this enzyme in the modulation of free GSH and/or spermidine in response to different environmental conditions, and possibly that GspdSH is more effective than GSH in protecting DNA from oxidative damage during stationary growth.

Explanatory Box 1: Spermidine, Spermine, and Related Polyamines

Some rather intricate amino-group-containing biomolecules, such as spermine and spermidine, find mention in this chapter. Although these nitrogen-based compounds are not redox active by themselves, they play an important role in biochemistry which also touches on numerous redox processes. We will therefore briefly depart from our focus on sulfur and take a closer look at these nitrogen species. Many of them occur in human biology, but also in some microorganisms and in plants. In some cases, their specific chemical properties seem to imply interesting pharmaceutical applications.

Spermidine is a small molecule containing two primary and one secondary amines. In chemistry, it is called 1,8-diamino-4-azaoctane or *N*-(3-aminopropyl)-1,4-diaminobutane and has the chemical formula $NH_2(CH_2)_3NH(CH_2)_4NH_2$. Spermidine is the biochemical precursor of spermine. The latter contains four nitrogen atoms, is known as *N,N′*-bis(3-aminopropyl)-1,4-diaminobutane, and has the chemical formula $NH_2(CH_2)_3NH(CH_2)_4NH(CH_2)_3NH_2$. Spermidine and spermine are positively charged at physiological pH and bind rather well to negatively charged biomolecules, including certain proteins and enzymes—whose function and activity may in part be affected by this binding—and in particular to DNA. Here, the two small nitrogen-containing molecules take on some of the roles of histones, especially in sperm cells (hence the presence of the term 'sperm' in the names).

Together with some other nitrogen-rich small molecules, spermidine, and spermine form a class of positively charged compounds which are also of potential pharmaceutical interest, as they can bind and hence interact rather strongly with many negatively charged biological molecules and also interact fairly well with metal ions and metalloproteins due to their multi-dentate ligand properties. Indeed, putrescine (1,4-butanediamine, $NH_2(CH_2)_4NH_2$), chemically speaking a truncated version of spermidine, resembles the classic bidentate ligand 1,2-diaminoethane (ethylene diamine, en, $NH_2(CH_2)_2NH_2$). Putrescine is a foul smelling compound generated in small quantities in the human body where it represents a precursor of spermidine (and hence also of spermine). Putrescine binds to the polyamine modulatory site of the NMDA (*N*-methyl-D-aspartate) receptor. Interestingly, putrescine also occurs as part of the breakdown of amino acids in dead bodies. Here, it is joined by another small amine-containing compound, which is appropriately called 'cadaverine' (1,5-diaminopentane, $NH_2(CH_2)_5NH_2$). Various *alpha,omega*-diaminoalkanes have recently been considered for their potential antimicrobial activity and also cytotoxicity. Here longer chain and analogs bolaamphiphilic are of particular interest, and also emanate less of a smell.

2.1 Trypanothione

Trypanothione ($T(SH)_2$) is the bis-glutathionyl conjugate of spermidine whose biosynthesis is a subtle extension of that for GspdSH (Fig. 4). $T(SH)_2$ is the predominant LMW thiol among a number of pathogenic protozoa of the *Trypanosoma* and *Leishmania* genera, which are causative agents of tropical diseases such as Leishmaniasis, Chagas disease, and African Sleeping Sickness. In these organisms, $T(SH)_2$ substitutes for many of the GSH-dependent redox functions of other organisms. The $T(SH)_2$-dependent antioxidant system is summarized in Fig. 5.

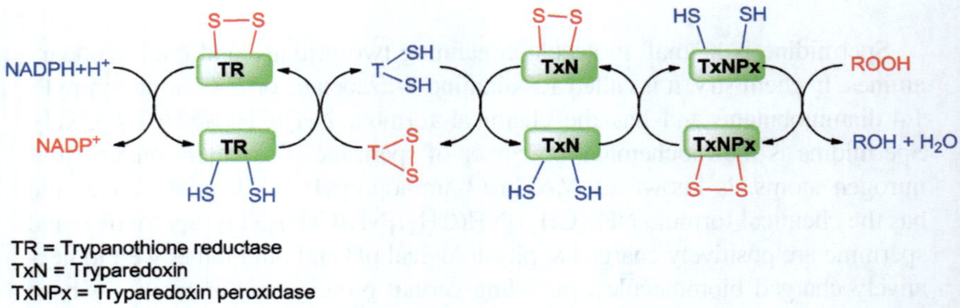

Fig. 5 The trypanothione-tryparedoxin redox system

Like GSH, $T(SH)_2$ helps to maintain an intracellular reducing environment and is oxidized to its intramolecular disulfide (TS_2) in the process. The NADPH-dependent trypanothione reductase (TR) catalyzes the regeneration of $T(SH)_2$ from TS_2 to maintain a high intracellular thiol-to-disulfide redox ratio. Although trypanosomatids also contain GSH (as an intermediate of $T(SH)_2$ biosynthesis), trypanothione reductase does not recognize GSSG as a substrate, hence GSH is maintained in its reduced state via chemical and thiol-*S*-transferase catalyzed thiol-disulfide exchange reactions between GSSG and $T(SH)_2$ (Moutiez et al. 1995). $T(SH)_2$ provides reducing equivalents to a tryparedoxin and tryparedoxin peroxidase redox couple in order to facilitate the enzymatic reduction of toxic peroxides. In this system, tryparedoxin is equivalent to the ubiquitous thioredoxin proteins found in most other organisms and is also able to provide reducing equivalents for ribonucleotide reductase (Dormeyer et al. 2001). Methyl glyoxal is a toxic carbonyl electrophile often produced as a by-product of glycolysis (Booth et al. 2003). In GSH-utilizing organisms, GSH reacts chemically with methylglyoxal to form a thiohemiacetal (Fig. 1, step-vi). Glyoxalase-I then catalyzes its isomerization to a GS-lactoyl thioester conjugate, which is then hydrolyzed (by glyoxalase-II) to liberate the nontoxic D-lactate and regenerate GSH. In *Trypanosoma cruzi* and *Leishmania major*, a similar $T(SH)_2$-dependent methyl glyoxal detoxification pathway has been identified, which exhibits a >200-fold selectivity for GspdSH and $T(SH)_2$ over GSH. The substrate specificities of the trypanothione-dependent redox and glyoxalase systems make them appealing drug targets in trypanothione-utilizing pathogens. Why these parasites have evolved to favor $T(SH)_2$ (over the biosynthetically more accessible GSH) as their major LMW thiol is intriguing. The lower thiol pK_a of $T(SH)_2$ (thiol $pK_a = 7.4$) (Moutiez et al. 1994) compared to GSH (thiol $pK_a = 8.9$) (Rabenstein 1973) implies that at physiological pH, significantly more $T(SH)_2$ than GSH resides in the thiolate form. This greatly enhances its reactivity with electrophiles and in thiol-disulfide exchange processes (Moutiez et al. 1994), reactions which contribute toward the ability of these parasites to survive in the extremely harsh (oxidative) environments that arise during host–pathogen interactions.

2.2 Gamma Glutamylcysteine

Several halobacteria (Newton and Javor 1985) and lactic acid bacteria (Kim et al. 2008) produce millimolar concentrations of γGC, but not GSH. An NADPH-dependent disulfide reductase has previously been purified from *Halobacterium halobium* that is able to reduce γGC-disulfide, but not GSSG (Sundquist and Fahey 1989). This supports the notion that γGC serves as a redox buffer in these microorganisms. The free cysteinyl carboxylate of γGC makes it more prone to metal-catalyzed auto-oxidation, suggesting it would be a poor substitute for GSH as a cellular redox buffer. In the presence of high salt concentrations (typical of halobacteria habitats), however, resistance of γGC to auto-oxidation is comparable to that of GSH (Sundquist and Fahey 1989). The linearity of the biosynthetic pathway from γGC to GSH to GspdSH to $T(SH)_2$ suggests a common phylogenetic ancestry among organisms that use these different biothiols. Alternatively, lactic acid bacteria and halobacteria may have obtained the γGC synthetase gene via gene transfer (Bayley 1982), or they may have evolved a γGC synthetase independently of other GSH-utilizing bacteria. The latter hypothesis is supported by the considerable diversity in the gene sequences encoding γGC synthetase among different organisms. Recently, forced evolution and bioinformatic studies have indicated that the L-proline biosynthetic pathway may have a secondary role in γGC biosynthesis in prokaryotes (Veeravalli et al. 2011).

2.3 Mycothiol

Mycothiol (MSH) is the predominant LMW thiol among Gram-positive bacteria of the Order Actinomycetales, including *Streptomyces*, *Mycobacteria* and *Corynebacteria*. MSH performs many of the functions observed for GSH in Gram-negative bacteria (Jothivasan and Hamilton 2008; Newton et al. 2008a). The first obligate step in MSH biosynthesis involves the UDP-GlcNAc-dependent retaining glycosyltransferase (MshA), which catalyzes the formation of the glycosidic linkage with the acceptor substrate L-*myo*-inositol-1-phosphate(Ins-1-P) (Fig. 6) (Newton et al. 2006b). The phosphate group of GlcNAc-Ins-1-P is then removed by a currently unknown phosphatase ('MshA2') (Newton et al. 2006b). A divalent metal-dependent *N*-acetyl hydrolase (MshB) then catalyzes the hydrolysis of the *N*-acetyl group to give the free aminosugar GlcN-Ins (Newton et al. 2000a; Huang et al. 2011) which is the substrate of an ATP-dependent cysteine ligase (MshC) that catalyzes the addition of the cysteine side chain (Bornemann et al. 1997). Finally, acetylation of the cysteinyl amine is mediated by the acetyl-CoA (CoASAc)-dependent acetyl-transferase MshD (Koledin et al. 2002). The crystal structures have been solved for MshA-D (see Fan et al. for a comprehensive review) (Fan et al. 2009). MshA and MshC are essential for MSH biosynthesis, but there is some redundancy in the function of MshB and MshD (as discussed later). In the absence of any external stimuli, MSH levels and MSH:MSSM ratios

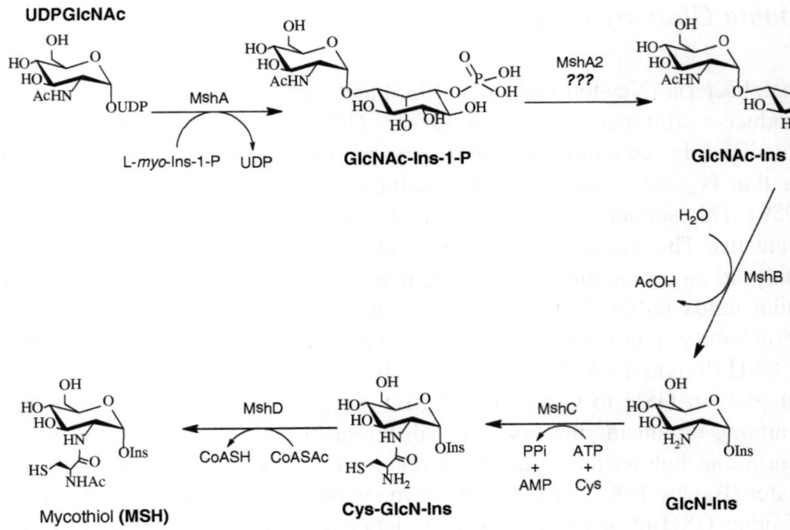

Fig. 6 Biosynthetic pathway of MSH

remain relatively constant during exponential growth, but can decrease during stationary phase (Buchmeier et al. 2006). They also vary in response to oxidative and osmotic stress (Newton et al. 2008a; Lee et al. 2005).

M. smegmatis and *M. tuberculosis mshD* deletion mutants produce only 1–2 % of wild type MSH levels, accompanied by a significant increase in Cys-GlcN-Ins concentrations. Significant levels of *N-des*-acetyl-*N*-formyl-mycothiol (fMSH) (~20 % of wild type MSH levels) and a small amount of *N-des*-acetyl-*N*-succinyl-mycothiol (sucMSH) are also produced (Fig. 7) (Newton et al. 2005; Buchmeier et al. 2006). The low levels of MSH and sucMSH are probably the products of chemical *N*-acylation of Cys-GlcN-Ins by acetyl-CoA and succinyl-CoA, respectively. It has been suggested that fMSH arises from enzymatic formylation of Cys-GlcN-Ins, catalyzed by N^{10}-formyl tetrahydrofolate methionyl-tRNA/ribonucleotide *trans*-formylases (Newton et al. 2005). The *M. smegmatis mshD* mutant was no more sensitive to H_2O_2 and alkyl peroxides than the wild type, indicating that fMSH can substitute for the redox buffer role of MSH in this strain. In contrast, the *mshD* knockout in *M. tuberculosis* displays increased sensitivity to H_2O_2 (Buchmeier et al. 2006) and is unable to grow in macrophages (Rengarajan et al. 2005). Evidently the MSH-dependent biochemistry (i.e., redox biochemistry and oxidant detoxification) of this slower growing pathogen is more critical for survival. It has been observed in a number of marine actinomycetes that the onset of secondary metabolism in early stationary phase growth is accompanied by a decrease in the biosynthesis of MSH and the formation of a MSH homolog where the *N*-acetyl group is replaced by a *N*-propionyl unit (pMSH) (Fig. 7) (Newton et al. 2008b). Propionyl-CoA serves as an acyl starter unit for polyketide

	MSH derivative	-NHR	RSH/RSSR[a]	RSH/RSSR[b]
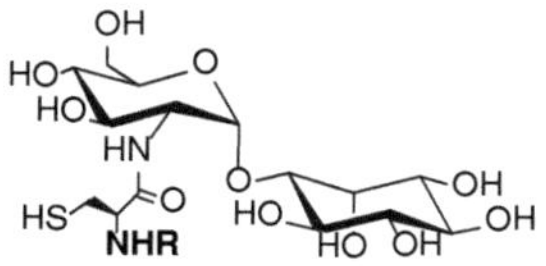	**MSH**	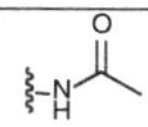	**500**	**40**
	fMSH	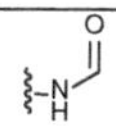	**110**	
	sucMSH	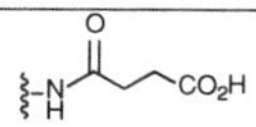	**90**	
	pMSH	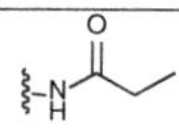		**10**

Fig. 7 Cellular redox ratios of different N-acyl mycothiol derivatives. [a] ΔmshD mutant of *M. smegmatis* during stationary phase (Newton et al. 2005). [b] 'Marinispora' CNQ703 (MAR2) during stationary phase (Newton et al. 2008b)

biosynthesis in the production of many secondary metabolites. High levels of propionyl-CoA, however, are potentially toxic, so the diversion of propionyl-CoA into pMSH production may serve to limit the accumulation of potentially toxic propionyl-CoA during degradative metabolism of odd chain and branched chain fatty acids and cholesterol. The cellular thiol/disulfide redox status of pMSH is comparable to that of cysteine, but is four times more oxidized than that of MSH (Newton et al. 2008b). This suggests that it may serve as a weaker mycothiol disulfide reductase substrate than MSSM, as has previously been observed for disulfides of fMSH and sucMSH (see Fig. 7) (Buchmeier et al. 2006).

The NADPH-dependent flavoprotein, mycothiol disulfide reductase (Mtr) maintains a high thiol/disulfide ratio by reducing MSSM back to MSH in an analog manner to the glutathione reductase redox pathway found in eukaryotes (Patel and Blanchard 1999, 2001). MSH levels in *M. smegmatis mtr* knockout mutants are the same as those in the wild type, which suggests some redundancy in the activity of Mtr, although the influence on MSH/MSSM ratios has not been reported (Holsclaw et al. 2011). A recent publication suggests that the thioredoxin system of *M. tuberculosis* (TrxR and TrxC) is able to serve as an MSSM reductase, which may account for the redundancy of Mtr (Attarian et al. 2009).

While the disulfides of fMSH and sucMSH are more slowly reduced by Mtr than MSSM, the rates of reduction are significantly enhanced in the presence of MSH, presumably due to the formation and accelerated reduction of the respective mycothiol mixed disulfides (Newton et al. 2005). Similar rate enhancements are observed when comparing the Mtr catalyzed reduction of the symmetrical aryl disulfide (Ellman's reagent) with that of its mixed disulfide derivative with an MSH mimetic (Stewart et al. 2008). These observations imply that only one half of the disulfide substrate makes a significant contribution toward substrate recognition by Mtr. It therefore seems plausible that Mtr could also function in the reduction of biologically relevant mycothiol-mixed disulfides (i.e., with Cys and/or CoA) in order to help maintain these biothiols in a reduced state. It is noteworthy that the glutathione reductase from *E. coli* does not recognize $(GspdS)_2$ as a substrate, but is able to reduce the GspdSSG mixed disulfide with an efficiency comparable to the one observed in the case of GSSG (Smith et al. 1995).

Analog to glutathiolation (Dalle-Donne et al. 2007; Ghezzi 2005), evidence is now emerging that protein mycothiolation functions as a reversible redox switch in the regulation of enzyme activity (Chi et al. 2013a). Two mycoredoxin enzymes (Mrx1 and Mrx2) have been characterized from *Corynebacterium glutamicum,* which form part of a MSH/Mtr-coupled redox system providing reducing equivalents to arsenate reductases to facilitate the reduction of arsenate (As(V)) to arsenite (As(III)) (Ordóñez et al. 2009; Villadangos et al. 2011). The ability of Mrx1 to reduce some mycothiolated proteins (Chi et al. 2013a) and small molecule MS-mixed disulfides (Van Laer et al. 2012) has also been demonstrated. A total of 25 different proteins are known to be mycothiolated in response to NaOCl stress in *C. glutamicum* (Chi et al. 2013a). These include antioxidant enzymes such as the thiol peroxidase (Tpx) whose peroiredoxin activity can be restored *in vitro* by Mrx1. Interestingly, one of the most predominantly mycothiolated proteins in this study was methionine synthase (MetE), which is a major target for protein bacillithiolation in *Bacilli* under the same NaOCl stress conditions (see later) (Chi et al. 2011a, 2012). As work in the mycoredoxin arena continues to mature, it will be interesting to compare similarities between protein-mycothiolation and protein glutathiolation redox regulation pathways in actinomycetes and Gram-negative bacteria, respectively.

MSH deficient bacteria show a significant increase in sensitivity to oxidative stress (Newton et al. 1999; Rawat et al. 2002) although MSH and Mtr requirements are not the same for all actinomycetes. An *mtr* null mutant of *M. smegmatis* shows no change in growth or cellular MSH levels compared to the wild type, however, the mutants show increased sensitivity to electrophiles, oxidants, and several antibiotics (Rawat et al. 2002, 2004, 2007). In addition, a two-fold reduction in MSH levels is observed on treatment of *M. smegmatis* with H_2O_2 (Holsclaw et al. 2011). With *M. tuberculosis*, it has proven to be possible, but much more challenging, to generate MSH-deficient mutants (Sareen et al. 2003; Buchmeier and Fahey 2006). The fact that MSH is more important for the viability of *M. tuberculosis* than for *M. smegmatis* (Rawat et al. 2002), *S. coelicolor* (Park et al. 2006) or *C. glutamicum* (Feng et al. 2006) indicates that the latter three microorganisms are biochemically better equipped to facilitate cell growth in the absence of MSH.

In GSH-utilizing organisms, many electrophilic xenobiotics are detoxified by a four-step process initiated by glutathione-*S*-transferases (GSTs), which can catalyze *S*-conjugation of GSH to electrophiles, such as α,β-unsaturated ketones, quinones, epoxides, and aryl halides (Hayes et al. 2005). In GSH-utilizing bacteria and plants, the GS-conjugates are then directly excreted from the cell (Kaluzna and Bartosz 1997; Martinoia et al. 1993; Müller et al. 1994). A more efficient detoxification pathway is utilized by actinomycetes in which mycothiol-*S*-conjugate amidase (Mca) hydrolyzes the glucosaminyl-amide bond of MS-conjugates to

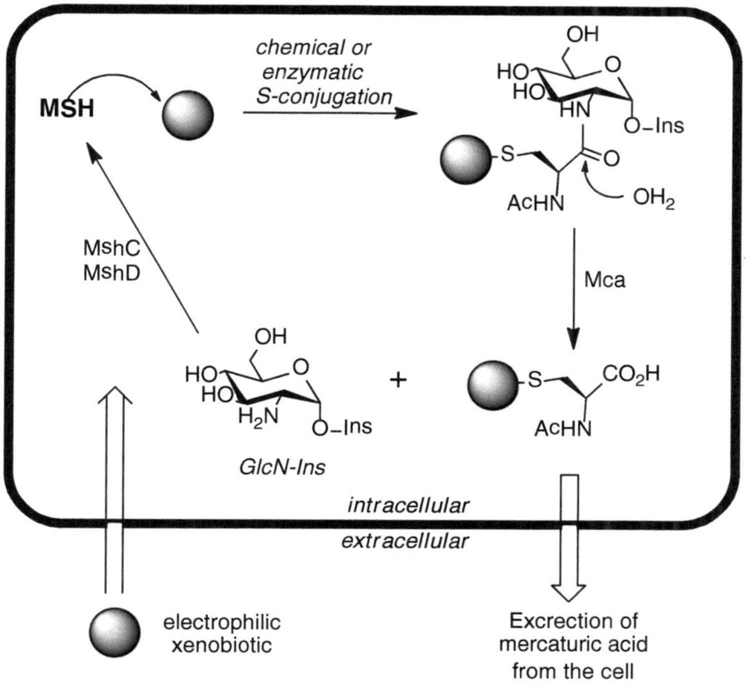

Fig. 8 Mca detoxification pathway

liberate the mercapturic acid labeled toxin and GlcN-Ins (Fig. 8). The mercapturic acid derivatives are then exported from the cell while GlcN-Ins is recycled back into the MSH biosynthetic pathway (Newton et al. 2000b).

Mca, which is a homolog of MshB (with 42 % sequence identity between Mca and MshB in *M. tuberculosis*) (Steffek et al. 2003) possesses zinc-dependent *N*-acyl hydrolase activity. The ability of Mca to *N*-deacetylate GlcNAc-Ins has been proposed to explain the redundancy of MshB, with Mca providing the compensatory GlcNAc-Ins deacetylase activity that retains ~10 % of the wild type MSH levels in *mshB* null mutants. Mca has a broad substrate tolerance for mycothiol conjugates, ranging from small, relatively non-polar, MS-conjugates (e.g., MS-acetophenone) to large macrocyclic antibiotics (MS-rifamycin) (Newton et al. 2006a). Mca also displays amidase activity with free MSH (Newton et al. 2006a). Based on this activity, the intracellular rate of MSH degradation to *N*-Acetyl-Cysteine and GlcN-Ins is estimated to be 20 % per hour suggesting that MSH serves as an intracellular storage reservoir for sulfur/cysteine that can be accessed via the Mca-catalyzed hydrolysis of MSH (Bzymek et al. 2007).

Mca null mutants exhibit increased sensitivity to a range of alkylating agents (iodoacetamide, *N*-ethylmaleimide), electrophilic oxidants (menadione, plumbagin), and antibiotics (e.g., streptomycin, rifamycin, tetracycline) (Rawat et al. 2004;

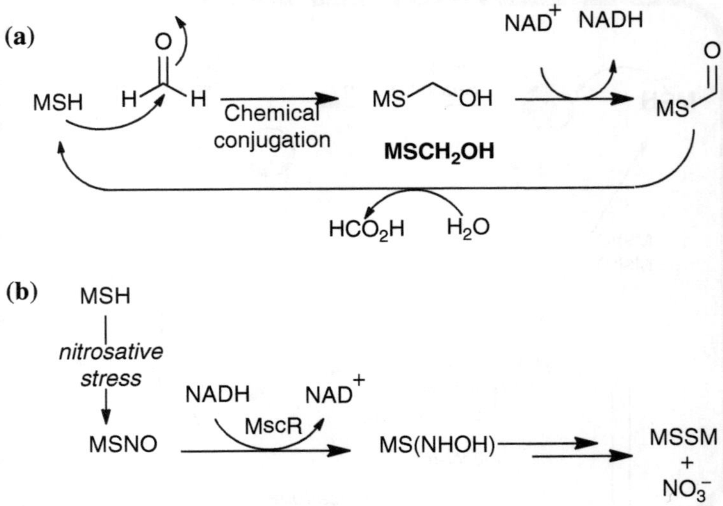

Fig. 9 **a** Formaldehyde dehydrogenase and **b** Mycothiol-S-nitroso reductase activity of MscR

Park and Roe 2008). In therapeutically relevant pathogens such as *M. tuberculosis*, Mca evidently plays an important role in resistance against electrophilic drugs such as the rifampicins, which are furnished with Michael acceptor motifs susceptible to *S*-conjugation with MSH. Mycothiol-*S*-transferases able to catalyze such MS-conjugation reactions have recently been identified, but have not yet been characterized in detail (Newton et al. 2011). Many actinomycetes produce a broad spectrum of antibacterial natural products, and mercapturic acid derivatives of such metabolites have previously been observed in the fermentation broths of these bacteria (Newton and Fahey 2002). This suggests that actinomyctes also use MSH and Mca to limit the intracellular toxicity and to regulate the intracellular concentrations of their own antimicrobial secondary metabolites. For this purpose, MS-conjugate transferase/amidase activities might only become physiologically relevant processes if the intracellular concentrations of such metabolites reach a critical level. This concept warrants further investigation that could provide valuable insights into the potential role(s) of MSH in secondary metabolite regulation and transport within actinomycetes.

Formaldehyde is a central metabolite in the metabolism of C_1 compounds, such as methane and methanol that can be further oxidized to CO_2 to provide a valuable source of energy for methylotrophic microorganisms. Formaldehyde is also toxic, however, even at low concentrations. NAD/MSH-dependent formaldehyde dehydrogenases (MscR) limit the intracellular accumulation of formaldehyde via its catalytic oxidation to formic acid (Fig. 9a) (Ophem et al. 1992; Eggeling and Sahm 1985; Vogt et al. 2003). MscR enzymes also display preferential NADH-dependent mycothiol-*S*-nitroso (MSNO) reductase activity and are likely to be implicated in the regulation of nitrosative stress (Fig. 9b) (Vogt et al. 2003).

(a)

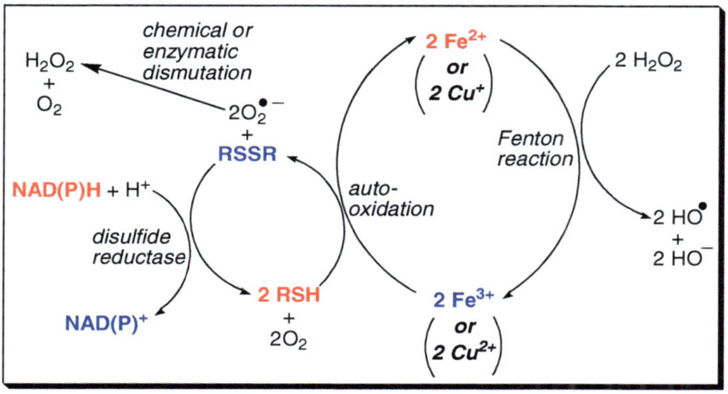

(b)

Cu^{2+}mediated auto-oxidation (relative rates)		Fe^{3+}mediated auto-oxidation (relative rates)	
Cys	100 (Newton et al. 1995)	Cys	100 (Park and Imlay 2003)
Cys-GlcN-Ins	11 (Newton et al. 2005)	GSH	10 (Park and Imlay 2003)
GSH	7 (Newton et al. 1995)		
MSH	1 (Newton et al. 1995)		

Fig. 10 **a** How biothiols fuel the Fenton reaction; **b** Metal-mediated auto-oxidation of thiols (relative rates in *parentheses*)

Studies have shown that nitric oxide sensitivity in *M. smegmatis* is increased in MSH-deficient mutants (Miller et al. 2007).

The thiol pK_a and redox potential of MSH are implicated in its functional efficiency both as a nucleophile and a redox buffer. Surprisingly, neither of these important parameters has yet been reported for MSH. In aerobic organisms, cellular thiols are susceptible to metal (e.g., Fe^{3+}, Cu^{2+}) mediated auto-oxidation (Akaria and Kamide 1975; Winterbourn 1993) which results in the production of toxic peroxide and concomitant production of disulfides (Fig. 10). Cysteine undergoes rapid auto-oxidation and its intracellular accumulation is cytotoxic under aerobic conditions. In most eukaryotes, GSH is used as a more stable cysteine reservoir which is at least ~ ten-fold more resistant than cysteine to such auto-oxidation processes (Newton et al. 1995). The relative rate of MSH auto-oxidation is another seven-fold slower than that of GSH (Newton et al. 2005). The reduced thiol auto-oxidation rates of GSH and MSH are partly attributed to blocking of the amino substituent which could otherwise enhance Cu^{2+} coordination and facilitate the auto-oxidation process. Sequential blocking of both the carboxyl and amino groups of cysteine (i.e., in Cys-GlcN-Ins and MSH) also results in the progressive reduction in the rate of Cu^{2+}-catalyzed auto-oxidation (Newton et al. 1995). The intracellular half-life of MSH is ~50 h (Bzymek et al. 2007), mediated by the continuous degradation of

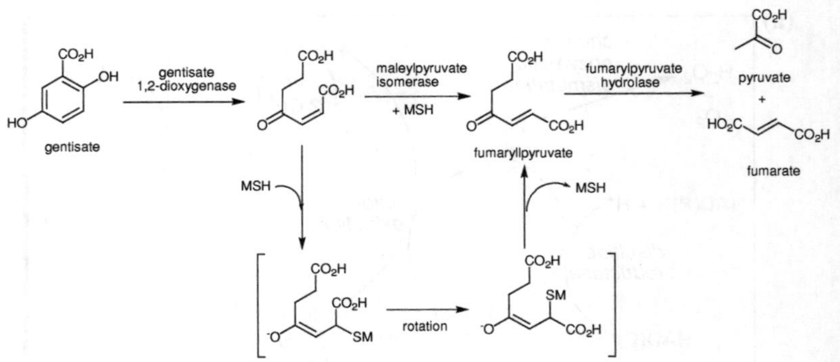

Fig. 11 Mycothiol-dependent gentisate metabolism

Mca which hydrolyzes MSH to give *N*-acetyl-cysteine and GlcN-Ins. The *N*-acetyl-cysteine product of MSH catabolism is rapidly *N*-deacetylated to cysteine. Cysteine and GlcN-Ins can be recycled back into MSH biosynthesis or utilized for other metabolic purposes as required. The constant Mca-catalyzed turnover of MSH provides a controlled, but continuous supply of cysteine (and sulfur) to serve other cellular needs while obviating the damaging effects of auto-oxidation.

Many aerobic bacteria are biochemically equipped to utilize aromatic substrates as an alternative carbon source, and gentisate and substituted gentisates are often key intermediates during the enzymatic degradation of such compounds. Maleylpyruvate is a key intermediate of the gentisate pathway in *C. glutamicum* and *Streptomyces* (Shen et al. 2005), whose subsequent catabolism is mediated by a MSH-dependent maleylpyruvate isomerase (Feng et al. 2006; Wang et al. 2007). A plausible isomerization mechanism has been proposed involving Michael addition of MSH onto the conjugated double bond followed by bond rotation and a subsequent retro-Michael addition to give fumarylpyruvate (Fig. 11).

MSH is evidently an essential metabolite in the *M. tuberculosis* pathogen and performs a number of important roles in other actinomycetes (e.g., *Corynebacteria*) of industrial significance. If MSH truly parallels the full range of metabolic functions served by GSH then there are still a number of MSH-dependent biochemical functions that are yet to be uncovered, such as redox regulation of protein function and methylglyoxal detoxification. Such knowledge could have a realistic and significant impact on future developments in anti-tubercular chemotherapy, antibiotic production, and bioremediation as well as providing opportunities to further elucidate some of the unique metabolic pathways of the actinomycetes.

2.4 Bacillithiol

In 2009, bacillithiol (BSH) (Fig. 2) was identified as an LMW thiol among many low G + C Gram-positive bacteria (Firmicutes) (Newton et al. 2009), which do not produce GSH or MSH. These include bacilli (e.g., *Bacillus anthracis,*

Fig. 12 Biosynthetic pathway for BSH

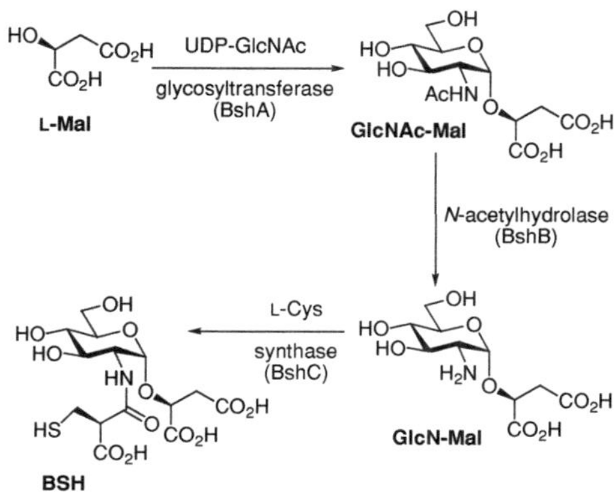

B. subtilis, B. cereus, B. megaterium, B. pumilis) and some, but not all staphylococci (e.g., *Staphylococcus aureus, S. saprophyticus*), and streptococci (*Streptococcus agalactiae*) (Newton et al. 2009). BSH biosynthesis is initiated by a retaining glycosyltransferase (BshA) that catalyzes the glycosylation of L-malic acid with UDP-GlcNAc to afford D-GlcNAc-L-Mal (Fig. 12) (Gaballa et al. 2010; Parsonage et al. 2010; Fang et al. 2013). An *N*-acetyl hydrolase (BshB) then liberates the free amine D-GlcN-L-Mal (Gaballa et al. 2010; Parsonage et al. 2010). Gene knockout studies in *B. subtilis* have identified a bacillithiol synthase (BshC) (Gaballa et al. 2010), which mediates the condensation of D-GlcN-L-Mal with L-cysteine to yield BSH. In the absence of this *bshC* gene, *B. subtilis* is unable to produce BSH, and the intracellular levels of its biosynthetic precursor (GlcNMal) have been shown to accumulate (Gaballa et al. 2010). This is very strong genetic evidence that this BshC candidate mediates the final stage of BSH biosynthesis. Attempts to demonstrate the BSH synthase activity of BshC *in vitro*, however, have so far been unsuccessful (Gaballa et al. 2010). In *B. subtilis*, the master regulator of disulfide (thiol depletion) stress (Spx) is an activator of the BSH biosynthesis genes (Gaballa et al. 2013).

BshA and BshC are essential for BSH biosynthesis, but the activity of BshB is shared with another *N*-deacetylase, similar to what is observed in MSH biosynthesis (Gaballa et al. 2010; Parsonage et al. 2010). This redundancy is due to a candidate bacillithiol-*S*-conjugate amidase (Bca) that is able to *N*-deacetylate GlcNAc-Mal with comparable efficiency to that observed with BshB (Fang et al. 2013). BSH deficient mutants of *B. subtilis* display increased sensitivity to oxidative, osmotic, and acid stress as well as a 100-fold reduction in sporulation efficiency (Gaballa et al. 2010). The mutants also show increased sensitivity to some antibiotics (e.g., fosfomycin).

Fosfomycin is a broad spectrum antibiotic, which serves as a covalent inhibitor of MurA (the first enzyme involved in peptidoglycan biosynthesis) via reaction of the epoxide motif with a substrate binding site cysteine. Thiol-*S*-transferase fosfomycin resistance genes (*fosB*) have been detected in many BSH-producing

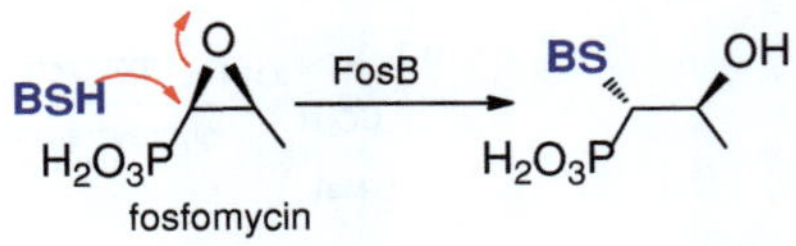

Fig. 13 The bacillithiol-*S*-transferase catalyzed inactivation of fosfomycin by FosB

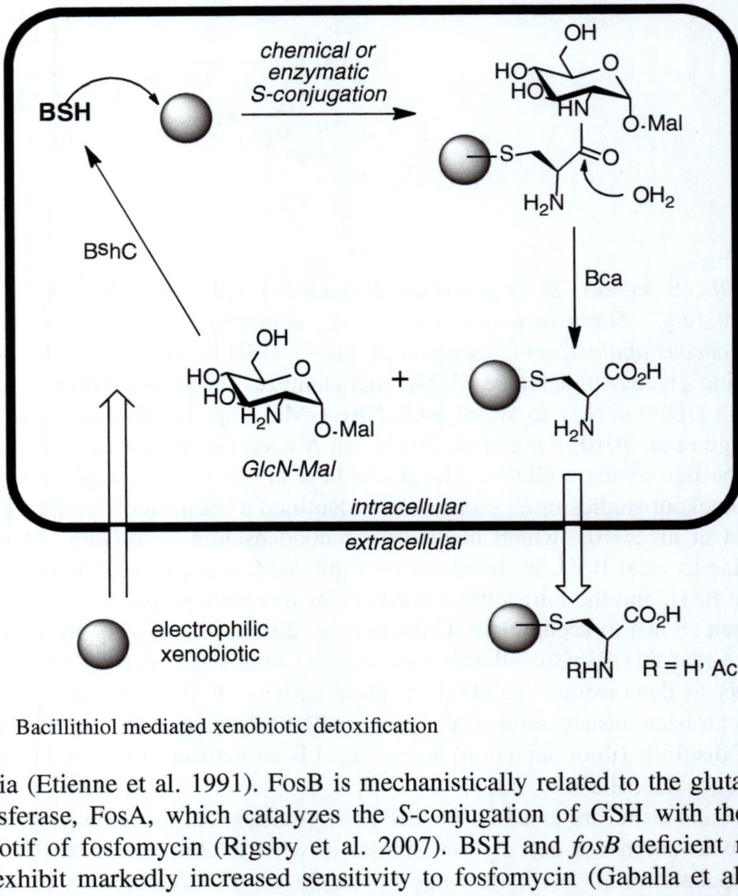

Fig. 14 Bacillithiol mediated xenobiotic detoxification

bacteria (Etienne et al. 1991). FosB is mechanistically related to the glutathione-*S*-transferase, FosA, which catalyzes the *S*-conjugation of GSH with the epoxide motif of fosfomycin (Rigsby et al. 2007). BSH and *fosB* deficient mutants both exhibit markedly increased sensitivity to fosfomycin (Gaballa et al. 2010; Parsonage et al. 2010; Poether et al. 2013; Rajkarnikar et al. 2013) and detailed kinetic studies of *S. aureus* FosB have clearly demonstrated it to be a bacillithiol-*S*-transferase (Fig. 13) (Roberts et al. 2013).

Another bacillithiol-*S*-transferase has very recently been isolated from *B. subtilis*, which appears to be present in a significant number of BSH-utilizing bacteria (Newton et al. 2011). Its bacillithiol-*S*-transferase activity has been demonstrated with the synthetic electrophile, monochlorobimane, but so far no physiologically relevant substrates have been identified. Evidence is beginning to emerge that the BSH-dependent xenobiotic detoxification pathways share similarities with those of MSH (Fig. 14). Cysteine- and *N*-acetylcysteine-*S*-conjugates of

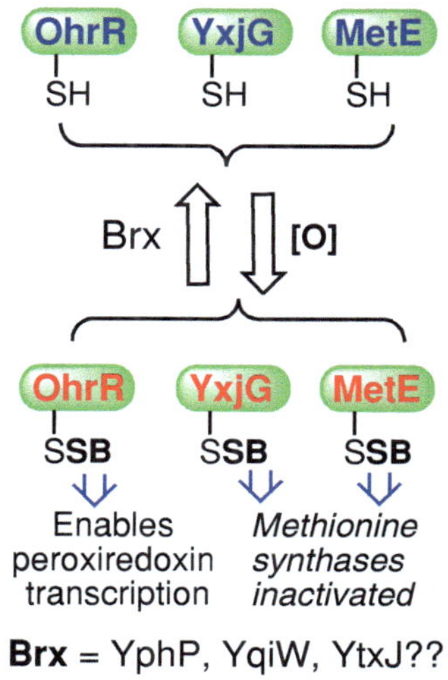

Fig. 15 Examples of the predominant reversible protein bacillithiolations observed during NaOCl treatment of *B. subtilis*. *OhrR* peroxiredoxin transcription regulator; *MetE* and *YxjG* methionine synthases

monochlorobimane are exported into the extracellular media of *S. aureus* when treated with this xenobiotic (Newton et al. 2011). This indicates that BSH utilizes a bacillithiol-*S*-conjugate amidase (Bca) pathway for the conversion of BS-conjugates into Cys(NAc)-conjugated derivatives in a manner similar to the characterized Mca pathway in MSH-utilizing microorganisms.

Intracellular BSH is predominantly present in its reduced form with BSH/BSSB redox ratios ranging from 40:1 (in *B. anthracis*) (Parsonage et al. 2010) to 400:1 (in *B. subtilis*) (Newton et al. 2009; Sharma et al. 2013). Reduced BSH is presumably maintained by an as yet unidentified BSSB reductase. BSH also forms mixed disulfides with protein thiols (bacillithiolations) which can regulate protein function. In *B. subtilis* the redox sensitive peroxiredoxin transcription regulator (OhrR) is bacillithiolated during cumene hydroperoxide stress suggesting a function for BSH in redox sensing (Lee et al. 2007; Soonsanga et al. 2008). Several other proteins are also bacillithiolated when *Bacilli* are exposed to NaOCl stress (Chi et al. 2011b, 2013b). These include two methionine synthases (MetE and YxjG), a pyrophosphatase (PpaC) and a 3-D-phosphoglycerate dehydrogenase (SerA). Growth of both *ohrR* and BSH deficient mutants are severely impaired by hypochlorite compared to the wild type. *S*-Bacillithiolation of OhrR leads to a 200-fold up-regulation of the OhrA peroxiredoxin that confers protection against NaOCl (Chi et al. 2011b). *S*-bacillithiolation of methionine synthases causes hypochlorite-induced methionine starvation/auxotrophy, which inhibits the initiation of protein translation (Chi et al. 2011b). This could serve to prevent the synthesis (and potential oxidative damage) of additional proteins while the cell detoxifies the oxidant and restores the thiol redox homeostasis.

Three candidate bacilliredoxin (Brx) proteins have been identified (YtxJ, YphP, and YqiW), which are widely conserved among BSH-producing Firmicutes and could potentially mediate the de-bacillithiolation of BS-proteins (Fig. 15)

(Gaballa et al. 2010). YtxJ is a thioredoxin family protein with a conserved cysteine residue in a TCPIS-motif reminiscent of those found in many monothiol glutaredoxins. YqiW and YphP are related (53 % sequence identity) thioredoxin-like proteins where the classical redox active CXXC motif found in thioredoxins, is replaced by an invariant CGC active site motif. Under NaOCl stress, *S*-bacillithiolation of YphP is observed at the more solvent exposed 'nucleophillic' Cys53 residue of its redox active CGC motif. This is a likely intermediate in the Brx catalytic cycle (Chi et al. 2011b). The reduction potential of YphP (−130 mV) (Derewenda et al. 2009) is significantly higher than those typically observed for thioredoxins and glutaredoxins (<−200 mV) (Åslund et al. 1997). The standard redox potential of BSH (−221 mV) (Sharma et al. 2013) (Fig. 3) implies it is a weaker redox buffer than GSH (−240 mV). However it should be noted that redox potentials are a thermodynamic property based on thiol-disulfide exchange equilibria and while the redox status can be maintained in a steady state in living cells, it is never at equilibrium. The actual redox buffering properties of BSH are likely driven by other factors (i.e., cellular abundance and catalytic efficiencies of BSH-specific redox enzymes).

In Gram-negative bacteria, intracellular concentrations of Cys (~0.4 mM) are usually much lower than GSH (up to 10 mM) (Newton et al. 1996). BSH levels in many Bacilli (e.g., *S. aureus*, *B. anthracis*) are more comparable to those of Cys (Newton et al. 2009). The significance of this is not yet known, but it is noteworthy that the unusually high CoA levels in these microorganisms often equal, or exceed, those of BSH and they also produce a CoA-disulfide reductase enzyme (see later), suggesting there may be some redundancy in the redox buffer functions of these two thiol cofactors. Interestingly, *B. subtilis* does not have a CoA disulfide reductase, but its BSH levels can increase from 0.5 to 5 mM as cell cultures progress from early-exponential growth to stationary phase (Sharma et al. 2013). It is not yet known how the catalytic efficiencies of BSH redox enzymes (i.e., bacilliredoxins and BSSB reductases) compare with their GSH counterparts. Further investigations to unravel the unique characteristics of this newly discovered redox system will address these fundamental questions on BSH redox biology.

2.5 Ergothioneine

Ergothioneine (ESH) is a 2-thiolhistidine (Figs. 2, 16) that has been detected (in millimolar concentrations) in bacteria, fungi, plants, animals, and humans. Animals obtain ESH from dietary sources, whereas plants are believed to obtain ESH from the soil. ESH is only biosynthesized by some fungi (Ishikawa et al. 1974) and actinomycetes (Seebeck 2010). In *M. smegmatis*, the ESH biosynthetic pathway has been elucidated (Fig. 16) (Seebeck 2010). Initial steps involve the tri-*N*-methylation of histidine by EgtD, followed by oxidative sulfurization of hercynine using γGC. A glutamine amidotransferase then hydrolyzes the γ-glutamyl

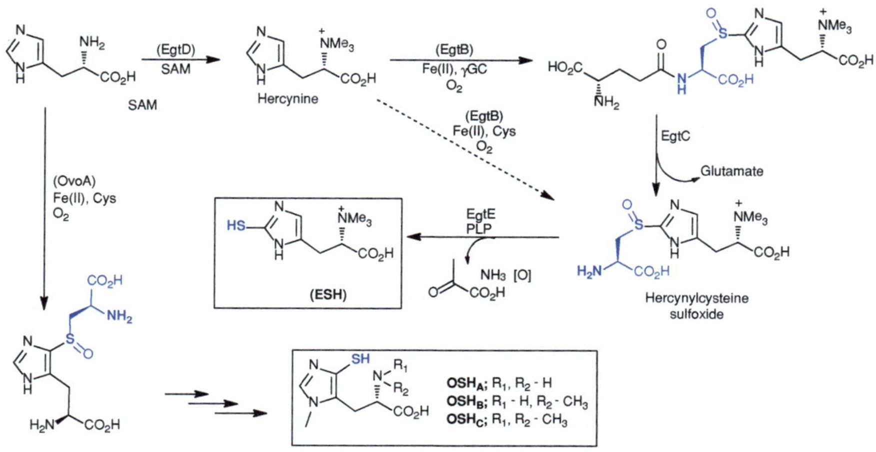

Fig. 16 Biosynthesis of ergothioneine and ovothiols

motif to give hercynylcysteine sulfoxide. A PLP-dependent *β*-lyase (EgtE) then converts hercynylcysteine sulfoxide into ESH. A shorter pathway was originally proposed for ESH biosynthesis in fungi (*Neurospora* crassa) (Ishikawa et al. 1974) with cysteine serving as the thiol substrate for oxidative sulfurization of hercynine, but this remains to be unequivocally proven. However, it is noteworthy that cysteine, not *γ*GC, is the substrate for the oxidative sulfurization step in ovothiol (OvSH) biosynthesis (Vogt et al. 2001; Braunshausen and Seebeck 2011).

ESH serves as an important redox modulator, which in higher organisms is primarily stored in tissues that are often exposed to oxidative stresses. ESH has a very high thiol redox potential (−60 mV) (Jocelyn 1972) (Fig. 3) with the disulfide (ESSE) being less thermodynamically stable due to the greater stability of the thione form of ESH (Fig. 17a). Under physiological conditions ESH exists entirely in its reduced state due to the greater stability of the thione form of ESH (Fig. 17a). The considerable nucleophilicity of the predominant amino-thione form of ESH at physiological pH is attributed to the mesomeric release of the lone pair electrons on the nitrogen atom of the amino group (Fig. 17b) (Misiti et al. 2001; Kawano et al. 1982). This enhanced reactivity of ESH could explain its notable antioxidant properties despite its high redox potential. ESH strongly influences GSNO decomposition, and can enhance the rate of decomposition 20-fold more than GSH (Misiti et al. 2001). ESH is also a powerful scavenger of $HO^{\bullet}$ (Akanmu et al. 1991; Asmus et al. 1996), HOCl (Akanmu et al. 1991), singlet oxygen ($^1O_2$) (Takeshima and Inoue 1996), and peroxynitrite (Jang et al. 2004) at biologically significant rates, suggesting it can play a relevant role in the detoxification of these oxidants. In contrast, it does not appear to react with H_2O_2 or $O_2^{\bullet-}$ (Akanmu et al. 1991). The relative formation constants of ESH-metal complexes are $Cu^{2+} > Hg^{2+} > Zn^{2+} > Cd^{2+} > Co^{2+} \geq Ni^{2+}$, with ESH behaving as a unidentate ligand that coordinates the metal through the sulfur atom (Motohash et al. 1974).

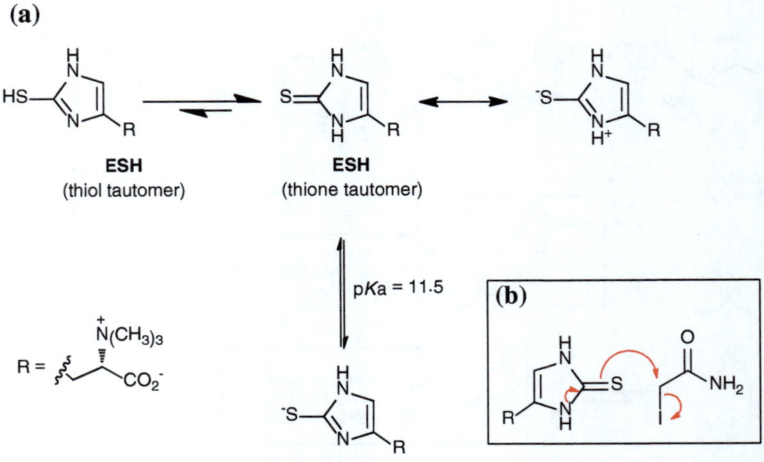

Fig. 17 **a** Thiol-thione tautomerism of ESH; **b** mechanism explaining the nucleophilicity of the uncharged thione form of ESH at physiological pH

Unlike other LMW thiols at physiological pH, ESH forms a stable complex with Cu^{2+} in a non-redox active form, which prevents auto-oxidation and hydroxy radical formation (Akanmu et al. 1991; Rowley and Halliwell 1982). To date, most studies of ESH function have been performed in eukaryotes, but a recent study of MSH deficient mutants of *M. smegmatis* (also an ESH producer) showed a 20- to 30-fold increase in ESH production compared to the wild type indicating that it may compensate toward the loss of MSH in these mutants (Ta et al. 2011). In the extensive *in vitro* and *in vivo* studies that have been reported, ESH appears to serve an important role as a biochemical antioxidant. So far, no enzymes have been identified that use ESH as a cofactor.

2.6 Ovothiol

Ovothiols A-C ($OSH_{A\text{-}C}$) form a group of structurally related 1-methyl-4-thiol-histidines with differing degrees of *N*-methylation (Figs. 2, 16). They are found alongside GSH in echinoderms (Turner et al. 1986) and a variety of other marine invertebrates (Palumbo et al. 1984; Rossi et al. 1985), some halo-tolerant cyanobacteria (Selmanreimer et al. 1991), as well as the $T(SH)_2$-producing parasites *Crithidia fasciculata* (Steenkamp and Spies 1994) and *Leshmania spp*. (Spies and Steenkamp 1994; Ariyanayagam and Fairlamb 2001). The thiol-histidine motif of OSH appears as a deceptively subtle regioisomer of ESH and there are evident similarities in the biosynthetic pathways (Vogt et al. 2001; Steenkamp et al. 1996) [although to date, only the first enzyme of OSH biosynthesis (OvoA) has been

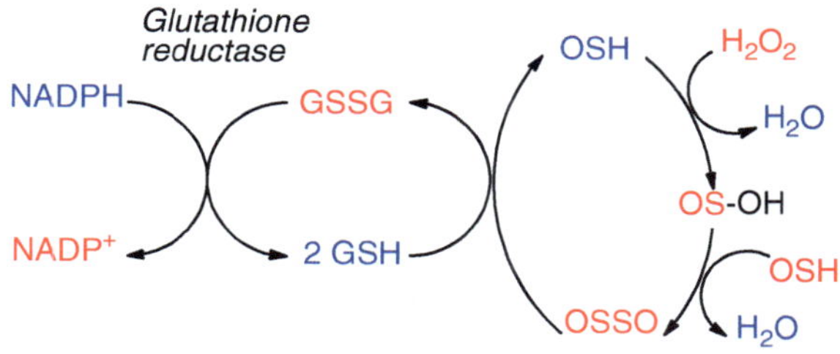

Fig. 18 The peroxidase activity of OSH

fully characterized] (Braunshausen and Seebeck 2011). Nonetheless, there are notable differences between the biophysical and biochemical properties of OSH and ESH. Interestingly, the redox potential of OSH is pH-dependent, although it is highest in the region of physiological pH (~ −60 mV) (Weaver and Rabenstein 1995) (Fig. 3), hence the disulfide is very unstable and OSH is swiftly regenerated via thiol-disulfide exchange with other LMW thiols (e.g., GSH). The high redox potential coupled with the acidity of the histidine thiol (pK_a ~ 2.3) means that, under physiological conditions, OSH exists exclusively in its thiolate anion form. Ovothiols play an important role in resistance to oxidative stress. For example, in fertilized sea urchin eggs, an oxidative burst produces large quantities of H_2O_2 to feed ovoperoxidase-catalyzed formation of the fertilization envelope. To prevent oxidative damage from excess H_2O_2, ~5 mM OSH_C is produced. This functions as a chemical equivalent of GSH peroxidase (Fig. 18) (Turner et al. 1988) and is even more effective than catalase in the detoxification of the H_2O_2 produced during egg fertilization. Recent progress in identifying some of the genes implicated in OSH biosynthesis should now enable the preparation of OSH deficient mutants of *Leishmania* parasites to be prepared. This will help to determine the importance of OSH for parasite viability, and may potentially validate OSH biosynthetic enzymes as new drug targets.

2.7 Coenzyme A

Coenzyme A (CoA) is an essential thiol cofactor in all living organisms that is widely implicated in central metabolic pathways including the Krebs cycle and fatty acid metabolism. Nonetheless, it merits some discussion since NAD(P) H-dependent CoA disulfide reductases (CoADR) have been isolated and biochemically characterized from *S. aureus* (delCardayre et al. 1998; Luba et al. 1999; Mallett et al. 2006), *B. anthracis* (Wallen et al. 2008), *Borrelia burgdorferi* (Boylan et al. 2006), and *Pyrococcus horikoshii* (Harris et al. 2005), indicating that CoA has intracellular redox functions in many bacteria. Bioinformatic analyses indicate that CoADR is broadly distributed among the bacterial and archaeal kingdoms (Mallett et al. 2006). Some bacteria (e.g., *B. anthracis, S. aureus*), which utilize the CoADR/CoA redox couple, also produce bacillithiol in similar quantities (Newton et al. 2009). Redundancy provided by these two thiol redox buffers may confer an advantage in virulence and/or survival of such pathogens (Parsonage et al. 2010). Little is known, however, about the redox functions of CoA in a physiological setting. Its redox potential (−234 mV) (Keire et al. 1992b) is comparable to that of GSH (Fig. 3), suggesting that CoA could make a significant contribution toward the electrochemical potential inside the cell. A thiol pK_a of 9.83 (Keire et al. 1992a) implies that CoA exists exclusively in its 'unreactive' thiol form at physiological pH. It is reasonable to speculate that enzymes must be required to enhance the reactivity of CoA in cellular redox processes, although currently none have been reported. In dormant spores of *B. megaterium, B. cereus*, and *Clostridium bifermentans* approximately 45 % of cellular CoA is linked to proteins as mixed disulfides (Setlow and Setlow 1977) and more than 75 % of these are cleaved from the proteins during germination. This clearly demonstrates a protein-thiol redox regulation and/or protective role for CoA, similar to that of GSH. The mechanism(s) by which these CoASS-proteins are reduced are not yet known.

3 Future Considerations

It is clearly evident that LMW non-protein thiols are key players in a wide range of biological redox processes. Despite the great advances in LMW thiol redox biochemistry in recent decades, there are still many important questions that remain unanswered. While LMW thiols are central players in an extensive range of redox regulation/metabolism processes, not all organisms use the same thiol cofactor(s) to this effect, as evidenced by the discovery of MSH and BSH among different Gram-positive bacteria. There are still many elements of MSH, BSH, ESH, and OSH biochemistry that remain to be explored. For BSH and MSH the bacilliredoxin/mycoredoxin mediated regulation of protein function is a newly emerging area that is ripe for exploration. It will also be interesting to see if there are any ESH- or OSH-dependent enzymes in producer organisms or whether the cellular functions of these thiol-histidine metabolites are purely driven by their unusual chemical properties. Another pertinent challenge resides in the identity of additional unknown LMW thiol(s), which are utilized by microorganisms that do not produce the aforementioned thiols. There are many microorganisms (for example among the Archaea) whose low molecular weight thiol redox buffer(s) have not yet been identified or explored (either bioinformatically or experimentally). Evidently there are many fundamental scientific questions in the LMW thiology arena that remain to be addressed, the answers to which will provide a more comprehensive understanding of their importance and multifaceted functions among different organisms.

Acknowledgments We would like to acknowledge the following for financial support: The EU FP-7 ITN grant number 215009 (M. Arbach) and the BBSRC grant number BB/H013504/1 (C. Hamilton). We also thank Dr. Alex Roberts for her critical reading of this chapter.

References

Akanmu D, Cecchini R, Aruoma OI, Halliwell B (1991) The antioxidant action of ergothioneine. Arch Biochem Biophys 288(1):10–16

Akaria H, Kamide H (1975) Autoxidation of cysteine catalyzed by copper in glycylglycine buffer. Chem Pharm Bull 23:1671–1676

Ariyanayagam MR, Fairlamb AH (2001) Ovothiol and trypanothione as antioxidants in trypanosomatids. Mol Biochem Parasitol 115(2):189–198

Åslund F, Berndt KD, Holmgren A (1997) Redox potentials of glutaredoxins and other thiol-disulfide oxidoreductases of the thioredoxin superfamily determined by direct protein-protein redox equilibria. J Biol Chem 272(49):30780–30786

Asmus KD, Bensasson RV, Bernier JL, Houssin R, Land EJ (1996) One-electron oxidation of ergothioneine and analogues investigated by pulse radiolysis: redox reaction involving ergothioneine and vitamin C. Biochem J 315:625–629

Attarian R, Bennie C, Bach H, Av-Gay Y (2009) Glutathione disulfide and S-nitrosoglutathione detoxification by *Mycobacterium tuberculosis* thioredoxin system. FEBS Lett 583(19):3215–3220

Bayley ST (1982) Problems in tracing the early evolution of cells as illustrated by the archaebacteria and particularly by the halobacteria. Zentralblatt Fur Bakteriologie Mikrobiologie Und Hygiene I Abteilung Originale C-Allgemeine Angewandte Und Okologische Mikrobiologie 3(1):65–68

Benesch RE, Benesch R (1955) The acid strength of the -SH group in cysteine and related compounds. J Am Chem Soc 77(22):5877–5881

Bollinger JM, Kwon DS, Huisman GW, Kolter R, Walsh CT (1995) Glutathionylspermidine metabolism in *Escherichia coli*. J Biol Chem 270(23):14031–14041

Booth IR, Ferguson GP, Miller S, Li C, Gunasekera B, Kinghorn S (2003) Bacterial production of methylglyoxal: a survival strategy or death by misadventure? Biochem Soc Trans 31(Pt 6):1406–1408

Bornemann C, Jardine MA, Spies HSC, Steenkamp DJ (1997) Biosynthesis of mycothiol: elucidation of the sequence of steps in *Mycobacterium smegmatis*. Biochem J 325:623–629

Boylan JA, Hummel CS, Benoit S, Garcia-Lara J, Treglown-Downey J, Crane EJ, Gherardini FC (2006) Borrelia burgdorferi bb0728 encodes a coenzyme A disulphide reductase whose function suggests a role in intracellular redox and the oxidative stress response. Mol Microbiol 59(2):475–486

Braunshausen A, Seebeck FP (2011) Identification and characterization of the first ovothiol biosynthetic enzyme. J Am Chem Soc 133(6):1757–1759

Buchmeier N, Fahey RC (2006) The mshA gene encoding the glycosyltransferase of mycothiol biosynthesis is essential in *Mycobacterium tuberculosis* Erdman. FEMS Microbiol Lett 264(1):74–79

Buchmeier NA, Newton GL, Fahey RC (2006) A mycothiol synthase mutant of *Mycobacterium tuberculosis* has an altered thiol-disulfide content and limited tolerance to stress. J Bacteriol 188(17):6245–6252

Bzymek KP, Newton GL, Ta P, Fahey RC (2007) Mycothiol import by *Mycobacterium smegmatis* and function as a resource for metabolic precursors and energy production. J Bacteriol 189(19):6796–6805

Chi BK, Gronau K, Mader U, Hessling B, Becher D, Antelmann H (2011a) S-bacillithiolation protects against hypochlorite stress in *Bacillus subtilis* as revealed by transcriptomics and redox proteomics. Mol Cell Proteomics MCP 10(11):M111.009506-M009111.009506

Chi BK, Gronau K, Maeder U, Hessling B, Becher D, Antelmann H (2011b) S-bacillithiolation protects against hypochlorite stress in *Bacillus subtilis* as revealed by transcriptomics and redox proteomics. Mol Cell Proteomics. doi:10.1074/mcp.M1111.009506

Chi BK, Roberts AA, Huyen TT, Bäsell K, Becher D, Albrecht D, Hamilton CJ, Antelmann H (2012) S-bacillithiolation protects conserved and essential proteins against hypochlorite stress in firmicutes bacteria. Antioxid Redox Signal 18:1273–1295

Chi BK, Busche T, Laer KV, Basell K, Becher D, Clermont L, Seibold GM, Persicke M, Kalinowski J, Messens J, Antelmann H (2013a) Protein S-mycothiolation functions as redox-switch and thiol protection mechanism in *Corynebacterium glutamicum* under hypochlorite stress. Antioxid Redox Signal (in press). doi: 10.1089/ars.2013.5423

Chi BK, Roberts AA, Tran Thi Thanh H, Baesell K, Becher D, Albrecht D, Hamilton CJ, Antelmann H (2013b) S-bacillithiolation protects conserved and essential proteins against hypochlorite stress in firmicutes bacteria. Antioxid Redox Signal 18(11):1273–1295

Clement GE, Hartz TP (1971) Determination of the microscopic ionization constants. J Chem Educ 48(6):395

Dalle-Donne I, Rossi R, Giustarini D, Colombo R, Milzani A (2007) S-glutathionylation in protein redox regulation. Free Radical Biol Med 43(6):883–898

del Cardayre SB, Stock KP, Newton GL, Fahey RC, Davies JE (1998) Coenzyme A disulfide reductase, the primary low molecular weight disulfide reductase from *Staphylococcus aureus*—purification and characterization of the native enzyme. J Biol Chem 273(10):5744–5751

Derewenda U, Boczek T, Gorres KL, Yu M, Hung L-w, Cooper D, Joachimiak A, Raines RT, Derewenda ZS (2009) Structure and function of *Bacillus subtilis* YphP, a prokaryotic disulfide isomerase with a CXC catalytic motif. Biochemistry 48(36):8664–8671

Dormeyer M, Reckenfelderbaumer N, Ludemann H, Krauth-Siegel RL (2001) Trypanothione-dependent synthesis of deoxyribonucleotides by *Trypanosoma brucei* ribonucleotide reductase. J Biol Chem 276(14):10602–10606

Duan X-H, Tang Z-H, Guo X-Y (2010) Biosynthesis and function of glutathione in plant. Bull Bot Res 30(1):98–105

Eggeling L, Sahm H (1985) The formaldehyde dehydrogenase of *Rhodococcus erythropolis*, a trimeric enzyme requiring a cofactor and active with alcohols. Eur J Biochem 150(1):129–134

Etienne J, Gerbaud G, Fleurette J, Courvalin P (1991) Characterization of staphylococcal plasmids hybridizing with the fosfomycin resistance gene fosB. FEMS Microbiol Lett 84(1):119–122

Fahey RC, Newton GL (1983) Occurence of low molecular thiols in biological systems. In: Larsson A, Holmgren A, Mannervik B, Orrenius S (eds) Functions of glutathione biochemical, physiological, toxicological and clinical aspects. Raven Press, New York, pp 251–260

Fairlamb AH, Cerami A (1992) Metabolism and functions of trypanothione in the kinetoplastida. Annu Rev Microbiol 46(1):695–729

Fan F, Vetting MW, Frantom PA, Blanchard JS (2009) Structures and mechanisms of the mycothiol biosynthetic enzymes. Curr Opin Chem Biol 13(4):451–459

Fang Z, Roberts AA, Weidman K, Sharma SV, Claiborne A, Hamilton CJ, Dos Santos PC (2013) Cross-functionalities of Bacillus deacetylases involved in bacillithiol biosynthesis and bacillithiol-S-conjugate detoxification pathways. Biochem J 454:239–247

Feng J, Che YS, Milse J, Yin YJ, Liu L, Ruckert C, Shen XH, Qi SW, Kalinowski J, Liu SJ (2006) The gene ncgl2918 encodes a novel maleylpyruvate isomerase that needs mycothiol as cofactor and links mycothiol biosynthesis and gentisate assimilation in *Corynebacterium glutamicum*. J Biol Chem 281(16):10778–10785

Gaballa A, Newton GL, Antelmann H, Parsonage D, Upton H, Rawat M, Claiborne A, Fahey RC, Helmann JD (2010) Biosynthesis and functions of bacillithiol, a major low-molecular-weight thiol in Bacilli. Proc Nat Acad Sci 107(14):6482–6486

Gaballa A, Antelmann H, Hamilton CJ, Helmann JD (2013) Regulation of *Bacillus subtilis* bacillithiol biosynthesis operons by Spx. Microbiology (Reading, England) 159(Pt 10):2025–2035

Ghezzi P (2005) Review: regulation of protein function by glutathionylation. Free Radical Res 39(6):573–580

Harris DR, Ward DE, Feasel JM, Lancaster KM, Murphy RD, Mallet TC, Crane EJ (2005) Discovery and characterization of a Coenzyme A disulfide reductase from *Pyrococcus horikoshii*. FEBS J 272(5):1189–1200

Hayes JD, Flanagan JU, Jowsey IR (2005) Glutathione transferases. Annu Rev Pharmacol Toxicol 45(1):51–88

Holsclaw CM, Muse Iii WB, Carroll KS, Leary JA (2011) Mass spectrometric analysis of mycothiol levels in wild-type and mycothiol disulfide reductase mutant *Mycobacterium smegmatis*. Int J Mass Spectrom 305(2–3):151–156

Huang X, Kocabas E, Hernick M (2011) The activity and cofactor preferences of N-Acetyl-1-d-myo-inosityl-2-amino-2-deoxy-α-d-glucopyranoside deacetylase (MshB) change depending on environmental conditions. J Biol Chem 286(23):20275–20282

Ishikawa Y, Israel SE, Melville DB (1974) Participation of an intermediate sulfoxide in enzymatic thiolation of imidazole ring of hercynine to form ergothioneine. J Biol Chem 249(14):4420–4427

Jang JH, Aruoma OI, Jen LS, Chung HY, Surh YJ (2004) Ergothioneine rescues PC12 cells from beta-amyloid-induced apoptotic death. Free Radical Biol Med 36(3):288–299

Jocelyn P (1972) Biochemistry of the SH group. Academic Press, London

Jothivasan VK, Hamilton CJ (2008) Mycothiol: synthesis, biosynthesis and biological functions of the major low molecular weight thiol in actinomycetes. Nat Prod Rep 25(6):1091–1117

Kaluzna A, Bartosz G (1997) Transport of glutathione S-conjugates in *Escherichia coli*. Biochem Mol Biol Int 43(1):161–171

Kawano H, Higuchi F, Mayumi T, Hama T (1982) Studies on ergothioneine. 7. Some effects of ergothioneine on glycolytic metabolism in red-blood-cells from rats. Chem Pharm Bull 30(7):2611–2613

Keire DA, Robert JM, Rabenstein DL (1992a) Microscopic protonation equilibria and solution conformations of coenzyme A and coenzyme A disulfides. J Org Chem 57(16):4427–4431

Keire DA, Strauss E, Guo W, Noszal B, Rabenstein DL (1992b) Kinetics and equilibria of thiol/disulfide interchange reactions of selected biological thiols and related molecules with oxidized glutathione. J Org Chem 57(1):123–127

Kim E-K, Cha C-J, Cho Y-J, Cho Y-B, Roe J-H (2008) Synthesis of gamma-glutamylcysteine as a major low-molecular-weight thiol in lactic acid bacteria Leuconostoc spp. Biochem Biophys Res Commun 369(4):1047–1051

Koledin T, Newton GL, Fahey RC (2002) Identification of the mycothiol synthase gene (mshD) encoding the acetyltransferase producing mycothiol in actinomycetes. Arch Microbiol 178(5):331–337

Lee EJ, Karoonuthaisiri N, Kim HS, Park JH, Cha CJ, Kao CM, Roe JH (2005) A master regulator sigma(B) governs osmotic and oxidative response as well as differentiation via a network of sigma factors in *Streptomyces coelicolor*. Mol Microbiol 57(5):1252–1264

Lee JW, Soonsanga S, Helmann JD (2007) A complex thiolate switch regulates the *Bacillus subtilis* organic peroxide sensor OhrR. Proc Natl Acad Sci USA 104(21):8743–8748

Luba J, Charrier V, Claiborne A (1999) Coenzyme A-disulfide reductase from *Staphylococcus aureus*: evidence for asymmetric behavior on interaction with pyridine nucleotides. Biochemistry 38(9):2725–2737

Mallett TC, Wallen JR, Karplus PA, Sakai H, Tsukihara T, Claiborne A (2006) Structure of coenzyme A-disulfide reductase from *Staphylococcus aureus* at 1.54 angstrom resolution. Biochemistry 45(38):11278–11289

Martinoia E, Grill E, Tommasini R, Kreuz K, Amrhein N (1993) ATP-dependent glutathione S-conjugate 'export' pump in the vacuolar membrane of plants. Nature 364(6434):247–249

Masip L, Veeravalli K, Georgioui G (2006) The many faces of glutathione in bacteria. Antioxid Redox Signal 8(5–6):753–762

Meister A (1988) On the discovery of glutathione. Trends Biochem Sci 13(5):185–188

Miller CC, Rawat M, Johnson T, Av-Gay Y (2007) Innate protection of *Mycobacterium smegmatis* against the antimicrobial activity of nitric oxide is provided by mycothiol. Antimicrob Agents Chemother 51(9):3364–3366

Misiti F, Castagnola M, Zuppi C, Giardina B, Messana I (2001) Role of ergothioneine on S-nitrosoglutathione catabolism. Biochem J 356(3):799–804

Motohash N, Mori I, Sugiura Y, Tanaka H (1974) Metal-complexes of ergothioneine. Chem Pharm Bull 22(3):654–657

Motohashi N, Mori I, Sugiura Y (1976) C-13-nuclear magnetic-resonance and Raman spectroscopic studies on ionization and mercury complex of ergothioneine. Chem Pharm Bull 24(8):1737–1741

Moutiez M, Mezianecherif D, Aumercier M, Sergheraert C, Tartar A (1994) Compared reactivities of trypanothione and glutathione in conjugation reactions. Chem Pharm Bull 42(12):2641–2644

Moutiez M, Aumercier M, Schoneck R, Mezianecherif D, Lucas V, Aumercier P, Ouaissi A, Sergheraert C, Tartar A (1995) Purification and characterization of a trypanothione-glutathione thioltransferase from *Trypanosoma cruzi*. Biochem J 310:433–437

Müller M, Meijer C, Zaman GJ, Borst P, Scheper RJ, Mulder NH, de Vries EG, Jansen PL (1994) Overexpression of the gene encoding the multidrug resistance-associated protein results in increased ATP-dependent glutathione S-conjugate transport. Proc Natl Acad Sci 91(26):13033–13037

Newton GL, Fahey RC (2002) Mycothiol biochemistry. Arch Microbiol 178(6):388–394

Newton GL, Javor B (1985) Gamma-Glutamylcysteine and thiosulfate are the major low-molecular-weight thiols in halobacteria. J Bacteriol 161(1):438–441

Newton GL, Bewley CA, Dwyer TJ, Horn R, Aharonowitz Y, Cohen G, Davies J, Faulkner DJ, Fahey RC (1995) The structure of U17 isolated from *Streptomyces clavuligerus* and its properties as an antioxidant thiol. Eur J Biochem 230(2):821–825

Newton GL, Arnold K, Price MS, Sherrill C, Delcardayre SB, Aharonowitz Y, Cohen G, Davies J, Fahey RC, Davis C (1996) Distribution of thiols in microorganisms: mycothiol is a major thiol in most actinomycetes. J Bacteriol 178(7):1990–1995

Newton GL, Unson MD, Anderberg SJ, Aguilera JA, Oh NN, delCardayre SB, Av-Gay Y, Fahey RC (1999) Characterization of *Mycobacterium smegmatis* mutants defective in 1-D-myo-inosityl-2-amino-2-deoxy-alpha-D-glucopyranoside and mycothiol biosynthesis. Biochem Biophys Res Commun 255(2):239–244

Newton GL, Av-Gay Y, Fahey RC (2000a) N-acetyl-1-D-myo-inosityl-2-amino-2-deoxy-alpha-D-glucopyranoside deacetylase (MshB) is a key enzyme in mycothiol biosynthesis. J Bacteriol 182(24):6958–6963

Newton GL, Av-Gay Y, Fahey RC (2000b) A novel mycothiol-dependent detoxification pathway in mycobacteria involving mycothiol S-conjugate amidase. Biochemistry 39(35):10739–10746

Newton GL, Ta P, Fahey RC (2005) A mycothiol synthase mutant of *Mycobacterium smegmatis* produces novel thiols and has an altered thiol redox status. J Bacteriol 187(21):7309–7316

Newton GL, Ko M, Ta PL, Av-Gay Y, Fahey RC (2006a) Purification and characterization of *Mycobacterium tuberculosis* 1D-myo-inosityl-2-acetamido-2-deoxy-alpha-D-glucopyranoside deacetylase, MshB, a mycothiol biosynthetic enzyme. Protein Expr Purif 47(2):542–550

Newton GL, Ta P, Bzymek KP, Fahey RC (2006b) Biochemistry of the initial steps of mycothiol biosynthesis. J Biol Chem 281(45):33910–33920

Newton GL, Buchmeier N, Fahey RC (2008a) Biosynthesis and functions of mycothiol, the unique protective thiol of Actinobacteria. Microbiol Mol Biol Rev 72(3):471–494

Newton GL, Jensen PR, MacMillan JB, Fenical W, Fahey RC (2008b) An N-acyl homolog of mycothiol is produced in marine actinomycetes. Arch Microbiol 1–11

Newton GL, Rawat M, La Clair JJ, Jothivasan VK, Budiarto T, Hamilton CJ, Claiborne A, Helmann JD, Fahey RC (2009) Bacillithiol is an antioxidant thiol produced in Bacilli. Nat Chem Biol 5(9):625–627

Newton GL, Leung SS, Wakabayashi JI, Rawat M, Fahey RC (2011) The DinB superfamily includes novel mycothiol, bacillithiol, and glutathione S-transferases. Biochemistry 50(49):10751–10760

Nicely NI, Parsonage D, Paige C, Newton GL, Fahey RC, Leonardi R, Jackowski S, Mallett TC, Claiborne A (2007) Structure of the type III pantothenate kinase from *Bacillus anthracis* at 2.0 & #197; Resolution: implications for coenzyme A-dependent redox biology. Biochemistry 46(11):3234–3245

Ophem PW, Beeumen J, Duine JA (1992) NAD-linked, factor-dependent formaldehyde dehydrogenase or trimeric, zinc-containing, long-chain alcohol dehydrogenase from *Amycolatopsis methanolica*. Eur J Biochem 206(2):511–518

Ordóñez E, Van Belle K, Roos G, De Galan S, Letek M, Gil JA, Wyns L, Mateos LM, Messens J (2009) Arsenate reductase, mycothiol, and mycoredoxin concert thiol/disulfide exchange. J Biol Chem 284(22):15107–15116

Palumbo A, Misuraca G, Dischia M, Donaudy F, Prota G (1984) Isolation and distribution of 1-methyl-5-thiol-L-histidine disulfide and a related metabolite in eggs from echinoderms. Comp Biochem Physiol B-Biochem Mol Biol 78(1):81–83

Park S, Imlay JA (2003) High levels of intracellular cysteine promote oxidative DNA damage by driving the fenton reaction. J Bacteriol 185(6):1942–1950

Park J-H, Roe J-H (2008) Mycothiol regulates and is regulated by a thiol-specific antisigma factor RsrA and & #x03C3;R in *Streptomyces coelicolor*. Mol Microbiol 68(4):861–870

Park JH, Cha CJ, Roe JH (2006) Identification of genes for mycothiol biosynthesis in *Streptomyces coelicolor* A3(2). J Microbiol 44(1):121–125

Parsonage D, Newton GL, Holder RC, Wallace BD, Paige C, Hamilton CJ, Dos Santos PC, Redinbo MR, Reid SD, Claiborne A (2010) Characterization of the N-Acetyl-Î ± -d-glucosaminyl l-malate synthase and deacetylase functions for bacillithiol biosynthesis in *Bacillus anthracis*. Biochemistry 49(38):8398–8414

Patel MP, Blanchard JS (1999) Expression, purification, and characterization of *Mycobacterium tuberculosis* mycothione reductase. Biochemistry 38(36):11827–11833

Patel MP, Blanchard JS (2001) Mycobacterium tuberculosis mycothione reductase: pH dependence of the kinetic parameters and kinetic isotope effects. Biochemistry 40(17):5119–5126

Poether D-C, Gierok P, Harms M, Mostertz J, Hochgaefe F, Antelmann H, Hamilton CJ, Borovok I, Lalk M, Aharonowitz Y, Hecker M (2013) Distribution and infection-related functions of bacillithiol in *Staphylococcus aureus*. Int J Med Microbiol 303(3):114–123

Pompella A, Visvikis A, Paolicchi A, De Tata V, Casini AF (2003) The changing faces of glutathione, a cellular protagonist. Biochem Pharmacol 66(8):1499–1503

Rabenstein DL (1973) Nuclear magnetic resonance studies of the acid-base chemistry of amino acids and peptides. I. Microscopic ionization constants of glutathione and methylmercury-complexed glutathione. J Am Chem Soc 95(9):2797–2803

Rajkarnikar A, Strankman A, Duran S, Vargas D, Roberts AA, Barretto K, Upton H, Hamilton CJ, Rawat M (2013) Analysis of mutants disrupted in bacillithiol metabolism in *Staphylococcus aureus*. Biochem Biophys Res Commun 436(2):128–133

Rawat M, Newton GL, Ko M, Martinez GJ, Fahey RC, Av-Gay Y (2002) Mycothiol-deficient *Mycobacterium smegmatis* mutants are hypersensitive to alkylating agents, free radicals, and antibiotics. Antimicrob Agents Chemother 46(11):3348–3355

Rawat M, Uppal M, Newton G, Steffek M, Fahey RC, Av-Gay Y (2004) Targeted mutagenesis of the *Mycobacterium smegmatis* mca gene, encoding a mycothiol-dependent detoxification protein. J Bacteriol 186(18):6050–6058

Rawat M, Johnson C, Cadiz V, Av-Gay Y (2007) Comparative analysis of mutants in the mycothiol biosynthesis pathway in *Mycobacterium smegmatis*. Biochem Biophys Res Commun 363(1):71–76

Rengarajan J, Bloom BR, Rubin EJ (2005) Genome-wide requirements for *Mycobacterium tuberculosis* adaptation and survival in macrophages. Proc Nat Acad Sci USA 102(23):8327–8332

Rigsby RE, Brown DW, Dawson E, Lybrand TP, Armstrong RN (2007) A model for glutathione binding and activation in the fosfomycin resistance protein, FosA. Arch Biochem Biophys 464(2):277–283

Roberts AA, Sharma SV, Strankman AW, Duran SR, Rawat M, Hamilton CJ (2013) Mechanistic studies of FosB: a divalent metal-dependent bacillithiol-S-transferase that mediates fosfomycin resistance in *Staphylococcus aureus*. Biochem J 451(1):69–79

Rossi F, Nardi G, Palumbo A, Prota G (1985) 5-thiolhistidine, a new amino-acid from eggs of *Octopus vulgaris*. Comp Biochem Physiol B-Biochem Mol Biol 80(4):843–845

Rowley DA, Halliwell B (1982) Superoxide-dependent formation of hydroxyl radicals in the presence of thiol compounds. FEBS Lett 138(1):33–36

Rubino FM, Pitton M, Caneva E, Pappini M, Colombi A (2008) Thiol-disulfide redox equilibria of glutathione metaboloma compounds investigated by tandem mass spectrometry. Rapid Commun Mass Spectrom 22(23):3935–3948

Sareen D, Newton GL, Fahey RC, Buchmeier NA (2003) Mycothiol is essential for growth of *Mycobacterium tuberculosis* Erdman. J Bacteriol 185(22):6736–6740

Seebeck FP (2010) In vitro reconstitution of mycobacterial ergothioneine biosynthesis. J Am Chem Soc 132(19):6632–+

Selmanreimer S, Duhe RJ, Stockman BJ, Selman BR (1991) L-1-N-methyl-4-mercaptohistidine disulfide, a potential endogenous regulator in the redox control of chloroplast coupling factor-I in Dunaliella. J Biol Chem 266(1):182–188

Setlow B, Setlow P (1977) Levels of acetyl coenzyme A, reduced and oxidized coenzyme A, and coenzyme A in disulfide linkage to protein in dormant and germinated spores and growing and sporulating cells of *Bacillus megaterium*. J Bacteriol 132(2):444–452

Sharma SV, Arbach M, Roberts AA, Macdonald CJ, Groom M, Hamilton CJ (2013) Biophysical features of bacillithiol, the glutathione surrogate of *Bacillus subtilis* and other firmicutes. ChemBioChem 14(16):2160–2168

Shen XH, Jiang CY, Huang Y, Liu ZP, Liu SJ (2005) Functional identification of novel genes involved in the glutathione-independent gentisate pathway in *Corynebacterium glutamicum*. Appl Environ Microbiol 71(7):3442–3452

Smith K, Borges A, Ariyanayagam MR, Fairlamb AH (1995) Glutathionylspermidine metabolism in *Escherichia coli*. Biochem J 312(2):465–469

Soonsanga S, Lee JW, Helmann JD (2008) Oxidant-dependent switching between reversible and sacrificial oxidation pathways for *Bacillus subtilis* OhrR. Mol Microbiol 68(4):978–986

Spies HSC, Steenkamp DJ (1994) Thiols of intracellular pathogens—Identification of ovothiol-A in *Leishmania donovani* and structural-analysis of a novel thiol from *Mycobacterium bovis*. Eur J Biochem 224(1):203–213

Stanovnik B, Tišler M (1964) Dissociation constants and structure of ergothioneine. Anal Biochem 9(1):68–74

Steenkamp DJ, Spies HSC (1994) Identification of a major low-molecular-mass thiol of the trypanosomatid crithidia-fasciculata as ovothiol-A. Facile isolation and structural-analysis of the bimane derivative. Eur J Biochem 223(1):43–50

Steenkamp DJ, Weldrick D, Spies HSC (1996) Studies on the biosynthesis of ovothiol A. Eur J Biochem 242(3):557–566

Steffek M, Newton GL, Av-Gay Y, Fahey RC (2003) Characterization of *Mycobacterium tuberculosis* mycothiol S-conjugate amidase. Biochemistry 42(41):12067–12076

Stewart MJG, Jothivasan VK, Rowan AS, Wagg J, Hamilton CJ (2008) Mycothiol disulfide reductase: solid phase synthesis and evaluation of alternative substrate analogues. Org Biomol Chem 6(2):385–390

Sundquist AR, Fahey RC (1989) The function of gamma-glutamylcysteine and bis-gamma-glutamylcystine reductase in *Halobacterium halobium*. J Biol Chem 264(2):719–725

Ta P, Buchmeier N, Newton GL, Rawat M, Fahey RC (2011) Organic hydroperoxide resistance protein and ergothioneine compensate for loss of mycothiol in *Mycobacterium smegmatis* mutants. J Bacteriol 193(8):1981–1990

Takeshima S, Inoue T (1996) The strong fluorescence-quenching effect of ergothioneine. Chem Pharm Bull 44(1):201–203

Turner E, Klevit R, Hopkins PB, Shapiro BM (1986) Ovothiol: a novel thiohistidine compound from sea-urchin eggs that confers NAD(P)H-O2 oxidoreductase activity on ovoperoxidase. J Biol Chem 261(28):3056–3063

Turner E, Hager L, Shapiro B (1988) Ovothiol replaces glutathione peroxidase as a hydrogen peroxide scavenger in sea urchin eggs. Science 242(4880):939–941

Van Laer K, Buts L, Foloppe N, Vertommen D, Van Belle K, Wahni K, Roos G, Nilsson L, Mateos LM, Rawat M, van Nuland NAJ, Messens J (2012) Mycoredoxin-1 is one of the missing links in the oxidative stress defence mechanism of Mycobacteria. Mol Microbiol 86(4):787–804

Veeravalli K, Boyd D, Iverson BL, Beckwith J, Georgiou G (2011) Laboratory evolution of glutathione biosynthesis reveals natural compensatory pathways. Nat Chem Biol 7(2):101–105
Villadangos AF, Van Belle K, Wahni K, Tamu Dufe V, Freitas S, Nur H, De Galan S, Gil JA, Collet J-F, Mateos LM, Messens J (2011) Corynebacterium glutamicum survives arsenic stress with arsenate reductases coupled to two distinct redox mechanisms. Mol Microbiol 82(4):998–1014
Vogt RN, Spies HSC, Steenkamp DJ (2001) The biosynthesis of ovothiol A (N1-methyl-4-mercaptohistidine). Eur J Biochem 268(20):5229–5241
Vogt RN, Steenkamp DJ, Zheng RJ, Blanchard JS (2003) The metabolism of nitrosothiols in the mycobacteria: identification and characterization of S-nitrosomycothiol reductase. Biochem J 374:657–665
Wallen JR, Paige C, Mallett TC, Karplus PA, Claiborne A (2008) Pyridine nucleotide complexes with *Bacillus anthracis* coenzyme A-disulfide reductase: a structural analysis of dual NAD(P)H specificity. Biochemistry 47(18):5182–5193
Wang W, Ballatori N (1998) Endogenous glutathione conjugates: occurrence and biological functions. Pharmacol Rev 50(3):335–356
Wang R, Yin YJ, Wang F, Li M, Feng J, Zhang HM, Zhang JP, Liu SJ, Chang WR (2007) Crystal structures and site-directed mutagenesis of a mycothiol-dependent enzyme reveal a novel folding and molecular basis for mycothiol-mediated maleylpyruvate isomerization. J Biol Chem 282(22):16288–16294
Weaver KH, Rabenstein DL (1995) Thiol/disulfide exchange reactions of ovothiol A with glutathione. J Org Chem 60(6):1904–1907
Winterbourn CC (1993) Superoxide as an intracellular radical sink. Free Radical Biol Med 14(1):85–90

Authors Biography

Chris J. Hamilton graduated from the University of Liverpool (1994) and obtained his PhD from the University of Exeter in 1997 working on the synthesis of biologically stable nor-carbovir triphosphate analogs as inhibitors of HIV reverse transcriptase. This was followed by postdoctoral appointments at the University of St Andrews (protein phosphatase enzymology), University of Dundee (synthesis and mechanistic studies of lunaria alkaloids as trypanothione reductase inhibitors), and the University of East Anglia (carbohydrate chemistry/biochemistry). Shortly after receiving a Wellcome Trust SHOWCASE award, he took up a post as Lecturer in Organic Chemistry at Queen's University Belfast in 2003 before returning to the University of East Anglia in 2007 where he is currently a reader in medicinal chemistry.

Chris' research interests reside at the Chemistry/Biology interface with particular interests in (1) Chemical and biochemical functions of low molecular weight thiols in Gram-positive bacteria; (2) Redox mechanisms of bioactive organosulfur metabolites from garlic; (3) Mechanistic enzymology.

Miriam Arbach (born 1987) has been trained as a biologist (specialized in plant physiology and plant biochemistry) at the RWTH Aachen University where she completed her Bachelor studies in 2009. Subsequently, in 2010, she became an Early Stage Researcher within the EU Marie Curie Initial Training Network "RedCat" and started her PhD program at the University of East Anglia (UEA) under the joint supervision of Dr. Chris Hamilton (UEA) and Dr. Murree Groom (ECOspray Ltd).

Miriam's PhD studies explore how garlic-derived diallyl polysulfane metabolites perturb cellular redox processes in soil bacteria and nematodes, and how such substances may find applications as nematicides in food crop protection. Miriam is currently about to complete her PhD thesis at the University of East Anglia.

Murree Groom obtained a joint honors degree in Chemistry and Biochemistry from UCNW in 1981 and was sponsored by the then MAFF Chief Scientist Group in the UK to pursue a PhD thesis on the cause and control of cavity spot in carrots. He completed his thesis in 1988 after 4 years of research combined at The Scottish Crop Research Institute and University of Dundee under the supervision of Prof. John Raven FRS. An appointment as a Chemist to the Diversey Corporate Technology Centre followed, where Murree worked on the formulation and manufacture of biocides for the dairy industry. Murree returned back to Norfolk as an agronomist for W.H. Knights and Sons where he developed and managed on farm R&D capabilities in pesticide analysis, entomology, nematology, plant pathology, and integrated pest management.

In 1997, Murree joined ECOspray and now directs research and development with specific interests in biopesticides produced from garlic extracts. ECOspray Ltd is the only company in the EU with regulatory dossiers that support authorization of garlic-based extracts as nematicides and insecticides. Since being with ECOspray Ltd, Murree has been instrumental in co-supervising several research students and securing three patents that all relate to the processing and chemistry of garlic. The present focus of Murree's work is construction of synthetic formulations based on garlic chemistry that could replace the materials derived from processing whole garlic plants.

Part XI
Connecting Section Between Chapters 11 and 12

After considering a range of intracellular thiol compounds and proteins, and highlighting various reactive sulfur species able to interact with these thiols, we will now explore how this kind of effective and often selective reactivity can be turned into a useful application. The next chapter will therefore consider the topic of redox active secondary metabolites from a rather different industrial perspective. We leave behind academic research for a moment and consider how research can be transformed into a commercially viable product. Here, we explore how products and insights obtained by research are protected by patents, how an economically viable product is developed, and subsequently licensed for practical uses, and what it means to set up a new company and subsequently stir it through the rocks and cliffs that are omnipresent during the first few years after establishing such a company. We also witness the rather merciless, competitive, and often global environment surrounding the natural products industry, and how it is possible for a small company to combine research with the continuous development of new and refined products in order to stay ahead of the game and in the market.

While it is possible to deal with these issues from the perspective of an economist, we have decided against yet another purely academic—and predictably complicated—contribution. For this chapter, we have rather chosen Awais Anwar and his colleagues from ECOSpray Ltd. UK as authors. Having been involved in 'company business' at ECOSpray for many years, these authors are able to provide a firsthand view on how natural products, in this case derived from garlic and onions, can be turned into a commercial success. While the story they tell is based on their own experiences in one specific company, many aspects are easily transferable to other companies dealing with different natural products. Indeed, we will consider a similar story related to a completely different natural-turned-commercial product later on, and the parallels will be obvious.

As we will see, one key aspect of all of these stories is the importance of conducting research and engaging with academia in order to gain an economic advantage. Indeed, this balance of mutual needs and benefits forms the bracket that holds natural products research and product development firmly together.

Chapter 12
How to Turn the Chemistry of Garlic into a 'Botanical' Pesticide

Awais Anwar, Murree Groom, Miriam Arbach and Chris J. Hamilton

Keywords ECOspray • EU market • Garlic • Organosulfur compounds • Plant protection

1 Introduction

In the near future the demand for greatly increased agricultural production and use of rural land for human development must be integrated with the rapidly increasing need for food for the current global population of 7 billion people (2011), expected to swell further by 2.5 billion in the next three decades. Achieving food sufficiency in a sustainable manner is a major and imperative challenge for farmers, agro-industries, researchers, and governments (Schillhorn Van Veen 1999). The intensification of agriculture to fulfill the need for food has increased the number of insect, fungal, and nematode pest species attacking different crops and their simultaneous resistance to many crop protection products. The combined effect of these two factors increases the risk of lost production, balanced to some extent by increased cycles of intensification. A more recent factor affecting food production, possibly linked to increased global human population growth, is the effect of climate change on resource availability and the more subtle impact that climate change is exerting on ecosystems that in the main support the present agricultural methodology.

A. Anwar (✉) · M. Groom · M. Arbach
ECOspray Limited, Grange Farm Hilborough, Thetford, Norfolk IP26 5BT, UK
e-mail: awais@ecospray.com

M. Arbach · C. J. Hamilton
School of Pharmacy, University of East Anglia, Norwich Research Park,
Norwich NR4 7TJ, UK

C. Jacob et al. (eds.), *Recent Advances in Redox Active Plant and Microbial Products*,
DOI: 10.1007/978-94-017-8953-0_12,

A very recent example of climate change affecting food production is the water restrictions imposed in southern and eastern England in 2012. Some farmers in the affected regions reduced areas of production to match the likely availability of water, or transferred production further north to areas where water was more readily available. At the same time, in March 2012, record temperatures were measured in Scotland on three consecutive days.[1]

Synthetic pesticides play a major role in crop protection programs and have immensely benefited mankind. Nevertheless, their indiscriminate and prophylactic use has resulted in the development of resistance by various pests (such as insects, weeds, etc.,), the resurgence and outbreak of new pests exploiting ecological niches left by perturbations in ecosystems, the toxicity to non-target organisms and hazardous effects on the environment, ultimately endangering the sustainability of entire ecosystems (Jeyasankar and Jesudasan 2005).

In recent years the EU has therefore implemented a fundamental reform of the Common Agricultural Policy (CAP) focusing on the environment, food safety, and animal welfare standards, imposing the rule that farming activity implements *integrated pest management* (IPM) that should deliver both, favorable agricultural and environmental outcomes (Schillhorn Van Veen 1999).

Due to unfavorable environmental side effects and health concerns, many synthetic carbamate, organophosphate, and organophthalide-based pesticides have therefore been banned (Council Directive 91/414/EEC) or are currently under re-evaluation (Regulation 2009/1107/EC & Directive 2009/128/EC).[2] The consequences of the sudden loss of such a wide range of crop protection products, with applications to many food crops, are unclear in the longer term.

A caveat in delivering this policy is that the agro-chemical industry does not equally distribute the costs for research and registration on all types of pesticides. As an example, direct research and expenditure development for many of the widely consumed fruits and vegetables in the EU is only a fraction of the expenditure on the global 'broad area' crops, such as cereals and rice. Lack of availability of many crop protection actives for minor crops has left some horticultural crops with only one active available to combat a particular pest or disease and in some situations no appropriate crop protection product at all. For example, much of the cabbage root fly control in the EU is dependent on chlorpyriphos. Carrot fly control is heavily reliant on Lambda-cyhalothrin and there are currently no suitable nematicides approved in Italy for use on carrots and only one appropriate nematicide (Vydate®) available for this purpose in the UK.

In fact, the development of new nematicides is rarely supported by the agrochemical industry even though in some cases, such as in the Netherlands, they represent more than 60 % of the total pesticides used in agriculture. This is due to the fact that nematodes are a rather difficult target to control and the economic cost of research and registration is an enormous hurdle for any prospective new synthetic

[1] UK Met Office (http://www.metoffice.gov.uk).

[2] Information about the Directives can be found on EC website in the area of EUR-Lex (Access to EULaw), http://eur-lex.europa.eu/en/index.htm.

nematicide to overcome. Using nematicide as an example, all the above facts come together and underline the increasing urgency for new and alternative pest control methods (Chitwood 2002).

An interesting approach of searching for bio-rational pesticides is screening naturally occurring compounds from plants (Isman 2006, 2008). Plants, as long-lived stationary organisms, must resist attackers over their lifetime, so they produce and exude constituents from plant secondary metabolism (PSMs) that play an important role in their defense (see also Chaps. 2, 9 and 10).

If developed for use as pesticides by the plants themselves, PSMs may have wider applications in weed and pest management, or they can be used as model compounds for the development of chemically synthesized derivatives. Many of such natural activities are 'environmentally friendly', pose few risks, if any, to humans and animals, have a selective mode of action, and avoid the emergence of resistant strains of pest species. As a result, these 'natural pesticides' are more likely to be used safely in Integrated Pest Management (IPM) (Isman 2006). Furthermore, they may also be suitable for organic food production.

There is extensive literature on the possible use of plant-derived preparations of crude or refined extracts (Brooks 1998; Copping 1999; Koul and Dhaliwal 2001) in various areas of crop protection (e.g., against insects, fungi, nematodes, bacteria, weeds). Not surprisingly, plant-derived fermentations, extracts, oils, and powders have long been used in pest management, in gardening, and (small scale) in agriculture. Nonetheless, it is now mandatory to attribute the efficacy of botanicals to specifically identified constituent compound(s) in order to delineate the mechanisms of bioactivity biologically and biochemically, and to provide adequate characterization, and in the field of medicine, to fully exploit the therapeutic potential of such extracts (Akhtar and Mahmood 1994). Recently, the European commission laid out guidelines concerning the data requirements for registering plant protection products, which focus on intensive characterization of material (SANCO/11470/2012– rev. 8).

Within this context, recent research has shown that naturally occurring sulfur compounds found in garlic (*Allium sativum*), in particular, provide a range of interesting leads for new antimicrobial and anticancer agents and may play a role as cancer cell selective cytotoxins and also as 'green'[3] pesticides (Cerella et al. 2009; Jacob 2006; Jacob and Anwar 2009).

2 Garlic and Its Chemistry

With a history going back over 4,000 years, garlic (*Allium sativum*) is well known as a herb and spice and is widely used around the world for its pungent flavor as a seasoning or condiment. Garlic is a fundamental component in many dishes

[3] The label 'green' in medicine and pesticide development therefore provides not (only) a social or political advantage, but also reflects the fact that natural products are often complicated chemicals well suited to fulfil a certain biological task. In many instances, they are even superior to 'artificial' pesticides or drugs.

of various regions around the globe. Garlic belongs to the family of *Alliums* that also includes other well-known plants like onions, leek, and chive. Typical for the family of *Alliums* is their high content of organosulfur compounds (OSCs) and, as a result, their characteristic taste and smell. The most abundant organosulfur metabolite produced by garlic plants is the non-protein organic amino acid alliin (*S*-allyl-L-cysteine sulfoxide), which is stored in large amounts in the cytosol of the garlic cells (5–14 mg/g fresh dry weight) (Block 2010).

Interestingly, garlic in its intact form, is odorless until it is crushed (or cells become otherwise damaged). Upon damaging the cellular structure in the plant, the enzyme alliinase is released from the vacuoles and catalyzes the rapid conversion of alliin into the chemically very reactive molecule allicin (1) (diallylthiosulfinate) (Cavallito and Bailey 1944). Allicin is an extraordinary bioactive molecule and is also ultimately responsible for the characteristic smell of freshly cut garlic (Chaps. 9 and 10 for more information).

From the perspective of the plant, this is a very effective defense mechanism against pathogens. At the same time, the formation of the active compound upon cell damage ensures that there is only a local formation of the active for a short period of time which ensures that allicin does not become toxic to the plant itself.

Because of its high chemical reactivity allicin (1) undergoes a cascade of further chemical rearrangements leading to other organosulfur molecules such as ajoene (9), dithiines (7, 8), and diallylpolysulfanes (2–6) whose chemical structures are shown in Fig. 1a (see also Chap. 10). In addition to large amounts of allicin, other sulfinates (methyl and propyl substituted) are also formed by the activity of alliinase, and, as a result, methyl and propyl polysulfanes are produced on degradation. All of these OSCs have been tested for various biological activities, and the highest activity has usually been found for the diallylpolysulfanes. Additionally, the diallyl substituted polysulfanes have a more pleasant smell than the analog dimethyl and allylmethyl compounds, which makes them a better candidate for the formulation of a natural nematicide.

The allylpolysulfanes occurring in aged garlic or garlic oils are mostly formed with two allyl groups at their ends and a varying number of sulfur atoms in a chain between them (see 2–6 in Fig. 1a). From diallylmonosulfide (2) up to the molecule containing nine sulfur atoms in the chain, all diallylpolysulfanes can be found and analyzed in garlic oil samples (Block 2010). Typically, diallyltrisulfide (4) and diallyltetrasulfide (5) are the most abundant, followed by diallyl disulfide (3) and relatively low concentrations of higher diallylpolysulfanes (6 and others). Having more sulfur atoms in the chain progressively destabilizes the molecule and there are stability issues with isolated diallyl penta sulfide and hexasulfide (longer chain diallylpolysulfanes have not been isolated as single molecules). This progressive instability makes high sulfur chain length polysulfanes difficult to handle as individual molecules. Interestingly, as the diallylpolysulfanes seem to form an equilibrium of molecules with different sulfur chain length (a hypothetical mechanism is shown in Fig. 1b), a mixture of these molecules is reasonably stable at room temperature. Such an 'equilibrated mixture' may, for instance, also be found in in garlic oils and in the formulated nematicides based on them.

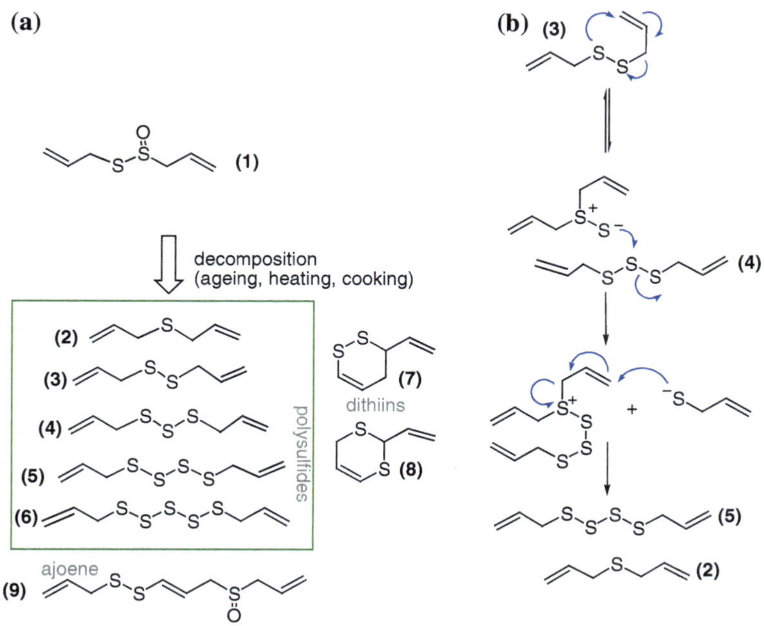

Fig. 1 **a** A selection of the sulfur chemistry found in garlic. Allicin (*1*), diallyl sulfide (DAS1 (*2*)), diallyl disulfide (DAS2 (*3*)), diallyltrisulfide (DAS3 (*4*)), diallyltetrasulfide (DAS4 (*5*)), diallylpentasulfide (DAS5 (*6*)), 3-vinyl-3,4-dihydro-1,2-dithiin (*7*), 2-vinyl-2,4-dihydro-1,3-dithiin (*8*), E-ajoene (*9*). **b** An example for a possible mechanism for rearrangements of diallylpolysulfanes to form an equilibrium of different chain length (poly)sulfanes

Another interesting feature of the diallylpolysulfanes is their progressive insolubility. Diallylpolysulfanes with higher chain lengths are very hydrophobic with low solubility in water (Schneider et al. 2011). This highlights another major challenge that has to be met when formulating the diallylpolysulfanes into an active nematicide, as solubility in water and controlled release of actives is critical for any agricultural applications.

2.1 *A Brief Look at Possible Modes of Action of Diallylpolysulfanes*

During the last decade research has shown that this unique pungent herb is very effective against different strains of pathogenic bacteria as well as different types of cancer, exhibiting increased activity for longer chain polysulfanes (4 > 3 > 2) with no data yet available for Diallylpentasulfide and diallylhexasulfide (Anwar et al. 2008; Cerella et al. 2009).

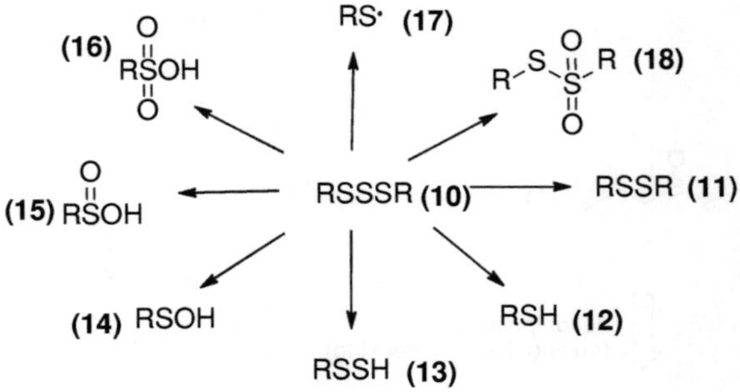

Fig. 2 Different RSS that can be formed from polysulfanes (e.g., a trisulfide (*10*)), either by reacting with thiols or by undergoing redox reactions ((*11*) disulfide (*12*) thiol (*13*) hydroperthiol (*14*) sulfenic acid (*15*) sulfinic acid (*16*) sulfonic acid (*17*) thiyl radical (*18*) thiosulfonate)

When considering possible modes of action of diallylpolysulfanes and their potential cellular targets (Munchberg et al. 2007) it is important to bear in mind that the sulfur atom can adopt many different oxidation states (from the most reduced state in thiols −2 to the most oxidized one in sulfate +6) and as a result sulfur forms part of many different molecules which can induce complex reaction cascades inside cells. Some of the Reactive Sulfur Species (RSS) that are formed by diallylpolysulfanes are shown in Fig. 2.

Similar to thiol/disulfide exchange reactions, diallylpolysulfanes are likely to react with cellular thiols, such as abundant low molecular weight (LMW) thiols (e.g., glutathione) or protein thiols, to form mixed disulfides. LMW thiols play a central role in cellular redox regulation, hence reactions with diallylpolysulfanes could potentially perturb the cellular redox status and induce oxidative stress (see Chap. 9). Likewise *S*-thiolation of cysteine residues of enzymes can result in altered protein function and changes to the cellular 'thiolstat' (Jacob 2011; Jacob and Ba 2011; Jacob et al. 2011).

Nevertheless, it is questionable if this mechanism is the main mode of action of the longer chain polysulfanes as it does not fully explain the increased activity of higher chain polysulfanes compared to diallyl disulfide, which is often observed.

Therefore an initial reaction with LMW thiol or protein thiols seems to be plausible, followed by a range of other reactions shown in Fig. 3. The resulting hydropersulfide (RSSH) (or hydropolysulfane RS_xH with $x \geq 3$) is mainly deprotonated at physiological pH, because of its lower acid dissociation constant. It is likely to react further with cellular thiols stimulating the release of hydrogen sulfide (H_2S), a vasoactive signaling molecule, as shown by Benavides et al. (Benavides et al. 2007). Additionally, the hydroperthiol may react as a strong reductant to convert dioxygen (O_2^-) to form the superoxide radical anion ($O_2^{\bullet -}$) and perthiyl radical ($RSS^\bullet$) which can dimerize to form a diallyltetrasulfide (Munday et al. 2003). This 'pseudo catalytic cycle' (which starts with DAS3 to form DAS4) would explain the higher activity of the higher diallylpolysulfanes, but needs further investigation. The formation of the superoxide radical anion in this cycle leads to further formation of reactive oxygen species (ROS) which ultimately may result in DNA damage and cell death (Chatterji et al. 2005).

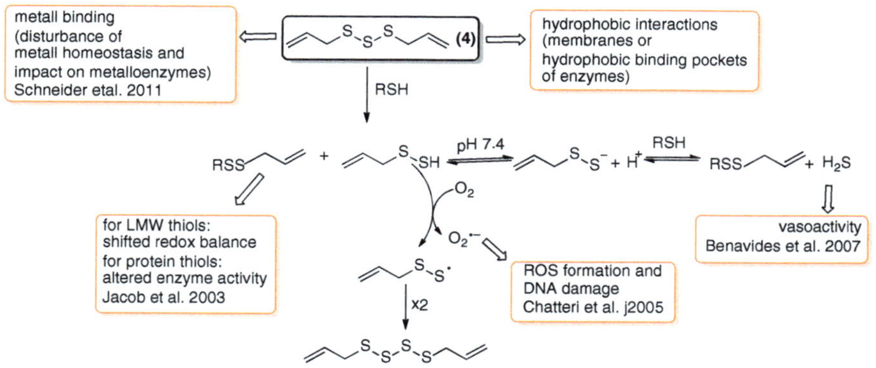

Fig. 3 Possible reactions of diallylpolysulfanes *in vivo* and various physiological consequences. The diagram shows an incomplete overview of possible reactions, cellular targets, and pathways and the given references provide more details about individual pathways

In addition to the redox activity of these molecules, their lipophilicity may contribute to their biological activity. The ability to interact with membranes, bind to hydrophobic binding pockets of proteins and to coordinate with certain metals was investigated by Schneider et al. (2011). In particular DAS3 and DAS4 (the more sulfur atoms in the chain, the more lipophilic) have an effect on the membranes of red blood cells and, as an example for an influence on metalloproteins, also possess the ability to coordinate to hemoglobin.

Clearly, the formation pathways and the physical and biochemical properties of garlic's bioactive organosulfur metabolites are very complex. Such complexities create many challenges that need to be addressed in their formulation, application and regulation within an agricultural setting. Such challenges, however, are worthwhile pursuing, because such food-derived bioactive molecules can provide more desirable alternatives when compared to many of the environmentally persistent agrochemical pesticides that are currently in use.

2.2 From Garlic Concentrate to a Potent Nematicide (ECOspray)

The extensively studied sulfur compounds derived from garlic represent excellent candidates for the formulation of 'green' nematicides or insecticides. By applying the literature available combined with a program of basic research starting in 2000, the UK-based company ECOspray has embarked on a journey to gain EU regulatory approval for the use of its garlic concentrate (i.e., containing a mixture of polysulfanes) in agriculture.

A plant extract may contain many biologically active molecules, the quantities and composition of which may vary with the cultivar used, environmental conditions during growth, and conditions used through processing. Such complexity is not easily addressed in many plant extracts within the present regulatory framework.

Here, the ECOspray garlic concentrate was prepared by integrating a range of different technologies usually applied to the production of plant derived food flavorings and colors to produce a stable mixture of biologically active OSCs. This concentrate was characterized chemically and physically to a level that satisfied regulatory compliance.

The key difficulty in registering botanical products is the intrinsic complexity of extracts or concentrates produced from a given plant. The EU regulatory framework, now broadly similar in all countries across the Union, has been developed to register single molecules as actives that have been produced synthetically with the associated rigorous quality control and possibly in large industrial facilities. This turns experimentation on toxicity, residue retention, breakdown in the environment, and on metabolism of synthetic molecules into a relatively straightforward, although expensive, undertaking.

To overcome some of the above difficulties, ECOspray utilized a vast body of literature when compiling the required regulatory dossiers. For instance, it employed data mostly derived from the literature on toxicity, residues, ecotoxicity, and fate and behavior in the environment. As the regulatory data package matured, more effort was focused on establishing the physical chemistry and efficacy of garlic. Within the current framework for regulating pest control products, these two areas are the most technically challenging to satisfy when registering botanical extracts. Alongside the development of chemical characterization techniques for garlic extracts, ECOspray developed a range of bioassays and initiated field trials. Here, from 2002 until present, ECOspray used a combination of food grade garlic juice concentrate with a granulation technology, based on mixing the garlic extract with a carrier matrix in a granulator to produce NEMguard®. NEMguard® is the first botanical nematicide from garlic registered in the EU (for cost-effective sources of natural products see Explanatory Box 1).

Explanatory Box 1: Agricultural Waste as Source of Redox Active Secondary Metabolites

This particular chapter considers the commercial use of redox active secondary metabolites. The materials in question contain mixtures of diverse organic sulfur compounds (OSCs), including various polysulfanes. Similar strategies have also been used to turn polyphenols from grape seeds or resveratrol from grape skins into commercial products. While the science and technology is in place to extract, distill, or otherwise process plant materials in order to derive at a commercial product, additional aspects need to be considered before such a product becomes a 'hot seller', a *commerciallyviable* product. Besides the usual regulatory issues, the palette and extent of potential applications (e.g., widespread use in agriculture versus niche market for parsley farmers), and the price users are inclined to pay for such a product, the question of raw materials is of paramount importance. Here, agricultural waste is clearly a very attractive option. It is cheap, readily available, and almost unlimited as vegetable or berry farming always generates waste, simply because of the usual deterioration of the product and the fact that some of the products 'don't look great' and hence are difficult to sell. Here, garlic, onions, but also grape seeds (from winemaking), mustard seeds, and a wide range of other products are readily available at a very competitive cost and may be used.

At the same time, the growing interest in such natural products as raw products for refined, commercially viable end-products may stimulate a wider production of such plants, berries, mushrooms, etc. Here, farming may shift from the mass production of cheap crops to the (limited) cultivation of produce rich in valuable secondary metabolites. Such a move toward planting and harvesting 'green gold' is particularly attractive to developing countries, where the demand for cheap yet ecologically friendly and healthy plant protective substances and therapeutic agents is high. Indeed, local growth, refinement, and application of such products may not only provide access to such products but also generate new wealth generating cycles, especially if waste products can be used. This strategy is illustrated nicely in the case of the grape, which can be used to manufacture a wide range of valuable products, including juices, wine, resveratrol, grape seed oil, and polyphenol-rich grape seed flour. As a beneficial side effect, such a move would also enrich the local flora and break with the kind of monoculture still present in many relevant Asian and African countries.

Further evidence for approval came from field trials where garlic-based NEMguard® was effective in reducing nematode attack and symptoms of nematode parasitism in carrot and parsnip roots. The work focused on various free living nematodes in the UK that damage carrot and parsnip crops, and root knot nematodes which damage carrot and tomato crops in Italy.

NEMguard® was as effective against both types of nematode as the synthetic nematicides used for comparison (Temik®, Nemacur® and Nemathorin®) (Figs. 4 and 5). NEMguard® is capable of delivering efficacy comparable to synthetic products such as Temik® (aldicarb), Vydate® (oxamyl) and Nemacur® (Figs. 4 and 5), all of which are under intense international regulatory scrutiny for adverse ecotoxicity and residues in treated crops.

The full range of ECOspray's garlic active ingredients was voted through the EU Commission for inclusion into Annex 1 of directive 91/414 in October of 2008. Finally, for ECOspray there was some light at the end of a very long tunnel.

ECOspray is now able to start developing broader commercial opportunities while still working on technical issues as approvals are being sought in many countries both within and outside the European Union, including the USA, Australia, and South Africa. ECOspray has in addition received a registration for ECOguard®,

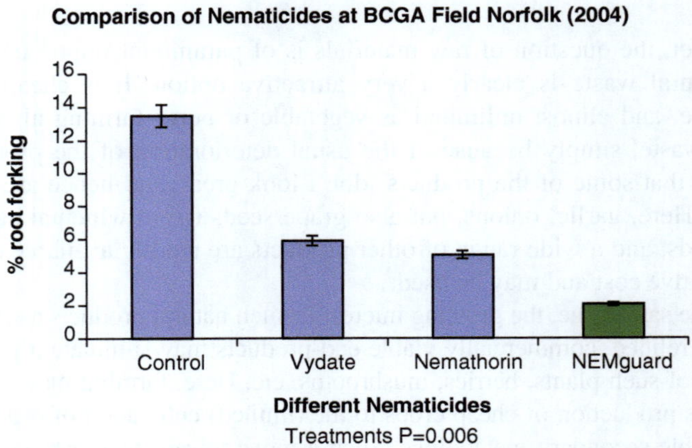

Fig. 4 A comparison of commercial nematicides against NEMguard®. There is a clear evidence that NEMguard® is comparable to synthetic commercial products with an extra advantage of being a natural product. © ECOspray Limited UK

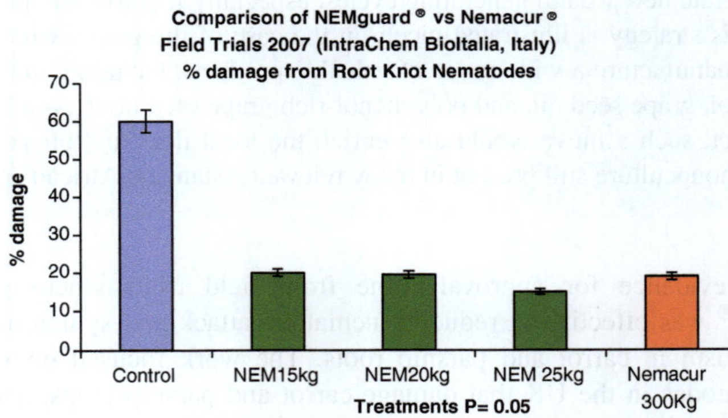

Fig. 5 Comparison of NEMguard®, different application rates, with commercial nematicide, Nemacur® in field trials. Please note that maximum application of NEMguard® is 25 kg/ha while Nemacur® is 300 kg/ha. © ECOspray Limited UK and IntraChem, Italy

a garlic-based insecticide registered for cabbage root fly control in Denmark in February 2006 (it received organic registration for the same product in September of the same year in Denmark). And in early 2008 it received a registration for ECOguard® also for cabbage root fly control in Norway. In 2010, ECOspray obtained regulatory approval in the UK for Eagle Green Care®, the only liquid nematicide approved for use on elite sports turf.[4] The timeline taken by ECOspray to register a single product from start to full approval is outlined in Fig. 6.

---

[4] MAAP No. 14989.

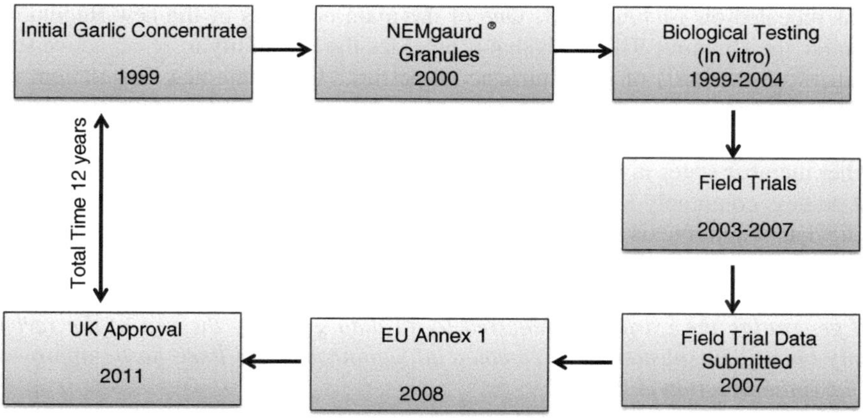

Fig. 6 Flow diagram showing the registration time frame taken by ECOspray for one product (NEMguard®)

3 Regulatory Requirements to Register New Pesticide/ Botanical Products for Crop Protection, Using Garlic as an Example

Crop protection products are some of the most thoroughly tested and regulated chemicals in the world. The process for bringing new pesticides to the market involves research, development (R&D), and registration. Before any pesticide can be used, sold, supplied, advertised, or stored it must be approved for use. In the UK, for example, these approvals are granted by the Chemicals Regulation Directorate (CRD)[5] on behalf of Ministers under a range of specific pesticide-related legislation.

In 1991 the European Directive, 91/414/EEC, was introduced as a basis for registration of crop protection products. It aimed to harmonize the registration of pesticides across the European Union by setting out unified basic data requirements and a process of evaluation for pesticide active substances at a European level. Directive 91/414/EC also required that all existing crop protection products should be re-registered using updated end points in order to remain in the market. For many crop protection products re-registration under 91/414/EC has been unsuccessful and many products that have been in the market for over 40 years are now no longer available. This fact alone has become a powerful driving force for new and widespread sustainable crop protection technologies. A new EU regulation 1107/2009 entered into force on 14 December 2009 and Directive 91/414/EEC

[5] CRD is a division of the Health and Safety Executive (HSE).

was repealed on 14 June 2011. One of the main elements of the new Regulation (unlike the Directive 91/414) is that it provides the possibility to *reject* active substances on the basis of their intrinsic properties.[6] Under the new Regulation, the EU is divided into three zones, Northern, Central, and Southern, with the concept that once a plant protection product approval is granted in one member state, the other member states in that zone are able to use the evaluation to grant an approval (a process commonly known as Mutual Recognition), as long as any national specific data requirements and risk assessments have been completed.

In total, close to 1,000 active substances are part of the review but it is likely that across Europe over 450 substances will no longer be sold because of the cost of generating the extra data required to fill data gaps. By the end of the review only about 350 substances of commercial significance are likely to be approved. This is the situation in 2012.

The active substances that are authorized at EU level are listed in Annex 1 of the Directive. The registration of pesticide products containing these active substances is the responsibility of individual Member States.

Irrespective of the regulatory framework, crop protection products are usually synthesized or chemically extracted from natural sources and then subjected to regulatory scrutiny. They are then tested using a series of increasingly complex screens to see if they have any 'biological activity' or potential as a pesticide.

This early research phase will include preliminary toxicological and environmental testing. If the molecule has good properties in all these areas and shows good commercial potential, it will then go on to the development phase. On average, only 1 in 70,000 compounds tested goes forward. The cost of this research phase averages £71 million for each new product. The subsequent development phase costs an average of £60 million per product and only 1 in 2 make it through to an application for registration.

The total time taken from first synthesis (discovery) to first sale averages over 9 years and only 1 in 139,000 molecules synthesized will make it to the market. The total cost per new synthetic crop protection product is estimated at £140 million.[7]

In contrast, a botanical extract being considered as a new crop protection active will still have to pass through the same regulatory review as a new synthetic substance, but may be able to address much of the technical data requests *by reference to published literature*. By adopting this approach, much of the costs of registration for a botanical extract can be mitigated. There are as yet very few examples of botanical pesticides where this principle has been applied. Garlic-based substances produced by ECOspray Ltd. are among the first. A detailed overview of the key points in the registration to approval process is presented in Fig. 7.

[6] A concept commonly known as 'hazard-based cut-off criteria'. Further information available on the PSD website http://www.pesticides.gov.uk.

[7] All costs mentioned above relate to the year 2000, Figures taken from the *Crop Protection Association* (CPA).

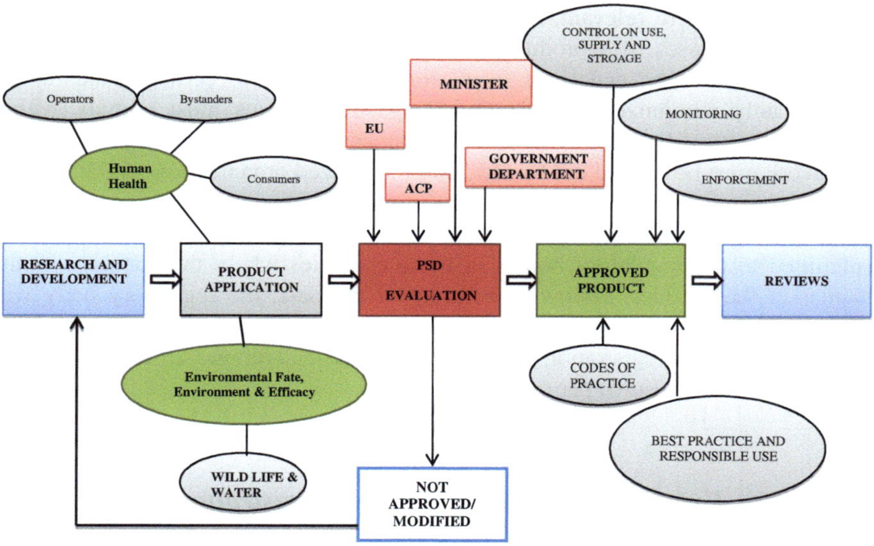

Fig. 7 Pesticide product route map highlighting different steps and authorities involved in the approval of such a product in the UK

The process of developing a new molecule or botanical extract into a pesticide product includes:

- Generating enough representative material for sustained biological and safety testing.
- Testing the material in a variety of formulations to find the safest and most effective way of using it in the field.
- Further biological testing to investigate the activity of the product against a variety of target pests, weeds or diseases in a number of crops under a variety of environmental conditions.
- Trials to determine the fate of the molecule and its metabolites (breakdown products) in the environment, soil, and plants.
- Safety testing both in terms of toxicology and environmental impact to meet the statutory requirements of regulatory bodies in the EU, USA, and elsewhere. Or a major literature review and construction of reasoned arguments to address regulatory data requests.

3.1 Registration Processes in the EU

3.1.1 Application for Authorization

To file an application for registration as a pesticide in the EU, the active must be included in Annex 1, a central database of acceptable active substances. Applications for approval are made in writing using the relevant application forms. Approval can only be granted once all the data and/or information on the safety, efficacy, and, where relevant, humaneness, of the pesticide are considered to be acceptable. All uses of a product must be supported by data or reasoned arguments. Data must be generated from work carried out to certain standards by appropriately recognized organizations (e.g., laboratories must comply within Good Laboratory Practice (GLP), or developed from the literature). Accredited field work must be carried out by officially recognized organizations and must represent the climactic zone that the product is intended to be used in.

The results of the developmental studies and data on key end points are then submitted to the regulatory bodies in those countries where the company wants to market the product. Following scrutiny of the data, the regulatory body will decide whether or not to register the product. The cost of preparing and compiling the application for registration of a synthetic product averages around £8 million (around 9.5 million Euro) per product.

3.2 Garlic Extract as an Example in the Registration Process: The ECOspray Experience

Fortunately, chemical actives derived from a botanical source may avoid the need to establish a plethora of new (and expensive) data, by reference to high quality scientific and technical literature which is already in the public domain. Even with this advantage, interpretation of scientific and technical literature in a format compliant with the regulatory review process takes considerable time and until very recently has tended to progress through review on an iterative basis. This was more a consequence of a very limited range of examples for regulators to work from within existing regulatory guidelines, than poor or inappropriate interpretations of data found within the literature, sometimes coupled with new experimental data on the plant extract.

Garlic is a plant with a long and voluminous historical account. Apart from the widely known culinary attributes of the plant as a 'vigorous and pungent' flavor enhancer, preparations of garlic, mostly as aqueous extracts or distillates in oil form have been shown to exhibit a wide range of beneficial properties in human medicine, and as plant protection agents. A significant volume of these more recent accounts has been conducted to a standard high enough to enable the data to be used at regulatory level by ECOspray Ltd.

The key milestones in the ECOspray journey, how they were approached and passed are described below.

- How did the market for bio-pesticides work in the year 2000?
- What has changed ten years later?
- Early engagement with regulators and dealing with a changing regulatory format (2000–2004).
- The first technical milestone (analysis of the core extract).
- Early indications of broad bioactivity.
- The first efficacy milestone passed as nematicide (2003) and insecticide (2005).
- Significant improved characterization of the product (2005–2010).

- Navigation through changing regulatory frameworks (2005–2010).
- The first major approvals (2006).
- R&D takes off and ECOspray enters the market with a bionematicide (2010).
- Sustaining approvals (2010–present). Transition from national to EU regulatory scale.

4 ECOspray Enters the EU Market, Smelling of Garlic

Identifying a change in the market, ECOspray, a UK-based business came into existence in 1997 focusing on the chemistry of garlic as a potential source of new pesticides from a benign botanical origin.

In the UK in the late 1990s, natural products for crop protection were of fringe interest with their primary entry point being the organic sector or as general plant stimulants. To avoid the stringent requirements of a regulatory system designed to regulate synthetic chemicals, these natural products were marketed with no assertive claims about pesticidal function on the product label. This tended to keep these products off the regulatory radar but as a consequence they suffered from low commercial credibility and had little impact in the mainstream of crop protection.

When ECOspray sought to bring garlic to the market as a pesticidal product it had no choice but to adopt the guidelines which applied to the chemistry of a new synthetic active under the Control of Pesticides Regulations (COPR 1986). The 91/414/EC directive had not been fully implemented at this time and ECOspray had to take full account of the biochemical complexity of garlic in order to plot a course through the chemical and manufacturing regulatory data requirements. For more than 50 years, the whole regulatory framework had been evolving around single molecule synthetic activities. For ECOspray Ltd., this involved: characterizing and developing quality control for a complex plant extract; developing new analytical technology and proving batch-to-batch consistency. This was not an easy task but was completed by ECOspray Ltd. around 2004.

Initially (1996–1998), several poor quality garlic-based materials entered the EU market, but results using these products were very inconsistent. On one occasion, two sequential deliveries of poor quality garlic extract were subjected to a quality control bioassay. One batch tested exhibited 100 % mortality on the test organism, while the other batch exhibited 0 % mortality. With this scale of variability, skepticism about the potential of these products persisted in both the market and UK regulatory bodies.

A critical mass for greater market penetration of natural products had still not been achieved by the early 2000s. There were, however, some initial signs that within the complex mixtures of garlic extracts, a viable biological activity as crop protection agents could be identified. Garlic and Neem-based products started to attract growing interest around this time and ECOspray identified a much more consistent source of garlic extract. This material has since been the basis for all approved ECOspray Ltd. crop protection products.

Most remarkable is the increasing evidence that a food grade extract of garlic can kill nematodes and other pests with the same efficacy as a material that started off as a chemical developed as a nerve gas during the second World War. Ultimately, the story just told is one of dogged determination from a very small company, pushing against closed doors for several years. These doors are now opening at a time when sustainability of food production techniques in the EU has become a major political issue.

5 Outlook

There has been considerable recent interest in the development of pesticides based on OSCs derived from garlic, onions and other *Allium* species since such food-based compounds should be environmentally benign, potentially widely available, and low in cost. It is known that *Allium* species are resistant to insects and fungi and that technologically advanced extracts and distillates of these plants can function as potent nematicides, insecticides, and repellents. Recent initiatives by the pesticide regulatory departments of European and North American governments have stimulated renewed interest in bio-pesticide technologies to replace toxic synthetic pesticides with more benign natural products. Considerable progress has been made recently with bringing botanical bio-pesticides to the market and the first well-researched examples of these products are starting to enter significant segments of the EU crop protection market.

A concerted effort in formulation development for bio-pesticides by multidisciplinary teams is still required to optimize bio-pesticide yield, efficacy, storage stability, and delivery to enable this technology to evolve further and to make a significant contribution to meeting today's agricultural and societal demands for safe and sustainable food production.

Acknowledgment We greatly acknowledge the EU FP-7 ITN grant number 215009 for supporting the project.

References

Akhtar M, Mahmood I (1994) Potentiality of phytochemicals in nematode control—a review. Bioresour Technol 48(3):189–201

Anwar A, Burkholz T, Scherer C, Abbas M, Lehr C-M, Diederich M, Jacob C (2008) Naturally occurring reactive sulfur species, their activity against Caco-2 cells, and possible modes of biochemical action. J Sulfur Chem 29(3):251–268

Benavides GA, Squadrito GL, Mills RW, Patel HD, Isbell TS, Patel RP, Darley-Usmar VM, Doeller JE, Kraus DW (2007) Hydrogen sulfide mediates the vasoactivity of garlic. Proc Natl Acad Sci USA 104:17977–17982

Block E (ed) (2010) Garlic and other Alliums—the lore and the science. RSC Publishing, Cambridge

Brooks GT (1998) Book review: phytochemicals for pest control. In: Hedin PA, Hollingworth RM, Masler EP, Miyamoto J, Thompson DG (eds) American Chemical Society, Washington DC, USA, Pestic Sci 52 (3):303–304. ISBN 0-8412-3488-4

Cavallito CJ, Bailey JH (1944) Allicin, the antibacterial principle of allium sativum. I. isolation, physical properties and antibacterial action. J Am Chem Soc 66(11):1950–1951
Cerella C, Scherer C, Cristofanon S, Henry E, Anwar A, Busch C, Montenarh M, Dicato M, Jacob C, Diederich M (2009) Cell cycle arrest in early mitosis and induction of caspase-dependent apoptosis in U937 cells by diallyltetrasulfide (Al2S4). Apoptosis 14(5):641–654
Chatterji T, Keerthi K, Gates KS (2005) Generation of reactive oxygen species by a persulfide (BnSSH). Bioorg Med Chem Lett 15(17):3921–3924
Chitwood DJ (2002) Phytochemical based strategies for nematode control. Annu Rev Phytopathol 40:221–225
Copping LG (1999) Insecticides of natural origin. In: Dev S, Koul O. Harwood Academic Publishers, The Netherlands, 1997. Pestic Sci 55(7):756–757. ISBN 3 7186 5913 1
Isman MB (2006) Botanical insecticides, deterrents, and repellents in modern agriculture and an increasingly regulated world. Annual Review of Entomology, vol 51, pp 45–66
Isman MB (2008) Botanical insecticides: for richer, for poorer. Pest Manag Sci 64(1):8–11
Jacob C (2006) A scent of therapy: pharmacological implications of natural products containing redox-active sulfur atoms. Nat Prod Rep 23(6):851–863
Jacob C (2011) Redox signalling via the cellular thiolstat. Biochem Soc Trans 39:1247–1253
Jacob C, Anwar A (2009) Sulfides in *Allium* vegetables. In: Knasmueller S, DeMarini D, Johnson I, Gerhaeuser C (eds) Chemoprevention of cancer and DNA damage by dietary factors. Wiley,GmbH & Co., Weinheim, pp 663–684
Jacob C, Ba LA (2011) Open season for hunting and trapping post-translational cysteine modifications in proteins and enzymes. Chem Bio Chem 12(6):841–844
Jacob C, Jamier V, Ba LA (2011) Redox active secondary metabolites. Curr Opin Chem Biol 15(1):149–155
Jeyasankar A, Jesudasan RWA (2005) Insecticidal properties of novel botanicals against a few lepidopteran pests. Pestology 29:42–44
Koul O, Dhaliwal GS (2001) Phytochemical biopesticides. Harwood Academic, Luxembourg
Munchberg U, Anwar A, Mecklenburg S, Jacob C (2007) Polysulfides as biologically active ingredients of garlic. Org Biomol Chem 5(10):1505–1518
Munday R, Munday JS, Munday CM (2003) Comparative effects of mono-, di-, tri-, and tetra-sulfides derived from plants of the *allium* family: Redox cycling in vitro and hemolytic activity and phase 2 enzyme induction in vivo. Free Radical Biol Med 34(9):1200–1211
Schillhorn Van Veen TW (1999) Agricultural policy and sustainable livestock development. Int J Parasitol 29(1):7–15
Schneider T, Ba LA, Khairan K, Zwergel C, Bach ND, Bernhardt I, Brandt W, Wessjohann L, Diederich M, Jacob C (2011) Interactions of polysulfanes with components of red blood cells. Med Chem Comm 2(3):196–200

Authors Biography

Awais Anwar obtained an MSc degree in Chemistry (2000) from the University of Peshawar, Pakistan. He then studied at the Quaid-I-Azam University Islamabad for an advanced MPhil degree in Biochemistry (2005) which was followed by a short-term research project at Eberhard Karls University of Tuebingen, Germany with Prof. Wolfgang Voelter. Awais obtained his PhD in 2009 from the University of Saarland under the supervision of Prof. Claus Jacob. In 2010 Awais has conducted research at the RWTH Aachen as part of his

Marie-Curie postdoctoral research fellowship. He is currently working for ECOspray Limited UK where he is involved in product development, formulation, regulatory data management, and field trials. At the same time, Awais is also investigating the possible application of Reactive Sulfur Species as novel anticancer therapeutics.

Murree Groom obtained a joint honours degree in Chemistry and Biochemistry from UCNW in 1981 and was sponsored by the then MAFF Chief Scientist Group in the UK to pursue a PhD thesis on the cause and control of cavity spot in carrots. He completed his thesis in 1988 after 4 years of research combined at The Scottish Crop Research Institute and University of Dundee under the supervision of Prof. John Raven FRS. An appointment as a Chemist to the Diversey Corporate Technology Centre followed, where Murree worked on the formulation and manufacture of biocides for the dairy industry. Murree returned back to Norfolk as an agronomist for W.H. Knights and Sons where he developed and managed on farm R&D capabilities in pesticide analysis, entomology, nematology, plant pathology, and integrated pest management.

In 1997, Murree joined ECOspray and now directs research and development with specific interests in biopesticides produced from garlic extracts. ECOspray Ltd is the only company in the EU with regulatory dossiers that support authorisation of garlic-based extracts as nematicides and insecticides. Since being with ECOspray Ltd., Murree has been instrumental in co-supervising several research students and securing three patents that all relate to the processing and chemistry of garlic. The present focus of Murree's work is construction of synthetic formulations based on garlic chemistry that could replace the materials derived from processing whole garlic plants.

Miriam Arbach (born 1987) has been trained as a biologist (specialised in plant physiology and plant biochemistry) at the RWTH Aachen University where she completed her Bachelor studies in 2009. Subsequently, in 2010, she became an Early Stage Researcher within the EU Marie Curie Initial Training Network "RedCat" and started her PhD program at the University of East Anglia (UEA) under the joint supervision of Dr. Chris Hamilton (UEA) and Dr. Murree Groom (ECOspray Ltd).

Miriam's PhD studies explore how garlic-derived diallyl polysulfane metabolites perturb cellular redox processes in soil bacteria and nematodes, and how such substances may find applications as nematicides in food crop protection. Miriam is currently about to complete her PhD thesis at the University of East Anglia.

Chris J. Hamilton graduated from the University of Liverpool (1994) and obtained his PhD from the University of Exeter in 1997 working on the synthesis of biologically stable nor-carbovir triphosphate analogs as inhibitors of HIV reverse transcriptase. This was followed by postdoctoral appointments at the University of St Andrews (protein phosphatase enzymology), University of Dundee (synthesis and mechanistic studies of lunaria alkaloids as trypanothione reductase inhibitors) and the University of East Anglia (carbohydrate chemistry/biochemistry). Shortly after receiving a Wellcome Trust SHOWCASE award, he took up a post as Lecturer in Organic Chemistry at Queen's University Belfast in 2003 before returning to the University of East Anglia in 2007 where he is currently a reader in medicinal chemistry.

Chris' research interests reside at the Chemistry/Biology interface with particular interests in (1) Chemical and biochemical functions of low molecular weight thiols in Gram positive bacteria; (2) Redox mechanisms of bioactive organosulfur metabolites from garlic; (3) Mechanistic enzymology.

Part XII
Connecting Section Between Chapters 12 and 13

With this glimpse on how to turn a particular redox active secondary metabolite into a practical application with considerable commercial success, we will leave the vast and rapidly expanding field of natural organic sulfur compounds. The next couple of chapters will consider other, often less well-known secondary metabolites that have recently attracted the attention of researchers for a number of reasons. Besides an often pronounced redox activity—or ability to interfere with cellular redox processes—these compounds exhibit interesting biological activities and highlight the diverse approaches that may be used to turn a natural substance into a (more) powerful applicable agent.

In the next chapter, we begin our considerations with the family of phytols. Phytols are chemically rather simple diterpenes. While terpenes are commonly associated with plants, especially conifers, (which indeed produce a terpene-rich resin), diterpenes are formed in and by many animals, including insects, and in ruminants as part of digestive degradation of chlorophyll. Together, diterpenes exhibit a wide spectrum of biological activities, ranging from their use as a chemical deterrent against predators in beetle to possible anti-inflammatory effects once applied to the human body (which itself is unable to produce such diterpenes by its own). While compounds such as phytol are not redox active by themselves, they are actually capable of inducing redox changes to the cell, for instance, by upregulating intracellular levels of ROS. Here they join species such as Zn^{2+} which also 'redox modulate' without being redox active under these conditions (see Chap. 4). As part of this chapter, Peter Olofson and his colleagues will focus on the production of diterpenes from (chlorophyll contained in) grass, generated as a digestive by-product or waste product in ruminants. This chapter will highlight some of the anti-inflammatory activities associated with these compounds, and outline possible future applications. At the same time, it will instigate a closer look at different methods currently available to turn a plant metabolite, such as chlorophyll, into a different, more active and hence more valuable substance. This line of thought will then be followed up by several of the subsequent chapters.

Chapter 13
Phytol: A Chlorophyll Component with Anti-inflammatory and Metabolic Properties

Peter Olofsson, Malin Hultqvist, Lars I. Hellgren and Rikard Holmdahl

Keywords Chlorophyll • Inflammatory diseases • Metabolism • Phytol • Reactive oxygen species

1 Introduction

The metabolic syndrome denotes a cluster of clinical conditions associated with obesity, which are strongly associated to the risk of subsequent development of Type 2 diabetes and cardiovascular diseases (CVD). Currently, an estimated 10 million US adults have diabetes and another 25 million have impaired glucose tolerance (IGT), an intermediate between insulin resistance and diabetes. By 2030, 40.5 % of the US population is predicted to suffer from some form of CVD. Each year, CVD claims more lives than the next four leading causes of death combined (Heidenreich et al. 2011; Koch-Henriksen and Sorensen 2010). The pathophysiologic are mechanisms known to increase CVD risk in individuals with insulin resistance include formation of advanced glycation endproducts, hypertension, proinflammatory and prothrombotic states, and dyslipidemia, including triglyceridemia and hypercholesterolemia, i.e., increased serum levels of triglycerides and cholesterol.

P. Olofsson (✉) · M. Hultqvist
Redoxis AB, Medicon Village, 223 81 Lund, Sweden
e-mail: peter.olofsson@redoxis.com

L. I. Hellgren
Center for Biological Sequence Analysis, DTU Systems Biology, Technical University of Denmark, Kongens Lyngby, Denmark

R. Holmdahl
Medical Inflammation Research, MBB, Karolinska Institutet, Stockholm, Sweden

C. Jacob et al. (eds.), *Recent Advances in Redox Active Plant and Microbial Products*,
DOI: 10.1007/978-94-017-8953-0_13, © Springer Science+Business Media Dordrecht 2014

Other types of common and global disorders are autoimmune diseases. Autoimmune diseases like multiple sclerosis (MS) (Koch-Henriksen and Sorensen 2010), rheumatoid arthritis (RA) (Scott et al. 2010), and Type 1 diabetes (Todd et al. 2011) all affect specific organs or tissues of the body. In RA, it is the cartilaginous synovial joints of hands and feet that are inflamed and destroyed. In MS, the spinal chord and brain are affected by inflammation, leading to neurotoxicity and destruction of the myelin sheath with symptoms such as impaired vision, peripheral inflammation, and disturbed balance. In type 1 diabetes, the insulin-producing beta cells within the Langerhans islets of the pancreas are attacked resulting in a defective insulin production. Despite decades of research, a multitude of pharmaceutical drugs developed and billions of US dollars spent on research for new and improved drugs, there is still no cure available for these kinds of chronic inflammatory diseases. Interestingly, it is also observed that the combination of autoimmunity and the so-called metabolic syndrome seems to increase all over the world (Gremese and Ferraccioli 2011; Pereira et al. 2009) due to better diagnostics and healthcare systems but also to a spreading of the westernized lifestyle that is often accompanied by increasing prevalence of weight-related diseases. In addition, the improved quality of life results in a longer life span for all humans followed by disorders that are connected with old age, yet are not lethal anymore and require prolonged treatment.

Metabolic and inflammatory diseases represent two major global health concerns for millions of patients, are of high socioeconomic importance and are two areas of highest relevance for medical research and efforts to identify novel pharmaceuticals. Observations of both anti-inflammatory and beneficial metabolic properties of the diterpene oil phytol are thus very interesting (for details on terpenes see Explanatory Box 1). In light of these findings, we will in this chapter present the natural 'by-derived' compound, phytol, with anti-inflammatory as well as metabolic properties as an alternative for improving health.

Explanatory Box 1: Terpenes

As part of this chapter, a rather diverse class of natural products called 'terpenes' finds extensive mention. Terpenes form a large class of organic compounds which are produced by a variety of plants, in particular coniferous plants, but also by certain animals and microorganisms. In plants, terpenes form a major component of resin. Turpentine and similar essential oils derived from various types of plants contain an often diverse spectrum of terpenes. Chemically speaking, terpenes are 'built' from various units of isoprene, which has the molecular formula C5H8 and, like certain sulfur-containing garlic-components, also possesses an allyl-group. Indeed, the biosynthesis of terpenes often involves isoprene building blocks. Yet there are exceptions. The phytols discussed as part of this chapter are not synthesized from smaller monomeric species but are generated via the 'breakdown' of a larger molecule, i.e., of chlorophyll. This underlines the high diversity of the terpenes when it comes to their chemical structures, properties, origin, and biosynthetic pathways. Indeed, terpenes are often not the end-products of biosynthesis but also provide the basis for the synthesis of a vast spectrum of other biologically relevant substances, which include steroids (e.g., cholesterol), hormones, vitamins (e.g., vitamin A, vitamin K) and certain pheromones.

Terpenes possess many practical applications. Because of their strong—and mostly pleasant—smell, they are used as part of various fragrances and perfumes. Terpenes are also used as flavor additives for food and have a long tradition in traditional or alternative Medicine. Here, they are employed as part of 'aromatherapy', but also against cancer and malaria. Indeed, many terpenes exhibit diverse antimicrobial, antifungal, antiprotozotic, and anthelmintic activities, which are currently under intense scientific investigation. Antiparasitic properties, in particular, are not only of interest in medicine but also in the field of agriculture. Such substances appear to be rather promising in the context plant protection, 'green' pesticide development and as repellents and pheromones. Terpenes have extraordinarily diverse chemical structures. Therefore, it is futile to try to discuss the biological activity of terpenes in general. Individual terpenes need to be considered separately, as each of them will exhibit a unique spectrum of activities.

2 Phytol

Phytol (CAS 150-86-7; 3,7,11,15-tetramethyl-2-hexadecen-1-ol); (Fig. 1) is a natural linear diterpene alcohol that is an oily liquid. It is close to insoluble in water, but soluble in most organic solvents. The chemical family of terpenes is derived biosynthetically from units of isoprene, which have the molecular formula C_5H_8 (see Explanatory Box 1). The basic molecular formulas of terpenes are multiples of that $(C_5H_8)n$, where n is the number of linked isoprene units. The isoprene units may be linked together 'head to tail' to form linear chains or they may be arranged to form rings. Therefore, one can consider the isoprene unit as one of nature preferred building blocks. Diterpenes, like phytol, are composed of four isoprene units and have the molecular formula $C_{20}H_{32}$. Besides phytol, other common examples of diterpenes include retinol, cafestol, kahweol, cembrene, and taxadiene.

In nature, phytol constitutes the aliphatic chain of the chlorophyll molecules and represents approximately one-third of the mass of both chlorophylls *a* and *b*, representing about 0.2 % of the wet weight of green plants. Although monogastric animals, like man, cannot hydrolyze the ester linkage between chlorophyll and phytol, it is released during rumenal digestion of green plants in ruminant animals (van den Brink and Wanders 2006). In fact, 0.1–0.3 % of the total lipid content in the

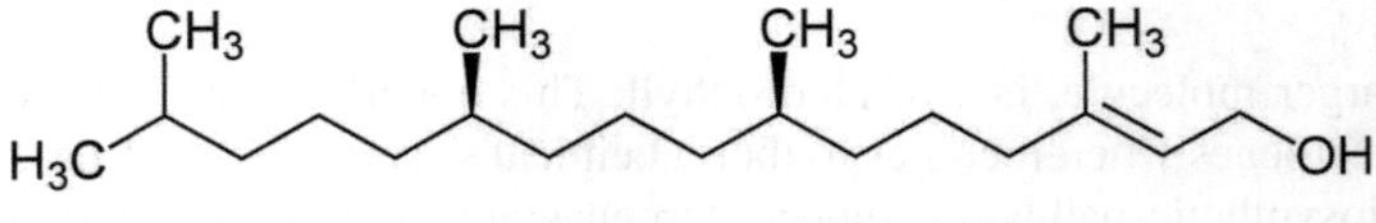

Fig. 1 Structure of phytol

lactating cow's rumen has been determined to be phytol (Patton and Benson 1966). Besides the possibility to be produced from natural chlorophyll sources, phytol can be derived chemically through a series of reactions starting from acetylene and acetone and is used among others in the production of vitamins E and K. From the synthetic phytol, all isomers can be derived while the naturally occurring phytol only consists of the *E*-isomer. The natural form of phytol can be produced from chlorophyll paste obtained from green plants like spinach using distillation procedures.

Phytol is readily absorbed in the small intestine of all mammals, and is metabolized to phytanic acid (3,7,11,15-tetramethylhexadecanoic acid), which is degraded via peroxisomal α-oxidation (Gloerich et al. 2007). The catabolism of phytol is not fully characterized at the molecular level, but includes the oxidation of the alcohol to an aldehyde (phytenal) by a yet unknown alcohol dehydrogenase. The aldehyde is oxidized further to a carboxylic acid (phytenic acid) by the fatty aldehyde dehydrogenase. These two reactions occur in the endoplasmic reticulum. In the peroxisome, phytenic acid is activated to phytenoyl-CoA, and the double bond is reduced by the peroxisomal *trans*-2-enoyl-CoA reductase, producing phytanic acid (van den Brink and Wanders 2006).

2.1 Phytanic Acid

Intake of phytol efficiently increases the concentration of phytanic acid in the circulation as well as in organs (for further details regarding so-called nutraceuticals see also Explanatory Box 2). When mice were given a diet containing 0.5 % phytol for 8 weeks, the hepatic phytol concentration increased from non-detectable in control mice fed a phytol-free diet, to 0.75 nmol/mg protein, while phytanic acid increased to almost 18 nmol/mg protein (Gloerich et al. 2005). In the same study, the intervention with 0.5 % phytol in the diet led to an increase in plasma phytanic acid from 0.8 to 49 μM.

Explanatory Box 2: Nutraceuticals

The expression 'nutraceutical' (or 'nutriceutical') is a combination of the words 'nutrition' and 'pharmaceutical' and refers to natural food ingredients that are also biologically (possibly pharmaceutically) active. It is based on a rather old concept in pharmaceutical sciences, and especially also in orthomolecular medicine, which in a nutshell considers food as a medicine. Nutraceuticals may be consumed as part of normal food, such harvested plants, berries or fruits (e.g., garlic), or in the form of functional food which is somehow fortified with the biologically active ingredients (garlic oils, milk enriched with vitamins, etc.). It is also possible to isolate such nutraceuticals and to apply them in form of herbal extracts or as food supplements (e.g., multivitamin pills). In most cases, the idea of nutraceuticals is neither new nor particularly inventive, as it has been known for millennia that certain edible plants, mushrooms, etc., seem to exert a beneficial effect on the human body which in modern days has been ascribed to one or more active ingredients found in these foodstuffs.

Interestingly, the field of nutraceuticals has recently witnessed a certain renaissance. This renewed interest in active ingredients found in edible plants and mushrooms is based on rather surprising findings, which link the consumption of rather low amounts of such compounds to pronounced effects. Compounds such as proanthocyanidins, for instance, seem to exert significant antimicrobial effects when passing through the gastrointestinal tract. They also seem to inhibit certain enzymes involved in the uptake of sugars, fats, and proteins and hence may be useful to reduce caloric intake and hence body weight. Other compounds, such as xanthohumol from hop, seem to induce considerable epigenetic changes.

Indeed, some of these findings have given rise to new fields of nutritional research, such as nutriepigenetics. The latter deals with the impact of food ingredients on epigenetic changes, often related to histone modifications. There is now a long list of (often chemopreventive) agents which (also) target the epigenome, including redox active selenium compounds, polyphenols, curcumin, resveratrol, coumarins, di- and poly-sulfanes from garlic and various isothiocyanates. These epigenome-changing compounds include many redox active secondary metabolites, which will be discussed in various chapters of this book.

Due to the relatively efficient oxidation of phytol to phytanic acid in cows, the dietary intake of phytanic acid is considerably larger than the intake of phytol. Although very few studies on the content of phytol in foodstuffs are present, available data indicates that there are 70–100 times higher concentrations of phytanic acid than phytol in dairy fat from herbivores like cows or goats. In most populations, dairy products are the main source of phytol and phytanic acid, although marine fats (e.g., fish oils) are also rich sources. It has, for example, been reported that fish oil supplements contain up to 750 mg phytanic acid per 100 g oil. Among dairy products, the content of phytanic acid varies substantially, depending on the feeding regime of the cow. Since chlorophyll is the precursor for the formation of phytol and phytanic acid, their respective concentrations in dairy products and ruminant meat are strongly correlated to the amount of green plant material in the food.

Thus, while Danish conventional butter only contains around 0.1 % phytanic acid in late wintertime, in late summer, it has been determined to reach up to 0.55 % in milk fat from cows that have been grazing on mountain pastures all summer (Drachmann and Hellgren, unpublished work).

The levels of phytol and phytanic acid in the circulation in humans, of course, varies with the food we eat. Thus the mean concentration of phytanic acid in plasma from nonvegetarians has been reported to be varying from 3 to about 6 μM (Allen et al. 2008; Werner et al. 2011). In a recent study, it was shown that intake of 45 g/day of dairy fat for 4 weeks significantly increased the plasma phytanic acid concentration, even when the concentration in the milk fat was as low as 0.15 %, Werner et al. (2011) demonstarting that changes in dietary habits rapidly alter the level of phytanic acid.

Due to the presence of the methyl-branches at C-3 in phytanic acid, the latter cannot be catabolized directly through β-oxidation, but is first oxidized to pristanic acid in the peroxisomal α-oxidation. Pristanic acid is further degraded in the peroxisomal β-oxidation (Mukherji et al. 2003).

3 Refsum's Disease

Heredopathia atactica polyneuritiformis (Refsum's disease) is a hereditary recessive disorder affecting the nervous system function and characterized by retinitis pigmentosa (RP), hypertrophic peripheral neuropathy and cerebellar ataxia (Refsum 1976). Patients suffering from this disease have a defect in the enzyme responsible for the α-oxidation of fatty acids and therefore lack the function to normally metabolize phytanic acid, resulting in its accumulation. The severity of Refsum's disease is correlated to the concentration of phytanic acid in the serum (Mize et al. 1966). The disease is usually diagnosed during childhood or young adulthood when visual problems become apparent. The first symptom of Refsum's disease is night blindness followed by gradual loss of peripheral vision.

The best way to treat Refsum's disease is to maintain a strict diet that excludes food with a high content of phytanic acid (or phytol). By maintaining a phytanic acid low diet (less than 10 mg/day) it is possible to keep the serum level of phytanic acid below 10 mg/100 ml and thus prevent the adverse effects characterizing the disease (Masters-Thomas et al. 1980).

4 Control of Serum Levels of Cholesterol and Triglycerides

Due to their hydrophobic nature, lipids are transported in the blood as part of lipoprotein complexes. The lipoprotein particles include low density lipoproteins (LDLs) and high density lipoproteins (HDLs), primarily involved in tissue cholesterol balance. Because LDL, which transports cholesterols into tissues, is involved in the formation of atherosclerotic plaques in humans, cholesterol in these particles is often considered to be the disease-mediating form of cholesterol. By contrast, HDL, which has the capacity to transport surplus cholesterol away from the tissues, is associated with a decreased risk of developing CVD. Increased levels of HDL, and increased cholesterol within these HDL particles, are thus considered to be advantageous. The most beneficial effect of controlling levels of cholesterol and triglycerides is considered to be related to the correlation to increased risk of CVD-like myocardial infarcts and strokes, which are linked to severe outcomes and eventually lethality. For a long time, the pharmaceutical development of triglyceride-lowering drugs, such as the fenofibrates that are used to treat hyperglycemia, (McKeage and Keating 2011) or cholesterol-lowering drugs, such as statins used to treat hypercholesterolemia. Lardizabal and Deedwania (2011) have focused on managing LDL and VLDL (very low density lipoprotein) levels (Sheng et al. 2012). As the relationship between HDL, LDL, and VLDL—as well as cholesterol metabolism—is far from understood, more focus has recently been drawn to the levels of HDL (Lin et al. 2010) and Lipoprotein A (Nordestgaard et al. 2010) as markers of CVD.

4.1 Phytol Administered to Animals

Rats fed on a diet containing 5 % phytol had the capacity to rapidly absorb and degrade phytol (Mize et al. 1966). Rats have a capacity to absorb as much as 0.2 g phytol/kg body weight into the intestinal lymphatic system after oral administration (Baxter and Steinber 1967; Baxter et al. 1967). The most important route of uptake is via the intestinal lymphatic system and it was observed that more than 50 % of orally administered phytol is absorbed. Dietary phytol given to animals (rat, mouse, rabbit, and chinchillas) at doses of 1–5 % in the food leads to accumulation in both liver and serum. Once phytol was removed from the food, however, the serum and liver concentrations of accumulated phytol rapidly normalized (Steinberg et al. 1966).

Since non-ruminant animals lack the capacity to release the phytol-moiety from chlorophyll during digestion, the absorption of phytol, when provided as part of chlorophyll, is much lower compared to when given as pure phytol. For example, spinach can contain as much as 1–2 % phytol but when administered orally to rats, only about 1–2 % of the phytol content is absorbed (Baxter and Steinber 1967). Hence, when phytol is consumed as chlorophyll it is mostly excreted with the feces.

4.2 Phytol Administered to Humans

Administration and kinetic studies of radiolabeled phytol have been performed in patients suffering from Refsum's disease (Baxter 1968; Baxter et al. 1967). It was also observed how exogenously administered phytol was incorporated into triglycerides and phospholipid fatty acids, after synthesis to phytanic acid (Kahlke and Wagener 1966). As mentioned, serum levels of phytanic acid in healthy individuals are at micromolar concentrations (between 3 and 10 μM, respectively) (ten Brink et al. 1992). In patients with Refsum's disease the plasma concentration of phytanic acid and pristanic acid can rise to 1,300 and 80 μM, respectively (Verhoeven et al. 1998).

4.3 Molecular Function of Phytol

Many of the biological effects of phytanic acid and phytol observed, have been assumed to be transduced through agonist activity on the nuclear receptors RXRα (retinoid X receptor) (Lemotte et al. 1996; McCarty 2001) and PPARα (Ellinghaus et al. 1999; Zomer et al. 2000). Phytanic acid, but not phytol, has been shown to be an RXR activator in concentrations similar to those found in plasma (≥4 μM) (Kitareewan et al. 1996), while both phytol and phytanic acid bind to and activate PPARα (Goto et al. 2005). Due to the importance of these transcription factors in metabolic control, it has been suggested that phytanic acid—containing food—might act as a nutraceutical in the prevention of obesity-related diseases (Hellgren 2010).

5 Inflammation Models

Traditional herbal remedies containing phytol have been suggested to have inflammation ameliorating properties. For example, *E*-Phytol extracted from Aoki (*Aucuba Japonica*) was shown to exhibit arthritis ameliorating properties when applied topically as a crèmes in milligram doses (1–10 mg/application) and resulted in 20–40 % inhibition of induced paw edema in rats (Shimizu and Tomoo 1994).

Parts of the anti-inflammatory effects of phytol have been attributed to its capacity to induce production of reactive oxygen species (ROS) from the phagocyte NADPH oxidase (NOX2) complex. Contrary to the general dogma that ROS have a damaging effect on tissues and cells, it was suggested that increased ROS production could actually have a disease ameliorating effect (Gelderman et al. 2007; Hultqvist et al. 2006; Olofsson et al. 2003). By studying arthritogenic effects of carbohydrates of different length we could separate adjuvant effects from ROS inducing effects (Hultqvist et al. 2006). The ROS inducing effect was separated from the adjuvant effect of alkanes since only alkane oils with carbon chains of more than 15 carbons induced arthritis, while shorter alkanes were more potent in inducing an oxidative burst. It was also found that phytol was a potent inducer of ROS production in mammalian cells. When phytol was administered to arthritis-prone rats, we found a restoration of the ROS-producing capacity, as well as prevention of the onset of arthritis. Further studies confirmed this dramatic effect on arthritiogenicity and adjuvant capacity related to small changes in structure. Interestingly, phytol therefore represents another example of a redox modulating compound which by itself is not redox active (see Chap. 4 and other examples, such as Zn^{2+}). Phytol did not only prevent development of arthritis if injected before the onset of the disease. Rats with acute or chronic arthritis also showed a reduction in the severity of the disease after phytol administration. Even if phytol mainly affects ROS production, it has been hypothesized that preventive effects are mediated by indirectly affecting arthritiogenic T cells shown to be the main players in experimental models of arthritis (Holmberg et al. 2006). T cells that are oxidized *in vivo* are less arthritiogenic and production of ROS have been shown to be of importance for this regulation (Gelderman et al. 2007). The mechanism by which phytol induces ROS production in mammalian cells, however, is still unknown.

In addition, Phytol shows a strong protective effect when used in vaccination experiments against *Staphylococcus aureus* in mice, (Lim et al. 2006a, b) suggesting a great impact on bacterial resistance that might also be attributable to its ROS-inducing capacity. Phytol and phytol-based adjuvants are safe, with a high benefit-to-toxicity ratio. Recently, the anti-inflammatory effect of phytol was investigated in the course of experimental autoimmune uveitis in mice. Rather disappointingly, there was no effect on disease development (Daudin et al. 2011).

6 Metabolic Models

As described above, both phytol and phytanic acid have been implicated in the prevention of metabolic diseases, due to their agonist activity on RXRα and PPARα.

There are no published *in vivo* intervention studies with phytanic acid in either animal models or in humans. Thus, the existing data on effects of phytanic acid are based on *in vitro* studies and feeding studies with phytol. Several rodent studies have shown that an enhanced supply of phytol in the diet reduces serum and hepatic triglyceride levels. Hence 3 weeks of feeding mice phytol (0.5 weight % of diet) resulted in a 40 % reduction of serum triacylglycerol, while serum cholesterol levels remained unaffected (Van den Branden et al. 1986). In another mouse study, the same concentration of phytol reduced hepatic triacylglycerol by almost 75 % (Hellgren 2010). The reduction in triacylglycerols has been attributed partly to a PPARα-induced activation of both mitochondrial and peroxisomal β-oxidation in the liver, Gloerich et al. (2005), Hashimoto et al. (2006) although Hashimoto et al. also have also shown that phytol-induced activation of hepatic fatty acid oxidation is not inhibited completely in PPARα KO-mice (Hashimoto et al. 2006).

Besides the well-characterized effects on lipid metabolism, phytol and phytanic acid may also have positive effects on metabolic control via other mechanisms of action. In primary rat hepatocytes cultured in the presence of 100 µM phytanic acid, non-insulin-dependent glucose uptake as well as the mRNA levels of the glucose-transporters, GLUT-1 and GLUT-2, were substantially enhanced (Heim et al. 2002). The authors did not include any data on post-absorptive glucose metabolism in the paper. It is therefore impossible to conclude whether the enhanced glucose uptake is to be considered as a positive outcome, due to enhanced glucose-clearance from the blood, or whether the increased rate of uptake will drive enhanced fatty acid synthesis and lipid accumulation.

Phytanic acid, but not phytol, was also shown to be a powerful inducer of uncoupler protein-1 (UCP-1) and other markers of brown adipocyte differentiation (Schluter et al. 2002a, b). Due to their ability of uncoupled mitochondrial electron-transport, brown adipocytes have the capacity to increase cellular energy expenditure (Nedergaard and Cannon 2010). Hence, increased differentiation of brown adipocytes could increase energy expenditure and reduce the risk of obesity. Interestingly, it was recently shown that the prevalence of brown adipose tissue in adult humans is inversely related to BMI and body fat percentage (Vijgen et al. 2011).

Hence, available data indicates that intake of phytol and/or phytanic acid might have various positive effects on energy expenditure, fatty acid and glucose metabolism. *In vivo* studies using physiologically relevant concentrations of both phytol and phytanic acid, however, are needed before any definite conclusions can be drawn.

7 Other Health-Related Properties

Phytol has been suggested to posses other potential health benefits in other disease models like muscular dystrophia, (Hidiroglou and Jenkins 1972) as well as anticonvulsant and anti-epileptic effects (Costa et al. 2012). Taken together with the better studied anti-inflammatory, adjuvant properties (Chowdhury and Ghosh 2012) and most promising serum lipid lowering functions, further studies on the matter of the health properties of phytol are both challenging and interesting.

8 Conclusions

A natural component like phytol, with apparent therapeutic effects on widespread and serious disorders, like autoimmunity and CVD, will in the future have an important role to play a part of nutritional products with health applications. Indeed, phytol can be produced from a range of agricultural products and by-products with high contents of chlorophyll. Phytol therefore represents an interesting food ingredient to be introduced into healthcare or functional food products with a natural origin, for the improvement of health and for anti-inflammatory effects.

References

Allen NE, Grace PB, Ginn A, Travis RC, Roddam AW, Appleby PN, Key T (2008) Phytanic acid: measurement of plasma concentrations by gas–liquid chromatography–mass spectrometry analysis and associations with diet and other plasma fatty acids. Brit J Nutr 99(3):653–659

Baxter JH (1968) Absorption of chlorophyll phytol in normal man and in patients with refsum's disease. J Lipid Res 9(5):636–641

Baxter JH, Steinber D (1967) Absorption of phytol from dietary chlorophyll in rat. J Lipid Res 8(6):615–620

Baxter JH, Steinberg D, Mize CE, Avigan J (1967) Absorption and metabolism of uniformly 14C-labeled phytol and phytanic acid by the intestine of the rat studied with thoracic duct cannulation. Biochim Biophys Acta 137(2):277–290

Chowdhury RR, Ghosh SK (2012) Phytol-derived novel isoprenoid immunostimulants. Front Immunol 3:49

Costa JP, Ferreira PB, De Sousa DP, Jordan J, Freitas RM (2012) Anticonvulsant effect of phytol in a pilocarpine model in mice. Neurosci Lett 523(2):115–118

Daudin JB, Monnet D, Kavian N, Espy C, Wang A, Chereau C, Goulvestre C, Omri S, Brezin A, Weill B, Batteux F, Nicco C (2011) Protective effect of pristane on experimental autoimmune uveitis. Immunol Lett 141(1):83–93

Ellinghaus P, Wolfrum C, Assmann G, Spener F, Seedorf U (1999) Phytanic acid activates the peroxisome proliferator-activated receptor alpha (PPAR alpha) in sterol carrier protein 2-/ sterol carrier protein x-deficient mice. J Biol Chem 274(5):2766–2772

Gelderman KA, Hultqvist M, Olsson LM, Bauer K, Pizzolla A, Olofsson P, Holmdahl R (2007) Rheumatoid arthritis: the role of reactive oxygen species in disease development and therapeutic strategies. Antioxid Redox Signal 9(10):1541–1567

Gloerich J, van den Brink DM, Ruiter JPN, van Vlies N, Vaz FM, Wanders RJA, Ferdinandusse S (2007) Metabolism of phytol to phytanic acid in the mouse, and the role of PPAR alpha in its regulation. J Lipid Res 48(1):77–85

Gloerich J, van Vlies N, Jansen GA, Denis S, Ruiter JPN, van Werkhoven MA, Duran M, Vaz FM, Wanders RJA, Ferdinandusse S (2005) A phytol-enriched diet induces changes in fatty acid metabolism in mice both via PPAR alpha-dependent and -independent pathways. J Lipid Res 46(4):716–726

Goto T, Takahashi N, Kato S, Egawa K, Ebisu S, Moriyama T, Fushiki T, Kawada T (2005) Phytol directly activates peroxisome proliferator-activated receptor alpha (PPAR alpha) and regulates gene expression involved in lipid metabolism in PPAR alpha-expressing HepG2 hepatocytes. Biochem biophys res commun 337(2):440–445

Gremese E, Ferraccioli G (2011) The metabolic syndrome: the crossroads between rheumatoid arthritis and cardiovascular risk. Autoimmun Rev 10(10):582–589

Hashimoto T, Shimizu N, Kimura T, Takahashi Y, Ide T (2006) Polyunsaturated fats attenuate the dietary phytol-induced increase in hepatic fatty acid oxidation in mice. J Nutr 136(4):882–886

Heidenreich PA, Trogdon JG, Khavjou OA, Butler J, Dracup K, Ezekowitz MD, Finkelstein EA, Hong YL, Johnston SC, Khera A, Lloyd-Jones DM, Nelson SA, Nichol G, Orenstein D, Wilson PWF, Woo YJ, Coordina AHAA, Council S, Inter CCR, Cardiology CC, Prevention CE, Thrombosi CA, C CCC, Nursing CC, Dis CKC, Anesthe CCS, Quality IC (2011) Forecasting the future of cardiovascular disease in the United States a policy statement from the american heart association. Circulation 123(8):933–944

Heim M, Johnson J, Boess F, Bendik I, Weber P, Hunziker W, Fluhmann B (2002) Phytanic acid, a natural peroxisome proliferator-activated receptor agonist, regulates glucose metabolism in rat primary hepatocytes. Faseb J 16(3):718–720

Hellgren LI (2010) Phytanic acid-an overlooked bioactive fatty acid in dairy fat? Ann Ny Acad Sci 1190:42–49

Hidiroglou M, Jenkins KJ (1972) Effect of phytol on the incidence of muscular dystrophy in rabbits. Can J Physiol Pharmacol 50(2):171–172

Holmberg J, Tuncel J, Yamada H, Lu S, Olofsson P, Holmdahl R (2006) Pristane, a non-antigenic adjuvant, induces MHC class II-restricted, arthritogenic T cells in the rat. J Immunol 176(2):1172–1179

Hultqvist M, Olofsson P, Gelderman KA, Holmberg J, Holmdahl R (2006) A new arthritis therapy with oxidative burst inducers. PLoS Med 3(9):e348

Kahlke W, Wagener H (1966) Conversion of H3-Phytol to phytanic acid and its incorporation into plasma lipid fractions in heredopathia atactica polyneuritiformis. Metab 15(8):687–693

Kitareewan S, Burka LT, Tomer KB, Parker CE, Deterding LJ, Stevens RD, Forman BM, Mais DE, Heyman RA, McMorris T, Weinberger C (1996) Phytol metabolites are circulating dietary factors that activate the nuclear receptor RXR. Mol Biol Cell 7(8):1153–1166

Koch-Henriksen N, Sorensen PS (2010) The changing demographic pattern of multiple sclerosis epidemiology. Lancet Neurol 9(5):520–532

Lardizabal JA, Deedwania P (2011) Lipid-lowering therapy with statins for the primary and secondary prevention of cardiovascular disease. Cardiol Clin 29(1):87–103

Lemotte PK, Keidel S, Apfel CM (1996) Phytanic acid is a retinoid X receptor ligand. Eur J Biochem 236(1):328–333

Lim SY, Bauermeister A, Kjonaas RA, Ghosh SK (2006a) Phytol-based novel adjuvants in vaccine formulation: 2. assessment of efficacy in the induction of protective immune responses to lethal bacterial infections in mice. J Immune Based Ther Vaccines 4:5

Lim SY, Meyer M, Kjonaas RA, Ghosh SK (2006b) Phytol-based novel adjuvants in vaccine formulation: 1. assessment of safety and efficacy during stimulation of humoral and cell-mediated immune responses. J Immune Based Ther Vaccines 4:6

Lin Y, Mousa SS, Elshourbagy N, Mousa SA (2010) Current status and future directions in lipid management: emphasizing low-density lipoproteins, high-density lipoproteins, and triglycerides as targets for therapy. Vasc Vascular health Risk Manage 6:73–85

Masters-Thomas A, Bailes J, Billimoria JD, Clemens ME, Gibberd FB, Page NG (1980) Heredopathia atactica polyneuritiformis (Refsum's disease): 1. Clinical features and dietary management. J Hum Nutr 34(4):245–250

McCarty MF (2001) The chlorophyll metabolite phytanic acid is a natural rexinoid—potential for treatment and prevention of diabetes. Med Hypotheses 56(2):217–219

McKeage K, Keating GM (2011) Fenofibrate a review of its use in dyslipidaemia. Drugs 71(14):1917–1946

Mize CE, Avigan J, Baxter JH, Fales HM, Steinberg D (1966) Metabolism of phytol-U-14C and phytanic acid-U-14C in the rat. J Lipid Res 7(5):692–697

Mukherji M, Schofield CJ, Wierzbicki AS, Jansen GA, Wanders RJ, Lloyd MD (2003) The chemical biology of branched-chain lipid metabolism. Prog Lipid Res 42(5):359–376

Nedergaard J, Cannon B (2010) The changed metabolic world with human brown adipose tissue: therapeutic visions. Cell Metab 11(4):268–272

Nordestgaard BG, Chapman MJ, Ray K, Boren J, Andreotti F, Watts GF, Ginsberg H, Amarenco P, Catapano A, Descamps OS, Fisher E, Kovanen PT, Kuivenhoven JA, Lesnik P, Masana L, Reiner Z, Taskinen MR, Tokgozoglu L, Tybjaerg-Hansen A, Conse EAS (2010) Lipoprotein(a) as a cardiovascular risk factor: current status. Eur Heart J 31(23):2844–2853

Olofsson P, Holmberg J, Tordsson J, Lu SM, Akerstrom B, Holmdahl R (2003) Positional identification of Ncf1 as a gene that regulates arthritis severity in rats. Nat Genet 33(1):25–32

Patton S, Benson AA (1966) Phytol metabolism in the bovine. Biochim Biophys Acta 125(1):22–32

Pereira RM, de Carvalho JF, Bonfa E (2009) Metabolic syndrome in rheumatological diseases. Autoimmun Rev 8(5):415–419

Refsum S (1976) Heredopathia atactica polyneurotiformis: therapeutic and pathogenetic aspects. Rivista di patologia nervosa e mentale 97(3):115–126

Schluter A, Barbera MJ, Iglesias R, Giralt M, Villarroya F (2002a) Phytanic acid, a novel activator of uncoupling protein-1 gene transcription and brown adipocyte differentiation. Biochem J 362(Pt 1):61–69

Schluter A, Giralt M, Iglesias R, Villarroya F (2002b) Phytanic acid, but not pristanic acid, mediates the positive effects of phytol derivatives on brown adipocyte differentiation. FEBS Lett 517(1–3):83–86

Scott DL, Wolfe F, Huizinga TWJ (2010) Rheumatoid arthritis. Lancet 376(9746):1094–1108

Sheng X, Murphy MJ, MacDonald TM, Wei L (2012) Effect of statins on total cholesterol concentrations and cardiovascular outcomes in patients with diabetes mellitus: a population-based cohort study. Eur J Clin Pharmacol 68(8):1201–1208

Shimizu M, Tomoo T (1994) Anti-inflammatory constituents of topically applied crude drugs. V. Constituents and anti-inflammatory effect of Aoki, Aucuba japonica Thunb. Biol Pharm Bulletin 17(5):665–667

Steinberg D, Avigan J, Mize CE, Baxter JH, Cammermeyer J, Fales HM, Highet PF (1966) Effects of dietary phytol and phytanic acid in animals. J Lipid Res 7(5):684–691

ten Brink HJ, Schor DS, Kok RM, Poll-The BT, Wanders RJ, Jakobs C (1992) Phytanic acid alpha-oxidation: accumulation of 2-hydroxyphytanic acid and absence of 2-oxophytanic acid in plasma from patients with peroxisomal disorders. J Lipid Res 33(10):1449–1457

Todd JA, Knip M, Mathieu C (2011) Strategies for the prevention of autoimmune type 1 diabetes. Diabet Med: J British Diabet Assoc 28(10):1141–1143

Van den Branden C, Vamecq J, Wybo I, Roels F (1986) Phytol and peroxisome proliferation. Pediatr Res 20(5):411–415

van den Brink DM, Wanders RJ (2006) Phytanic acid: production from phytol, its breakdown and role in human disease. Cell Mol Life Sci: CMLS 63(15):1752–1765

Verhoeven NM, Wanders RJ, Poll-The BT, Saudubray JM, Jakobs C (1998) The metabolism of phytanic acid and pristanic acid in man: a review. J Inherit Metab Dis 21(7):697–728

Vijgen GH, Bouvy ND, Teule GJ, Brans B, Schrauwen P, van Marken Lichtenbelt WD (2011) Brown adipose tissue in morbidly obese subjects. PLoS One 6(2):e17247

Werner LB, Hellgren LI, Raff M, Jensen SK, Petersen RA, Drachmann T, Tholstrup T (2011) Effect of dairy fat on plasma phytanic acid in healthy volunteers—a randomized controlled study. Lipids Health Dis 10:95

Zomer AW, van Der Burg B, Jansen GA, Wanders RJ, Poll-The BT, van Der Saag PT (2000) Pristanic acid and phytanic acid: naturally occurring ligands for the nuclear receptor peroxisome proliferator-activated receptor alpha. J Lipid Res 41(11):1801–1807

Authors Biography

Peter Olofsson (born 1972) obtained his research-based Master degree in Biochemistry and Molecular Biology at the end of the 1990s. To further advance his skills in research he continued with a doctoral education and received a PhD from the University of Lund in 2003 with a focus on genetics and the inheritance of complex genetic disorders. Peter's main scientific interest and continued research is within the field of immunology and autoimmunity. Within this field, his research during the last 10 years has been devoted to the role of redox regulating enzymes and the role of Reactive Oxygen Species as regulators of the innate immune system. Since 2003, Peter has shifted from academic research to the industry and development of therapies for severe autoimmune conditions. Here, he contributes his extensive experience of preclinical models of autoimmunity, drug discovery, and project development. Besides academic training and research, Peter has also an education in finance and law and an executive MBA from Gothenburg University. Peter has held project-leading positions at companies such as Arexis AB and Biovitrum AB (publ.). He is engaged as principal investigator of several EU-funded research collaborations and is co-author of more than 30 peer-reviewed articles and patents. Since 2007, Peter has been the CEO of the drug development company, Redoxis AB.

Malin Hultqvist (born 1979) holds a degree in Pharmaceutical Biomedicine from Gothenburg University in Sweden. She has started her scientific career as a PhD student in the laboratory of Professor Rikard Holmdahl in 2003, at the University of Lund. In 2007 Malin defended her PhD thesis in the field of redox regulation of immunology, with a thesis entitled "The Role of Reactive Oxygen Species in Animal Models of Autoimmunity". Since 2008, Malin is working in the company Redoxis, dealing with the development of small molecules for the treatment of autoimmune diseases by targeting the phagocyte NADPH oxidase pathway. She is currently the CSO of the company.

Malin has a strong interest in redox regulation of the immune system and has over 25 publications in this field.

Lars I. Hellgren (born 1962) graduated with an M.Sc. in Biology from the University of Gothenburg in the early 1990s, and continued his research education with PhD studies in Plant Physiology at the same university. During his PhD, he developed an interest in the role of lipid metabolism in cellular development and regulation, and its relevance for physiological performance, both at organ and whole organism level—and not only in plants but also in relation to human nutrition and medicine. After having defended his PhD thesis in 1996, Lars took up a position in Nutritional Biochemistry of Lipids at the Department of Biochemistry and Nutrition at the Technical University of Denmark. During the last 10 years, Lars' research has focused on dysfunctions in lipid metabolism in the development of metabolic diseases, and how dietary fatty acid intake can prevent or prime this development. As part of these studies, he has a particular interest in naturally occurring PPAR-agonists, such as phytanic acid, and has led a project studying whether dairy fat with increased concentrations of phytanic acid could be protective against the development of insulin resistance and non-alcoholic fatty liver disease.

During the last couple of years, the role of altered lipid fluxes into tissue-resident immune cells during the development of metabolic diseases has gained much of his research interest, and he is now group leader for the group "Systems Biology of Immune Regulation" at the Center for Biological Sequence Analysis, Department of Systems Biology, Technical University of Denmark. To date, Lars has published over 50 peer-reviewed publications, 34 of them since 2007.

Rikard Holmdahl (born 1953) has been trained in immunology at Uppsala University, from where he obtained a PhD in 1985 and an MD in 1987. In 1993 he moved to Lund University as Professor in Medical Inflammation Research. He has been a Professor at the Finnish National Academy, located in Turku, from 2007 until 2011 (50 %) and Professor at Karolinska Institutet (Stockholm) from 2008 until now.

Rikard is an expert in the genetics and immunology of animal models for autoimmune diseases. He is currently the Head of the Medical Inflammation Research Laboratory at the Karolinska Institutet and Turku University, which is a leading research center for the studies of experimental animal models for autoimmune diseases. He is also assisting several scientific journals (associated editor for Arthritis Research & Therapy, an executive committee member of the European Journal of immunology and in the editorial board of Scand J Immunol). He is a partner in and has been a coordinator of several European grant consortia. He is also a founder of and advisor for several biopharmaceutical companies.

Rikard leads an active research team focusing on unravelling the genetic and environmental control of autoimmune diseases, with emphasis on models for rheumatoid arthritis (RA) and multiple sclerosis (MS). His team has identified the major loci in models of RA and MS and positioned some of the most important genes. One of these genes is Ncf1, which was identified to control an oxidation pathway regulating autoimmunity and chronic inflammation. He has also positioned the MHC class II genes in animal models and has conducted studies on humanised MHC class II mice. His team is currently focusing on the etiologic and pathogenic events directed by antigen-specific T and B cells causing and regulating autoimmune disease.

Rikard has authored over 450 publications in Medline listed journals and has currently over 19,000 citations and a Hirsch-index of 76 (according to Google Scholar).

Part XIII
Connecting Section Between Chapters 13 and 14

The use of ruminants, such as cows, to turn a common, widely available plant metabolite into a valuable diterpene product associated with potential applications in Nutrition and Medicine, is clearly a very elegant approach toward the refinement of metabolites. Indeed, most practical applications in the field of Medicine or Agriculture tend to rely initially on readily available metabolites from plants, fungi or bacteria. After the first success of the natural substance, however, the desire to modify the core structure of the active ingredient in order to obtain an improved activity and/or less side effects (i.e., higher selectivity) becomes a driving force for further research and development. There are many historical examples that bear witness to this particular development away from the natural compound and toward more amenable (semi-)synthetic derivatives. One of the best known examples is perhaps the development of aspirin, acetyl-salicylic acid, yet there are many other examples as well.

Nowadays, there are various methods that enable researchers—and industry—to deviate from the chemical structure provided initially by the natural source, and to arrive at tailored derivatives. These methods include more traditional techniques, such as digestion (as in the case of the diterpenes produced by cows), degradation (e.g., in the case of garlic, where chemical decomposition converts allicin into polysulfanes), fermentation (e.g., for the production of alcohol), maceration (in the case of resveratrol extraction from grape skins), or enzymatic conversion (e.g., stereo-selective oxidation of educts by haloperoxidases, see Chap. 17). Increasingly, however, biotechnology and chemical synthesis are used to address the growing demand for 'derivatives' of natural products. Both methods promise a variety of new substances 'derived from Nature' in good yields and high purity, and can be applied on an industrial scale.

We will first consider the emerging power of biotechnology. The last couple of decades have seen enormous progress in this field, and it is now feasible to map out the biosynthetic pathways of most secondary metabolites. Having obtained such knowledge regarding the individual enzymes involved in the biosynthesis of a given natural product, it then becomes possible to increase the efficiency of production, and even to change this 'production line' in order to obtain new products by targeted mutation of the specific enzymes involved. Once such mutations are in place—and the organism is still viable and the respective assembly line still functioning—a large-scale production of suitably altered metabolites becomes feasible. Indeed, as such mutated plants, fungi or bacteria can be cultivated, an almost endless supply of the desired natural and 'almost natural' products may become available.

The next chapter by Frédéric Bourgaud will highlight this approach using the furanocoumarin synthesis in higher plants as an example. This chapter will provide insight into modern research in molecular genetics, and show how previous 'black boxes' involved in the biosynthesis of a rather complex natural metabolite are consecutively being filled with newly discovered enzymes. Eventually, this kind of research illuminates the production line of the furanocoumarin biosynthesis and provides the basis for an enhanced biosynthetic efficiency and even further product development via genetic engineering.

Chapter 14
Recent Advances in Molecular Genetics of Furanocoumarin Synthesis in Higher Plants

Frédéric Bourgaud, Alexandre Olry and Alain Hehn

Keywords Furanocoumarin • Phenylpropanoidpathway • Psoralen • Umbelliferone

1 Introduction

Furanocoumarins represent a subclass of polyphenolics which are principally found in four families of higher plants: *Apiaceae* (celery, parsnip, Ammi, parsley, etc.,), *Fabaceae* (Psoralea, Cullen, Bituminosa, Coronilla, etc.,), *Rutaceae* (Ruta, Citrus, Zanthoxylum, Fagara, etc.,), and *Moraceae* (Ficus) (Bourgaud et al. 1989). They are structurally characterized by the occurrence of a furan ring attached to the carbons 6 and 7 (i.e., linear furanocoumarins) or 7 and 8 (i.e., angular furanocoumarins) of a benzo-α-pyrone nucleus (Fig. 1).

Several dozens of furanocoumarins have been reported in plants (Fig. 1) and have been assigned to important ecological functions in the plant kingdom. They have been described as potent allelochemicals bearing various properties and acting as phytoalexins (Beier and Oertli 1983), anti-insect (Schuler 2011), and allelopathic agents (Baskin et al. 1967; Hale et al. 2004; Razavi 2011). In the context of humans, furanocoumarins are used in traditional and modern medicine, exhibiting pharmacological properties such as anticancer (Luo et al. 2011; Plumas et al. 2003) or anti-psoriatic properties (Adisen et al. 2008).

The characterization of the furanocoumarin pathway in plants was achieved during the period of 1960–1990 thanks to the use of radiolabeled compounds (Bourgaud et al. 2006). Until recently, however, there was no information available

F. Bourgaud (✉) · A. Olry · A. Hehn
Agronomie et Environment, Université de Lorraine (Nancy-Colmar), UMR 1121, 54505 Vandœuvre-lès-Nancy, France
e-mail: frederic.bourgaud@univ-lorraine.fr

C. Jacob et al. (eds.), *Recent Advances in Redox Active Plant and Microbial Products*,
DOI: 10.1007/978-94-017-8953-0_14,

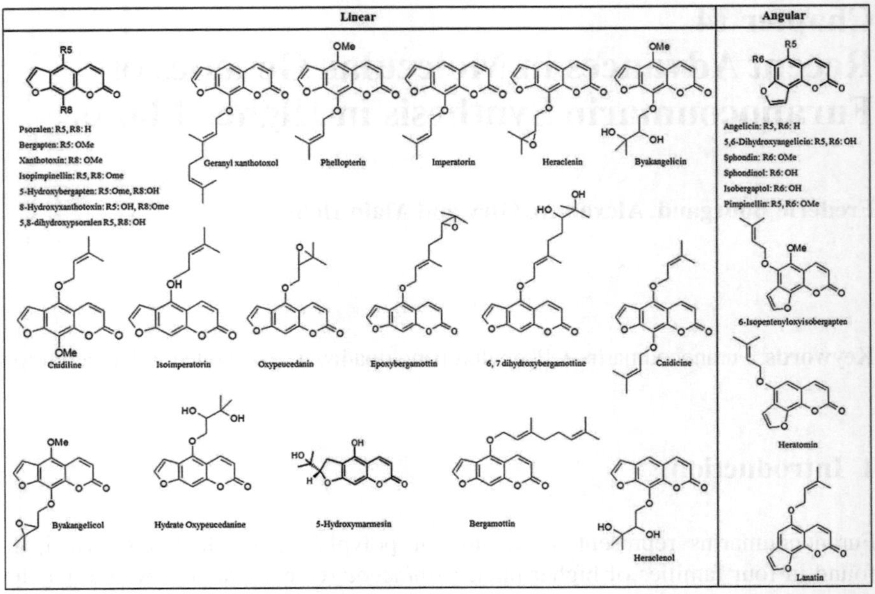

Fig. 1 Overview of several substructures furanocoumarins

regarding the molecular genetics behind this synthesis in higher plants. The purpose of this chapter is to provide an update on recent results published over the last years and that concern the first plant genes functionally characterized in this pathway. The biosynthesis of furanocoumarins that occurs in plants can be divided into several steps (Fig. 2).

1. **General phenylpropanoid pathway**. These molecules derive from the general phenylpropanoid pathway after the conversion of phenylalanine into cinnamic acid by phenylalanine ammonia lyase (PAL) and its subsequent transformation into *p*-coumaric acid by the cinnamate-4-hydroxylase enzyme (C4H).
2. **Synthesis of umbelliferone**. This important step is characterized by the *ortho*-hydroxylation of *p*-coumaric acid by a cinnamoyl-CoA 2′hydroxylase enzyme (C2′H) and, until more recently, has remained the last unknown hydroxylation step operating on the cinnamic acid nucleus, after the previous discoveries of C4H (Teutsch et al. 1993) and C3′H (*p*-coumaroyl-quinic or shikimic 3′-hydroxylase) (Schoch et al. 2001).
3. **Synthesis of demethylsuberosin (DMS) and osthenol**. These steps correspond to the prenylation of umbelliferone by a dimethylallyl pyrophosphate moiety, which occurs respectively on carbon 6 for DMS and carbon 8 for osthenol. This key step is operated by prenyltransferase enzymes that allow the opening of the routes to linear and angular furanocoumarins, for DMS and osthenol, respectively.

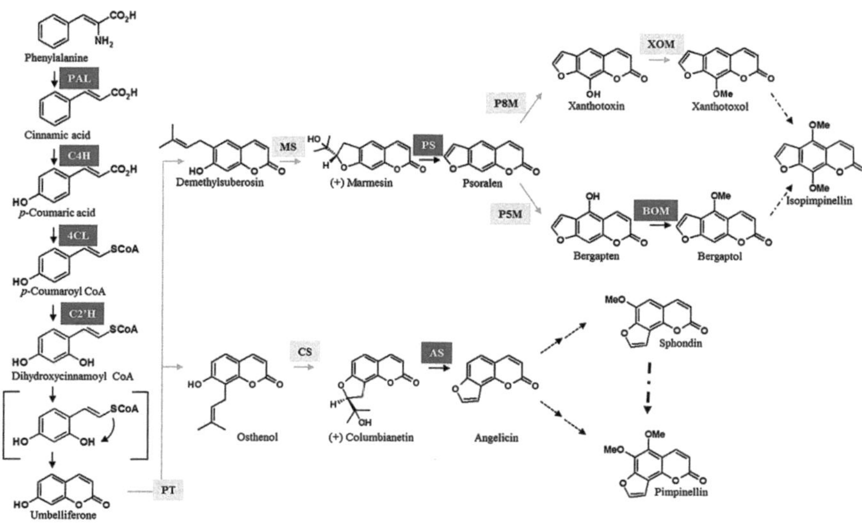

Fig. 2 Biosynthesis of furanocoumarins

4. **Synthesis of psoralen and angelicin**. Psoralen and angelicin are formed in a strict parallel manner, respectively, after the conversion of DMS into marmesin, and osthenol into columbianetin by a set of cytochrome P450 enzymes (P450).
5. **Synthesis of hydroxylated forms of psoralen and angelicin**. Psoralen and angelicin are frequently transformed into more hydroxylated products, such as 5-hydroxypsoralen (bergaptol) and 8-hydroxypsoralen (xanthotoxol) from psoralen. These enzymatic steps are governed by cytochrome P450 enzymes.
6. **Synthesis of prenylated furanocoumarins**. These prenylated furanocoumarins are frequently found in Citrus plants, like 5-geranyloxypsoralen (bergamottin, Fig. 1) a major component of bergamot essential oil.
7. **Synthesis of methoxylated products**. Hydroxylated forms of psoralen and angelicin are subsequently transformed into methoxylated products, such as 5-methoxypsoralen (bergapten), 8-methoxypsoralen (xanthotoxin), 5,8-dimethoxypsoralen (isopimpinellin), or 6-methoxyfuro[2,3-h]chromen-2-one (sphondin). This last step of transformation is achieved by methyl-transferase enzymes.

2 General Phenylpropanoid Pathway

At present, there is no report of PAL activity specifically related to the furanocoumarin pathway. The function of four different isozymes of PAL was investigated in *Petroselinum crispum* but no difference could be highlighted regarding their kinetic constants (Appert et al. 1994). PAL3 was shown to be preferentially activated in wounded leaves (Lois and Hahlbrock 1992). It is unknown, however, if the entry of phenylalanine into the furanocoumarin pathway is controlled by a special subclass of PAL enzymes, like PAL3 in parsley. If the furanocoumarin pathway were organized in metabolons (Jorgensen et al. 2005; Winkel 2004) as is the case for the lignin or flavonoid pathway in *Arabidopsis*, it would be reasonable to expect a furanocoumarin-dedicated PAL isoform, which might operate specifically in this pathway.

The synthesis of *p*-coumaric acid from cinnamic acid is realized by a P450, the enzyme C4H, which has received much attention in the literature (Fahrendorf and Dixon 1993; Gravot et al. 2004; Hubner et al. 2003; Koopmann et al. 1999; Pierrel et al. 1994). The first C4H from a furanocoumarin producing plant was isolated from parsley, and was demonstrated to be highly transcribed upon induction by UV, fungal elicitation, or wounding (Koopmann et al. 1999). The activities of C4H were also compared between furanocoumarin producing plants and plants that do not synthesize these molecules. *Ruta graveolens*, a *Rutaceae* plant that synthesizes linear furanocoumarins, displays a C4H with kinetic constants very comparable (i.e., K_m around 1 μM) to the *Arabidopsis* C4H, a plant that presents a single P450 involved in this activity (Gravot et al. 2004).

A basic biochemical property of furanocoumarins is represented by their capacity to act as mechanism-based inhibitors (MBI) of P450 enzymes (Gravot et al. 2004). This biochemical property has been studied extensively in humans where the absorption of furanocoumarins can dramatically interfere with drugs. Casual uptake of furanocoumarins via the human diet can eventually prevent the normal drug clearance process by disrupting the pool of P450s (Girennavar et al. 2007; Gorgus et al. 2010). Because plants contain numerous P450s, this process of mechanism-based inhibition is likely to occur not only in humans but also in plants which synthesize furanocoumarins. P450 inhibition by furanocoumarins was compared for several C4Hs originating from furanocoumarin producing and non-furanocoumarin producing plants (Gravot et al. 2004). It was demonstrated that C4Hs from furanocoumarin containing plants (*Ruta graveolens* and parsley) were much more resistant to MBI than C4Hs, which were cloned from plants devoid of furanocoumarins (*Arabidopsis thaliana* and Jerusalem artichoke). These results might reflect an adaptation process that would have taken place in plants which synthesize furanocoumarins, rendering their C4Hs more capable to operate in a cellular environment rich in furanocoumarins without being disrupted (Gravot et al. 2004).

3 Synthesis of Umbelliferone

Umbelliferone (7-hydroxycoumarin) is a ubiquitous molecule found in plants. It is not only involved in the synthesis of furanocoumarins, but also in the formation of general methoxylated coumarins such as scopoletin and its glycoside derivative, scopolin, two important phytoalexins in plants (Kai et al. 2006). This synthesis of umbelliferone was a missing link in the general phenylpropanoid pathway until it was recently described in *Ruta graveolens* (Vialart et al. 2012). This synthesis is controlled by a C2′H, which hydroxylates *p*-coumaroyl-CoA at the *ortho* position, and not by a P450, as was speculated previously (Ellis and Amrhein 1971).

The first C2′H described by Vialart and co-workers is a bifunctional enzyme that not only synthesizes umbelliferone in *Ruta graveolens*, but also converts feruloyl-CoA into scopoletin. This latter activity, corresponding to a feruloyl-CoA 6′ hydroxylation (F6′H), was described previously in *Arabidopsis* where this enzyme is also present and responsible for the formation of scopoletin (Kai et al. 2006). This F6′H found in *Arabidopsis*, however, is not capable of synthesizing umbelliferone from *p*-coumaroyl-CoA, but only converts feruloyl-CoA into scopoletin, to the difference of the bifunctional C2′H found in *R. graveolens*. Ruta C2′H and *Arabidopsis* F6′H share a large percentage of identity at the amino acid level (58.9 %), strongly suggesting that the bifunctional C2′H might have evolved from the monofunctional F6′H in the course of plant evolution.

Ruta C2′H mRNA were found to be actively transcribed upon elicitation by UV light (Vialart et al. 2012). Another interesting feature of Ruta C2′H is related to its inhibition by psoralen: this enzyme is inhibited competitively by psoralen, and not by other furanocoumarins, suggesting that the psoralen concentration could act as a negative regulator of the enzyme's activity. This inhibition mechanism might prevent excessive concentrations of psoralen from accumulating in the plant cell, which might turn out to be toxic for the plant, due to psoralen's general toxicity (Vialart et al. 2012).

It is worth noting that C2′H does not operate on *p*-coumaric acid itself, but on its CoA derivative. Therefore, this C2′H enzyme is likely to work in tandem with a *p*-coumarate CoA ligase. Notably, a *p*-coumarate CoA ligase was previously described in *R. graveolens* and the corresponding gene was shown to be preferentially transcribed in parallel with the synthesis of furanocoumarins, turning this *p*-coumarate CoA ligase into an ideal candidate to work in conjunction with the Ruta C2′H (Endler et al. 2008).

4 Synthesis of Demethyl-suberosin and Osthenol

Umbelliferone is subsequently transformed into DMS or osthenol after respective prenylation at C6 or C8 by a dimethylallyl pyrophosphate (DMAPP) substrate. This enzymatic step is of pivotal importance to the furanocoumarin pathway because it opens the path to either linear or angular furanocoumarins.

There have been several *C*-prenyltransferases described over the past few years that operate on flavonoids (Sasaki et al. 2008; Shen et al. 2012; Yazaki et al. 2009). Until now, however, the molecular characterization of a prenyltransferase enzyme responsible for the prenylation of umbelliferone at C6 or C8 is still missing. Notably, it is unknown if a single enzyme is responsible for both prenylations at C6 and C8, or if several specific enzymes are required.

Although the question of the umbelliferone DMAPP prenyltransferase is still unresolved, the metabolic origin of the DMAPP moiety was investigated. The use of deuterated glucose unequivocally established that in celery DMAPP derives from the methyl erythrose phosphate (MEP) pathway and not from the mevalonate pathway (Stanjek et al. 1999). This is an indication that the umbelliferone DMAPP prenyltransferase is likely to operate at the chloroplastic level, the assigned location of the MEP pathway, as was already demonstrated for flavonoid *C*-prenyltransferases (Sasaki et al. 2008).

5 Synthesis of Psoralen and Angelicin

Psoralen is formed from DMS after the action of two consecutive P450s: the marmesin synthase and the psoralen synthase (Hamerski and Matern 1988a, b).

Psoralen synthase was the first P450 committed to the furanocoumarin pathway which was characterized at the molecular level, and is referred to as CYP71AJ1 in the international P450 nomenclature (http://drnelson.uthsc.edu/CytochromeP450.html). The CYP71AJ1 cDNA was isolated from elicitor-treated cells of *Ammi majus* (*Apiaceae*), and the corresponding enzyme was demonstrated to be responsible for a very atypical reaction for a P450, which corresponds to a carbon–carbon bond cleavage with a, in the plant kingdom, unique release of one molecule of acetone (Larbat et al. 2007). Interestingly, it was demonstrated that CYP71AJ1 is resistant to mechanism-based inhibition by furanocoumarins, suggesting that this P450 enzyme can operate without major difficulties in a cellular environment rich in furanocoumarins (Larbat et al. 2007). CYP71AJ1 is an exclusive psoralen synthase and cannot form angelicin from columbianetin. Columbianetin, however, was shown to act as a competitive inhibitor of marmesin (Larbat et al. 2007). This result demonstrates that columbianetin can somehow enter and fit well into the active site of CYP71AJ1, challenging marmesin, the real substrate, in this matter. Several orthologous genes have been isolated from other *Apiaceous* plants. These enzymes display the enzymatic characteristics of CYP71AJ1 (Larbat et al. 2009). An additional orthologous gene was isolated from *Psoralea corylifolia*, but in this case, the function has not been characterized so far (Parast et al. 2011).

Angelicin synthase was subsequently discovered in *Pastinaca sativa* (parsnip, *Apiaceae*), a plant that synthesizes both linear and angular furanocoumarins (Larbat et al. 2009). This enzyme is referred to as CYP71AJ4 and was compared to CYP71AJ3, a psoralen synthase orthologous to CYP71AJ1, which was also isolated from *P. sativa* (Larbat et al. 2009). CYP71AJ4 and CYP71AJ3 were shown to share a high percentage of identity at the amino acid level (73.8 %). The two protein sequences were also compared for a restricted number of peptides referred to as substrate recognition sites (SRS), which correspond to amino acids implicated in critical interactions with the substrates. Interestingly, this percentage of identity dropped to 40 % once the SRS were concerned. These results on CYP71AJ3 and CYP71AJ4 suggest that the two psoralen and angelicin synthases from *P. sativa* could have derived from a common ancestor gene and one of the two copies would have evolved through duplication and neo-functionalisation into angelicin synthase.

At present, there are 15 different *CYP71AJ* genes recorded in the cytochrome P450 databank maintained at the University of Tennessee Health Science Center (http://drnelson.uthsc.edu/CytochromeP450.html). All the 15 accessions were isolated from *Apiaceae* plants but only four of them are, to date, functionally characterized. These functions were reported as psoralen synthase for CYP71AJ1, CYP71AJ2, and CYP71AJ3 (Larbat et al. 2009, 2007) and angelicin synthase for CYP71AJ4 (Larbat et al. 2009). Interestingly, CYP71AJ5 and CYP71AJ14 were isolated from *Thapsia garganica* and CYP71AJ6 and CYP71AJ12 from *Thapsia villosa,* which are both *Apiaceae* plants which do not synthesize furanocoumarins but only coumarins (Larsen and Sandberg 1970; Mendez and Rubido 1980). This occurrence of CYP71AJ in a plant devoid of furanocoumarins provides an indication that this CYP71AJ subfamily might also be involved in the biosynthesis of other molecules, presumably coumarins in the case of *Thapsia* species.

Marmesin synthase is another enzyme involved in the furanocoumarin pathway. No molecular evidence is available so far but Hamersky and Matern showed that incubating microsomes prepared from elicitated *A. majus* cells in the presence of dimethyl [3-C^{14}] suberosin, NADPH and oxygen leads to the appearance of (+)marmesin. They also demonstrated that this reaction could be inhibited by carbon monoxide and that the inhibition could be reversed by blue light, events that are characteristic of P450s (Hamerski and Matern 1988b). Finally, according to the strict parallel organization of linear and angular furanocoumarin pathways, another P450 enzyme is likely to be involved in the transformation of osthenol into columbianetin, the direct precursor of angelicin. To date, there is no report of such enzymes, or their corresponding genes, at the molecular level.

6 Hydroxylation of Psoralen and Angelicin

Psoralen and angelicin are frequently transformed further into methoxylated forms, such as bergapten. These methoxylated forms of furanocoumarins necessarily require an initial series of hydroxylations.

Until now, there is only one report of a hydroxylase enzyme responsible for the transformation of furanocoumarins, involving the psoralen 5-monooxygenase which was isolated from *Ammi majus* (Hamerski and Matern 1988a). This enzyme belongs to the cytochrome P450 family but there is no report of it at the molecular level.

7 Prenylation of Furanocoumarins

Citrus and other *Rutaceae* plants synthesize linear furanocoumarins, which are frequently transformed into prenylated compounds. For instance, bergamot or lime essential oils contain large quantities of bergamottin (see Fig. 1) and its corresponding C8 isomer (8-geranyloxypsoralen).

Recently, a coumarin-specific prenyltransferase was identified in lemon peel which converts bergaptol into bergamottin (Munakata et al. 2012). Although the corresponding gene was not characterized by the authors, its identification soon should provide a first *O*-prenyltransferase capable to operate on furanocoumarins.

8 Synthesis of Methoxylated Furanocoumarins

Hydroxylated furanocoumarins, such as bergapten or xanthotoxol, are only found in very low quantities in plants (Bourgaud et al. 1989): these molecules are subsequently transformed into methoxylated products, bergapten, and xanthotoxin for bergaptol or xanthotoxol, respectively.

A bergaptol-*O*-methyltransferase isolated from elicitor-treated cells of *A. majus* reflected the first cDNA of a gene from the furanocoumarin pathway ever functionally characterized (Hehmann et al. 2004). The corresponding enzyme was shown to be highly specific for bergaptol and does not accept xanthotoxol, the C(8) corresponding isomer. This finding corroborates previous reports on the separation of bergaptol and xanthotoxol methyltransferases from *R. graveolens* (Sharma et al. 1979) or *P. crispum* (Hauffe et al. 1986). Recently, two other bergaptol-*O*-methyltransferases were isolated and characterized from *Glehnia littoralis* (Ishikawa et al. 2009) and *Angelica dahurica* (Lo et al. 2012), which share high sequence similarities with the *A. majus* bergaptol-*O*-methyltransferase.

9 Degradation of Furanocoumarins in Plants

Furanocoumarins are long and well known to inhibit P450 enzymes (Guo and Yamazoe 2004) and their catabolism has been studied extensively in insects. In these organisms, cytochrome P450s belonging to the CYP6 family are involved in the degradation of furanocoumarins, and contribute to the chemical warfare between insects and plants (Berenbaum and Zangerl 2008; Schuler 2011). Besides this example, little is known regarding the catabolization of furanocoumarins in higher plants. Kruse and collaborators showed that xanthotoxin can be transformed into 5-hydroxy-8-methoxypsoralen in by *Arabidopsis* plants overexpressing two different P450s: CYP82C2 and CYP82C4 (Kruse et al. 2008). Because *Arabidopsis* does not synthesize furanocoumarins, however, it is difficult to speculate on the identity of enzymes that might be involved in the degradation of furanocoumarins, in a furanocoumarin producing plant.

10 Conclusions and Prospects

Publications on the cDNA of genes from the furanocoumarin pathway have increased over the last decade, and we have now the following enzymes and genes available in publications, classified in order of appearance of the corresponding enzyme in the pathway (Fig. 2):

- A *p*-coumaroyl CoA ester 2′-hydroxylase from *R. graveolens* (synthesis of umbelliferone) (Vialart et al. 2012).
- Three psoralen synthases from *A. majus* (Larbat et al. 2007), *Apium graveolens* and *Pastinaca sativa* (Larbat et al. 2009).
- An angelicin synthase from *P. sativa* (Larbat et al. 2009).
- Three bergaptol-*O*-methyltransferases isolated and characterized from *Ammi majus* (Hehmann et al. 2004), *Glehnia littoralis* (Ishikawa et al. 2009) and *A. dahurica* (Lo et al. 2012).

The molecular characterization of the pathway, however, is incomplete, with important genes and enzymes still missing, such as:

- Prenyltransferases which operate on umbelliferone to yield DMS and osthenol
- Marmesin and columbianetin synthases, two important cytochrome P450 enzymes involved in the linear and angular furanocoumarin pathways
- Psoralen and angelicin monooxygenases which are cytochrome P450 enzymes involved in the hydroxylation of basic furanocoumarin structures.
- Xanthotoxol-*O*-methyltransferase and other *O*-methyltransferases operating both on xanthotoxol, 5-hydroxyangelicin, (isobergaptol) and 6-hydroxyangelicin (sphondinol) to deliver the methoxylated derivatives.

Besides the incomplete molecular characterization of the pathway, it is worth noting that we still have a very limited number of genes and enzymes described, mainly isolated from *Apiaceae* plants. The isolation of orthologous genes from other plant families such as *Rutaceae* (Citrus, Ruta), *Fabaceae*, or *Moraceae* will open the avenue to a better understanding of the evolutionary status of furanocoumarins in higher plants. For now, it is presently unknown if this pathway has appeared in an ancestor plant and subsequently was spread over different plant groups, or if it is the result of a parallel and independent evolutionary process occurring in several groups of plants.

References

Adisen E, Karaca F, Oztas M, Gurer MA (2008) Efficacy of local psoralen ultraviolet a treatments in psoriasis, vitiligo and eczema. Clin Exp Dermatol 33(3):344–345

Appert C, Logemann E, Hahlbrock K, Schmid J, Amrhein N (1994) Structural and catalytic properties of the 4 phenylalanine ammonia-lyase isoenzymes from parsley (Petroselinum-crispum Nym). Eur J Biochem 225(1):491–499

Baskin JM, Murrell JT, Wolf FT (1967) Psoralen in seeds of psoralea psoralioides (Leguminosae). Phyton 24(2):85

Beier RC, Oertli EH (1983) Psoralen and other linear furocoumarins as phytoalexins in celery. Phytochemistry 22(11):2595–2597

Berenbaum MR, Zangerl AR (2008) Facing the future of plant-insect interaction research: le retour a la "raison d'etre". Plant Physiol 146(3):804–811

Bourgaud F, Allard N, Guckert A, Forlot P (1989) in *Psoralens, past, present and future of Photochemoprotection and other Biological Activities*. Fitzpatrick T, Forlot P, Pathak MA., Urbach F ed. J. Libbey Eurotext, Paris pp. 301–306

Bourgaud F, Hehn A, Larbat R, Doerper S, Gontier E, Kellner S, Matern U (2006) Biosynthesis of coumarins in plants: a major pathway still to be unravelled for cytochrome P450 enzymes. Phytochem Rev 5(2–3):293–308

Ellis BE, Amrhein N (1971) Nih-shift during aromatic orthohydroxylation in higher plants. Phytochemistry 10(12):3069–3072

Endler A, Martens S, Wellmann F, Matern U (2008) Unusually divergent 4-coumarate : CoA-ligases from *Ruta graveolens* L. Plant Mol Biol 67(4):335–346

Fahrendorf T, Dixon RA (1993) Stress responses in alfalfa (*Medicago sativa L.*). XVIII: molecular cloning and expression of the elicitor-inducible cinnamic acid 4-hydroxylase cytochrome P450. Arch Biochem Biophys 305(2):509–515

Girennavar B, Jayaprakasha GK, Patil BS (2007) Potent inhibition of human cytochrome P450 3A4, 2D6, and 2C9 isoenzymes by grapefruit juice and its furocoumarins. J Food Sci 72(8):C417–C421

Gorgus E, Lohr C, Raquet N, Guth S, Schrenk D (2010) Limettin and furocoumarins in beverages containing citrus juices or extracts. Food Chem Toxicol 48(1):93–98

Gravot A, Larbat R, Hehn A, Lievre K, Gontier E, Goergen JL, Bourgaud F (2004) Cinnamic acid 4-hydroxylase mechanism-based inactivation by psoralen derivatives: cloning and characterization of a C4H from a psoralen producing plant-*Ruta graveolens*-exhibiting low sensitivity to psoralen inactivation. Arch Biochem Biophys 422(1):71–80

Guo LQ, Yamazoe Y (2004) Inhibition of cytochrome P450 by furanocoumarins in grapefruit juice and herbal medicines. Acta Pharmacol Sin 25(2):129–136

Hale AL, Meepagala KM, Oliva A, Aliotta G, Duke SO (2004) Phytotoxins from the leaves of *Ruta graveolens*. J Agric Food Chem 52(11):3345–3349

Hamerski D, Matern U (1988a) Biosynthesis of psoralens. Psoralen 5-monooxygenase activity from elicitor-treated *Ammi majus* cells. FEBS Lett 239(2):263–265

Hamerski D, Matern U (1988b) Elicitor-induced biosynthesis of psoralens in ammi-majus L suspension-cultures—microsomal conversion of demethylsuberosin into (+)marmesin and psoralen. Eur J Biochem 171(1–2):369–375

Hauffe KD, Hahlbrock K, Scheel D (1986) Elicitor-stimulated furanocoumarin biosynthesis in cultured parsley cells—S-adenosyl-L-methionine: bergaptol and S-adenosyl-L-methionine—xanthotoxol O-methyltransferases. Z Naturforsch C 41(1–2):228–239

Hehmann M, Lukacin R, Ekiert H, Matern U (2004) Furanocoumarin biosynthesis in *Ammi majus* L.—Cloning of bergaptol O-methyltransferase. Eur J Biochem 271(5):932–940

Hubner S, Hehmann M, Schreiner S, Martens S, Lukacin R, Matern U (2003) Functional expression of cinnamate 4-hydroxylase from *Ammi majus* L. Phytochemistry 64(2):445–452

Ishikawa A, Kuma T, Sasaki H, Sasaki N, Ozeki Y, Kobayashi N, Kitamura Y (2009) Constitutive expression of bergaptol O-methyltransferase in *Glehnia littoralis* cell cultures. Plant Cell Rep 28(2):257–265

Jorgensen K, Rasmussen AV, Morant M, Nielsen AH, Bjarnholt N, Zagrobelny M, Bak S, Moller BL (2005) Metabolon formation and metabolic channeling in the biosynthesis of plant natural products. Curr Opin Plant Biol 8(3):280–291

Kai K, Shimizu B, Mizutani M, Watanabe K, Sakata K (2006) Accumulation of coumarins in *Arabidopsis thaliana*. Phytochemistry 67(4):379–386

Koopmann E, Logemann E, Hahlbrock K (1999) Regulation and functional expression of cinnamate 4-hydroxylase from parsley. Plant Physiol 119(1):49–55

Kruse T, Ho KL, Yoo HD, Johnson T, Hippely M, Park JH, Flavell R, Bobzin S (2008) In planta biocatalysis screen of P450s identifies 8-methoxypsoralen as a substrate for the CYP82C subfamily, yielding original chemical structures. Chem Biol 15(2):149–156

Larbat R, Hehn A, Hans J, Schneider S, Jugde H, Schneider B, Matern U, Bourgaud F (2009) Isolation and functional characterization of CYP71AJ4 encoding for the first P450 monooxygenase of angular furanocoumarin biosynthesis. J Biol Chem 284(8):4776–4785

Larbat R, Kellner S, Specker S, Hehn A, Gontier E, Hans J, Bourgaud F, Matern U (2007) Molecular cloning and functional characterization of psoralen synthase, the first committed monooxygenase of furanocoumarin biosynthesis. J Biol Chem 282(1):542–554

Larsen PK, Sandberg F (1970) Constituents of umbelliferous plants. XV. Coumarins from thapsia garganica L. The structure of new coumarin. Acta Chem Scand 24(3):1113

Lo SC, Chung PE, Wang CS (2012) Molecular cloning and functional analysis of bergaptol-O-methyltransferase from *Angelica dahurica* (Bai Zhi) and using it to efficiently produce bergapten in E. coli. Bot Stud 53(2):197–206

Lois R, Hahlbrock K (1992) Differential wound activation of members of the phenylalanine ammonia-lyase and 4-coumarate—CoA ligase gene families in various organs of parsley plants. Z Naturforsch C 47(1–2):90–94

Luo KW, Sun JG, Chan JYW, Yang L, Wu SH, Fung KP, Liu FY (2011) Anticancer effects of imperatorin isolated from angelica dahurica: induction of apoptosis in HepG2 cells through both death-receptor- and mitochondria-mediated pathways. Chemotherapy 57(6):449–459

Mendez J, Rubido J (1980) Coumarins and phenols of thapsia-villosa fruits. Planta Med 38(2):177–179

Munakata R, Inoue T, Koeduka T, Sasaki K, Tsurumaru Y, Sugiyama A, Uto Y, Hori H, Azuma J, Yazaki K (2012) Characterization of coumarin-specific prenyltransferase activities in citrus limon peel. Biosci Biotech Bioch 76(7):1389–1393

Parast BM, Chetri SK, Sharma K, Agrawal V (2011) In vitro isolation, elicitation of psoralen in callus cultures of psoralea corylifolia and cloning of psoralen synthase gene. Plant Physiol Bioch 49(10):1138–1146

Pierrel MA, Batard Y, Kazmaier M, Mignottevieux C, Durst F, Werckreichhart D (1994) Catalytic properties of the plant cytochrome-P450 Cyp73 expressed in yeast—substrate-specificity of a cinnamate hydroxylase. Eur J Biochem 224(3):835–844

Plumas JL, Drillat P, Jacob MC, Richard MJ, Favrot MC (2003) Extracorporeal photochemotherapy for treatment of clonal T cell proliferations. B Cancer 90(8–9):763–770

Razavi SM (2011) Plant coumarins as allelopathic agents. Int J Biol Chem 5:86–90

Sasaki K, Mito K, Ohara K, Yamamoto H, Yazaki K (2008) Cloning and characterization of naringenin 8-prenyltransferase, a flavonoid-specific prenyltransferase of sophora flavescens. Plant Physiol 146(3):1075–1084

Schoch G, Goepfert S, Morant M, Hehn A, Meyer D, Ullmann P, Werck-Reichhart D (2001) CYP98A3 from *Arabidopsis thaliana* is a 3 '-hydroxylase of phenolic esters, a missing link in the phenylpropanoid pathway. J Biol Chem 276(39):36566–36574

Schuler MA (2011) P450s in plant-insect interactions. Bba-Proteins Proteom 1814(1):36–45

Sharma SK, Garrett JM, Brown SA (1979) Separation of the S-adenosylmethionine: 5- and 8-hydroxyfuranocoumarin O-methyltransferases of *Ruta graveolens* L. By general ligand affinity chromatography. Z Naturforsch C 34C(5–6):387–391

Shen G, Huhman D, Lei Z, Snyder J, Sumner LW, Dixon RA (2012) Characterization of an isoflavonoid-specific prenyltransferase from lupinus albus. Plant Physiol 159(1):70–80

Stanjek V, Piel J, Boland W (1999) Synthesis of furanocoumarins: mevalonate-independent prenylation of umbelliferone in *Apium graveolens* (*Apiaceae*). Phytochemistry 50(7):1141–1145

Teutsch HG, Hasenfratz MP, Lesot A, Stoltz C, Garnier JM, Jeltsch JM, Durst F, Werckreichhart D (1993) Isolation and Sequence of a cdna-encoding the Jerusalem-Artichoke cinnamate 4-hydroxylase, a major plant cytochrome-P450 involved in the general phenylpropanoid pathway. P Natl Acad Sci USA 90(9):4102–4106

Vialart G, Hehn A, Olry A, Ito K, Krieger C, Larbat R, Paris C, Shimizu B, Sugimoto Y, Mizutani M, Bourgaud F (2012) A 2-oxoglutarate-dependent dioxygenase from *Ruta graveolens* L. exhibits p-coumaroyl CoA 2′-hydroxylase activity (C2′H): a missing step in the synthesis of umbelliferone in plants. Plant J Cell Mol Biol 70(3):460–470
Winkel BS (2004) Metabolic channeling in plants. Annu Rev Plant Biol 55:85–107
Yazaki K, Sasaki K, Tsurumaru Y (2009) Prenylation of aromatic compounds, a key diversification of plant secondary metabolites. Phytochemistry 70(15–16):1739–1745

Authors Biography

Frédéric Bourgaud PhD in Agronomy, is a specialist of plant natural compounds (secondary metabolites). He started his career as a scientist at Laboratoires Goupil SA (Paris, France, 1988–1990). Then he moved for a postdoctoral position at the University of Padua (Italy, 1990–1991). He became Assistant Professor in Agronomy at Nancy University-INPL (National Polytechnical Institute, 1991–1999) and then was nominated full Professor at University of Lorraine (Nancy, France, 1999–present) where he has been teaching biotechnologies. He is currently heading a joint research unit between INRA (National Institute for Agronomical Research in France) and the University of Nancy, specialized in the study of plant secondary metabolites. He is also Vice-Chairman in charge of science at "Plant Advanced Technologies," a company that he co-founded in 2005. He has authored 50 peer-reviewed papers and book chapters.

Alexandre Olry PhD in Molecular Enzymology has a strong experience in protein biochemical and enzymatical characterization. He started his career at the laboratory of Maturation des ARNs and molecular enzymology (MAEM) (University of Lorraine, Nancy, France) under the direction of Pr Guy Branlant and got a postdoctoral position at the Institut de Biologie Moléculaire des Plantes (IBMP), (University Louis Pasteur, Strasbourg, France) in the department of Plant metabolic responses under the direction of Dr Werck-Reichhart. He worked in the field of P450 involved in the hydroxylation of fatty acids for cutin and suberin synthesis in Arabidopsis. In 2008, he was nominated as Research Engineer at University of Lorraine in the laboratory Agronomie et Environnement (LAE) under the direction of Pr Frédéric BOURGAUD. His research is focusing on the biochemical and Enzymatical characterization of the enzymes involved in the furanocoumarin pathway.

Alain Hehn obtained a PhD in Molecular and Cellular Biology 1995 at the University Louis Pasteur in Strasbourg (France) in the field of molecular virology. After a first postdoctoral fellowship in the Max Planck Institut fuer Zuechtungsforshung in Koeln (Germany) he joined the group of Danièle Werck Reichhart at the Institut of Plant Molecular Biology in Strasbourg for studying the involvement of cytochrome P450s in the plant secondary metabolism. In 2001, he obtained an engineer position in the laboratory Agronomie et Environnement (Nancy, France) and was subsequently nominated as an Associate Professor in 2007 at the University of Lorraine. He is currently teaching molecular biology, biotechnology, and plant breeding and the research topic he developed since 2001 concerns the identification and functional characterization of genes involved in various secondary metabolite biosynthetic pathways.

Alain Hehn obtained a PhD in Molecular and Cellular Biology 1995 at the University Louis Pasteur in Strasbourg (France) in the field of plant cellular biology. After a first postdoctoral fellowship at the Max Planck Institut für Züchtungsforschung in Köln (Germany) he joined the group of Danièle Werck-Reichhart at the Institut of Plant Molecular Biology in Strasbourg for studying the involvement of cytochrome P450s in the plant secondary metabolism. In 2001 he obtained an engineer position in the laboratory Agronomie et Environnement (Nancy, France) and was subsequently nominated as an Associate Professor in 2007 at the University of Lorraine. He is currently teaching molecular biology, [illegible] and plant breeding and the research topic he developed since 2001 [illegible] the identification and functional characterization of genes involved in [illegible] [illegible]ary metabolite biosynthetic pathways.

Part XIV
Connecting Section Between Chapters 14 and 15

We have mentioned previously that traditional techniques and modern biotechnology represent the means to obtain derivatives of natural products. Synthetic chemistry, obviously, is another powerful method, which enables researchers and Industry to generate a wealth of molecules, which often resemble natural products, yet contain certain alterations or additions to enable such substances to better fulfill their practical purposes. Indeed, the origins of organic chemistry are firmly rooted in the desire to synthesize natural products without the help of a mysterious *vis vitalis*, and Friedrich Woehler's (1800–1882) keystone organic syntheses, i.e., the conversion of cyanogen to oxalic acid in 1824, and of ammonium cyanate to urea in 1828, are not only highlights of an emancipating organic chemistry as a science free of vitalism, but also examples of a successful natural product synthesis.

Synthetic organic chemistry has obviously advanced considerably since those early days. For almost two centuries now, chemistry has been occupied with the synthesis of natural products and their derivatives, and has witnessed considerable successes in this field. The total synthesis of complex secondary metabolites, such as diverse vitamins, proanthocyanidins, and plant hormones, in particular, has raised the interest of organic chemists. The total synthesis of such molecules is often rather difficult and in some instances has taken years to accomplish. Nonetheless, once an appropriate synthetic procedure is in place, it is then possible to produce the natural product in question in considerable amounts and adequate purity—and also to embark on the synthesis of desired derivatives. As such, a total synthesis is often a major task. More modest approaches have also been considered. Here, the partial synthesis of new compounds derived from isolated natural products synthetically by comparably small and straightforward modifications (e.g., via esterification of a free acid or alcohol, via methylation of a hydroxyl or amine group) has become rather popular.

The following chapter will consider an elegant approach of such a partial synthesis, which 'combines' chemically readily available redox active plant secondary metabolites, such as coumarins, quinones, curcumin, and fatty acids to yield novel *hybrid molecules*. These hybrids bring together the chemistry—and often also the biological activities—of the individual compounds they are based upon, yet in many aspects (such as solubility, bioavailability, and metabolic stability) surpass their individual 'parts'. Importantly, the synthesis of such hybrid molecules does not necessarily have to involve simple 'coupling' or 'conjugation' of existing building blocks, e.g., via a di- or polysulfane bridge. As the next chapter will show, it is also possible to 'create' the second part of such molecules synthetically by modifying functional groups that are present in the first one. As part of this chapter, Gilbert Kirsch and Segio Valente will consider conjugates of a range of (redox active) secondary metabolites, including curcumin, caffeic acid, resveratrol, kojic acid, flavanoids, fatty acids, retinoic acids, and alkaloids. While some of these conjugation reactions leading to hybrid molecules are straightforward chemically, others require more eloquent synthetic pathways. Nonetheless, the biological activities associated with many of the resulting hybrid molecules are often highly simulating and compensate for the synthetic difficulties, which may have been encountered before.

Chapter 15
Natural Products Molten Together: Toward Multifunctional Hybrid Molecules with Specific Activities and Applications

Gilbert Kirsch and Sergio Valente

Keywords Bioconjugation • Chemical coupling methods • Curcumin • Fatty acids • Hybrid molecules

1 Introduction

It happens in Nature that compounds with interesting biological activities are formed of individual blocks of different structures that are generally active by themselves. For example, some chromone and flavanoid alkaloids (Fig. 1) have been found in many different plants (Houghton 2000; Khadem and Marles 2012; Jain et al. 2012; Vlietinck et al. 1998; Ismail et al. 2009). Similarly, different acids such as quinic acid or shikimic acid, are found to be conjugated to caffeic acid.

In these cases the combination is the result of a biosynthetic process. At the same time, chemists were interested in melting together natural products known for specific activity in a way to obtain a multiple shot for the biological effect or also, sometimes, to enhance *in vivo* distribution. We are presenting here some examples of the research carried out in this field with a special focus on redox active compounds. A literature search for such 'hybrid' compounds suffers from a nonspecific design for such derivatives. They are sometimes named conjugates, also bioconjugates and mixed conjugates or bioconjugate compounds.

G. Kirsch (✉)
SRSMC UMR 7565 (former Laboratoire d'Ingénierie Moléculaire et Biochimie Pharmacologique), Université de Lorraine, 57070 Metz, France
e-mail: gilbert.kirsch@univ-lorraine.fr

S. Valente
Facoltà di Farmacia e Medicina, Sapienza Università di Roma, P.le Aldo Moro 5, 00185 Rome, Italy

C. Jacob et al. (eds.), *Recent Advances in Redox Active Plant and Microbial Products*, 379
DOI: 10.1007/978-94-017-8953-0_15, © Springer Science+Business Media Dordrecht 2014

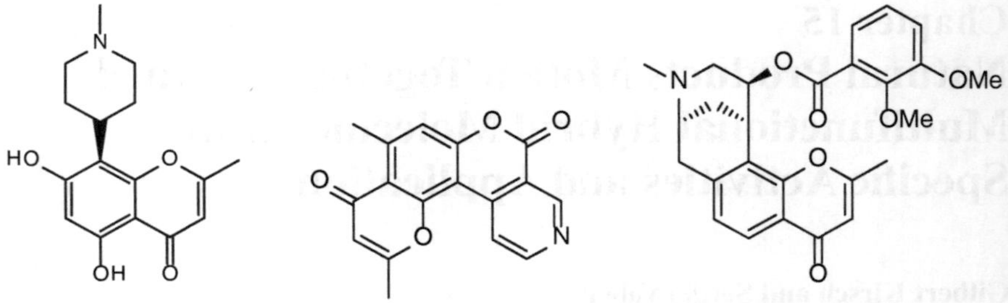

Fig. 1 A few selected examples of flavoalkaloids (flavopyridols) composed of two substructures with individual activity

We will therefore present here different classes of these conjugated compounds. Some classes have been studied more than others in the past. In fact, very often, synthetic biologically active biomolecules have been conjugated to some natural compounds with the aim of enhancing their bioavailability or also to target certain biological sites with higher precision (see Explanatory Box 1 for classical chemical coupling methods). Conjugation of two or even more natural compounds is less common in the literature. In fact, our simple literature search was unable to identify such a conjugation of three (or more) natural compounds in one molecule. This offers the possibility to the chemist to design many new compounds based on this 'melting' strategy, to generate new types of multifunctional hybrid molecules.

Explanatory Box 1: Chemical coupling reactions

This chapter highlights the idea of combining two or more natural products in one molecule in order to improve efficiency, widen the spectrum of activity, and also generate synergistic effects, for instance, an improved solubility and/or membrane permeability. One particular stimulating example from pharmaceutical research is the antibacterial drug sultamicillin, a hybrid co-drug which consists of two individual drugs, namely ampicillin and sulbactam, which are coupled together via a linker. Once reaching its target, the ester-linkages in the co-drug are cleaved and the two individual drugs are released to attack their individual targets. While the penicillin ampicillin targets the bacterial cell wall synthesis, the β-lactamase inhibitor sulbactam takes on the role of ampicillin's 'bodyguard' by preventing its inactivation by cellular β-lactamases. Sultamicillin is superior to a simple mixture of ampicillin and sulbactam because (a) both drugs are liberated together and hence reach their target together, therefore avoiding potential complications caused by different pharmacokinetic profiles of the individual drugs and (b) the hybrid has some improved solubility and permeability properties compared to the individual agents.

Whilst this approach of using hybrid molecules as 'co-drugs' is rather intriguing, it often requires some intricate chemistry in order to succeed. Here, it is not always straightforward to simply 'couple', 'link', 'conjugate', or even 'integrate' two complex molecules to yield the desired hybrid. As most natural products contain several reactive groups, such coupling procedures are far from trivial and often rely on extensive protection and reaction conditions, which in the worst case may even chemically ruin the units which are coupled together. We will therefore briefly consider some of the most promising and widely used coupling methods, of course without staking any claim on comprehensiveness.

One of the most straightforward and even reversible coupling methods is the one between two thiols which can be oxidized to a disulfide, which then holds together the two molecules in question. This kind of coupling is even used in Nature, for instance in the case of insulin, which is composed of an A-chain and B-chain held together by two disulfide bridges. The major advantages of this particular coupling method include a high selectivity for thiols, mild, oxidizing conditions for the formation of the disulfide bridge and reversibility of the coupling, as the disulfide can be reduced and hence opened up under mildly reducing conditions. One major disadvantage is the fact that the coupling occurs between two identical functional groups, (i.e., thiols) and hence may not result exclusively in the desired products, especially when two (or more) different units are supposed to be linked. In this case, various homo- and heterodimers are likely to occur unless the coupling itself is under enzymatic control (as is the case for most biochemical dimers).

It is therefore often advisable to use two different functional groups for coupling. Assuming that such reactions should occur under comparably mild conditions, a few coupling methods stand out. First of all, it is possible to couple an alcohol (or phenol) to an acid to form an ester. Many activated acids, coupling agents, and strategies are available to achieve such an ester linkage. As the latter can be cleaved under alkaline conditions and also by many enzymes, esters are often the preferred coupling methods in pharmaceutical research, especially if the individual units are supposed to be 'released' later on, for instance once inside a specific body, tissue or cell. Esters are also good candidates for pro-drugs. Similarly, acids can also be coupled with amines, which generates a amide bond. The latter is a lot more stable than the ester bond and amides are often the coupling products of choice if a stable coupling is desired. The same applies to sulfonamides that are based on amines coupled to a sulfonic acid. Amines can also be coupled to aldehydes, which yields imines. The latter are often reduced further to (secondary) amines, which also provide a fairly stable type of 'linkage'. Less straightforward but equally effective are simple substitution reactions, such as nucleophilic attack on molecules bearing a good leaving group, such as a halide or tosyl leaving group. This kind of coupling usually provides good yields, but may require more extreme reaction conditions.

Related reactions, which may proceed under milder conditions, involve catalyzed coupling methods. The latter include various types of the traditional 'Suzuki coupling', which employs a palladium (0) complex to couple together a boronic acid bearing partner with a halide bearing partner (Nobel Prize in Chemistry in 2010). Importantly, both functional groups involved in the coupling process, i.e., the boronic acid and the halide, are eliminated during this coupling (which is different from the methods mentioned above), and the resulting product contains a firm carbon–carbon single bond linking the two original units together.

More 'exotic' but equally effective and often specifically activated coupling methods include addition reactions, such as the 1,3-cycloaddition. A specific 1,3-dipolar cycloaddition involving an azide and alkyne (Huisgen cycloaddition) is used part of the 'click'-chemistry which leads to a fairly stable, cyclic addition product, a 1,2,3-triazole. As neither azides nor alkynes normally occur in biological systems, and this particular coupling method can be triggered by metal catalysts and light, the 'click' coupling has become extremely popular in biological chemistry, and many azide- and alkyne-bearing 'building blocs' are now commercially available. This kind of coupling can also be used, of course, to link two natural products together to form a new hybrid molecule.

In some instances, it is not possible to combine two individual natural products directly. This situation occurs, for instance, if the two units are rather bulky, and a direct coupling is not possible for steric reasons. Often, the two units also contain the same functional groups (e.g., two alcohols, two amines, two acids), and direct coupling is therefore not possible either. The use of a specifically designed 'spacer' may therefore be required to keep the two units apart and also to provide a specific coupling chemistry at both of its ends. Indeed, such a 'spacer' has been used to connect the two acid functions of ampicillin and sulbactam in sultamicillin. For the reasons mentioned above, adequate bifunctional spacers have also been developed to conjugate specific proteins and enzymes, and to attach certain molecules to antibodies. Here, we find some rather interesting types of 'coupling chemistry', including the coupling of cysteine residues to maleimides and of lysine residues to specifically designed activated acids.

While the chemical coupling of proteins with each other has recently received some rather stiff competition from so-called 'fusion proteins', the continuous development of effective and selective coupling methods still represents a major area of modern bioorganic chemistry, and also stimulates drug development and the formation of hybrid natural products.

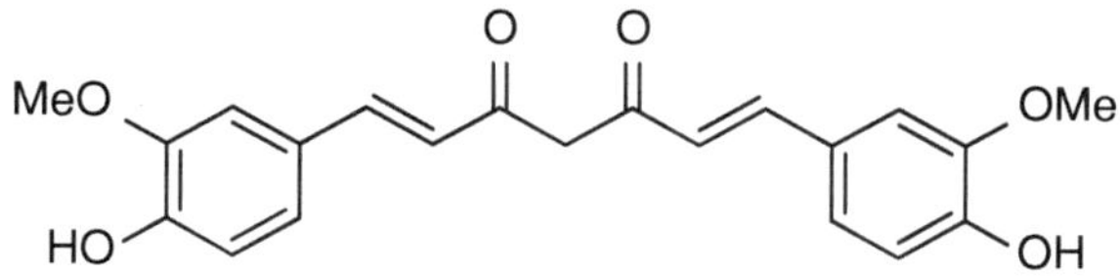

Fig. 2 Chemical structure of curcumin

2 Polyphenol Bioconjugates

Curcumin is among the most studied polyphenolic compounds, while for other polyphenolic derivatives, studies have appeared on cinnamic acid derivatives (caffeic acid and related structures), and resveratrol has also been considered.

2.1 Bioconjugation of Curcumin

Although curcumin (Fig. 2) displays many beneficial properties in preventing or perhaps even curing cancer, diabetes, inflammation, cataracts, rheumatoid arthritis, HIV, etc., it has not been possible so far to assign a definitive drug profile to this compound because of low bioavailability, fast metabolism, poor adsorption, and lack of targeted delivery (see also Chap. 16).

Bioconjugating curcumin to other molecules can in this way be looked at from two distinct angles:

1. For enhancing its bioavailability;
2. For enhancing bioavailability and targeted bioactive properties.

Concerning just bioavailability, various modulations have been achieved using the different possibilities of the phenolic groups present to be alkylated, arylated, heteroarylated, or esterified (Al-Wabli et al. 2012). This kind of work has been carried out to explore the possible use of such curcumin derivates as possible anticancer agents. For the same purpose, curcumin has also been conjugated to cyclodextrin (Salmaso et al. 2007). Different modifications of curcumin have also been performed in order to improve antibacterial activity. These modifications took account of the possible reaction of the remaining phenolic groups as well as of the reactivity of the central methylene group able to be transformed into a carbanion and hence allowing a substitution reaction. For example, amino acids were introduced at the three possible reaction sites and in this way the antibacterial activity was enhanced (Kumar et al. 2001; Pandey et al. 2011). The best results were obtained with piperoylglycine-curcumin (Fig. 3) (Mishra et al. 2005b).

In contrast, biological activity was not enhanced when a thymidyl group was grafted onto the central position (Kumar et al. 2001). Vitamin B3 (nicotinic acid) was conjugated to curcumin and the activity was improved (Pandey et al. 2011).

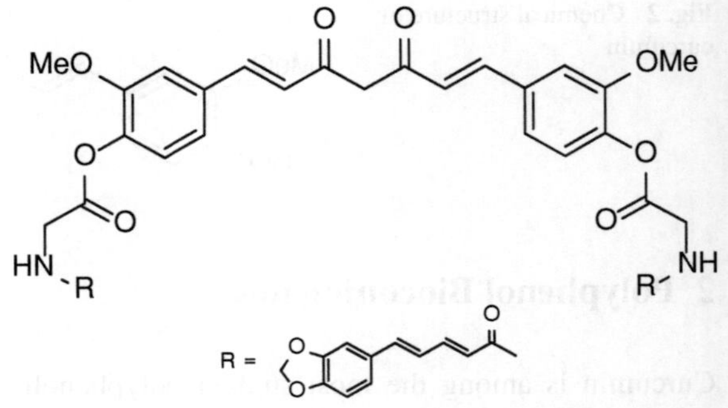

Fig. 3 Piperoylglycine-curcumin

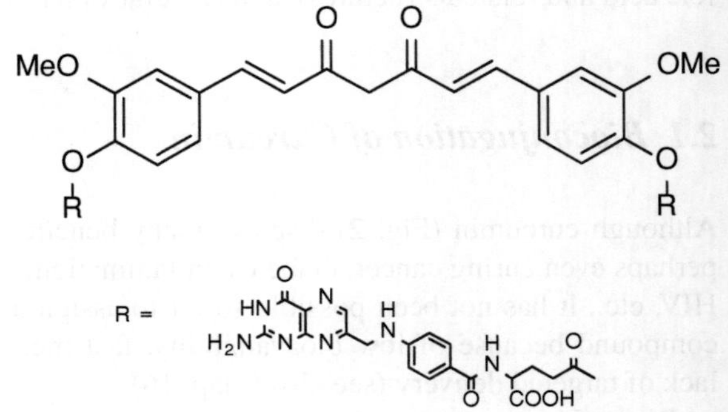

Fig. 4 Curcumin-folic acid conjugate

Similar results were also obtained when curcumin was esterified with piperic acid (Mishra et al. 2005b; Dubey et al. 2008). Structural modifications with fatty acids, dipeptides, and folic acid have also been described and such derivatives have been tested for antibacterial and antiviral properties (Mythri et al. 2011). A condensation with folic acid has been carried out on the phenol function and also on the central carbon atom through a linker (Fig. 4).

Good results were obtained for several of these bioconjugated systems in the context of antibacterial activity. In the case of antiviral activity, however, only the enhancement of lipophylicity due to esterification with fatty acids or dipeptides was efficient (Singh et al. 2010).

As curcumin has been shown to induce apoptosis in tumor cells through generation of reactive oxygen radicals, bioconjugates of curcumin with cysteine were synthesized (Bhaumik et al. 1999; Mishra et al. 2005a) Two cysteines were attached at each end of the curcumin molecule (via an ester linkage to the circumins-free OH groups) and a cyclic molecule resulted as as an octyl bridge was attached to each sulfur atom of the cysteines (Fig. 5).

Dipiperoyl and diglycinoyl derivatives showed higher apoptotic activity at lower concentration than curcumin itself. On the other hand, diglycinoyl-dipiperoyl and the disulfide had lost their apoptotic activity significantly. The apoptotic activity correlated well with the generation of ROS by the tumor cells, whereas GSH levels remained unaltered.

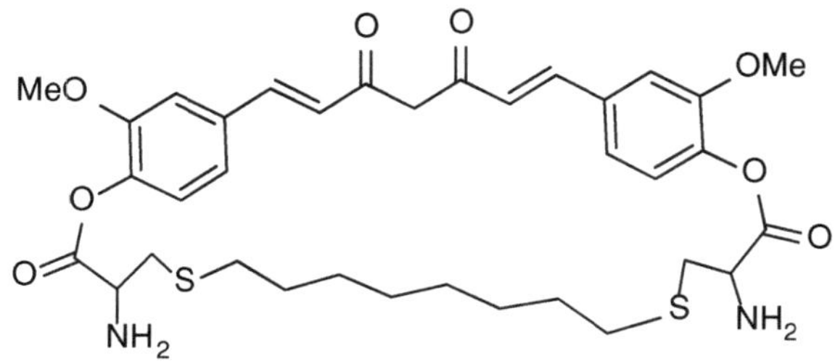

Fig. 5 Curcumin cyclized via a symmetric cysteine-based sulfur linked bridge

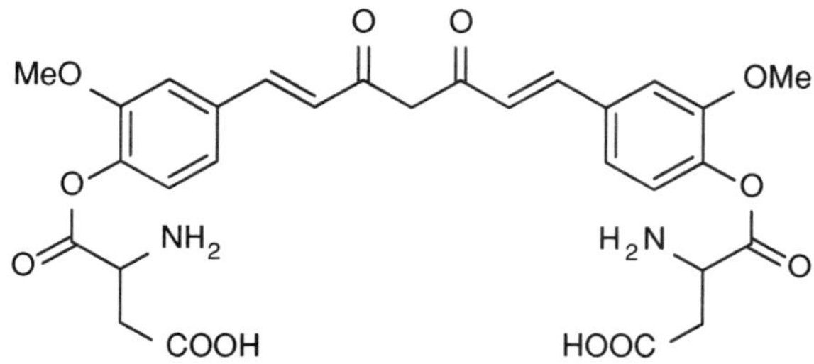

Fig. 6 Curcumin–glutamic acid conjugate

Parkinson's disease is a neurodegenerative disorder due to loss of dopaminergic neurons in the *Substantia Nigra*. Oxidative stress and GSH depletion precede mitochondrial dysfunction and dopamine loss and therefore are considered as early triggers for neurodegeneration. It has been shown that curcumin protects mitochondria against nitrosative and oxidative stress (Harish et al. 2010; Mythri et al. 2011). The bioavailability of curcumin, however, is limited and bioconjugation could overcome this obstacle (see also Chap. 19). To improve the activity of curcumin in neuronal cells, it was conjugated with demethylenated piperic acid, valine, and glutamic acid (Harish et al. 2010; Mythri et al. 2011). It could be shown in these studies that all derivatives were equally active or even more efficient than curcumin itself. Glutamoylcurcumin in particular (Fig. 6) showed an improved protection against neuronal cell death.

Some triterpenoids (serratans) from Japanese pines have shown to decrease the size of adenomas and total tumor numbers in rat models (Yamaguchi et al. 2008). Two derivatives of curcumin fused to methoxy serrat-14-en-21-ß-ol were studied and presented antitumor activity. Serratol moieties were conjugated to curcumin using a succinic acid linker between the remaining free hydroxyl groups on the curcumin and the serratol (Fig. 7) (Tsujii et al. 2010).

In this way mono- and bis-serratanes were prepared. Disappointingly, when considering these compounds as chemopreventive agents no positive results could be obtained (Tanaka et al. 2011). Similarly, testing them for anti-HIV activity showed no reverse transcriptase inhibitory effect either (Tanaka et al. 2009).

A possibility to obtain a better activity on the site of a tumor is to associate the chosen compound with a peptide targeting the (tumor) cell surface. LHRH (luteinizing hormone releasing hormone) is a decapeptide targeting pancreatic cell lines

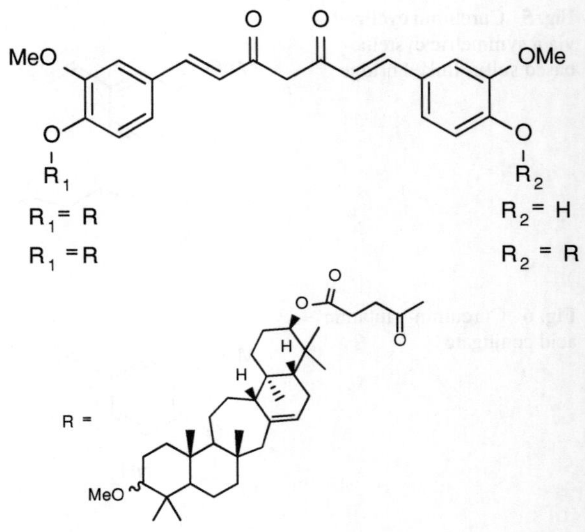

Fig. 7 Curcumin–serratol conjugates

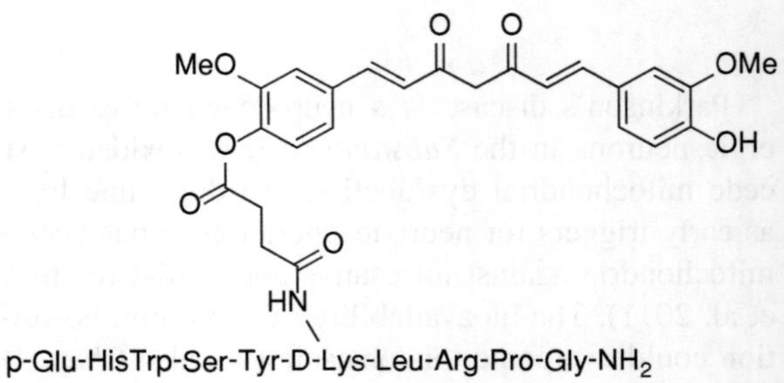

Fig. 8 Curcumin–LHRH conjugate

as well as others. Curcumin was thus conjugated to LHRH to yield the [DLys-6]-LHRH–curcumin conjugate and used against pancreatic cancer *in vitro* and *in vivo*. Chemically, conjugation was achieved by coupling the curcumin glutarate monoester with the decapeptide at lysine in position 6 (Fig. 8) (Aggarwal et al. 2011).

The conjugate induced apoptosis in pancreatic cancer cells and prevented the growth of pancreatic cancer cell xenografts in nude mice.

2.2 Caffeic Acid Conjugates

Caffeic acid (Fig. 9) was found to exhibit distinct antiproliferative effects (Gomes et al. 2003).

Caffeic acid has mostly been modified by esterification or amidification, and rarely been fused to other natural products. Of special interest is the esterification with quinic and chicoric acid, which has resulted in interesting derivatives for

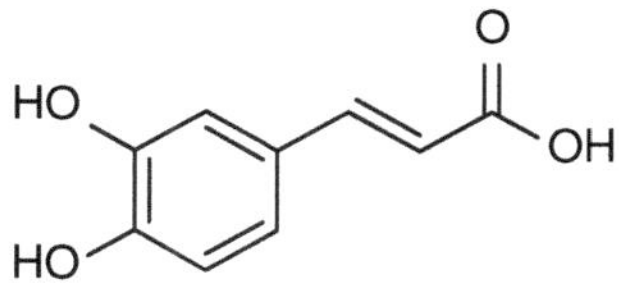

Fig. 9 Caffeic acid

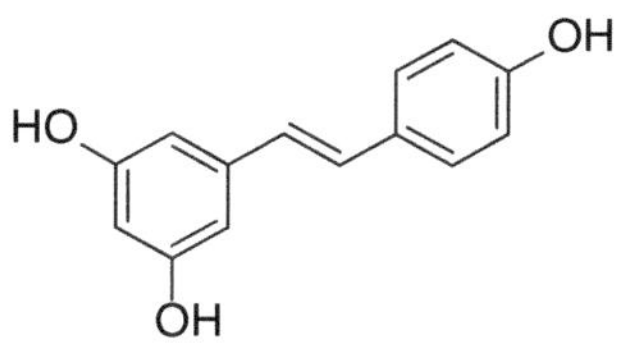

Fig. 10 Resveratrol

HIV integrase inhibition (McDougall et al. 1998). Conjugating caffeic acid with glutathione at the aromatic ring of the acid was performed to study the inhibition of nitration, yet the conjugated product showed lower inhibition than the starting compounds. Reduced dihydrolipoic acid was conjugated via the sulhydryl group to the ortho-quinone obtained from caffeic acid. A review article on caffeic acid as pharmacophore summarizes all the literature regarding these studies (Touaibia et al. 2011). These conjugated compounds were tested against hepatocellular carcinoma (Guerriero et al. 2011).

2.3 *Resveratrol and Conjugated Compounds*

Resveratrol (Fig. 10) is well known for its antioxidant properties. Few resveratrol bioconjugated compounds, however, have been prepared and tested in Biology. Various hybrids are described, yet these compounds do not correspond to 'molten together' natural products.

For applications in cosmetics (see also Chap. 16), resveratrol has been coupled with lipoic acid (Fig. 11) or tocopherol (Fig. 12). To prepare the monolipoate or the monotocopherolate, only one hydroxyl group of resveratrol had to be reacted out of the three hydroxyl groups present. The diacetate of resveratrol could be prepared by an enzymatic deprotection of its triacetate (Fig. 11). Esterification could then be performed, with a linker in the case of tocopherol (Gelo-Pujic et al. 2008).

For the conjugation of resveratrol with tocopherol, a succinyl linker has been attached between the hydroxyl groups of the two phenolic compounds (Fig. 12).

Both compounds presented good chemical stability when exposed to heat- and photoaging conditions. Human *Stratum Corneum* enzymes efficiently hydrolyzed the resveratrol conjugate releasing the active molecules under *in vitro* conditions after 72 h at neutral pH. The tocopherol conjugate was only partially hydrolyzed. In this case the resveratrol conjugate could be used in studies evaluating possible topical applications (Delaire et al. 2009).

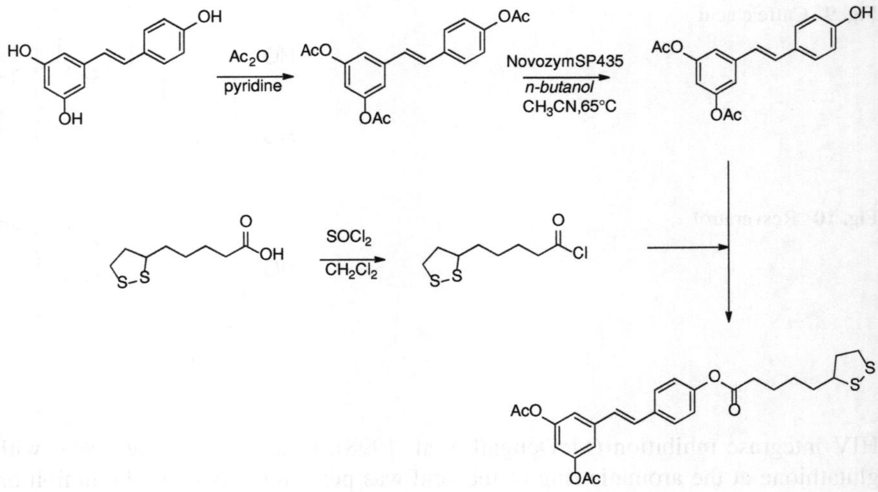

Fig. 11 Synthesis of resveratrol-lipoic acid conjugate

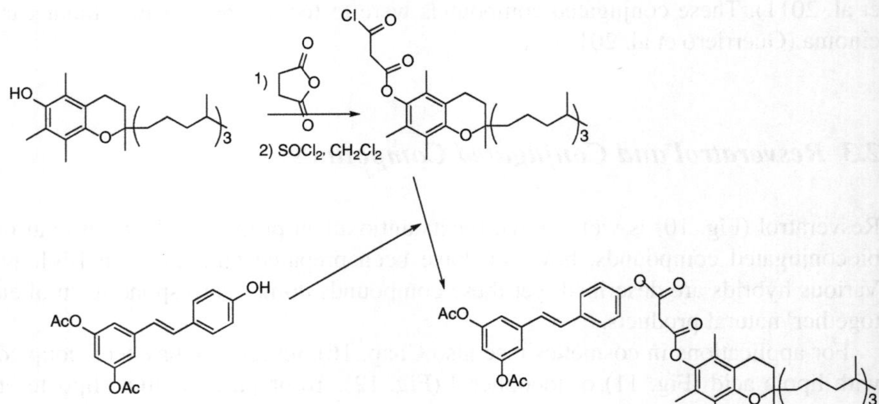

Fig. 12 Synthesis of resveratrol-tocopherol conjugate

3 Kojic Acid Conjugates

Kojic acid (Fig. 13), a compound produced from sugar through an enzymatic process, shows interesting tyrosinase inhibition properties that render it usable for cosmetic applications, especially for skin depigmentation. It has also good chelation properties for Cu^{2+} and Fe^{3+} ions and acts as a radical scavenger.

Fig. 13 Kojic acid

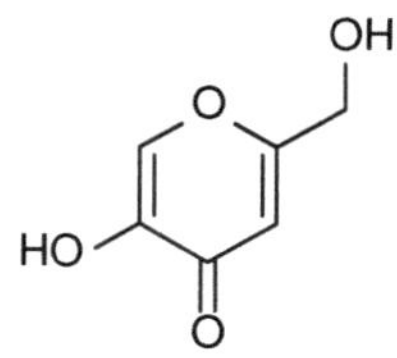

Coupling with different amino acids (phenylalanine, tryptophane, tyrosine, and histidine) improved the mushroom tyrosinase inhibition ability of kojic acid in a non-competitive manner (Noh et al. 2009). Kojic acid has also been coupled one or two methoxy-serratol moieties through a succinic acid linker (Tsujii et al. 2010). These compounds show potential as anti-HIV agents.

4 Flavanoid Bioconjugates

Some naturally occurring flavanoids have also been coupled with natural compounds. This was the case for some serratans composed of serratols and quercetine, baicalein, naringenin, hesperidine, genistein, hesperidine, and daidzein (Tanaka et al. 2009, 2011; Tsujii et al. 2010). The condensation between these two classes of natural products was realized using a succinic acid linker between the alcoholic function of the serratols and the phenolic group in the flavanoids. The purpose of this combination was to improve antioxidant and chemopreventive properties as well as anti-HIV activity by reverse transcriptase (RT) inhibition. Ultimately, however, none of the conjugated derivative exhibited interesting activity on the desired biological targets. Even the monosuccinoyl derivative of the starting triterpenoid was more active. These studies underline the potential drawbacks associated with merging two compounds into one.

Tocopherol (Vitamin E) is well known for its antioxidant properties, while natural triterpenoids have been tested for their anticancer/anti-HIV properties (Vasilevsky et al. 2011). Considering these properties, authors have prepared mixed compounds between these two classes of compounds. For example, condensation between betulonic and betulinic acids from the lupane series and tocopherol have been achieved using a dipeptide linker (Spivak et al. 2010; Alakurtti et al. 2006). Unfortunately, results concerning biological activity are still warranted.

5 Fatty Acid Conjugates

Many fatty acids have been conjugated with other active natural compounds, generally in order to enhance the lipophilicity or bioavailability of the natural compound (e.g., curcumin).

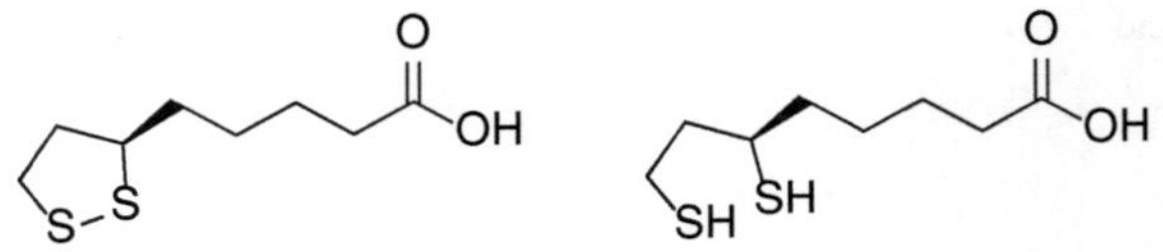

Fig. 14 Lipoic acid and its dihydro-derivative

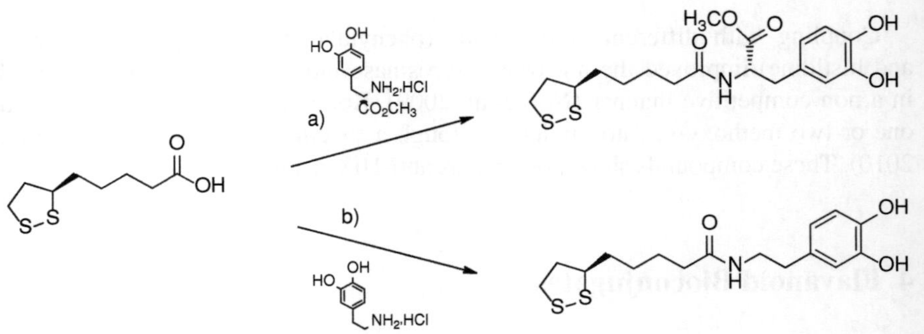

Fig. 15 Conjugation of lipoic acid with L-DOPA (**a**) and dopamine (**b**)

5.1 Lipoic Acid Conjugates

Lipoic acid and its dihydro-derivative (Fig. 14) have been used in conjugation reactions as discussed before in connection with resveratrol and tocopherol. Lipoic acid has also been conjugated to other molecules in order to obtain multifunctional co-drugs, for instance with L-Dopa and Dopamine (Di Stefano et al. 2006). The ability of lipoic acid to cross the blood–brain barrier and to accumulate in neuronal cells is well known (Di Stefano et al. 2006). Once inside neurons, lipoic acid is reduced to dihydrolipoic acid which lowers the redox activity of non-protein bound iron and copper ions. Conjugation with L-Dopa and Dopamine was achieved through an amide bond using classical reaction conditions (Fig. 15).

These compounds were thought to deliver the antioxidant moiety of lipoic acid directly to the specific group of cells where stress is related to pathology, i.e., normal aged cells and cells affected by neurodegenerative disorders. Indeed, the combination of lipoic acid with L-DOPA resulted in improved antioxidant properties compared to L-Dopa on its own.

The concept behind several depigmentation agents has been based on the conjugate reaction between sulhydryl compounds and DOPA-quinone. The reduced form of lipoic acid reacts with DOPA-quinone and the derivatives formed show a pronounced depigmenting effect by inhibition of DOPA-chrome formation. Unfortunately, dihydrolipoic acid is easily oxidable and poorly water soluble. To overcome this problem a zinc salt using its thiol groups as ligands was prepared and the lipoic acid was coupled via an amide bond to histidine. Salification of histidine resulted in good water solubility. The sodium zinc salt of

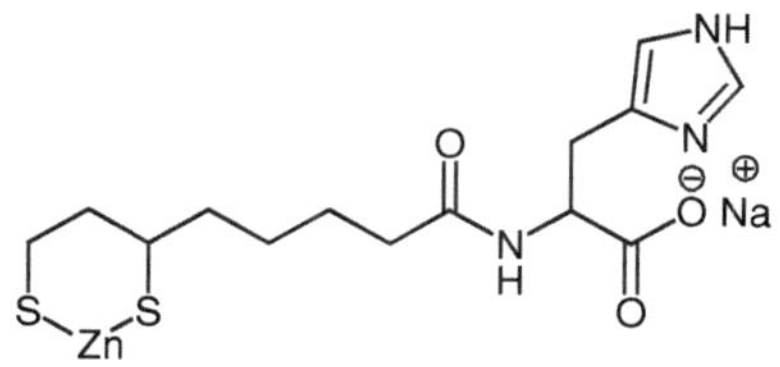

Fig. 16 Dihydrolipoic acid conjugate with histidine (sodium zinc salt)

dihydrolipoylhistidine formed retained its reactivity toward DOPA quinone and also significantly inhibited melanin synthesis. Hence this derivative is a strong candidate for use as a skin-lightening agent (Fig. 16) (Tsuji-Naito et al. 2007).

To fight or prevent diseases or conditions related to or caused by free radicals, a combination of lipoic acid and carnitine has been described using a hydrolyzable linker (Singh et al. 2007). Proline conjugated to lipoic acid via an amide bond has been tested for COX-2 inhibition (Taher et al. 2008). Also, diosgenin has been coupled to lipoic acid through an amino-sugar linker and the cytotoxicity of the hybrid has been studied in the context of neuroblastoma and breast cancer cell lines (Kaskiw et al. 2008). A complete review of lipoic acid and all types of lipoic acid conjugates can be consulted for further information (Koufaki et al. 2009).

5.2 Retinoic Acid Conjugates

Retinoic acid (Fig. 17) has been coupled with a vitamin B analog and this resulted in a mixture of eight compounds as a yellow oil, of which six were identified. This combination of retinoid derivatives was then tested for antiviral and anti-HIV properties (Kesel 2003).

Retinoids and especially retinoic acid was shown to repair chronological skin aging and photoaging (Fluhr et al. 1999). Retinoic acid, however, is fat soluble, a skin irritant and causes many side effects (Kim et al. 2008) such as skin-dryness, wounds, and scraping. These factors are limiting medicinal and cosmetic applications of the acid itself. These side effects are mostly due to the acid function which at the same time leads to chemical modification. Retinol, vitamin A, has also received considerable attention because of its anti-aging activity as part of long-term treatment. Like retinoic acid, retinol can hardly be used in cosmetic formulations due to its instability when exposed to light and oxygen. The combination of retinoic acid and retinol to form retinoylretinate, however, provides an elegant solution to most of these problems (Fig. 18). It has the advantage of hiding the carboxylic acid function of retinoic acid and, at the same time, stabilizing the retinol moiety (Kim et al. 2008).

Anti-aging properties of the individual building blocks were retained in this case and microspheres of biodegradable polymers (polylactic acid) were used as delivery vehicle for this poorly water-soluble anti-wrinkle agent (Kim et al. 2012).

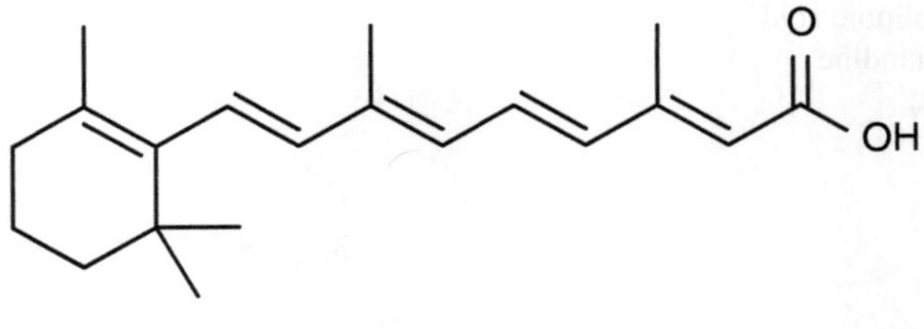

Fig. 17 Chemical structure of retinoic acid

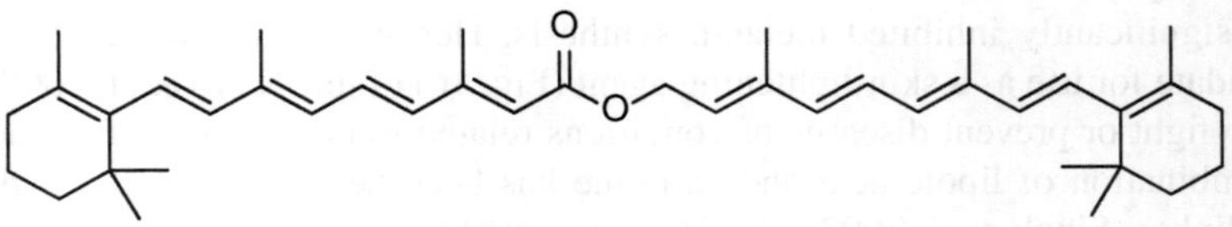

Fig. 18 Chemical structure of retinoylretinoate

6 Alkaloid Conjugates

Alkaloid conjugates have already been mentioned briefly in the introduction to this chapter. They very often exist as (naturally occurring) flavopyridols. Tubulin binding properties of noscapine (L-narcotine), an opium alkaloid, were revealed in 1998 (Ye et al. 1998). Conjugation of noscapine with folic acid was then carried out and the resulting compound was patented under the name "Targetin" for cancer treatment. A subsequent publication explained how molecular modeling had led to the idea of conjugating noscapine to folic acid, as the folate FR α receptor is overexpressed in ovarian epithelial cancers. Conjugation was achieved through an amide bond between folic acid and an amino-noscapine derivative (Fig. 19) (Staswick 2009).

It appears that not many other attempts have been reported so far to conjugate alkaloids with other natural compounds, hence leaving open a considerable spectrum of potential hybrid molecules for further research and development.

7 Miscellaneous Conjugates

Tryptophane has been conjugated to jasmonic and indole-3-acetic acid and studied as an inhibitor of of the effects of indole-3-acetic acid (auxin) (Staswick 2009). Conjugates between catecholamines and glutathione or cysteine have also been prepared to study the influence of oxygen-based radicals and other ROS on neuronal death in Parkinson's disease. As already mentioned, most of the neuronal damage may be related to the oxidation of L-DOPA or dopamine (Spencer et al. 1998).

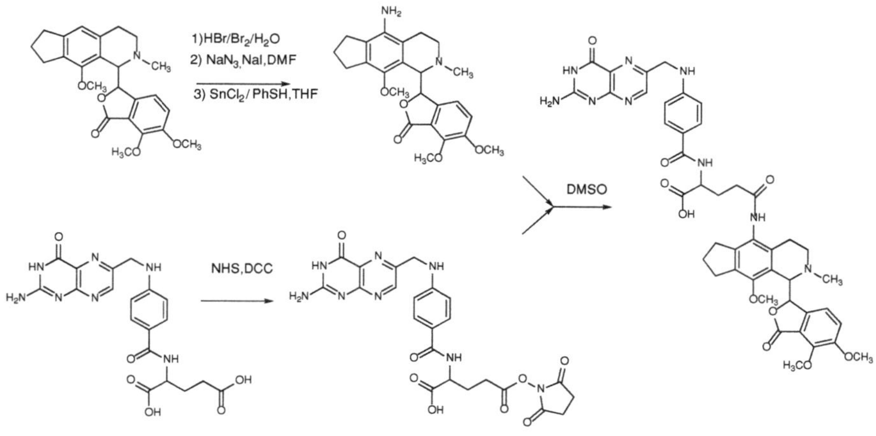

Fig. 19 Synthesis of noscapine-folic acid conjugate (targetin)

To overcome potential toxicity of drug carrying polymer nanocarriers (see also Chap. 19), conjugates of heparin and aminated folic or retinoic acid have been prepared using a microfluidic technology (Tran et al. 2012).

The last decades have also invested many efforts to identify useful carriers for intracellular delivery of beta-lactam antibiotics. Antibiotic-loaded liposomes have extensively been studied. These liposomes are often based on polyalkylcyanoacrylates and have shown some toxicity and side effects. In order to overcome these problems, it was proposed to link penicillin to slightly modified squalene skeletons. The bioconjugates formed were found to assemble spontaneously into nanoparticles and those have shown fast and significant activity against *Staphylococcus aureus* infected J774 cells (Semiramoth et al. 2012).

8 Conclusions

The trend to prepare hybrid molecules of several natural compounds 'molten together' in such comparably simple structures has appeared quite recently in the literature. The reasons for such conjugations or fusions can differ: enhancing bio-disponibility of some active compounds where bio-availability is low; targeting more precisely some receptors in the way to traffic the natural active compounds directly to the target-side, and last but not least, aiming for a double effect related to the biological activity of each building block. The principle of melting natural compounds together seems to gain importance and today is considered quite frequently when synthetic chemists are looking for—improved—bioactive molecules.

References

Aggarwal S, Ndinguri MW, Solipuram R, Wakamatsu N, Hammer RP, Ingram D, Hansel W (2011) [DLys(6)]-luteinizing hormone releasing hormone-curcumin conjugate inhibits pancreatic cancer cell growth in vitro and in vivo. Int j cancer J int du cancer 129(7):1611–1623

Al-Wabli RI, AboulWafa OM, Youssef KM (2012) Synthesis of curcumin and ethylcurcumin bioconjugates as potential antitumor agents. Med Chem Res 21(6):874–890

Alakurtti S, Makela T, Koskimies S, Yli-Kauhaluoma J (2006) Pharmacological properties of the ubiquitous natural product betulin. Eur J Pharm Sci 29(1):1–13

Bhaumik S, Anjum R, Rangaraj N, Pardhasaradhi BVV, Khar A (1999) Curcumin mediated apoptosis in AK-5 tumor cells involves the production of reactive oxygen intermediates. FEBS Lett 456(2):311–314

Delaire S, Adao A, Desmurs J-R, Gelo-Pujic M, Saint-Jalmes L, Kassem T (2009) Polyphenolic Bioprecursors

Di Stefano A, Sozio P, Cocco A, Iannitelli A, Santucci E, Costa M, Pecci L, Nasuti C, Cantalamessa F, Pinnen F (2006) L-Dopa- and dopamine-(R)-alpha-lipoic acid conjugates as multifunctional codrugs with antioxidant properties. J Med Chem 49(4):1486–1493

Dubey SK, Sharma AK, Narain U, Misra K, Pati U (2008) Design, synthesis and characterization of some bioactive conjugates of curcumin with glycine, glutamic acid, valine and demethylenated piperic acid and study of their antimicrobial and antiproliferative properties. Eur J Med Chem 43(9):1837–1846

Fluhr JW, Vienne MP, Lauze C, Dupuy P, Gehring W, Gloor M (1999) Tolerance profile of retinol, retinaldehyde and retinoic acid under maximized and long-term clinical conditions. Dermatology 199:57–60

Gelo-Pujic M, Desmurs JR, Kassem T, Delaire S, Adao A, Tawil D (2008) Synthesis of new antioxidant conjugates and their in vitro hydrolysis with Stratum corneum enzymes. Int J Cosmet Sci 30(3):195–204

Gomes CA, da Cruz TG, Andrade JL, Milhazes N, Borges F, Marques MP (2003) Anticancer activity of phenolic acids of natural or synthetic origin: a structure-activity study. J Med Chem 46(25):5395–5401

Guerriero E, Sorice A, Capone F, Costantini S, Palladino P, D'Ischia M, Castello G (2011) Effects of lipoic acid, caffeic acid and a synthesized lipoyl-caffeic conjugate on human hepatoma cell lines. Molecules 16(8):6365–6377

Harish G, Venkateshappa C, Mythri RB, Dubey SK, Mishra K, Singh N, Vali S, Bharath MM (2010) Bioconjugates of curcumin display improved protection against glutathione depletion mediated oxidative stress in a dopaminergic neuronal cell line: Implications for Parkinson's disease. Bioorg Med Chem 18(7):2631–2638

Houghton PJ (2000) Chemistry and biological activity of natural and semi-synthetic chromone alkaloids. In: Bioactive Natural Products (Part B), vol 21

Ismail IS, Nagakura Y, Hirasawa Y, Hosoya T, Lazim MI, Lajis NH, Shiro M, Morita H (2009) Chrotacumines A-D, chromone alkaloids from *Dysoxylum acutangulum*. J Nat Prod 72(10):1879–1883

Jain SK, Bharate SB, Vishwakarma RA (2012) Cyclin-dependent kinase inhibition by flavoalkaloids. Mini Rev Med Chem 12(7):632–649

Kaskiw MJ, Tassotto ML, Th'ng J, Jiang ZH (2008) Synthesis and cytotoxic activity of diosgenyl saponin analogues. Bioorg Med Chem 16(6):3209–3217

Kesel AJ (2003) Synthesis of retinoid vitamin A-vitamin B6 conjugate analogues for antiviral chemotherapy. Biochem Biophys Res Commun 300(3):793–799

Khadem S, Marles RJ (2012) Chromone and flavonoid alkaloids: occurrence and bioactivity. Molecules 17(1):191–206

Kim H, Kim B, Kim H, Um S, Lee J, Ryoo H, Jung H (2008) Synthesis and in vitro biological activity of retinyl retinoate, a novel hybrid retinoid derivative. Bioorg Med Chem 16(12):6387–6393

Kim H, Kim M, Quan Y, Moon T, Mun J, Cho H, Park N, Moon W, Lee K, Kim H, Lee J, Ryoo H, Jung H (2012) Novel anti-wrinkle effect of cosmeceutical product with new retinyl retinoate microsphere using biodegradable polymer. Skin Res Technol 18(1):70–76

Koufaki M, Detsi A, Kiziridi C (2009) Multifunctional lipoic acid conjugates. Curr Med Chem 16(35):4728–4742

Kumar S, Narain U, Tripathi S, Misra K (2001) Syntheses of curcumin bioconjugates and study of their antibacterial activities against beta-lactamase-producing microorganisms. Bioconjug Chem 12(4):464–469

McDougall B, King PJ, Wu BW, Hostomsky Z, Reinecke MG, Robinson WE Jr (1998) Dicaffeoylquinic and dicaffeoyltartaric acids are selective inhibitors of human immunodeficiency virus type 1 integrase. Antimicrob Agents Chemother 42(1):140–146

Mishra S, Kapoor N, Ali AM, Pardhasaradhi BVV, Kumari AL, Khar A, Misra K (2005a) Differential apoptotic and redox regulatory activities of curcumin and its derivatives. Free Radical Bio Med 38(10):1353–1360

Mishra S, Narain U, Mishra R, Misra K (2005b) Design, development and synthesis of mixed bioconjugates of piperic acid-glycine, curcumin-glycine/alanine and curcumin-glycine-piperic acid and their antibacterial and antifungal properties. Bioorg Med Chem 13(5):1477–1486

Mythri RB, Harish G, Dubey SK, Misra K, Bharath MMS (2011) Glutamoyl diester of the dietary polyphenol curcumin offers improved protection against peroxynitrite-mediated nitrosative stress and damage of brain mitochondria in vitro: implications for Parkinson's disease. Mol Cell Biochem 347(1–2):135–143

Noh JM, Kwak SY, Seo HS, Seo JH, Kim BG, Lee YS (2009) Kojic acid-amino acid conjugates as tyrosinase inhibitors. Bioorg Med Chem Lett 19(19):5586–5589

Pandey A, Gupta RK, Bhargava A, Agrawal B (2011) Antibacterial Activities of Curcumin Bioconjugates. Int J Pharmacol 7(8):874–879

Salmaso S, Bersani S, Semenzato A, Caliceti P (2007) New cyclodextrin bioconjugates for active tumour targeting. J Drug Target 15(6):379–390

Semiramoth N, Di Meo C, Zouhiri F, Said-Hassane F, Valetti S, Gorges R, Nicolas V, Poupaert JH, Chollet-Martin S, Desmaele D, Gref R, Couvreur P (2012) Self-assembled squalenoylated penicillin bioconjugates: an original approach for the treatment of intracellular infections. ACS Nano 6(5):3820–3831

Singh IP, Singh S, Noujaim A, Hirsche BD, Tam DS (2007) Cleavable carnitine-linker-lipoic acid conjugate,and their therapeutic use. PCT Int Appl, WO 2007095760

Singh RK, Rai D, Yadav D, Bhargava A, Balzarini J, De Clercq E (2010) Synthesis, antibacterial and antiviral properties of curcumin bioconjugates bearing dipeptide, fatty acids and folic acid. Eur J Med Chem 45(3):1078–1086

Spencer JP, Jenner P, Daniel SE, Lees AJ, Marsden DC, Halliwell B (1998) Conjugates of catecholamines with cysteine and GSH in Parkinson's disease: possible mechanisms of formation involving reactive oxygen species. J Neurochem 71(5):2112–2122

Spivak AY, Mufazzalova RR, Shakurova ER, Odinokov VN, Dzhemilev UM (2010) Synthesis of novel bioconjugates of alpha-tocopherol with lupane triterpenoids. Russ Chem B + 59 (1):241–250

Staswick PE (2009) The tryptophan conjugates of jasmonic and indole-3-acetic acids are endogenous auxin inhibitors. Plant Physiol 150(3):1310–1321

Taher N, Nureldin R, Anwar R (2008) Novel conjugates of polyunsaturated fatty acids with amines for the treatment of various inflammatory disease or disorder or cancer. PCT Int Appl, WO2008075366

Tanaka R, Tsujii H, Yamada T, Kajimoto T, Amano F, Hasegawa J, Hamashima Y, Node M, Katoh K, Takebe Y (2009) Novel 3 alpha-methoxyserrat-14-en-21 beta-ol (PJ-1) and 3 beta-methoxyserrat-14-en-21 beta-ol (PJ-2)-curcumin, kojic acid, quercetin, and baicalein conjugates as HIV agents. Bioorgan Med Chem 17(14):5238–5246

Tanaka R, Tsujii H, Yamada T, Kajimoto T, Tokuda H, Arai T, Suzuki N, Hasegawa J, Hamashima Y, Node M (2011) Conjugates of 3 alpha-methoxyserrat-14-en-21 beta-ol (PJ-1) and 3 beta-methoxyserrat-14-en-21 beta-ol (PJ-2) as cancer chemopreventive agents. Eur J Med Chem 46(8):3368–3375

Touaibia M, Jean-Francois J, Doiron J (2011) Caffeic Acid, a versatile pharmacophore: an overview. Mini Rev Med Chem 11(8):695–713
Tran TH, Nguyen CT, Kim DP, Lee YK, Huh KM (2012) Microfluidic approach for highly efficient synthesis of heparin-based bioconjugates for drug delivery. Lab Chip 12(3):589–594
Tsuji-Naito K, Hatani T, Okada T, Tehara T (2007) Modulating effects of a novel skin-lightening agent, alpha-lipoic acid derivative, on melanin production by the formation of DOPA conjugate products. Bioorg Med Chem 15(5):1967–1975
Tsujii H, Yamada T, Kajimoto T, Tanaka R, Tokuda H, Hasegawa J, Hamashima Y, Node M (2010) Hybrids of 3 alpha-methoxyserrat-14-en-21 beta-ol (PJ-1) and 3 beta-methoxyserrat-14-en-21 beta-ol (PJ-2) and various anti-oxidants as cancer chemopreventive agents. Eur J Med Chem 45(6):2191–2197
Vasilevsky SF, Govdi AI, Sorokina IV, Tolstikova TG, Baev DS, Tolstikov GA, Mamatuyk VI, Alabugin IV (2011) Rapid access to new bioconjugates of betulonic acid via click chemistry. Bioorg Med Chem Lett 21(1):62–65
Vlietinck AJ, De Bruyne T, Apers S, Pieters LA (1998) Plant-derived leading compounds for chemotherapy of human immunodeficiency virus (HIV) infection. Planta Med 64(2):97–109
Yamaguchi C, Wanibuchi H, Kakehashi A, Tanaka R, Fukushima S (2008) Chemopreventive effects of a serratane-type triterpenoid, 3 alpha-methoxyserrat-14-en-21 beta-ol (PJ-1), against rat lung carcinogenesis. Food Chem Toxicol 46(6):1882–1888
Ye KQ, Ke Y, Keshava N, Shanks J, Kapp JA, Tekmal RR, Petros J, Joshi HC (1998) Opium alkaloid noscapine is an antitumor agent that arrests metaphase and induces apoptosis in dividing cells. P Natl Acad Sci USA 95(4):1601–1606

Authors Biography

Gilbert Kirsch (born 1947) has been trained as an organic chemist at the Universities of Strasbourg and Metz. He started his academic carrier in 1973 at the University of Metz (now University of Lorraine) where he currently holds the position of Professor of Organic Chemistry. He has been a postdoc at Oak Ridge National Laboratory (TN) in the Nuclear Medicine Group and was also an invited scientist at Kodak (Rochester, NY) as well as invited professor at the University of Minho (Portugal) and Emory University (Atlanta, GA).

Gilbert's interests reside in heterocyclic chemistry, especially in the field of five-membered aromatic systems (thiophenes, selenophenes, tellurophenes, thiazoles, selenazoles) and their benzo-condensed derivatives. Lately, he developed synthetic work in the field of coumarins, looking at biological activities (CDC25 phosphatase inhibition). From his research, he published over 200 papers, wrote different chapters in books, like Patai's Functional Group Series, in Houben-Weyl, in Chemistry of Heterocyclic Compounds (J. Wiley Interscience), and in Springer's Selenium and Tellurium Chemistry. He holds also a few patents in the field of heart imaging and sulfur-containing tire additives. Gilbert has coordinated an Interreg program (acronym "Corena") about natural compounds for Medicine and Agriculture, and has participated in the EU ITN Marie Curie program "RedCat". He is participating in the regional programs Bioprolor and Biocaptech and in a national French ANR program on para-hydrogen.

Sergio Valente (born 1977) holds a cum laude Master degree in Pharmacy (2003) and an European PhD in "Pasteur Science" (2007) from the Department of Drug Chemistry and Technologies, "Sapienza" University of Rome. His thesis in the field of medicinal chemistry is entitled "Design, synthesis and biological validation of epigenetic modulators of histone/protein deacetylation and methylation". From July 2009 to May 2011, Sergio carried out postdoctoral research at LIMBP, Université de la Lorraine-Metz, France, under the supervision of Prof. Gilbert Kirsch and as part of the "RedCat" European Initial Training Network. Since October 2011, Sergio is Assistant Professor in Medicinal Chemistry, at the Department of Drug Chemistry and Technologies, Sapienza University of Rome, Italy.

Sergio's main research activity relates to the design, synthesis, and development of small epigenetic modulators of HDACs, SIRTs, HMTs, KDMs, DNMTs, in order to develop new approaches for anticancer chemotherapy, neurodegenerative, and metabolic pathologies. Moreover, he has developed an interest in coumarin-based inhibitors of CDC25 phosphatases endowed with anticancer activity. So far, Sergio is author/co-author of 55 scientific papers. He has been invited to deliver seven oral communications at national meetings on medicinal chemistry and he is presenting author of almost 30 poster communications at national and international congresses. He is also a participant in ten funded (national and European) projects.

Part XV
Connecting Section Between Chapters 15 and 16

The previous chapter briefly considered the advantage of employing hybrid molecules instead of individual compounds because of a better solubility and improved bioavailability. Indeed, many natural products are only poorly soluble in water and are also unable to cross cellular membranes. As a result, the bioavailability of such substances is inherently low in the case of more complex organisms, and especially in humans. Indeed, there are numerous examples of redox active secondary metabolites, including many flavonoids, which possess a considerable biological activity in cell culture, yet are virtually impossible to administer in humans in sufficient amounts because of a notoriously low solubility (see Chap. 2). Indeed, compounds such as oligomeric or polymeric proanthocyanidins are hardly able to enter the human body via the oral route, at least in their non-metabolized forms (see Chap. 8). It is therefore questionable if such substances actually reach the bloodstream and their target organs in sufficient amounts when taken up via food, and hence may be relevant for human health at all. This particular drawback associated with many secondary metabolites can be overcome by various 'tricks'. It is obviously possible to encapsulate such compounds into specifically designed carrier systems, such as micelles, vesicles, or liposomes. While this may solve the initial problem of poor solubility in water, it may not enable these compounds to cross lipophilic barriers. Alternatively, a small chemical modification, often in the form of a partial synthesis, may help. Here, solubility and lipophilicity of the substance in question are 'fine-tuned' by the modification of certain functional groups (e.g., esterification, methylation), or by conjugation to a more hydrophilic or lipophilic molecule (e.g., a fatty acid). This particular approach has been illustrated briefly as part of the previous chapter and results in a new molecule, which de facto is 'not quite' natural any more—with all the chemical, biochemical, biological, and regulatory implications such an artificial derivative brings with it.

The following chapter will therefore take a rather different approach. Solubility, and also the ability of compounds to cross biological barriers, does not necessarily imply isolated molecules in solution. Small aggregates of molecules, such as micelles, are also 'in solution' and certainly capable of crossing such barriers. In fact, micellation is often used to solubilize otherwise poorly soluble materials (e.g., in the case of detergents) and vesicles are employed regularly to traffic molecules across biological barriers. Interestingly, the same considerations also apply to nanoparticles. Not surprisingly, therefore, the idea of converting otherwise poorly soluble and biounavailable secondary metabolites into nanoparticles and hence render them 'available', has recently attracted considerable interest. Many of these rather exciting developments fall within the emerging field of applied biological nanotechnology and hence seem to have bypassed the mainstream of natural product research, at least so far. Cornelia M. Keck, Rainer H. Mueller, and Karl-Herbert Schaefer will therefore highlight some of the latest developments in this rapidly expanding field. They apply basic ideas of nanotechnology to redox active secondary metabolites and show how nanocrystals of compounds such as hesperidin exhibit unique solubility and release properties, which are of considerable interest for practical applications of these metabolites, and indeed have already paved the way for specific applications of such antioxidant nanocrystals, especially in the fields of Medicine and Cosmetics.

Chapter 16
Nanotaxis for Antioxidants

Cornelia M. Keck, Rainer H. Mueller and Karl-Herbert Schaefer

Keywords Antioxidants • Biopharmaceutics • Nanocrystal • Nanotaxi • Nanosuspension

1 Introduction

Many plants produce secondary plant metabolites that possess health beneficial properties. Examples include flavonoids, which act as antioxidants and thus can be isolated and used to support medical treatment (e.g., curcumin for colon cancer) or even to treat several diseases (Vauzour et al. 2010; Visioli et al. 2011). For an efficient supportive use or therapy, however, the compound must be bioavailable to the body. This means that upon administration, a sufficient amount of the compound must be taken up by the body (e.g., after oral administration into the blood or after dermal administration into the skin/epidermis and dermis). As only dissolved molecules can be taken up, the major prerequisite for bioavailability is a sufficient solubility of the compound at the site of administration (e.g., in the gastrointestinal

C. M. Keck (✉)
Applied Pharmacy, Campus Pirmasens, University of Applied Sciences Kaiserslautern, Carl-Schurz-Str. 10-16, 66953 Pirmasens, Germany
e-mail: ck@ckc-berlin.de

K.-H. Schaefer
Campus Zweibruecken, University of Applied Sciences Kaiserslautern, Amerikastr. 1, 66482 Zweibruecken, Germany

C. M. Keck · R. H. Mueller
Department of Pharmaceutics, Biopharmaceutics and NutriCosmetics, Freie Universitaet Berlin, Kelchstr. 31, 12169 Berlin, Germany

K.-H. Schaefer
Clinic of Pediatric Surgery, Workgroup Enteric Nervous System, University of Hospital Mannheim, Theodor-Kutzer-Ufer 1-3, 68167 Mannheim, Germany

C. Jacob et al. (eds.), *Recent Advances in Redox Active Plant and Microbial Products*,
DOI: 10.1007/978-94-017-8953-0_16,

tract or the water phase of a dermal creams). Unfortunately, for natural plant actives very poor solubility is very common, they possess very low saturation solubilities (e.g., rutin 125 μg/ml). Hence, even though these compounds possess high potential of bioactivity, the potential cannot be exploited due to their poor solubility in aqueous media (see also Explanatory Box 1) (Visioli et al. 2011). A solution to this problem is pharmaceutical nanotechnology. The principle of nanotechnology is that materials obtain new physicochemical properties when they are transferred from the macroscopic into the nanodimension, i.e., to a size of below approximately 1,000 nm down to a few nm (=nano range). Below a size of approximately 1,000 nm—besides other properties—the kinetic saturation solubility of materials strongly increases. Therefore a simple but effective strategy to overcome poor solubility is the transformation of the coarse plant actives, i.e., the coarse crystalline powder material, into nanocrystals.

Explanatory Box 1: Biopharmaceutical Classification System (BCS) of Drugs

This chapter addresses one of the most important issues in pharmacokinetics and drug delivery. In order to reach the human bloodstream and target organs, pharmaceutically active substances have to fulfill a number of very important criteria. Some general requirements on the physicochemical properties of such compounds are summarized in Lipinski's 'Rule of Five'. Here, the balance between hydrophilicity and lipophilicity of a given compound is particularly important. On the one hand, compounds need to be soluble in water in order to reach the walls of the gastrointestinal tract. On the other hand, they also need to cross various membranes and hence need to show a certain lipophilicity (there are, of course exceptions, which include compounds able to move through a channel, highjack a transport system or enter via endocytosis). The biopharmaceutical classification system therefore classifies compounds according to their solubility in aqueous media and their ability to cross membranes, i.e., their permeability. The resulting matrix contains four fields or classes, and each of these classes contains examples of important drugs.

Class I, for instance, contains the 'ideal' compounds with high solubility and permeability. It is represented by the beta-blocker Metoprolol®, which is used to treat disorders of the cardiovascular system. Class II represents compounds which are poorly soluble in water but still highly permeable, such as the anti-diabetic drug Glibenclamide®. It is mirrored by Class III which stands for highly soluble compounds which are poorly permeable. The histamine H_2-receptor antagonist cimetidine used to treat stomach ulcers is a representative of this class. And finally, Class IV contains drugs of low solubility and permeability. These compounds are rather poor from a pharmacokinetic perspective, yet some of them, such as the diuretic hydrochlorothiazide used to treat certain cardiovascular disorders (among others), are in clinical use.

While this classification is fairly useful as a first guide to pharmacokinetic properties, the real situation is often more complex. A fair number of drugs change their charge(s) and hence hydrophilicity depending on pH. Using pH as a control instance, it is possible to fine-tune solubility and permeability according to the pH in the gastrointestinal tract, and therefore also to determine where the drug becomes permeable and hence is taken up during its or-an-al journey. A similar strategy is employed in the case of produgs, which can be designed as water-soluble, yet poorly permeable substances which upon activation in the gastrointestinal tract change their solubility and permeability and hence become bioavailable.

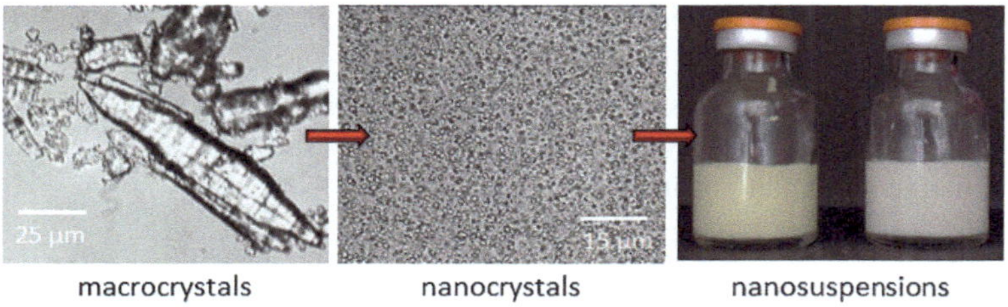

Fig. 1 Nanocrystals (*middle*) are obtained from coarse material (*left*). Often, due to the production process nanocrystals are dispersed in a liquid, the formulation is called a nanosuspension (*right*). Nanosuspensions are low viscous and 'milky like'. In contrast to macrosuspensions, nanosuspensions do not show sedimentation effects

Nanocrystals are nanoparticles composed of 100 % active (e.g., nutraceutical or drug). The particle size is in the nano range, typically 200–600 nm. Nanocrystals are produced by nanonizing coarse material being in the µm-size range, many powders are approx 50–200 µm (left and middle). This can be performed by several processes, e.g., precipitation, bead/pearl milling, and high pressure homogenization. Combination processes are also used, e.g., the combination of precipitation and high pressure homogenization (H 69), or the combination of ball milling and high pressure homogenization (CT process). All these production processes are 'wet' techniques, i.e., they require the dispersion of the material in a liquid. The liquid is in most cases water, an aqueous solution (e.g., isotonic water), a lipophilic liquid (e.g., oils, paraffin, waxes), or polyethylene glycol (PEG). Thus the product containing the nanocrystals is a liquid in which the nanocrystals are suspended, the nanocrystals are stabilized by a surrounding stabilizer layer (identical to macrosuspensions), i.e., surfactants, polymeric stabilizers, or simply viscosity enhancers. The product is therefore a nanosuspension (Fig. 1 right).

The nanocrystals can be used as a delivery system (=nanotaxi) to deliver active compounds (nutraceuticals, cosmetics, drugs) to different sites of the body. The target site determines the route of administration: oral, dermal, or parenteral (e.g., intravenously (*i.v.*), ocular). Depending on the route of administration, either the nanosuspension itself is used (e.g., *i.v.* injection) or transferred into another suitable dosage form, e.g., a tablet or capsule for oral and a crèmes/gel for dermal applications.

This chapter describes the special physicochemical properties of nanocrystals important for turning antioxidants biologically active. In addition, the main routes of administration and the respective dosage forms are briefly reviewed. Very briefly, the possibility for large-scale industrial production—the essential prerequisite for marketed products—is also covered. Last but not least, the aspect of nanotoxicology/nanotolerability is discussed, a point which is becoming increasingly important for manufacturers but also in the perception of the consumers.

2 Properties of Nanocrystals

Due to the decrease in particle size the physicochemical properties of the material change. The most important parameters, e.g., for the delivery of antioxidants, which change upon nanonization are:

- increase in dissolution velocity,
- increase in solubility,
- increased adhesiveness on surfaces or interfaces (e.g., biological membranes),
- increased bioavailability/uptake.

The increase in dissolution velocity can be explained by the increase in the total surface area of the particles. Based on the Noyes Whitney equation (Eq. 1), the dissolution velocity dc/dt is proportional to the surface area A, the diffusion constant D and to the concentration gradient $c_s - c_x/h$.

$$\mathbf{dc/dt = A \cdot D \cdot (c_s - c_x/h)} \tag{1}$$

c_s is the saturation solubility, c_x the concentration in the bulk surrounding a dissolving particle, h is the height of the boundary layer. Due to nanonization the total surface area A of the particles increases (Fig. 2a). In addition, the concentration gradient $c_s - c_x/h$ increases. This effect is more complex. The first prerequisite for an increased concentration gradient is the saturation solubility of the compound. Due to nanonization the curvature of the particles increases. This leads to a higher dissolution pressure (similar to the vapor pressure of liquids in smaller droplets in a gas phase). Based on the Kelvin equation a higher dissolution pressure leads to a higher kinetic solubility of the material (Fig. 2b). In addition, the diffusional distance h decreases with decreasing particle size. h can be calculated using the Prandtl equation (Fig. 2c).

Due to their small size, nanocrystals and all nanomaterials possess a much smaller mass to surface ratio compared to larger particles. This leads to an increased adhesion on surfaces or interfaces, because less weight is attached per attaching point (Fig. 3a).

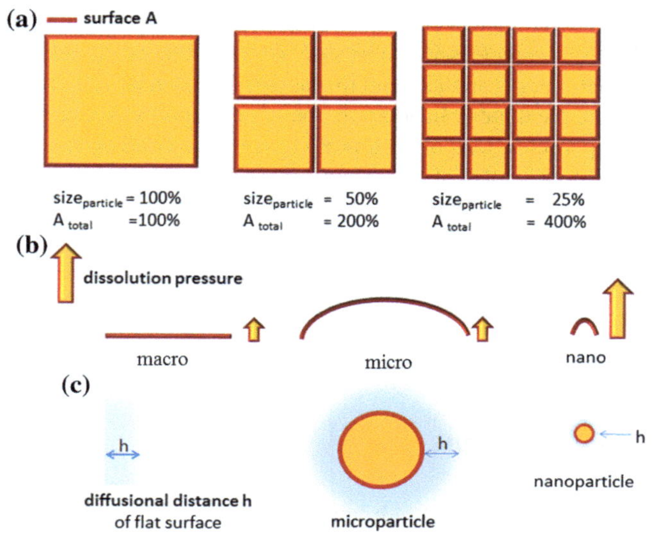

Fig. 2 Scheme of the changing properties of nanocrystals due to diminution leading to an increased dissolution velocity. **a** increase in the total surface area A; **b** increase in saturation solubility c_S due to increase in curvature, **c** decrease in diffusional distance h

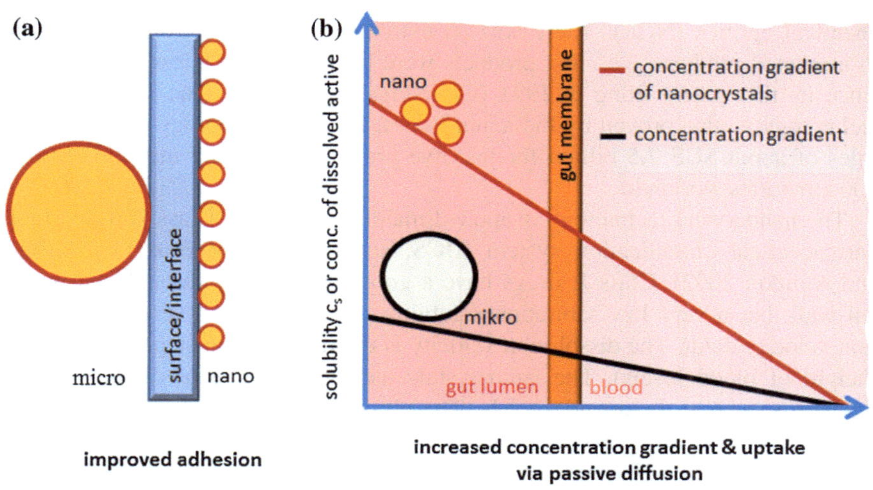

Fig. 3 **a** In comparison to macro- or microparticles, nanocrystals (and other nanomaterials) possess an increased adhesion on surfaces or interfaces, because less weight is attached per adhesion point. Having the same mass, one macroparticle corresponds to hundreds of nanoparticles, i.e., the same mass is attached via hundreds of adhesion points. **b** Due to increased solubility nanocrystals possess a higher concentration gradient, e.g., on membranes (gut or skin). This increases the uptake of the active via passive diffusion and thus the bioactivity

All these improved parameters contribute to an improved bioavailability/uptake upon application. Due to the increased solubility c_s the concentration gradient is increased (Fig. 3b). As many actives are taken up by passive diffusion a high concentration gradient will promote the uptake. In addition, the improved adhesion prolongs the retention time of the particles, leading to a prolonged time for the uptake. The improved bioactivity and the principles of the enhancement of the bioavailability of nanocrystals are reviewed in more detail e.g., in Mueller et al. 2000, 2011a, b; Peters and Mueller 1996).

3 Applications of Nanocrystals

3.1 Oral Administration

The properties of nanocrystals are mainly exploited in pharmaceutical industry, preferentially in oral products. The nanocrystals are used for improvement of the oral delivery of poorly soluble actives (drugs). Product examples are Emend (aprepitant, company Merck), Rapamune (drug sirolimus, company Wyeth), or Tricor (fenofibrate, company Abbott). In 2010, the nanocrystal technology was voted as the most successful drug delivery technology innovation (NanoCrystal Wins Technology Innovation Award 2010). This decision was based on the great success of this technology, as it was designed to overcome the important issues associated with poor water solubility, which concerns about 90 % of the new chemical entities (NCE). There was an extremely short time from patent filing to marketed products, i.e., first products were launched within less than 10 years since its first patent filing in 1991 (Liversidge et al. 1992). The success of this technology is also proven by the annual sales, i.e., the technology creates annual sales of about $US 7.5 billion for just five products which are marketed in over 100 territories worldwide.

The nanocrystal technology is applied mainly to drugs of Class II of the biopharmaceutical classification system (BCS, see Explanatory Box 1) (Martinez and Amidon 2002). Class II drugs have a good permeability through the intestinal wall, but a very low saturation solubility c_s and a related very low dissolution velocity dc/dt. The dissolution velocity is the rate limiting step for absorption (height of blood levels). The nanocrystals increase the rate of dissolution sufficiently to obtain therapeutic levels. The absorption is further promoted by the higher saturation solubility c_s of nanocrystals compared to microcrystals. This increase in c_s can be 10 times, but also up to 100 times or more—depending on the nature of the compound (increase is compound-specific). The increased c_s leads to a higher concentration gradient between gastrointestinal lumen and the blood, and a higher diffusive flow to the blood (Fig. 3b).

Many antioxidants are poorly soluble and can be classified as Class II compounds of the BCS. A typical example is coenzyme Q_{10}, the bioavailability is

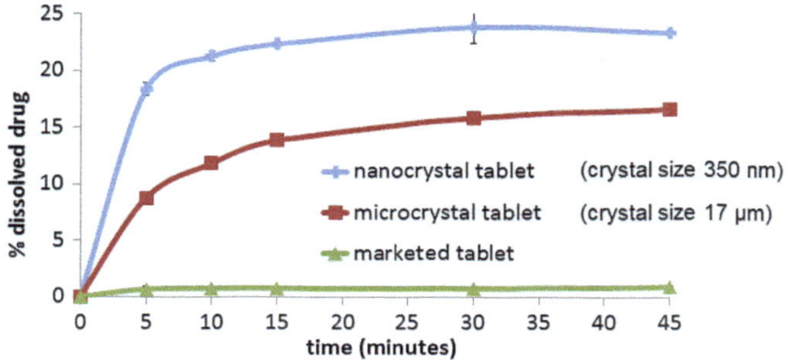

Fig. 4 *In vitro* dissolution of hesperidin from a marketed tablet versus a microcrystal and nanocrystal formulation (tablets with crystals size of 17 µm and 350 nm, respectively). Plotted is the percentage of released drug for the % released from dose in a Levy beaker as function of time [from p. 206 (Mauludin 2008)]

for most marketed oral products less than 4 % (Barakat et al.). Similar to Class II drugs, many antioxidant saturation solubilities are also very low, e.g., 20 µg/ml for hesperidin. Transfer to nanocrystals increased the solubility three-fold to 72 µg/ml (Mauludin 2008). In addition, the dissolution velocity increases distinctly by transfer to nanocrystals. In pharmacy the velocity and extent of dissolution is determined using the paddle method by the United States Pharmacopeia (USP, method 2). The powder or tablet is placed in a Levy beaker together with 900 ml release medium (water, buffer etc.,), and dissolution is determined at body temperature. The dissolved concentration is recorded by sampling as a function of time. This allows ofcomparing *in vitro* of the dissolution abilities of different technological formulations and also marketed products. Dissolution is a prerequisite for absorption and bioavailability, therefore if a marketed product does not dissolve *in vitro*, no bioavailability can be expected *in vivo*. Figure 4 shows the *in vitro* dissolution of hesperidin from a marketed product versus a microcrystal and a nanocrystal tablet. The release and dissolution from the marketed product is close to zero, very little bioavailability can be expected. The nanocrystal formulation is clearly manifold superior showing nicely the potential of the technology for nutraceuticals.

Besides a broad range of 'anti-'(inflammatory, cancer, etc.,) effects, antioxidants can have a strong neuroprotective effect. This is known for a variety of flavonoids such as quercetin. Several studies demonstrated impressively the supportive influence upon the central and peripheral nervous system. In cerebral and cerebellar tissues, quercetin administration leads to an improved protection against oxidative stress. In a rat model for a transient focal cerebral ischemia, quercetin administration prior and after the occlusion reduced significantly the infarction area and increased the functional capabilities (Ahmad et al. 2011). Similar effects are to be seen in the cerebellum (Arredondo et al. 2010). In a recent study, quercetin was administered in a dose of 200 mg/kg body weight in rats treated with streptozotocin (Lopes et al. 2012). Streptozotocin is used to induce diabetes. This leads to a loss of neurons and glial cells in the enteric nervous system. Quercetin

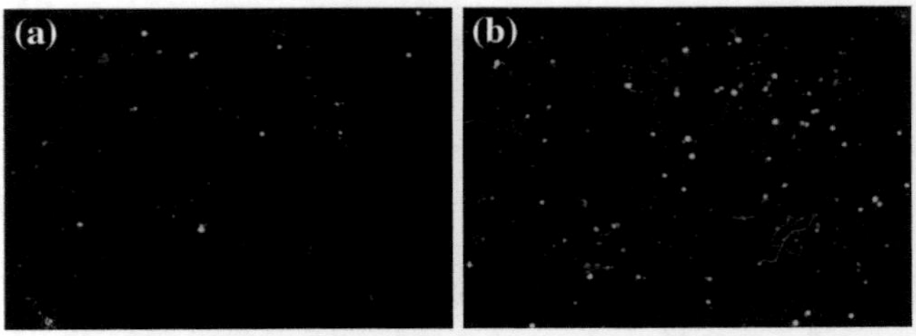

Fig. 5 Dissociated dorsal root ganglia from newborn rats, cultured in the presence of amyloid ß. Neurons were stained with a neuronal marker in *green*. While in the cultures with amyloid only most of the neurons die (**a**), cultures supplemented with antioxidant nanocrystals showed a much higher rate of surviving neurons (**b**). Moreover, the neurons had much longer neuritis as the neurons treated with amyloid alone

administration reduces neuronal and glial cell loss, and also reduces cell hypertrophy in at least a part of the enteric neurons. The individual neurons seem to benefit in a different amount from the quercetin treatment. This might mean that antioxidants either act only on specific neurons and glial cells, or that some of the cells are just more resistant against oxidative stress and do not need an additional protection. So far, it is not yet fully clear whether there is a pure protective effect, or whether there is also a stimulating effect upon the neural stem cell niche, which results in increasing numbers of neurons after neural stem cell differentiation, as it is known of resveratrol fatty acids (Hauss et al. 2007).

Due to the significant problems in dissolving most of the antioxidants, nanomodification might help to increase their uptake and so the bioavailability (Ghosh et al. 2009). This also might increase the efficiency of the antioxidants and turn them into very potent prophylactic agents also for neurodegeneration in general. In a pilot experiment, dorsal root ganglia were cultivated in the presence of natural amyloid beta protein to simulate an Alzheimer disease *in vitro*. Nanocrystalized antioxidants were added to the culture medium which reduced neuronal cell death significantly (Keck et al.). Figure 5 represents one result of this study.

3.2 Dermal Application

Besides the pharmaceutical use nanocrystals are exploited in dermal cosmetic products. Similar to the pharmaceutical use they are used to improve the bioavailability/bioactivity of poorly soluble cosmetics compounds. Examples of poorly soluble cosmetic actives are natural pant metabolites, e.g., flavonoids that possess a high antioxidative potential (Evans and Johnson 2010). In cosmetics, antioxidants are used in anti-aging products, because with age the amount of free radicals increases. Free radicals are associated with accelerated skin aging. However, free radicals are also created upon light exposure. This is not only due to UV light exposure, but also due to visible or infrared light exposure (Darvin et al. 2010; Zastrow et al. 2009).

The skin damaging effect of UV light is widely known and can be decreased by the application of UV absorbers and/or by the application of light scattering particles (e.g., titanium dioxide, physical protection). To circumvent damages due to the exposure to IR light today only the application of radical scavengers is known to be effective. Unfortunately, many isolated antioxidants, i.e., vitamin A or vitamin E, which are widely used in cosmetic products, were shown to possess also pro-oxidative effects under specific circumstances and thus should be avoided in such products (Petersen 2006; Thiele et al. 2001). Therefore, there is a tendency suggesting that antioxidants should be used only in physiological concentrations and moreover they should be of natural nature. Therefore flavonoids, which are found in many plants, e.g., citrus fruits, red grapes, etc., have become highly interesting. They are not known to possess pro-oxidative properties. These actives, however, are poorly soluble. Hence, they cannot dissolve and thus cannot penetrate into the skin.

In a first attempt these molecules were 'made soluble' by chemical synthesis (see also Chap. 15). This was achieved by synthesizing water soluble derivatives, e.g., by adding a highly water soluble sugar to the poorly soluble molecule. An example is Rutin, which was modified to rutin glycoside. Chemical synthesis is not a universal approach, i.e., for each molecule a new synthesis is required. In contrast, diminution is possible with each material. Thus the production of nanocrystals is a universal formulation approach and in principle can be applied to every poorly soluble compound. Besides the more convenient way of increasing water solubility of a compound by nanonization, a further benefit is that the original compound, i.e., the molecule itself is not changed. Hence by chemical addition of water soluble functional groups the properties, e.g., the activity of the molecule can change. In cosmetics, this was observed first for rutin. In an *in vivo* study the activity of the water soluble derivative rutin glycoside was compared to the activity of rutin nanocrystals. The active concentration of rutin glycoside was 5 % (w/w) in the final formulation but the concentration of dissolved rutin nanocrystals was only 0.01 %. Both formulations were applied to the skin (forearm) of the test individuals. The treated skin was exposed to UV light and the time needed to develop an erythema (minimal erythema dose, MED) was analyzed and compared to the MED of untreated skin. Formulations containing the water soluble derivative increased the MED, i.e., the sun protection by about 30 % when compared to the control. Rutin nanocrystals increased the sun protection by about 60 %. Hence the activity of nanocrystals was found to be 1,000-fold higher compared to the water soluble derivative, i.e., a 500 times lower concentration of active led to a two-fold higher sun protection (Petersen 2006). The increased activity of nanocrystals in comparison to water soluble derivatives can be explained by four facts:

1. Increased solubility leads to an increased concentration gradient and thus to an increased penetration of active into the skin.
2. The original molecule is more lipophilic than the water soluble derivative and penetrates better into the skin than the very hydrophilic derivative.

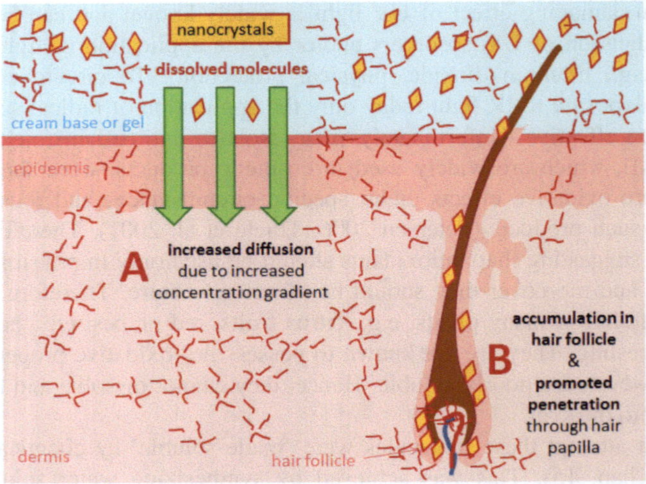

Fig. 6 Nanocrystals improve the penetration of actives upon dermal application. There are two reasons for this: *A* increased solubility of the active leads to an increased concentration gradient and thus to an increased penetration via passive diffusion of the active into the skin. *B* Nanoparticles, especially with a size of around 600 nm, can penetrate into hair follicles, from which penetration is increased further, e.g., through the more permeable papilla of the hair

3. The activity of the original molecule and/or the binding affinity in the skin cells is assumed to be higher than that of the derivative.
4. Nanoparticles, especially with a size of around 600 nm, were shown to penetrate into hair follicles, from which penetration is increased further.

The penetration enhancing effects of nanocrystals are depicted in Fig. 6. The special properties of nanocrystals upon dermal application are exploited in several cosmetic products. Examples are the products "Juvedical Renewing Eye Optimizer" (rutin nanocrystals, company Juvena), "ageLine®wo/man one Eye Lifting Serum" (rutin nanocrystals, company ageLine cosmeceuticals), "Edelweiss Wrinkle fighter" (rutin nanocrystals, company Audorasan), or platinum rare (hesperidin nanocrystals, company La Prairie).

3.3 Parenteral Administration

Parenteral administration is defined as administration circumventing the 'enteral' route (gastrointestinal tract). From this very different dosage forms are covered, from injectables to pulmonary, ocular etc., delivery. Some pharmacopeia define 'parenterals' in a strict sense as injectables only. Nanocrystals can be injected intravenously as nanosuspensions. Such suspensions are under development in the pharmaceutical industry, for example by Baxter Healthcare (NANOEDGE technology) (Kipp et al. 2003, 2005). These nanosuspensions need to be sterile and isotonic. In case of supportive therapy by nutraceuticals (e.g., antioxidants) the *i.v.* route appears attractive to generate higher and constant levels in blood (e.g., via infusion). This is of high interest for actives with a very short half-time, where elimination might be as fast as absorption, resulting in very low levels in the blood after oral administration.

Also, other *parenteral administration* routes are thinkable, e.g., to the eye or to other mucosal topical surfaces (e.g., cavity of the mouth, delivery as spray; vaginal mucosa in form of gel or fast-spreading suspension). The nanocrystal technology should be considered for various administrative routes when poorly soluble compounds are to be delivered.

4 Lab Scale and Industrial Production of Nanocrystals

There are basically two approaches for nanocrystal production, the bottom-up process and the top-down process (see also Explanatory Box 2). Detailed reviews can be found in Keck and Mueller 2006; Mueller et al. 2011a, b; Shegokar and Muller 2010. In the bottom-up process one starts from the dissolved molecule, which is aggregated to form the crystals. That means it is a typical precipitation process. The active is dissolved in a solvent, the solvent added to a non-solvent and the active precipitates in the form of fine crystals. For pharmaceutical applications, Prof. Sucker developed pharmaceutical *crystalline* nanoparticles (=nanocrystals) by precipitation, the so-called hydrosols (List and Sucker 1988; Sucker 1998). In the area of nutrition, BASF developed a process to generate *amorphous* nanoparticles of actives by a special precipitation process (Auweter et al. 2002). They are used, as e.g., additives to soft drinks (e.g., beta-carotene CWD). This technology is also offered for the delivery of poorly soluble pharmaceutical drugs as NanoMorph by the company Soliqs in Ludwigshafen, the drug delivery company of Abbott. As a precipitation process, it can easily be performed on a small scale but also on a large scale (e.g., using static blendes for mixing solvent and non-solvent).

Explanatory Box 2: Milling Vanilla

During the last decade or two, nanotechnology has witnessed an almost breathtaking rise from a specialized branch of material sciences and physical chemistry to a major field of modern, interdisciplinary research. While nanoparticles are hardly new, and indeed have been around since the first volcanoes erupted billions of years ago, the field of nanotechnology is still gathering steam. In a biological and pharmaceutical contexts, nanoparticles promise highly improved properties, including better solubility, solubilization, release, and permeability. Nanoparticles may even exhibit entirely new properties notably absent in their macroscopic brothers and sisters. These considerations also apply to numerous redox active plant secondary metabolites, which are difficult to extract, hardly soluble in aqueous media, and only poorly bioavailable. In this area of research and product development, nanotechnology may indeed provide many new answers to numerous old questions, and at last catapult a whole field of natural product research into the open, i.e., to provide access for nutritional, medical, cosmetic, and agricultural applications to name just a few.

As nanotechnology is at the forefront of many developments in natural sciences, the demand for nanoparticles is considerable. Hence one question often posed rather naïvely enquires how to manufacture such particles. The answer to this question often seems to be rather complicated. We will therefore briefly—and in simple words—consider some of the methods available, without any claim at all on completeness.

The expression ‘nanoparticle’ is usually applied to particles between 1 and 100 nm in diameter. In the pharmaceutical field particles below 1 μm in diameter in size are defined as nanoparticles. Larger particles are obviously microparticles and smaller ones are no longer particles in the strict sense but single molecules or small aggregates thereof. As nanoparticles are somehow particles of ‘intermediate size’, it is not surprising that such particles can be manufactured by two major avenues. On the one hand, one can start with single molecules and form a precipitate. Here, the particles grow from small to large(r), and it is up to the researcher or manufacturer to stop the growth process once the correct ‘nano-size’ has been reached. On the other hand, one may start with larger particles and employ—rather sophisticated—milling techniques to reduce the particles in size. Again, it is up to the operator to terminate this breaking-down process once the correct size of particles has been reached. While both approaches appear rather straightforward on paper, the praxis is rather complicated. Precipitations are not easy to control, and most traditional mills are not fine enough to mill down to the nanoscale. Hence the generation and the characterization of nanoscopic particles is a science by itself.

As part of the ‘bottom-up’ approach, the method of nanoprecipitation has gained considerable prominence. As part of this technique, a lipophilic active component, such as a water-insoluble natural product, is dissolved in an organic solvent and the solution is then injected rapidly into a hydrophilic phase to form an emulsion. The lipophilic phase is then evaporated to yield the respective particles. While this approach of rapid precipitation can be applied widely, it requires a certain solubility of the material in an organic solvent. This solubility is not always given, as the example of elemental sulfur nicely illustrates. Here, a different ‘bottom-up’ approach is used, which relies on the *in situ* synthesis of the product and its rapid precipitation. For instance, nanoparticles of elemental sulfur are synthesized by reacting sulfide and sulfite. In contrast, nanoparticles of elemental selenium are generated by reducing selenite with cysteine (selenium nanoparticles are also generated by certain bacteria, such as *Escherichia coli* and *Thauera selenatis*). Rod-shaped nanoparticles of tellurium can be obtained by the reduction of tellurium dioxide with hydrazine.

A special case of the 'bottom-up' approach is used for the manufacture of particles derived from combustible materials. As part of this method, a vaporous precursor (liquid or gas) is forced through an orifice at high pressure and burned. The resulting solid, i.e., an oxide or some kind of soot, is then purified. This kind of combustion or pyrolysis often results in aggregates and agglomerates and not necessarily in well-defined, uniform particles.

Most of the 'bottom-up' approaches employ solutions and liquids to generate particles which subsequently are used directly (in solution) or purified further. In contrast, the 'top-down' approach is usually based on mechanic processes which do not require the 'help' of any solvents. Here, various mills, such as a ball mill or a planetary mill are used to 'grind down' particles to a small size. Whilst some of these mills can be used to produce good quantities of uniformly sized particles, the balls used to 'crush' the particles may also cause contamination. This kind of contamination is particularly unpleasant when the particles are designed for biological applications. The use of hardened milling pearls, i.e., yttrium stabilized zirconium beads can help to overcome this problem. Alternatively, high pressure homogenization can be employed for nanonization. The coarse particles are passed through a small gap at high pressure. Based on the Bernoulli Priniciple cavitation occurs, which is strong enough to break the particles. Nanocrystals with improved properties (e.g., improved physical stability, smaller size) can be obtained by combining different techniques, e.g., bead milling with high pressure homogenization. Particles with sizes well below 100nm can be produced by combining, e.g., freeze drying and high pressure homogenization.

In the top-down processes one starts from large crystals in the μm range and reduces them in size by a milling process, typically wet milling because jet milling in the gas phase is not efficient enough to create nanocrystals. The main processes are bead milling and high pressure homogenization. The bead mill process for pharmaceutical nanocrystals was developed in 1991 by Liversidge and co-workers (Liversidge and Cundy 1995; Liversidge et al. 1992) at the company Nanosystems (later owned by élan, now Alkermes). A macrosuspension of the active is added to a milling container filled with milling beads (e.g., zirconium oxide, size 0.2–0.4 mm) and the beads are moved by an agitator. The crystals are ground between the moving beads yielding a nanosuspension. Most of the pharmaceutical products on the market are produced by bead milling.

High pressure homogenization (HPH) is widely used in the nutrition market (e.g., homogenized milk) but also in the pharmaceutical industry (e.g., production of parenteral nanoemulsions such as Lipofundin). The technology of making nanocrystals by HPH was developed by R.H. Mueller and co-workers (Müller et al. 1999). The principle of HPH is that a coarse dispersion (emulsion or suspension) is pressed under high pressure through a very tiny gap with a size of, e.g., 5–20 μm. In the gap the dispersion liquid has a very high streaming velocity (e.g., 500 m/s) which leads to cavitation (in case of water as dispersion medium) disintegrating droplets or crystals to the nano size. Cavitation means that the very fast streaming dispersion medium starts boiling at room temperature, gas bubbles form creating pressure waves that disintegrate particles. In addition, particle collision and friction of the liquid contributes to the size reduction. Typical pressures are 1,000–1,500 bar and 10–20 passes (cycles) through the homogenizer. Trade name for these nanocrystals is DissoCubes® (improved dissolution from cuboid crystals). Used are piston-gap homogenizers and microfluidizers (so-called IDD (insoluble drug delivery) technology) (Majuru and Oyewumi 2009). The pharmaceutical product on the market is Triglide, produced by SkyePharma. Baxter Healthcare is also using high pressure homogenization for its intravenous nanosuspensions. The production process is aseptic.

Recently the smartCrystal® technology was developed (Keck and Mueller 2010; Mueller and Keck 2008; Petersen 2006). The smartCrystal® technology represents a family of processes (=tool box) yielding nanocrystals with special properties. The technology is owned by Abbott US for pharma products, and by PharmaSol Berlin for non-pharma products (incl. nutraceuticals, cosmetics). The smartCrystals® are considered as the second generation of nanocrystals because of their improved properties. They possess a higher physical stability against electrolytes in the gut when the particles are produced by a combination of bead milling and subsequent high pressure homogenization (CT process). This is important because electrolytes in the body reduce the zeta potential of the nanocrystals; they aggregate and lose their special nanoproperties, such as increased solubility, adhesion, and bioavailability increase. Another smartCrystal® process is the creation of nanocrystals below 100 nm, which possess a very pronounced increase in solubility (H96 process, combination of freeze-drying and HPH). The smartCrystals® appear as most suitable formulation for orally and dermally applied antioxidants.

5 Nanocrystals: A Proven Effective Delivery System

Basic questions for each delivery system are:

- Is the system really effective?
- Is the system safe and well tolerated by the user?

The first question arises especially when delivery systems are used in areas in which no proof of efficiency has to be delivered by producers, e.g., in cosmetics but also with regard to nutraceuticals. The ingredients need to be listed on the package, but there is no requirement to prove that the nutraceutical active of a tablet or capsule really reaches the blood and an effective blood level is achieved (see Chap. 2). Classical examples are nutraceutical products with the poorly water soluble compound coenzyme Q_{10}. A recent review revealed that most of the products have an oral bioavailability of 1–4 % (Barakat et al.). The fraction absorbed was $\leq$ 3.5 % for about 93 % of the products reviewed (26 out of 28). Just one product had a fraction absorbed by 6.2 % and one by 12.5 %. In contrast, the pharmaceutical nanocrystals are tested in clinical studies proving that the technology is effective in enhancing the oral bioavailability. In addition to the clinically tested pharma products on the market, animal studies with other drugs also prove the effectiveness. For example, the oral bioavailability of the drug danazol could be increased from 5.2 % for the marketed suspension Danocrine to 82.3 % in case of the nanosuspension (Liversidge and Cundy 1995). Nanocrystals are an efficient system for increasing the transport of actives across biological membranes, ranging from the wall of the gastrointestinal tract to the skin and the blood–brain barrier (Dunay et al. 2004; Schöler et al. 2001).

6 Nanocrystals: A Proven Safe, Body Friendly Nanotechnology System

There is increasing concern of the public regarding the safety of nanotechnology. An objective reason is the limited data available about many nanosystems. But there is also a subjective perception by the public, mainly supported by uncritical reports in the popular press, which is simplifying or misinterpreting things to generate dramatic headlines. Certainly, the safety of nanosystems needs to be assessed on a reliable scientific basis. According to the new EU regulations, cosmetic products containing nanoparticles need to be labeled as nano products. Definition of a nanoparticle according to these regulations are only nanoparticles $\leq$ 100 nm, not from >100–1,000 nm! The EU definition is (European Commission 2011).

> A natural, incidental or manufactured material containing particles, in an unbound state or as an aggregate or as an agglomerate and where, for 50 % or more of the particles in the number size distribution, one or more external dimensions is in the size range 1–100 nm.

The logic that only nanoparticles $\leq$100 nm are covered by the regulation is that they have a much higher toxic risk than nanoparticles larger than 100 nm. The small nanoparticles can be internalized by any cell of the body via endocytosis, therefore they are riskier. Particles larger than approximately 100 nm can only be internalized by a very limited number of cells in the body, e.g., the macrophages in liver, spleen and lung. In addition, the larger particles have no or little access to these cells.

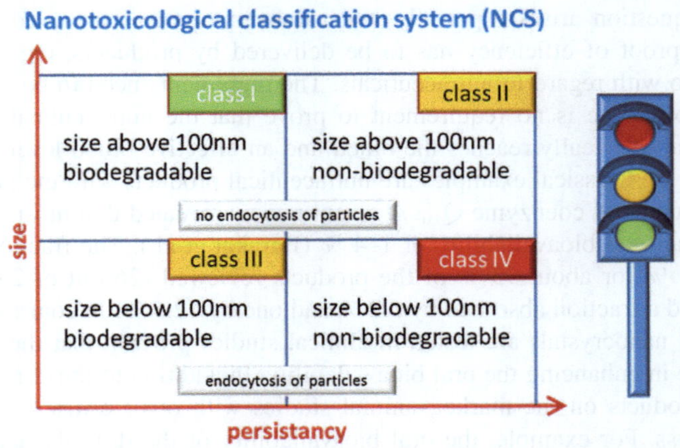

Fig. 7 Nanotoxicological classification system (*NCS*) dividing the nanoparticles into four classes of no/little risk to potentially (but not necessarily) high risk (similar to the traffic light system from *green to red*). Figure reproduced with permission from (Mueller et al. 2011a)

Another factor determining the tolerability is the degradability/non-degradability of the particles in the body. A priori, particles that are easily biodegradable are less risky than non-biodegradable particles. The classical examples for the latter are the fullerenes and the carbon nanotubes (CNT). Nevertheless, there are still proposals to use them for delivery of actives. However, they are non-biodegradable and will just stay in the body forever—an exhibit permanent toxic effects (e.g., CNT seem to act similar to asbestos fibers) (Poland et al. 2008).

Based on these considerations, a nanotoxicological classification system (NCS) was proposed (Keck submitted; Mueller et al. 2011b) based on the two facts:

- size: <100 nm and >100–1,000 nm, and
- degradability: yes /no

This classification system divides the nanoparticles into four classes of no/little risk to potentially (but not necessarily) high risk, according to a traffic light system from green via amber to red (Fig. 7):

- Class I (green): >100–1,000 nm, biodegradable
- Class II (amber): >100–1,000 nm, *non-biodegradable*
- Class III (amber): <100 nm, but still biodegradable
- Class IV (red): <100 nm, *non-biodegradable*

Nanocrystals are biodegradable (they just dissolve with time when enough water gets in touch with the crystal). The nanocrystals >100 nm belong therefore to the class 'green' of the NCS. Their safety is also confirmed by the regulatory drug authorities (e.g., FDA in US) which have registered the oral and injectable drug nanocrystal products.

Nanocrystals below 100 nm belong to Class III (amber), but finally they will dissolve which makes them also low /no risk. In addition, it should be kept in mind that each 1,000 nm nanocrystal shrinks during its dissolution and will finally be below 100 nm before it completely dissolves. Therefore, the registered pharma products with nanocrystals will all lead finally to nanocrystals <100 nm in the body, but they are classified as safe by the regulatory authorities.

Apart from this: mankind was living with oral nanoparticles since its existence. Each fat droplet in food in the gastrointestinal tract will shrink to the nm range during its surface digestion. The same applies to each drug powder particle in a tablet swollen, e.g., each Aspirin. Form this respect, dissolvable nanocrystals are a proven safe system for oral administration, and also for dermal use because critical cells in the body cannot be accessed (e.g., in the liver).

7 Perspectives

Pharmaceutical nanocrystals were invented in 1991, and in less than 10 years the first pharmaceutical product was on the market (Rapamune 2000). This is a record considering that most of the pharmaceutical technologies take rather 20 years and more to reach the market (e.g., liposomes, invented around 1965, first marketed products around 1990, i.e., a quarter of a century to the market). This proves that the nanocrystal technology is simple, thus industry friendly and effective (pharma products are only registered with proven effect).

However, this smart delivery technology is not yet, or only little, exploited in other areas such as cosmetics, nutraceutics, and agronomy. First, cosmetic products have meanwhile been placed worldwide on the market via PharmaSol Berlin, but there are no nutraceutical products at all based on this nanotechnology. Here is a very big potential in putting really effective nutraceutical products on the market, not just claiming but having an effective bioavailability of problematic poorly soluble actives such as many antioxidants. The nanocrystal concept allows also a 'market differentiation' compared to manifold competitor products. This nanotechnology can potentially be a milestone toward a new class of nutraceuticals.

References

Ahmad A, Khan MM, Hoda MN, Raza SS, Khan MB, Javed H, Ishrat T, Ashafaq M, Ahmad ME, Safhi MM, Islam F (2011) Quercetin protects against oxidative stress associated damages in a rat model of transient focal cerebral ischemia and reperfusion. Neurochem Res 36(8):1360–1371

Arredondo F, Echeverry C, Abin-Carriquiry JA, Blasina F, Antunez K, Jones DP, Go YM, Liang YL, Dajas F (2010) After cellular internalization, quercetin causes Nrf2 nuclear translocation, increases glutathione levels, and prevents neuronal death against an oxidative insult. Free Radical Biol Med 49(5):738–747

Auweter H, Bohn H, Heger R, Horn D, Siegel B, Siemensmeyer K (2002) Precipitated water-insoluble colorants in colloid disperse form. USA Patent

Barakat A, Shegokar R, Dittgen M, Müller RH Coenzyme Q10 oral bioavailability: effect of formulation type

Darvin ME, Haag S, Meinke M, Zastrow L, Sterry W, Lademann J (2010) Radical production by infrared a irradiation in human tissue. Skin Pharmacol Physiol 23(1):40–46

Dunay IR, Heimesaat MM, Bushrab FN, Muller RH, Stocker H, Arasteh K, Kurowski M, Fitzner R, Borner M, Liesenfeld O (2004) Atovaquone maintenance therapy prevents reactivation of toxoplasmic encephalitis in a murine model of reactivated toxoplasmosis. Antimicrob Agents Chemother 48(12):4848–4854

European Commission (2011) Definition of a nanomaterial. ec.europa.eu/environment/chemicals/nanotech/index.htm

Evans JA, Johnson EJ (2010) The role of phytonutrients in skin health. Nutrients 2(8):903–928

Ghosh A, Mandal AK, Sarkar S, Panda S, Das N (2009) Nanoencapsulation of quercetin enhances its dietary efficacy in combating arsenic-induced oxidative damage in liver and brain of rats. Life Sci 84(3–4):75–80

Hauss F, Liu JW, Michelucci A, Coowar D, Morga E, Heuschling P, Luu B (2007) Dual bioactivity of resveratrol fatty alcohols: differentiation of neural stem cells and modulation of neuroinflammation. Bioorg Med Chem Lett 17(15):4218–4222

Keck CM, Chen R, Sauer M, Schaefer K-H Antioxidative nanocrystals for the prevention of the toxic effects of ß-Amyloid in Alzheimers`s disease

Keck CM, Mueller RH (2006) Drug nanocrystals of poorly soluble drugs produced by high pressure homogenisation. Eur J Pharm Biopharm 62(1):3–16

Keck CM, Mueller RH (2010) SmartCrystals—review of the second generation of drug nanocrystals. In: Torchilin V, Amiji MM (eds) Handbook of materials for nanomedicine. Pan Stanford Publishing Pte Ltd, London, pp 555–579

Kipp JE, Wong JCT, Doty MJ, Rebbeck CL (2003) Microprecipitation method for preparing submicron suspensions. USA Patent

Kipp JE, Wong JCT, Doty MJ, Werling J, Rebbeck CL, Brynjelsen S (2005) Method for preparing submicron particle suspensions. USA Patent

List M, Sucker H (1988) Pharmaceutical colloidal hydrosols for injection. GB Patent

Liversidge GG, Cundy KC (1995) Particle-size reduction for improvement of oral bioavailability of hydrophobic drugs. 1. absolute oral bioavailability of nanocrystalline danazol in beagle dogs. Int J Pharm 125(1):91–97

Liversidge GG, Cundy KC, Bishop JF, Czekai DA (1992) Surface modified drug nanoparticles. USA Patent

Lopes CRP, Ferreira PEB, Zanoni JN, Alves AMP, Alves EPB, Buttow NC (2012) Neuroprotective effect of quercetin on the duodenum enteric nervous system of Streptozotocin-induced diabetic rats. Dig Dis Sci 57(12):3106–3115

Majuru S, Oyewumi MO (2009) Nanotechnology in drug development and life cycle management. In: de Villiers PAaGSK MM (ed) Nanotechnology in drug delivery: 10 (Biotechnology: Pharmaceutical Aspects), vol 597–620. American Association of Pharmaceutical Scientists, Springer

Martinez MN, Amidon GL (2002) A mechanistic approach to understanding the factors affecting drug absorption: a review of fundamentals. J Clin Pharmacol 42(6):620–643

Mauludin R (2008) Nanosuspensions of poorly soluble drugs for oral administration. Freie Universität Berlin, Berlin

Müller RH, Becker R, Kruss B, Peters K (1999) Pharmaceutical nanosuspensions for medicament administration as systems with increased saturation solubility and rate of solution. USA Patent

Mueller RH, Boehm BHL, Grau MJ (2000) Nanosuspensions—a formulation approach for poorly soluble and poorly bioavailable drugs. In: Wise DL (ed) Handbook of pharmaceutical controlled release. Marcel Dekker, New York, pp 345–357

Mueller RH, Gohla S, Keck CM (2011a) State of the art of nanocrystals—special features, production, nanotoxicology aspects and intracellular delivery. Eur J Pharm Biopharm 78(1):1–9

Mueller RH, Keck CM (2008) Second generation of drug nanocrystals for delivery of poorly soluble drugs: smartcrystal technology. Eur J Pharm Sci 34(1):S20–S21

Mueller RH, Shegokar R, Gohla S, Keck CM (2011b) Nanocrystals: production, cellular drug delivery, current and future products. In: Prokop A (ed) Intracellular delivery: fundamentals and applications, fundamental biomedical technologies, vol 5. Springer Science + Business Media B.V., pp 411–432

NanoCrystal Wins Technology Innovation Award (2010). Drug discovery and development magazine

Peters K, Mueller RH (1996) Nanosuspensions as oral and parenteral formulations for poorly soluble drugs. Eur J Pharm Sci 4(1):S44

Petersen RD (2006) Nanocrystals for use in topical fomulations and method of production thereof. Germany Patent

Poland CA, Duffin R, Kinloch I, Maynard A, Wallace WAH, Seaton A, Stone V, Brown S, MacNee W, Donaldson K (2008) Carbon nanotubes introduced into the abdominal cavity of mice show asbestos-like pathogenicity in a pilot study. Nat Nanotechnol 3(7):423–428

Schöler N, Krause K, Kayser O, Muller RH, Borner K, Hahn H, Liesenfeld O (2001) Atovaquone nanosuspensions show excellent therapeutic effect in a new murine model of reactivated toxoplasmosis. Antimicrob Agents Chemother 45(6):1771–1779

Shegokar R, Muller RH (2010) Nanocrystals: industrially feasible multifunctional formulation technology for poorly soluble actives. Int J Pharm 399(1–2):129–139

Sucker H (1998) Hydrosole, eine Alternative für die parenterale Anwendung von schwer wasserlöslichen Wirkstoffen. In: Mueller RH, Hildebrand GE (eds) Pharmazeutische technologie: moderne arzneiformen, vol 2. Wissenschaftliche Verlagsgesellschaft mbH, Stuttgart, pp 383–391

Thiele JJ, Schroeter C, Hsieh SN, Podda M, Packer L (2001) The antioxidant network of the stratum corneum. Curr Probl Dermatol 29:26–42

Vauzour D, Rodriguez-Mateos A, Corona G, Oruna-Concha MJ, Spencer JP (2010) Polyphenols and human health: prevention of disease and mechanisms of action. Nutrients 2(11):1106–1131

Visioli F, De La Lastra CA, Andres-Lacueva C, Aviram M, Calhau C, Cassano A, D'Archivio M, Faria A, Fave G, Fogliano V, Llorach R, Vitaglione P, Zoratti M, Edeas M (2011) Polyphenols and human health: a prospectus. Crit Rev Food Sci Nutr 51(6):524–546

Zastrow L, Groth N, Klein F, Kockott D, Lademann J, Renneberg R, Ferrero L (2009) The missing link—light-induced (280–1,600 nm) free radical formation in human skin. Skin Pharmacol Physiol 22(1):31–44

Authors Biography

Cornelia M. Keck studied Biology at the Technical University Dresden (1994–1996) and Pharmacy at the Freie Universitaet (FU) Berlin (1996–2001). In 2003 she obtained her Postgraduate Certificate in Pharmacy from Otago University, Dunedin, New Zealand. She received her PhD in Pharmaceutics from the FU Berlin in 2006. Between 2006 and 2011 she worked for both pharmaceutical and cosmetic industry and continued research in the field of nanocarriers (nanocrystals and nanolipids) for improved delivery of poorly soluble actives. In 2011 she obtained her "venia legendi" from the FU Berlin and was also appointed as a Professor of pharmacology and pharmaceutics at the University of Applied Sciences Kaiserslautern. Cornelia is the founding coordinator of the degree program in 'Applied Pharmacy', which started in 2012 as the first course of its kind in Germany. It is designed to train pharmacists especially for pharmaceutical industry.

Cornelia's research focuses on pharmaceutical nanotechnology, nanoanalytics and the development, and characterization of pharmaceutical and cosmeceutical formulations for improved delivery of actives. The main focus is the formulation of poorly soluble actives for improved bioavailability in the context of dermal and oral delivery. Current developments also involve the design and evaluation of faster and cost-effective methods for the production of drug nanocrystals. So far, Cornelia has published two books, about 50 journal articles and book chapters, more than 200 proceedings and abstracts, and she is an inventor on three patents.

Rainer H. Mueller studied pharmacy and obtained his PhD in Pharmaceutics in 1983 (Kiel University, North Germany). He subsequently worked in the area of pharmaceutical nanotechnology at Nottingham University/UK from 1984–1988, and then for one year at the "Université Paris-Sud" in France. In 1991, he became professor of "Pharmaceutics and Biopharmaceutics" at the Freie Universitaet (FU) Berlin, and established his research group in "Pharmaceutical Nanotechnology". In 1999 he founded his research & development company PharmaSol GmbH in Berlin. Rainer Mueller is the main inventor of solid lipid nanoparticles (SLN®) and nanostructured lipid carriers (NLC®), and of two generations of drug nanocrystals (DissoCubes®, smartCrystal®). He introduced lipid nanoparticles and nanocrystals loaded with cosmetic actives (e.g., antioxidants) into cosmetic dermal products worldwide. To date, Rainer Mueller has published 20 books, about 450 journal articles and book chapters, more than 900 proceedings and abstracts, and he is an inventor on about 20 patent families.

Karl-Herbert Schaefer studied Biology (1979–1980) and Medicine (1980–1986) at the University of Saarland. After having worked as a research assistant at the department of anatomy, he started training in general surgery at the St. Joseph Hospital in Neunkirchen/Saar (1988–1991). Karl-Herbert finished his doctoral thesis at the department of Anatomy on the topic of "Morphology of GABA-ergic neurons in the medial preoptic area of the rat" in 1993. In 1992 and 1993 he spent 6 months at University College London in the groups of Kristjan Jessen, Rhona Mirsky, and Geoffrey Burnstock, starting with his work on the Enteric Nervous System (ENS). Back in Germany, he subsequently worked on his habilitation, which he completed in the year 2000 on the "Isolation and Cultivation of myenteric plexus from human and rats". Within that time he also obtained his qualification as anatomist. In 1999 he changed to the Clinic of Pediatric Surgery at the University Mannheim/Heidelberg., where he could intensify his studies on the human ENS. There he established and led until now the working group on ENS. In the year 2001, he was appointed to the Professorship in Biotechnology at the University of Applied Sciences in Kaiserslautern. Here he was responsible for the biomedical training of engineers in microsystems technology. From these activities, he developed a completely new study course, which combined biomedical and pharmaceutical aspects with micro- and nanotechnological approaches. The degree program "Applied Life Sciences" started with a Bachelor program in 2006, which was followed by a consecutive Master program in the year 2009. K.-H. Schaefer was the head of the Bachelor program until 2009 and is still the head of the Master program until now. In 2008 he spent 6 months as a visiting professor at the University of Stanford.

The focus of Karl-Herbert's research is pointing to the ENS, its development, plasticity as well as its changes within diseases. Currently, the role of the ENS within neurodegenerative diseases is being investigated. The research approaches used in his laboratory stretch from single cell culture systems, via neural stem cells to microelectrode array applications and whole gut organ bath systems.

Part XVI
Connecting Section Between Chapters 16 and 17

The previous discussion of nanocrystals and their possible applications in the fields of medicine and cosmetics has enabled us to consider rather specific and—scientifically as well as economically—viable practical applications of redox active secondary metabolites. Together with our discussion of garlic- and onion-derived products rich in organic sulfur compounds and the potential use of phytols in the treatment of inflammatory diseases, these nanocrystals provide yet another example where modern research and technology have enabled secondary metabolites to progress from the ingredients of rather primitive food supplements and herbal tinctures to rather sophisticated and scientifically sound products. Indeed, these are examples that refute the common skepticism held by many scientists against natural products. After proper research and development, such redox active secondary metabolites no longer belong to folk medicine and to the realm of make-believe, but find numerous applications and play on a level field with synthetic or partially synthetic compounds.

The following chapter will consider yet another application of a redox active natural product. Here, we have chosen the example of a rather unusual, vanadium-containing redox enzyme to illustrate that (a) not all useful natural products are necessarily small molecules, (b) a number of 'exotic' redox centers and associated activities can be found in biomolecules, and (c) more robust industrial applications are also possible besides the more common applications of natural products in nutrition, drug development, cosmetics, and as part of an eco-friendly agriculture. Jennifer A. Littlechild and Michail Isupov consider the use of haloperoxidase enzymes in redox reactions, which are important for industrial biocatalysis. Here, a sound knowledge of enzyme structure, stability, and catalytic activity is paramount before such enzymes can be used on a larger scale and for demanding industrial processes. At the same time, this chapter shows how genetic engineering of such enzymes may improve the expression yield of enzymes expressed in bacteria, their catalytic efficiency, stability, substrate specificity, and even the range of enzymatic products possible. Ultimately, the joint effort of enzymologists, molecular biologists, crystallographers, biochemists, organic, and technical chemists leads to new bio-catalytic processes, which are competitive with or even supersede the ones provided by classical organic synthesis, especially in the context of stereo-selective modifications.

Chapter 17
Haloperoxidase Enzymes as 'Redox Catalysts' Important for Industrial Biocatalysis

Jennifer Littlechild and Michail Isupov

Keywords Biocatalysis • *Corallina* • Enzyme structure and mechanism • Flavin • Vanadium haloperoxidase

1 Introduction to Industrial Biocatalysis

The use of enzymes as industrial catalysts is becoming increasingly important since they are able to carry out chemistry in an environmentally friendly and sustainable manner. (Woodley 2008) Such natural (and modified) catalysts are replacing many conventional chemistry processes and have the added advantage of carrying out many of their reactions in a stereospecific manner (see also Explanatory Box 1). One problem that can hinder enzyme use is their limited stability under the conditions of the industrial process. The enzymes can be used as 'whole cell' biocatalysts which may circumvent some of these stability issues. This, however, can lead to additional problems with transporting substrate(s) into the cells and the purity of the resultant product which can also be converted further by other cellular enzymes. In addition, downstream processing to recover the product is more demanding. In most cases the biocatalytic reaction is therefore carried out with a purified enzyme preparation which is often stabilized by immobilization.

Explanatory Box 1: Enzymes at Work

As the name already indicates, industrial biotechnology employs biotechnology on an industrial scale. Here, cells—or parts thereof—are used to drive large-scale processes. This branch of biotechnology, which in Europe is also

J. Littlechild (✉) · M. Isupov
Henry Wellcome Building for Biocatalysis, Biosciences, College of Life and Environmental Sciences, University of Exeter, Stocker Road, Exeter EX4 4QD, UK
e-mail: J.A.Littlechild@exeter.ac.uk

C. Jacob et al. (eds.), *Recent Advances in Redox Active Plant and Microbial Products*,
DOI: 10.1007/978-94-017-8953-0_17,

known as 'white biotechnology', has a long tradition which is rooted in fermentation processes, such as brewing, wine making, baking, or the production of cheeses and vinegar. In the last couple of decades, industrial biotechnology has expanded and diversified considerably, not least because of the availability of genetically engineered organisms, proteins, and enzymes. Apart from the obvious uses of organisms and enzymes in fermentation and food production, such processes are also used increasingly in the field of biomass processing, energy generation (e.g., biogas production), and removal of waste (e.g., in waste processing and water treatment plants). As part of this chapter, we will learn also more about the potential practical large-scale uses of individual enzymes. Within this context, enzymes are used increasingly as part of synthetic processes, as they enable regio- and stereoselective conversions. Enzymes are highly effective and selective, and hence can drive chemical conversions even in rather crude and 'dirty' mixtures of compounds.

Besides large-scale applications in the chemical industry, enzymes are also found in more trivial, yet equally important situations. Many washing powders, for instance, contain enzymes able to break down 'stains', such as lipases against fat and grease, amylases against starch, and proteases against protein-based contaminants. The next chapter will consider a more sophisticated use of enzymes. Here, a pineapple-derived cocktail of enzymes is employed for the treatment of certain injuries. Indeed, some enzymes are also potent pharmaceutical agents. The enzyme asparaginase, for instance, is found in many bacteria. As part of the drug Elspar®, it is used to treat certain types of leukemia, such as acute lymphoblastic leukemia (ALL). This enzyme hydrolyzes the amino acid asparagine to aspartic acid. As certain tumor cells have lost the ability to synthesize asparagine by themselves, this enzyme has a particularly deteriorating, almost selective impact on those cells without causing too much damage to healthy cells. Besides asparaginase, which is already used in the clinic, other amino acid-degrading enzymes are also currently considered as potential therapeutic agents, including the enzyme arginase, which degrades the amino acid arginine. Importantly, such enzymes need to be applied by injection, as an oral uptake of enzymes generally is still not possible because of poor bioavailability and digestion in the gastrointestinal tract.

Certain enzymes are also able to damage the cell wall of certain bacteria and hence have gained some prominence as natural antibiotics. The enzyme lysozyme, for instance, occurs among other sources also in many dairy products, especially in eggs. It is used on a large scale in the food industry to protect and preserve certain hard and semi-hard cheeses from bacterial and fungal degradation. Indeed, food industry often employs enzymes, and besides the 'antimicrobial' enzyme lysozyme, the enzyme transglutaminase has recently gained a certain prominence. This enzyme is able to cross-link proteins, and hence is used as a'meat glue' in order to improve the texture and appearance of processed, protein-rich meat products. It is used, for instance, in the production of imitation crabmeat, fish balls, and certain types of ham. Indeed, transglutaminase is often praised for its ability to convert pressed meat to a product with the same strength and appearance as whole uncut muscle.

Many natural compounds are halogenated and the enzymes that carry out these reactions have been characterized and used in biocatalytic reactions for the synthesis of new drug intermediates. Indeed, many new drugs entering the market are halogenated as well and this modification changes their biological activity. An example is the naturally chlorinated anti-cancer agent salinosporamide from a marine actinomycete *Salinispora tropica*, which shows increased activity in comparison to its non-chlorinated form Feling et al. (2003). Approximately 20 % of all pharmaceuticals are halogenated (Yarnell 2006) and this number is increasing further. Since it is often difficult to carry out the required halogenation reactions using conventional chemical routes, halogenases represent important biocatalysts in the pharmaceutical industry (Eustaquio et al. 2008).

Halogenated natural products have been found both in marine and terrestrial environments (Gribble 1998) and various haloperoxidases have been identified that are capable of forming a range of halogenated compounds, including volatile halogenated hydrocarbons such as bromoform (Ohsawa et al. 2001), halogenated fatty acids (Dembitsky and Srebnik 2002) halogenated terpenes, and halogenated indoles (Carter-Franklin and Butler 2004). Halogenating enzymes were discovered already in the 1950s when the enzymatic chlorination of a β-ketoadipic acid to δ-chlorolevulinic acid was observed during investigations into the biosynthesis of the antibiotic caldariomycin (Shaw and Hager 1959). This heme-containing haloperoxidase was isolated from the terrestrial fungus *Caldariomycetes fumago.* Haloperoxidases have now been identified in many organisms including mammals, birds, the brown, and red algae *Laminaria saccharina, Fucus distichus, Corallina officinalis* and *Corallina pilulifera,* molds, and terrestrial lichens such as *Xantharia parietina* as well as plants, bacteria, and fungi including *Curvularia inaequalis.* More recently, there have also been several general reviews on halogenating enzymes in the literature (Blasiak and Drennan 2009; Fujimori and Walsh 2007; Hill and Littlechild 2012; Vaillancourt et al. 2006).

This chapter will provide a brief overview of the different types of halogenating enzymes. It will concentrate specifically on the vanadium haloperoxidases with regard to their structure and mechanism, and their uses in industrial biocatalysis.

2 Haloperoxidase Enzymes

2.1 Heme Haloperoxidases

The class of heme haloperoxidases includes enzymes which generate hypohalous acid (HOX) by the direct hydrogen peroxide-dependent oxidation of halide ions (Hager et al. 1966). One of the first enzymes to be studied from this class was the chloroperoxidase (CPO) from the fungus *C. fumago* (Sundaramoorthy et al. 1998). HOX contains a positively oxidized halogen atom, also designated as halonium ion (X^+) and is a strongly electrophilic, halogenating agent. Heme-dependent haloperoxidases can oxidize chloride (Cl^-), bromide (Br^-), and iodide (I^-), but not fluoride (F^-). They contain a heme prosthetic group in the active site which is essential for their activity which entails coordination between the iron center and an imidazole nitrogen of a proximal histidine amino acid residue in the enzyme. The heme group, however, can also be coordinated between the iron centre and the sulfur of a cysteine amino acid as in the so-called heme-thiolate haloperoxidases (Hofrichter and Ullrich 2006). A heme-thiolate, haloperoxidase, for instance has been recognized in the fungus *Agrocybe aegerita* strain A1. The *A. aegerita* peroxidase (AaP) was originally thought to be a lignin peroxidase. The enzyme can now be produced in useful yields from cultures of *A. aegerita* grown in soy based media in stirred bioreactors. This has allowed the purification and further characterization of AaP. Spectral studies of this enzyme showed similarities to CPO, and to cytochrome P450 enzymes, and it is likely that the catalytic cycle involves the same transitions of the iron center. Investigations into the catalytic activities of AaP revealed it to possess strong brominating and weak chlorinating and iodinating abilities, as well as a wide array of oxygenase activities. For instance, it can convert phenol to 4- and 2-bromophenol in the presence of Br^- in a 4:1 ratio, with negligible polybromination. In this respect, it differs from CPO, which causes a high degree of polybromophenols (Onega et al. 2010). The structure of a fungal heme chloroperoxidase from *Leptoxyphium fumago* is shown in Fig. 1 (Kuhnel et al. 2007).

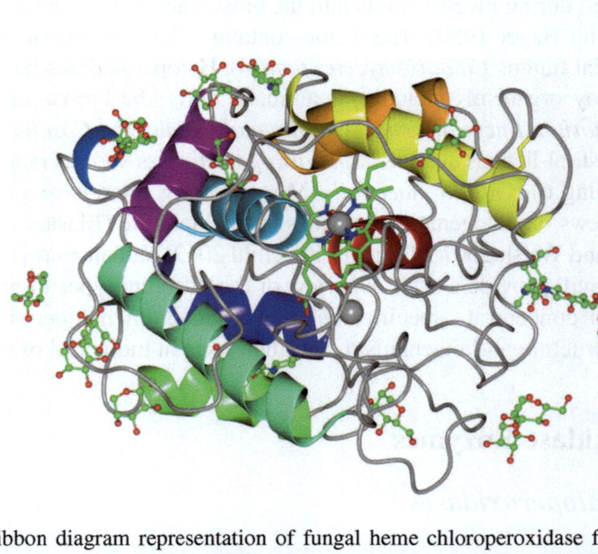

Fig. 1 A ribbon diagram representation of fungal heme chloroperoxidase from *Leptoxyphium fumago* (pdb 2j5 m; Kuhnel et al. 2007). Different α-helices are shown in *different colors*. The heme group is shown as *cylinders* and glycosylating sugar molecules are shown as *ball*-and-*stick* models. The heme iron and the manganese ion are shown as *spheres*. The figure was prepared by ccp4mg (McNicholas et al. 2011)

In addition to catalyzing halogenation reactions, the heme haloperoxidases are also able to catalyze a number of other oxidative transformations, including *N*-dealkylations, sulfoxidations, epoxidations of alkenes, and benzylic hydroxylations (Wojaczynska and Wojaczynski 2010). The epoxidation reactions of some CPO enzymes occur enantio-specifically, producing only one optical form of a given compound (Zaks and Dodds 1995).

A down-side to the biocatalytic oxidation reactions described above is that one of the substrates, i.e., hydrogen peroxide (H_2O_2) can inactivate the heme haloperoxidases by degradation of the heme moiety. It can also react in an undesired fashion with the substrate outside of the enzyme active site leading to non-specific by-products (Littlechild 1999; Vaillancourt et al. 2006; Butler and Sandy 2009; Wojaczynska and Wojaczynski 2010). To circumvent these problems, and to improve enzyme stability, H_2O_2 can be added slowly to the reactor. Alternatively, H_2O_2 can be generated *in situ*. Glucose oxidase (GOX), for instance, is able to mediate the oxidation of glucose to gluconic acid liberating H_2O_2 as a by-product (Karmee et al. 2009). In order to use this enzymatically generated H_2O_2, GOX has been immobilized with CPO in a polyurethane foam matrix. With the oxidation of methyl phenyl sulfide to the sulfoxide, this system showed an increase in the stability of the enzymes and increased optical purity (*ee.* 99 %) of the product (Wojaczynska and Wojaczynski 2010). The stereospecific oxidation of methyl phenyl sulfide has also been reported using CPO with H_2O_2 generated *in situ* using the catalytic reaction of hydrogen and oxygen in the presence of a palladium catalyst thereby linking a chemical and biocatalytic reaction in a cascade process (Karmee et al. 2009).

The heme oxygenase activities of these haloperoxidases resemble those of cytochrome P450 monooxygenases. Unlike these monooxygenases, however, the haloperoxidases do not hydroxylate aromatic rings.

Many of the reactions that are carried out by the heme haloperoxidases occur in a stereospecific manner and produce compounds which are of interest to both the biotechnology and pharmaceutical industries. One important example is the quantitative yields obtained in the oxidation of indoles to oxindoles, which are potent anti-inflammatory compounds (Littlechild 1999).

2.2 Non-heme Iron-Dependent Halogenases

Another class of halogenase enzymes includes the non-heme iron containing halogenases which are similar to non-heme iron oxygenases. These enzymes contain an iron coordinated between two histidine residues; in the oxygenases, an additional ligand is provided by the carboxylate side-chain groups of an aspartic or glutamic acid residue. In the halogenases of this class, the carboxylate ligand is replaced by a halide ion, such as Cl^- or Br^- (Blasiak and Drennan 2009). These enzymes are found in the biosynthetic pathways of compounds which are halogenated on inactivated carbon sites. Examples of such compounds include syringomycin E,

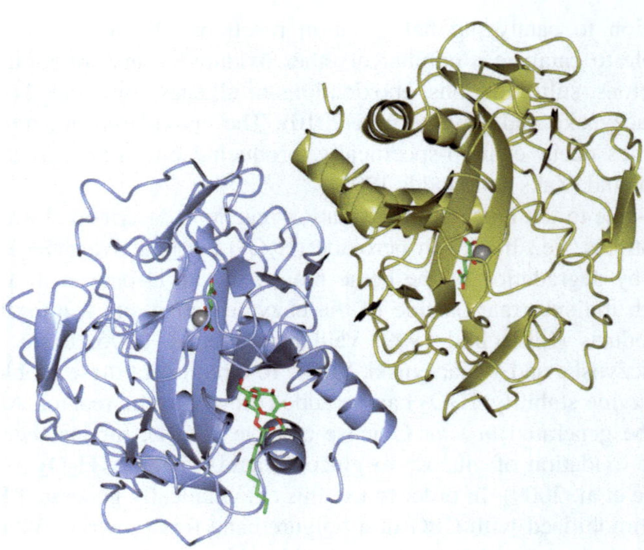

Fig. 2 The ribbon dimeric model of the non-heme iron halogenase SyrB2 from *P. syringae* (pdb 2fct; Blasiak et al. 2006) viewed along the molecular dyad. Oxoglutarate molecules and the inhibitor DSU ((2R,3S,4S,5S)-3,4-Dihydroxy-5-(hydroxymethyl)-5-((2R,3S,4S,5S,6R)-3,4,5-trihydroxy-6-methoxy-tetrahydro-2H-pyran-2-yloxy)-tetrahydrofuran-2-yl)methyl nonanoate) are shown as *stick* models. Iron atoms are shown as *spheres*. The figure was prepared by ccp4mg (McNicholas et al. 2011)

barbamide, and the kutznerides. The first example of a halogenase discovered from this group is the enzyme SyrB2 from *Pseudomonas syringae*, which is responsible for the chlorination step in the biosynthesis of syringomycin E. The structure of this protein is shown in Fig. 2.

2.3 *Flavin-Dependent Halogenases*

The flavin-dependent halogenases form a class of halogenating enzymes which require both oxygen and flavin adenine dinucleotide (FAD) for activity. The overall reactions of the two-component system are:

1) $FAD + NAD(P)H + H^+ \rightarrow FADH_2 + NAD(P)^+$ [Reductase]
2) $FADH_2 + O_2 + Cl^- + L\text{-}Trp + H^+ \rightarrow FAD + 7\text{-}Cl\text{-}L\text{-}Trp + 2H_2O$ [Halogenase].

The X-ray structure of PrnA (pdb 2aqj), a tryptophan-7-halogenase from *Pseudomonas fluorescens* has been reported in 2005 by Dong et al. (2005). This structure suggests a specific mechanism for regioselective chlorination (Dong et al. 2005) and is shown

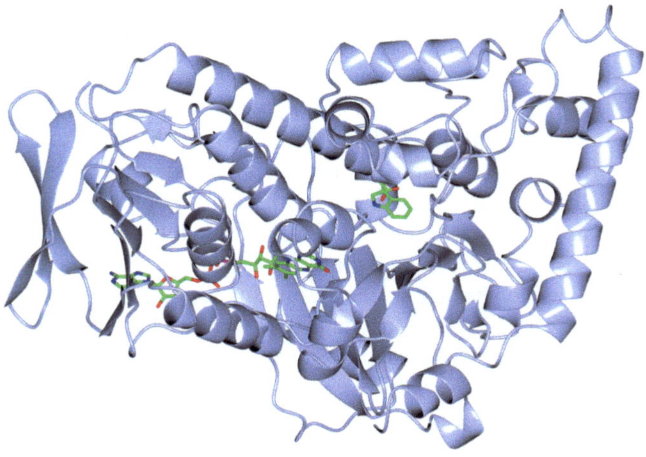

Fig. 3 The ribbon diagram of tryptophan 7-halogenase (PrnA) from *P. fluorescens* (pdb 2aqj; Dong et al. 2005) with the substrate tryptophan and its cofactor FAD shown as *stick* models. The figure was prepared by ccp4mg (McNicholas et al. 2011)

in Fig. 3. The structure was solved to 1.8 Å resolution and shows the protein to be a homodimer. The monomers are single domain proteins arranged into two modules, the larger flavin-binding module and the smaller tryptophan-binding module. The flavin-binding module consists of a four α-helix bundle flanked by two β-sheets, whereas the tryptophan module is entirely α-helical in secondary structure. The flavin moiety is shown to bind in a solvent-exposed groove, with the planar isoalloxazine ring effectively trapping a bound chloride ion within the groove. This structure has helped to elucidate the nature of the chlorinating species. The flavin moiety and chloride ion are located away from the tryptophan substrate with a 10 Å distance between the two binding sites. This strongly suggests that the halogenating species is a diffusible hypochlorite ion (OCl^-) rather than an FAD-bound hypochlorite (FAD-OCl). The tryptophan substrate is located in a pocket and forms stacking interactions with Trp455, His101 and Phe103. The $FADH_2$ is located in another pocket together with the chloride ion. The hypochlorite formed is prevented from diffusing into solution by the protein structure, and as such diffuses along a tunnel lined by unreactive amino acid side chains. It has been proposed that the hypochlorite ion is then coordinated, and orientated via hydrogen bonding with the protonated amino side chain of Lys79, whereby the electrophilic chlorine species is able to regioselectively attack the 7-position of the tryptophan ring. The arenium ion intermediate is stabilized via an interaction with the carboxylate of Glu346. Deprotonation of the arenium ion by Glu346 causes rearomatization of the ring. The active site is then regenerated by deprotonation of Glu346 by Lys79. The importance of Lys79 is demonstrated by the loss of activity in the K79A mutant, as reported by Dong et al. (2005).

Flavin-dependent halogenases also act upon other substrates and are used in the biosynthesis of chloramphenicol. The structure of CmlS, a flavin dependent halogenase, which is responsible for the formation of the dichloroacetyl moiety of chloramphenicol has been reported (Podzelinska et al. 2010). The structure reveals the enzyme to share the common flavin-binding domain, but also shows a novel 'winged helix' domain. The structure clearly demonstrates the flavin co-factor to be covalently bound via an ester linkage to the peptide through Asp277, unlike the prosthetic flavin co-factors found in other members of this class of halogenase. The 'winged helix' domain tops the flavin-binding module and generates a cover over the active site. The active site is lined predominantly with hydrophobic residues, and shows the conserved Lys required for chloramine formation. There is, however, also a notable absence of the stabilizing Glu residues seen in the tryptophan halogenases. The proposed mechanism is that the hypochlorite generated by the use of the oxidized flavin co-factor is reacted with Lys71 forming the lysine chloramine. This potentially chlorinates the enolate of acetoacetyl-CoA, with Tyr350 stabilizing the enolate intermediate. Glu44 is proposed to act as an acid catalyst in the neutralization of the nitranion; this process would occur in tandem to produce a dichloroacetoacetyl-CoA. The dichloroacetyl-CoA is released by hydrolysis of the terminal acetyl group (Podzelinska et al. 2010).

2.4 *S-adenosyl-L-methionine Dependent Chlorinases and Fluorinases*

S-adenosyl-L-methionine (SAM)-dependent halogenases have only recently been acknowledged as a new class of halogenating enzymes (Dong et al. 2004; Deng and O'Hagan 2008). These enzymes catalyze halogenation reactions by utilizing nucleophilic halide ions to attack the polar carbon-sulfur bond of SAM, displacing L-methionine, and generating a 5′-halogenated ribose moiety of adenosine. These reactions are examples of rare occurrences of S_N2 mechanisms in enzymology. They also explain why it is possible to use fluoride as part of a halogenation reaction. Unlike Cl^-, Br^-, or I^-, F^- cannot be oxidized to an 'F^+' species, such as HOF, since fluorine is the most electronegative element found in the Periodic Table. It is therefore not possible to employ an electrophilic fluorine species, and the substitution involving this element therefore necessarily has to proceed via a nucleophilic mechanism involving F^-. The crystal structure of 5′-fluoro-5′-deoxyadenosine synthase (5′-FDAS) was solved in 2004 (Dong et al. 2004). The protein forms a homotrimer as shown in Fig. 4. The monomeric subunits are single polypeptide chains with a large *N*-terminal domain (residues 8–180) and a smaller C-terminal domain (residues 195–298). The *N*-terminal domain consists of seven β-sheets that are sandwiched between α-helices, whereas the C-terminal domain is composed of β-sheet folds. A notable characteristic of 5′-FDAS is the 14-residue loop. This is a unique characteristic of this enzyme, and is involved in the trimer contacts stabilizing the trimer.

The active site residues are located at the interface of the monomeric subunits. SAM binds to a region between the *C*-terminal domain of one monomer and the

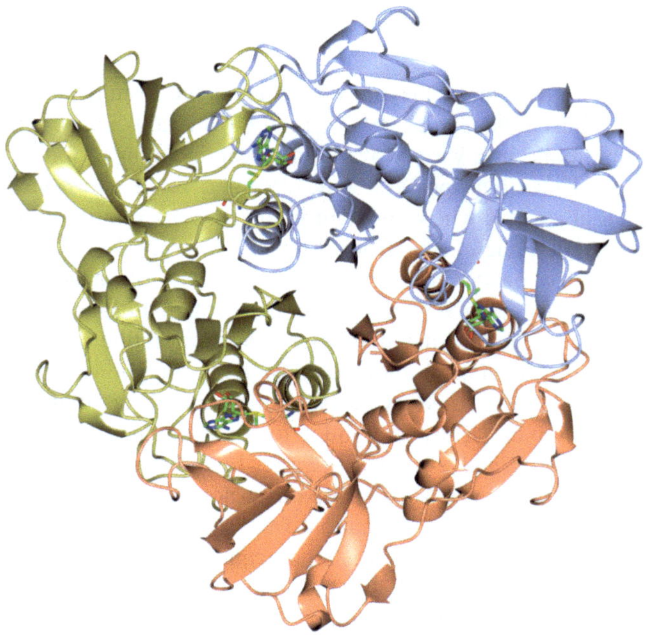

Fig. 4 A ribbon diagram representation of trimeric molecule of bacterial fluorinating enzyme 5′-fluoro-5′-deoxyadenosine synthase (pdb 1rqp; Dong et al. 2004) viewed approximately along three-fold molecular axis. Subunits are shown in *different colors*. The substrate and S-adenosylmethionine molecules are shown as *stick* models. The figure was prepared by ccp4mg (McNicholas et al. 2011)

N-terminal domain of the neighboring monomer, with three substrate molecules binding per trimer. Upon binding, SAM is buried within the protein structure, forming numerous hydrogen bonds, and hydrophobic interactions with residues of both domains.

It is believed that an open structure of the enzyme loosely binds fluoride ions. Upon binding, to SAM, the protein undergoes a conformational change, expelling water from the active site. This results in the desolvation of fluoride coordinated to residues in the active site.

2.5 Vanadium Haloperoxidases

Vanadium is an essential element for the activity of this group of haloperoxidase enzymes. These enzymes have been of increasing interest to both biologists and inorganic chemists with regard to their structure and mechanism. Therefore vanadium haloperoxidases form the main focus of this chapter and will be considered now in more detail.

3 Vanadium Haloperoxidases

A vanadium chloroperoxidase enzyme was first isolated from the terrestrial fungus *C. inaequalis* (Vanschijndel et al. 1993). In contrast, vanadium bromoperoxidase enzymes can be found in a variety of marine algae including *Ascophylum*, *Fucus*, *Laminaria*, and *Corallina* species. More recently, related enzymes have been found in bacterial species where they are involved, for example, in the napyradiomycin biosynthetic gene cluster from *Streptomyces* sp. CNQ-525 (Winter et al. 2007). A minireview on the chemistry and the biology of vanadium haloperoxidases has been published by Winter and Moore (Winter and Moore 2009).

The role of the vanadium haloperoxidases in their native environment is still under debate. The *Corallina* bromoperoxidase has been shown to be involved in the biosynthesis of brominated cyclic sesquiterpenes through *in vitro* chemoenzymatic conversion of (*E*)-(+)-nerolidol to yield the marine natural products α-snyderol, β-snyderol, and γ-snyderol (Carter-Franklin and Butler 2004). These enzymes will also brominate a range of other nonnatural compounds. The stereospecificity of bromohydrin formation from (*E*)-4-phenyl-buten-2-ol by the bromoperoxidase from *C. officinalis* is significantly different from the equivalent chemical reaction (Coughlin et al. 1993). Other reactions appear to be non-stereospecific but this varies with the substrate and reaction conditions.

3.1 Enzyme Structure

Several of the vanadium chloroperoxidase and bromoperoxidase enzymes have been studied at a structural level and in different mechanistic states with halogen and substrates bound. The main structural feature of all these enzymes is their α-helical bundle structure. The structure of the fungal *C. inaequalis* chloroperoxidase enzyme is shown in Fig. 5. In this enzyme, the active form is a monomer with the complete active site located in one monomer, whereas in the bromoperoxidases the active unit is a dimer with each subunit contributing to the other subunit to form two active sites.

The oligomeric state of the bromoperoxidase enzymes varies with the *Ascophylum* bromoperoxidase being the minimal active form composed of a single dimer, whereas the *Corallina* bromoperoxidase enzymes are dodecamers. The dimeric *Ascophylum* bromoperoxidase is stabilized by the formation of three disulfide bonds between the two monomers (Weyand et al. 1999). In the *Corallina* enzymes these cysteine residues are not conserved and the structure is a large arrangement of six dimers forming an ordered dodecamer as shown in Fig. 6 (Brindley et al. 1998; Isupov et al. 2000). The *N*-terminal regions of each monomer are organized into the center of the structure to form a cavity of 26 Å in diameter which does not have specific charge or hydrophobic properties. This makes it unlikely that the cavity is designed to bind metals and it is thought to play a more structural role. There are 12 monomers, each consisting of 19 α-helices which are 6 to 26 residues in length, eight 3_{10} helices and 14 β-strands which are mainly involved in β-hairpins. One surface of the monomer is flat and upon dimerization this surface forms the central region resulting in two

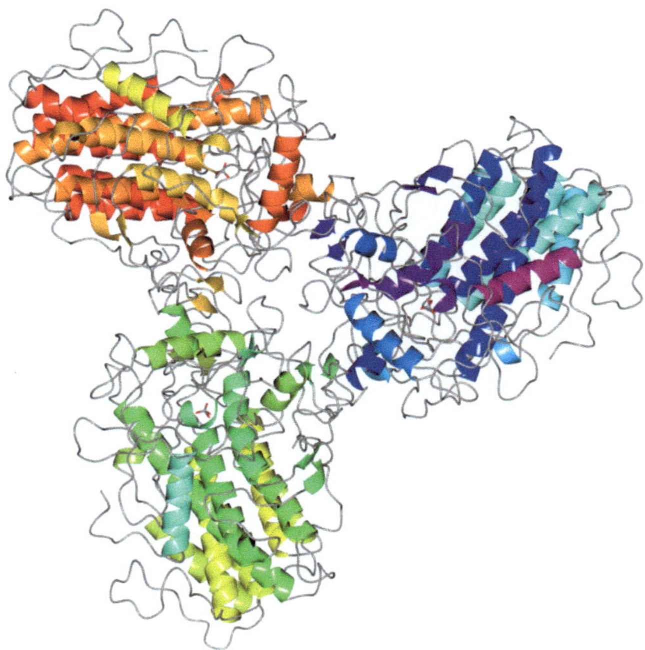

Fig. 5 A ribbon diagram representation of trimeric molecule of vanadium chloroperoxidase from *C. inaequalis* (pdb 1idq; Messerschmidt et al. 1997) viewed approximately along the three-fold molecular axis. The subunits are shown in *different colors*. The vanadate ions are shown as *ball*-and-*stick* models. The figure was prepared by ccp4mg (McNicholas et al. 2011)

four-helical bundles at the center of each dimer. The active site cleft uses amino acid residues from both monomers, with the residues of one predominantly being responsible for the bottom of the active site which binds the vanadate or phosphate, while the other constitutes the top region of the active site cleft. The dimers then interact to form the dodecamer which is approximately 150 Å in diameter, with a 23 cubic point group symmetry. The binding site for vanadate can also bind phosphate and resembles the phosphate binding site in acid phosphatases (Hemrika et al. 1997; Littlechild et al. 2002). Indeed, some crystal structures of the vanadium bromoperoxidase (Brindley et al. 1998; Littlechild and Garcia-Rodriguez 2003) show phosphate bound in the active site due to it displacing vanadate under the high phosphate concentrations used to crystallize the enzyme.

The active site residues that bind vanadate are conserved between the different haloperoxidase enzymes. In algal bromoperoxidases and fungal chloroperoxidases these residues include Arg545, His551, Arg406, Ser483, Lys398, and Gly484 (*Corallina* vanadium bromoperoxidase labeling). The phenylalanine proposed to bind to Cl^- in chloroperoxidase is replaced by His478, tryptophan is replaced by Arg395, and glycine is replaced by His485 in the *Corallina* bromoperoxidase enzyme. The structure of the *Corallina* vanadium bromoperoxidase has been solved with the natural vanadate as well as with phosphate bound. The general

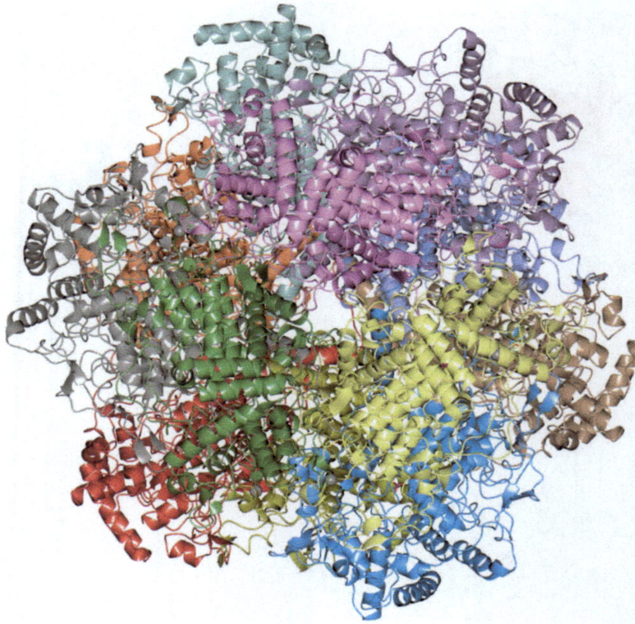

Fig. 6 A ribbon diagram representation of dodecameric vanadium bromoperoxidase from the algae *C. officinalis* (pdb 1qhb; Isupov et al. 2000) viewed approximately along a molecular threefold axis. Different subunits are shown in *different colors*. The inorganic phosphate ions are shown in *stick* mode and the calcium ions are shown as *spheres*. The figure was prepared by ccp4mg (McNicholas et al. 2011)

conformation of the active site remains unchanged in the natural vanadate form except that histidine 553 coordinates directly with the vanadate group with a displacement of the side chain nitrogen by 0.63 Å.

In the vanadate bound structure, the vanadium (V) coordination is in a trigonal bipyramidal geometry with one of the vanadate oxygens and the imidazole of the histidine in axial positions and three vanadate oxygens in equatorial positions. The conformation of vanadate is not stable and varies between the coordinated and the non-coordinated form. The distances between the vanadium atom and the oxygen atoms in the oxoanion are longer than those reported in other vanadium haloperoxidase structures.

3.2 Enzyme Mechanism

Crystallographic studies have provided some information to help us understand the mechanism of the vanadium dependent haloperoxidases. The nature of the brominating agent for these enzymes has been described as a 'Br^+-like intermediate' being either a free halogenating agent such as HOBr, Br_2, or Br_3^- or an enzyme-Br complex or a more 'exotic' V_{enz}-OBr species (Butler 1997). Extended X-ray absorption fine structure (EXAFS) studies (Christmann et al. 2004) with the related *Ascophyllum* vanadium bromoperoxidase have shown that the bromide binds within the active site of this enzyme in proximity to the vanadate center. This would permit the bromide to directly attack the peroxyvanadate center formed by the initial attack of H_2O_2 at the vanadate center in the first step of the halide oxidation. It has been debated as to whether the vanadium haloperoxidases bind their substrates, or if they catalyze only the formation of the brominating or chlorinating agent and this then interacts in a spontaneous chemical reaction with the substrate in the active site channel of the enzyme. This mechanism was used to explain the lack of stereospecificity which has been observed with many substrates. It has also been demonstrated for the bromoperoxidase from *A. nodosum* that there is a competition between indole and Phenol Red bromination (Tschirretguth and Butler 1994). Nonetheless, the same group has demonstrated that the indole was binding directly to the active site as shown by fluorescence quenching experiments. Furthermore, the *A. nodosum* bromoperoxidase enzyme was shown to catalyze a regiospecific indole bromination reaction (Martinez et al. 2001). These results provide direct evidence that the reaction is occurring within the enzyme active site. More recently, the *Corallina* bromoperoxidase was crystallized in the presence of bromide which can be detected directly in the crystal of the enzyme using data collected at a Synchrotron source at the bromine anomalous edge. These data have shown unambiguously that the bromide ion is bound to the enzyme between the vanadate and residue Arg397 at a hydrogen bonding distance of 2.8–3.1 Å. Movements of the active site residues are seen upon bromide binding. The displacement of Leu337 (terminal carbons move by 4.7 Å) and Phe373 is also seen. 'Breathing' of the enzyme reduces the size of the active site cavity between the phosphate and vanadate /bromide-bound structures. This provides a more hydrophobic environment for the catalysis to occur as shown in Fig. 7 (Littlechild et al. 2009).

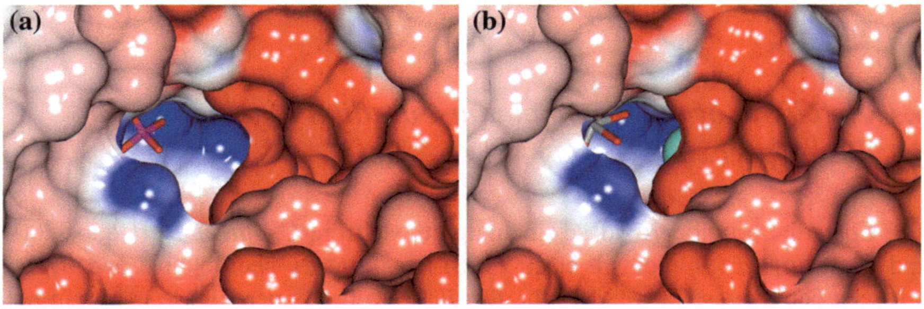

Fig. 7 An electrostatic surface potential of the active site cavity of vanadium bromoperoxidase from *C. pilulifera*, with **a** inorganic phosphate ion bound in the active site (pdb 1up8; Garcia-Rodriguez et al. 2005, Littlechild et al. 2009) and **b** vanadate and bromide bound in the active site (Garcia-Rodriguez 2005). Partial closure of the active site cavity is visible upon bromide binding. The figure was prepared by ccp4mg (McNicholas et al. 2011)

Synthetic and computational modeling studies have been carried out with synthetic small molecule analogs of the vanadium (V) active site to functionally mimic the oxidation of bromide and thioethers and density function theory has been used to understand the precise mechanism(s) (Kravitz and Pecoraro 2005; Zampella et al. 2006). Based on these studies, the Zampella group proposes that the resting form of vanadium reacts with H_2O_2 according to a mechanism that implies formation of an aqua complex, release of the apical water molecule according to a dissociative pathway and binding of H_2O_2 to vanadium. This leads to a side-on peroxo species that has been observed in a crystallographic complex with the chloroperoxidase enzyme. An acid–base catalyst accelerates the conversion of the side-on peroxo form (Zampella et al. 2006). The acid base shuttle is modeled as water and this lowers the activation barrier for intramolecular proton transfer.

The field of chemically producing small molecule mimics of the vanadium haloperoxidases has been very active in recent years. In collaboration with the group of Jim Tucker at Exeter (Tapper et al. 2006) we have been able to construct two tripodal receptors that have been shown to bind phosphate and vanadate anions in organic solvents through H-bonding interactions. The compounds were synthesized in a one-step reaction from their corresponding tetraamines, tris(2-aminoethyl)amine, and tris(3-aminopropyl)amine, respectively. Binding studies were carried out using the tetrabutylammonium salts of dihydrogen phosphate and dihydrogen vanadate as guest species. This study has provided the first successful complexation of vanadate by simple tripodal receptors in organic solvents and the first direct evidence of vanadate-binding by H-bonding interactions, similar to the ones observed at the active sites of the vanadium haloperoxidase enzymes. In addition to these binding studies involving simple mimics, the characterization of the halide binding site in the crystal structure of the *Corallina* vanadium haloperoxidase in close proximity to the vanadate site (see above) offers further scope for receptor design.

After this discussion of the structure and mechanism of the vanadium haloperoxidases, we will now turn our attention to overexpression and mutagenesis of the *Corallina* enzymes.

3.3 Refolding of the Recombinant Corallina officinalis Bromoperoxidase

The *Corallina* bromoperoxidases including those from *C. officinalis* (Coupe 2004) and *C. pilulifera* (Shimonishi et al. 1998) have already been cloned and overexpressed. The expression yield of the *C. pilulifera* bromoperoxidase is low in both *Saccharomyces cerevisiae* and *Escherchia coli* at 5.3 and 0.3 mg/l, respectively (Shimonishi et al. 1998). This is also the situation encountered with the related *C. officinalis* enzyme (Coupe 2004). In all cases, the majority of the enzyme is expressed in inclusion bodies and the alteration of induction of expression does not improve solubility. The high propensity for inclusion body formation is thought to be due to the size and quaternary structure of the dodecameric enzyme.

A method to refold the recombinant *E. coli* over-expressed *C. officinalis* enzyme has been developed (Coupe et al. 2007) using a systematic screen of refolding conditions. This has yielded active enzyme from protein without a His-tag. In contrast, it has been impossible to refold the tagged protein which can be explained by analysis of the crystal structure described above. Here, it is likely that the histidines forming the tag are unable to pack into the central core of the dodecameric protein, hence preventing refolding. Large scale refolding of the enzyme resulted in a yield of 40 mg of enzyme, therefore representing a significant increase from the protein yields achieved by simply purifying the soluble recombinant protein. In fact this yield represents a 59 % recovery rate resulting in enzyme that does not require any further purification as the majority of contaminants have been removed by the inclusion body washing process. The activity of this enzyme is also comparable to the one of the enzyme isolated directly from the algae when assayed using monochlorodimedone (MCD).

The refolded protein has been characterized and compared to the native enzyme and was shown to be stable at temperatures of 80 °C, over a pH range from 5.5 to 10 and in organic solvents such as ethanol, acetonitrile, methanol, and acetone. Circular dichroism studies have indicated that both the native and refolded enzyme have a similar structural conformation with the α-helical content showing a negative ellipticity at 230 nm, this being the most predominant feature of the spectra (Coupe 2004).

3.4 *Construction of a Truncated Dimeric Corallina officinalis Mutant Protein*

Several mutant enzymes were constructed which lack an increasing number of amino acids from the *N*-terminal end of the native *Corallina* bromoperoxidase protein. These were modeled on the crystal structure and constructs that resulted in no significant exposure of hydrophobic patches were chosen and investigated. The most successful mutant was one that had over 200 residues removed from the *N*-terminus. This mutant lacked residues up to the mid-region of α-helix 8. This length was chosen as α-helix 9 has been implicated in the active site cleft so the truncation did not extend further into this region. Despite this mutation being severe, i.e., removing a large region in between the active site channel and the distal region of the protein comprised of α-helices 14–16, the corresponding mutant proved to be the most successful one. It retained activity in the standard MCD assay and could also be overexpressed in *E. coli* in a soluble form (Coupe 2004). This dimeric form of the bromoperoxidase is being studied extensively as to its substrate specificity since it remains a favored form of the enzyme for commercial applications in biocatalysis.

3.5 Mutants of *Corallina* Vanadium Haloperoxidase to Change the Halide Specificity

There is also a considerable interest to understand the halide specificity of different vanadium haloperoxidase enzymes which is due to subtle changes of the amino acids around the active site. To modify the halogen specificity, Arg397 was targeted in the *Corallina* bromoperoxidase to see if an enzyme could be constructed that would carry out chloroperoxidase activity based on the fact that this residue varies between the fungal chloroperoxidase and the *Corallina* bromoperoxidase. Also, the bromide ion in the crystal structure is located close to this residue as shown in Fig. 8. A mutant of the *C. pilulifera* bromoperoxidase has been constructed, which also exhibits chloroperoxidase activity (Ohshiro et al. 2004). When arginine (R) is changed to tryptophan (W) or phenylalanine (F) then the mutant enzyme shows significant chloroperoxidase activity as well as bromoperoxidase activity. Furthermore, residue R397 has been substituted by the other 19 amino acids and the resulting mutant enzymes have been purified and their properties investigated. The specific chloroperoxidase activities of the R397W and R397F enzymes were 25.1 and 32.0 units/mg, and the K_m values for Cl^- were 780 and 670 mM, respectively. Unlike the native enzyme, both mutant enzymes were inhibited by NaN_3. In the case of the R397W enzyme, the incorporation rate of vanadate into the active site was also low, compared with the R397F mutant and the wild type enzyme. These results provide further evidence for the existence of a

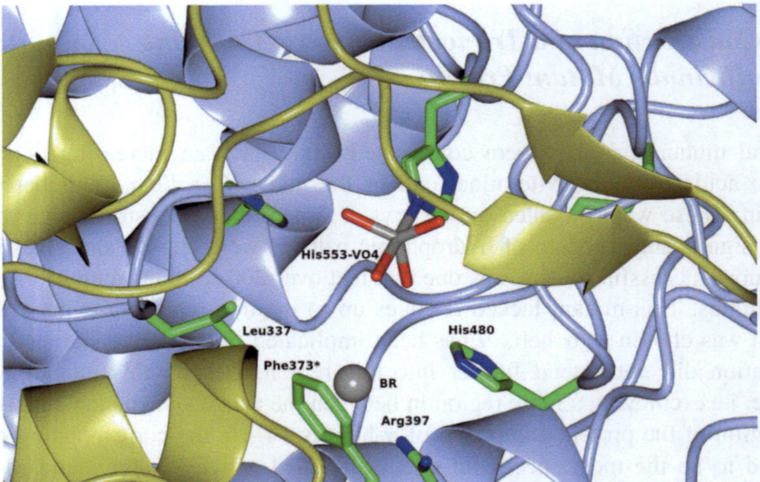

Fig. 8 The active site of the *C. pilulifera* bromoperoxidase complex with vanadate and bromide bound in the active site (Littlechild et al. 2009, Isupov and Littlechild 2014 manuscript in preparation). The vanadate-histidine adduct and active site residues are shown as *sticks* and the bromide ion is shown as a *sphere*. The figure was prepared by ccp4mg (McNicholas et al. 2011)

specific halogen binding site within the catalytic cleft of vanadium bromoperoxidase enzymes. As described above, the latter has now been observed in the crystal structure of the *Corallina* enzyme.

Other mutagenesis studies have been carried out with the *Curvularia* vanadium chloroperoxidase enzyme to understand the role of specific amino acid residues in the activity of the enzyme (Hemrika et al. 1999; Macedo-Ribeiro et al. 1999; Renirie et al. 2000).

3.6 Use of Vanadium Haloperoxidases in Biocatalysis

As already mentioned, if isolated enzymes are being used in biotransformation processes, then the stability of the enzymes is often a major issue. The vanadium haloperoxidase enzymes are structurally more robust and more resistant to H_2O_2 compared to many other (halo-)peroxidases. Often, these enzymes lack stereospecificity in their reaction. This is seen by the ability of the *Corallina* bromoperoxidase to cause multiple bromination of Phenol Red to Bromophenol Blue. Substrate specificity also varies between different vanadium haloperoxidases and this is thought to be due to the different residues lining the active site pocket. Five residues in the active site cleft are conserved between *Corallina* and *Ascophyllum* vanadium bromoperoxidases, which are not involved in vanadate binding. The *Corallina* enzyme has three charged amino acids within 7.5 Å of the vanadate site and the *Ascophyllum* enzyme has three hydrophilic amino acids, and no charged amino acids in this region. None of these amino acids is structurally conserved in the fungal *Curvularia* vanadium haloperoxidase. The vanadium bromoperoxidase enzymes have been shown to carry out the bromonium-assisted cyclization of terpenes and ethers (Butler and Carter-Franklin 2004; Carter-Franklin and Butler 2004). It is now accepted that the vanadium haloperoxidases found in red algae such as *Corallina*, *Laurencia pacifica* and *Plocamium cartilagineum* are able to catalyze the asymmetric bromination /cyclization reactions converting the sesquiterpene nerolidol to the snyderol family of natural marine products. These enzymes can also carry out sulfoxidation reactions of commercial interest, but not epoxidations (Littlechild 1999). Some of these sulfoxidation reactions can be stereospecific (Coughlin et al. 1993). The biotransformation of indene to *1S,2R*-indene oxide with an extract of a *Curvularia* fungus is used, for instance in the synthesis of the HIV protease inhibitor Crixivan® (Merck) (Zhang et al. 1999) (Fig. 9).

It is clear that the 'natural' role of the vanadium bromoperoxidases is to synthesize complex brominated marine molecules, many of which show anti-bacterial and anti-cancer properties. The fact that many new drugs entering the market are halogenated, turns such haloperoxidases into interesting and increasingly important natural tools which may be used in commercial biotransformations.

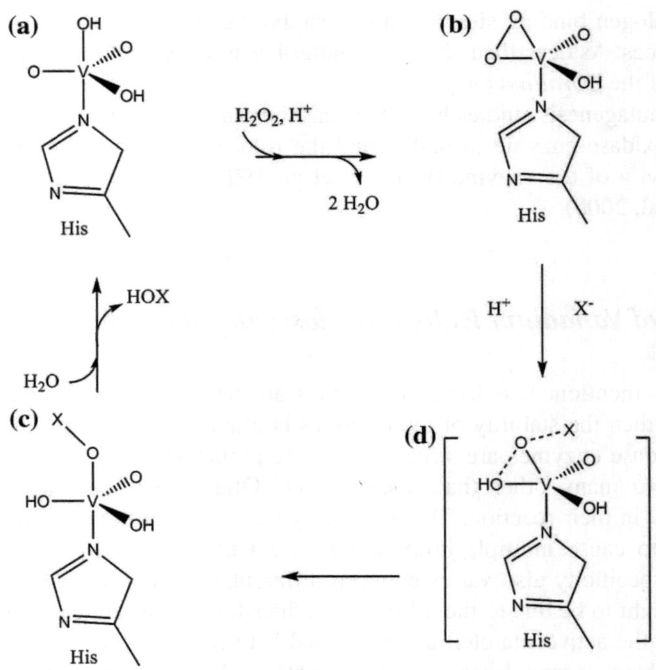

Fig. 9 A proposed reaction scheme for the intermediates formed at the vanadate center in vanadium-dependent bromoperoxidases, adapted from Zampella et al. 2006. The depicted complexes have been predicted by DFT (Density functional theory) studies and complex (**a**) and (**b**) by X-ray studies (Messerschmidt et al. 1997)

Acknowledgments The authors acknowledge the support of BBSRC project grants and funding of PhD studentships associated with the vanadium bromoperoxidase project and the support of European Community's 7th Framework Programme [FP7/2007-2013] under grant agreement No: 215009 (RedCat).

References

Blasiak LC, Drennan CL (2009) Structural perspective on enzymatic halogenation. Acc Chem Res 42(1):147–155

Blasiak LC, Vaillancourt FH, Walsh CT, Drennan CL (2006) Crystal structure of the non-haem iron halogenase SyrB2 in syringomycin biosynthesis. Nat 440:368–371

Brindley AA, Dalby AR, Isupov MN, Littlechild JA (1998) Preliminary X-ray analysis of a new crystal form of the vanadium-dependent bromoperoxidase from *Corallina officinalis*. Acta Crystallogr D 54:454–457

Butler A (1997) Vanadium-dependent redox enzymes. In: Vanadium haloperoxidase comprehensive biological catalysis. Academic Press, New York, pp 427–437

Butler A, Carter-Franklin JN (2004) The role of vanadium bromoperoxidase in the biosynthesis of halogenated marine natural products. Nat Prod Rep 21(1):180–188

Butler A, Sandy M (2009) Mechanistic considerations of halogenating enzymes. Nature 460(7257):848–854

Carter-Franklin JN, Butler A (2004) Vanadium bromoperoxidase-catalyzed biosynthesis of halogenated marine natural products. J Am Chem Soc 126(46):15060–15066

Christmann U, Dau H, Haumann M, Kiss E, Liebisch P, Rehder D, Santoni G, Schulzke C (2004) Substrate binding to vanadate-dependent bromoperoxidase from *Ascophyllum nodosum*: a vanadium K-edge XAS approach. Dalton T 16:2534–2540

Coughlin P, Roberts S, Rush C, Willetts A (1993) Biotransformation of alkenes by haloperoxidases—regiospecific bromohydrin formation from cinnamyl substrates. Biotechnol Lett 15(9):907–912

Coupe EE (2004) Phd thesis, Exeter, UK

Coupe EE, Smyth MG, Fosberry A, Hall RM, Littlechild JA (2007) The dodecameric vanadium-dependent haloperoxidase from the marine algae *Corallina officinalis*: cloning, expression, and refolding of the recombinant enzyme. Protein Expres Purif 52(2):265–272

Dembitsky VM, Srebnik M (2002) Natural halogenated fatty acids: their analogues and derivatives. Prog Lipid Res 41(4):315–367

Deng H, O'Hagan D (2008) The fluorinase, the chlorinase and the duf-62 enzymes. Curr Opin Chem Biol 12(5):582–592

Dong CJ, Flecks S, Unversucht S, Haupt C, van Pee KH, Naismith JH (2005) Tryptophan 7-halogenase (PrnA) structure suggests a mechanism for regioselective chlorination. Science 309(5744):2216–2219

Dong CJ, Huang FL, Deng H, Schaffrath C, Spencer JB, O'Hagan D, Naismith JH (2004) Crystal structure and mechanism of a bacterial fluorinating enzyme. Nature 427(6974):561–565

Eustaquio AS, Pojer F, Noe JP, Moore BS (2008) Discovery and characterization of a marine bacterial SAM-dependent chlorinase. Nat Chem Biol 4(1):69–74

Feling RH, Buchanan GO, Mincer TJ, Kauffman CA, Jensen PR, Fenical W (2003) Salinosporamide A: a highly cytotoxic proteasome inhibitor from a novel microbial source, a marine bacterium of the new genus Salinospora. Angew Chem Int Edit 42(3):355–357

Fujimori DG, Walsh CT (2007) What's new in enzymatic halogenations. Curr Opin Chem Biol 11(5):553–560

Garcia-Rodriguez E (2005) PhD thesis, University of Exeter

Garcia-Rodriguez E, Ohshiro T, Aibara T, Izumi Y, Littlechild J (2005) Enhancing effect of calcium and vanadium ions on thermal stability of bromoperoxidase from *Corallina pilulifera*. J Biol Inorg Chem 10:275–282

Gribble GW (1998) Naturally occurring organohalogen compounds. Acc Chem Res 31(3):141–152

Hager LP, Morris DR, Brown FS, Eberwein H (1966) Chloroperoxidase. II. Utilization of halogen anions. J Biol Chem 241(8):1769–1777

Hemrika W, Renirie R, Dekker HL, Barnett P, Wever R (1997) From phosphatases to vanadium peroxidases: a similar architecture of the active site. Proc Natl Acad Sci USA 94(6):2145–2149

Hemrika W, Renirie R, Macedo-Ribeiro S, Messerschmidt A, Wever R (1999) Xray crystal structures of active site mutants of the vanadium-containing chloroperoxidase from the fungus *Curvularia inaequalis*. J Biol Chem 274:23820–23827

Hill A, Littlechild J (2012) Oxidation and haloperoxidases. In: Comprehensive Chirality, vol 7. Elsevier Ltd., pp 329–349

Hofrichter M, Ullrich R (2006) Heme-thiolate haloperoxidases: versatile biocatalysts with biotechnological and environmental significance. Appl Microbiol Biot 71(3):276–288

Isupov MN, Dalby AR, Brindley AA, Izumi Y, Tanabe T, Murshudov GN, Littlechild JA (2000) Crystal structure of dodecameric vanadium-dependent bromoperoxidase from the red algae *Corallina officinalis*. J Mol Biol 299(4):1035–1049

Isupov M, Littlechild J (2014) Manuscript in preparation

Karmee SK, Roosen C, Kohlmann C, Lutz S, Greiner L, Leitner W (2009) Chemo-enzymatic cascade oxidation in supercritical carbon dioxide/water biphasic media. Green Chem 11(7):1052–1055

Kravitz JY, Pecoraro VL (2005) Synthetic and computational modeling of the vanadium-dependent haloperoxidases. Pure Appl Chem 77(9):1595–1605

Kuhnel K, Derat E, Terner J, Shaik S, Schlichting I (2007) Structure and quantum chemical characterization of chloroperoxidase compound 0, a common reaction intermediate of diverse heme enzymes. Proc Natl Acad Sci USA 104(1):99–104

Littlechild J (1999) Haloperoxidases and their role in biotransformation reactions. Curr Opin Chem Biol 3(1):28–34

Littlechild J, Garcia-Rodriguez E (2003) Structural studies on the dodecameric vanadium bromoperoxidase from Corallina species. Coordin Chem Rev 237(1–2):65–76

Littlechild J, Garcia-Rodriguez E, Dalby A, Isupov M (2002) Structural and functional comparisons between vanadium haloperoxidase and acid phosphatase enzymes. J Mol Recognit 15(5):291–296

Littlechild J, Rodriguez EG, Isupov M (2009) Vanadium containing bromoperoxidase—Insights into the enzymatic mechanism using X-ray crystallography. J Inorg Biochem 103(4):617–621

Macedo-Ribeiro S, Hemrika W, Renirie R, Wever R, Messerschmidt A (1999) X-ray crystal structures of active site mutants of the vanadium-containing chloroperoxidase from the fungus *Curvularia inaequalis*. J Biol Inorg Chem 4(2):209–219

Martinez JS, Carroll GL, Tschirret-Guth RA, Altenhoff G, Little RD, Butler A (2001) On the regiospecificity of vanadium bromoperoxidase. J Am Chem Soc 123(14):3289–3294

McNicholas S, Potterton E, Wilson KS, Noble MEM (2011) Presenting your structures: the CCP4mg molecular-graphics software. Acta Cryst D67:386–394

Messerschmidt A, Prade L, Wever R (1997) Implications for the catalytic mechanism of the vanadium-containing enzyme chloroperoxidase from the fungus*Curvularia inaequalis* by X-ray structures of the native and peroxide form. J Biol Chem 378:309–315

Ohsawa N, Ogata Y, Okada N, Itoh N (2001) Physiological function of bromoperoxidase in the red marine alga, *Corallina pilulifera*: production of bromoform as an allelochemical and the simultaneous elimination of hydrogen peroxide. Phytochemistry 58(5):683–692

Ohshiro T, Littlechild J, Garcia-Rodriguez E, Isupov MN, Iida Y, Kobayashi T, Izumi Y (2004) Modification of halogen specificity of a vanadium-dependent bromoperoxidase. Protein Sci 13(6):1566–1571

Onega M, Domarkas J, Deng H, Schweiger LF, Smith TAD, Welch AE, Plisson C, Gee AD, O'Hagan D (2010) An enzymatic route to 5-deoxy-5-[F-18]fluoro-D-ribose, a [F-18]-fluorinated sugar for PET imaging. Chem Commun 46(1):139–141

Podzelinska K, Latimer R, Bhattacharya A, Vining LC, Zechel DL, Jia ZC (2010) Chloramphenicol biosynthesis: The structure of CmlS, a flavin-dependent halogenase showing a covalent flavin-aspartate bond. J Mol Biol 397(1):316–331

Renirie R, Hemrika W, Wever R (2000) Peroxidase and phosphatase activity of active-site mutants of vanadium chloroperoxidase from the fungus *Curvularia inaequalis*—Implications for the catalytic mechanisms. J Biol Chem 275(16):11650–11657

Shaw PD, Hager LP (1959) Biological chlorination. III. beta-Ketoadipate chlorinase: a soluble enzyme system. J Biol Chem 234:2565–2569

Shimonishi M, Kuwamoto S, Inoue H, Wever R, Ohshiro T, Izumi Y, Tanabe T (1998) Cloning and expression of the gene for a vanadium-dependent bromoperoxidase from a marine macro-alga, *Corallina pilulifera*. FEBS Lett 428(1–2):105–110

Sundaramoorthy M, Terner J, Poulos TL (1998) Stereochemistry of the chloroperoxidase active site: crystallographic and molecular-modeling studies. Chem Biol 5(9):461–473

Tapper S, Littlechild JA, Molard Y, Prokes I, Tucker JHR (2006) Anion binding tripodal receptors as structural models for the active site of vanadium haloperoxidases and acid phosphatases. Supramol Chem 18(1):55–58

Tschirretguth RA, Butler A (1994) Evidence for organic substrate-binding to vanadium bromoperoxidase. J Am Chem Soc 116(1):411–412

Vaillancourt FH, Yeh E, Vosburg DA, Garneau-Tsodikova S, Walsh CT (2006) Nature's inventory of halogenation catalysts: oxidative strategies predominate. Chem Rev 106(8):3364–3378

Vanschijndel JWPM, Vollenbroek EGM, Wever R (1993) The chloroperoxidase from the fungus Curvularia-inaequalis—a novel vanadium enzyme. Biochim Biophys Acta 1161(2–3):249–256

Weyand M, Hecht HJ, Kiess M, Liaud MF, Vilter H, Schomburg D (1999) X-ray structure determination of a vanadium-dependent haloperoxidase from Ascophyllum nodosum at 2.0 angstrom resolution. J Mol Biol 293(3):595–611

Winter JM, Moffitt MC, Zazopoulos E, McAlpine JB, Dorrestein PC, Moore BS (2007) Molecular basis for chloronium-mediated meroterpene cyclization - Cloning, sequencing, and heterologous expression of the napyradiomycin biosynthetic gene cluster. J Biol Chem 282(22):16362–16368

Winter JM, Moore BS (2009) Exploring the chemistry and biology of vanadium-dependent haloperoxidases. J Biol Chem 284(28):18577–18581

Wojaczynska E, Wojaczynski J (2010) Enantioselective synthesis of sulfoxides: 2000–2009. Chem Rev 110(7):4303–4356

Woodley JM (2008) New opportunities for biocatalysis: making pharmaceutical processes greener. Trends Biotechnol 26(6):321–327

Yarnell A (2006) Nature's X-factors. Chem Eng News 84(21):12–14

Zaks A, Dodds DR (1995) Chloroperoxidase-catalyzed asymmetric oxidations—substrate-specificity and mechanistic study. J Am Chem Soc 117(42):10419–10424

Zampella G, Fantucci P, Pecoraro VL, De Gioia L (2006) Insight into the catalytic mechanism of vanadium haloperoxidases. DFT investigation of vanadium cofactor reactivity. Inorg Chem 45(18):7133–7143

Zhang J, Roberge C, Reddy J, Connors N, Chartrain M, Buckland B, Greasham R (1999) Bioconversion of indene to trans-2S, 1S-bromoindanol and 1S, 2R-indene oxide by a bromoperoxidase/dehydrogenase preparation from Curvularia protuberata MF5400. Enzyme Microb Tech 24(1–2):86–95

Authors Biography

Prof. Jennifer Littlechild is Professor of Biological Chemistry and Director of the Henry Wellcome Centre for Biocatalysis at Exeter. She carried out her PhD in the Biophysics Laboratory, Kings College, London University, UK followed by a postdoctoral fellowship at the Biochemistry Department at Princeton University, USA. In 1975 she became a group leader at the Max-Planck Institute for Molecular Genetics in Berlin, Germany. In 1980 she returned to the UK to Bristol University and in 1991 to Exeter. Her current research grants are from UK research councils, BBSRC, EPSRC, and the EU and large and SME industries.

Jennifer's research interests involve the structural and mechanistic characterization of the C-C bond forming enzymes transketolase and aldolase, vanadium haloperoxidases, Baeyer-Villiger monooxygenases, aminoacylases, novel esterases and lipases, gamma lactamases, alcohol dehydrogenases, dehalogenases, transaminases, and other enzymes from thermophilic bacteria and archaea. Many of these enzymes are used in combination with conventional chemical synthesis for the production of new optically pure drugs of interest to pharmaceutical companies. To date, Jennifer has published over 130 publications in refereed high impact journals and presented her research work internationally.

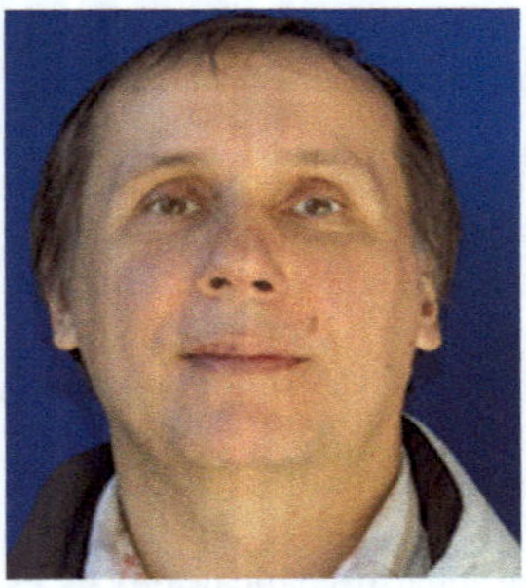

Dr. Michail Isupov (born 1963) has been trained in molecular and chemical physics at the Moscow Institute of Physics and Technology. He received his PhD in Physics and Mathematics from the Moscow Institute of Crystallography. In 1994 he started his postdoctoral studies at the University of Exeter in the UK, where he is now a Senior Research fellow.

Michail is working in the field of protein crystallography of macromolecules and is interested in structure–function relationships of proteins. He is also involved in the development of new methods for macromolecular structure solution. His particular interest concerns enzymes involved in biocatalysis, including pyridoxal phosphate-dependent enzymes, enzymes involved in halogenation and dehalogenation, hydrolases, and also oxidative stress-response proteins. To date Michail has published over 60 publications and deposited around 60 structures in the Protein Data Bank.

Part XVII Connecting Section Between Chapters 17 and 18

While haloperoxidases form one class of enzymes with promising applications in industrial redox processes, there are numerous other enzymes that can also be employed for practical uses, either as extracts or in their isolated form. In the field of redox active metabolites, enzymes such as alliinase from *Allium* plants (e.g., garlic and onions), or myrosinase from *Brassicaceae* plants (e.g., mustard, garden cress, rape, wasabi, broccoli) come to mind (see Chaps. 9 and 10). Indeed, alliinase is a highly interesting C-S-lyase enzyme, which converts its rather harmless sulfoxide substrate alliin into the powerful thiol-modifying thiosulfinate allicin. It is now possible to employ this enzymatic conversion as part of a two-component approach in order to generate highly reactive thiosulfinates *in situ*, i.e., at the location and time of their desired action. For this purpose, alliinase itself can be isolated from garlic cloves in good amounts, and its substrates are easily accessible via chemical synthesis starting from L-cysteine. Such two-component 'binary' systems have been suggested for practical uses in the fight against certain types of cancer and also as plant protectants in an eco-friendly agriculture. Similar practical applications are currently also emerging for the myrosinase enzyme, which converts its harmless glucosinolate substrates into highly cytotoxic (and generally toxic) thiocyanates, isothiocyanates and nitriles. The latter cause a barrage of damages on proteins and enzymes. In contrast to the thiosulfinates, which react exclusively with thiol groups, these reactive electrophilic species are less picky when it comes to their targets, and tend to modify amines as well as thiols, i.e., primarily but not exclusively lysine and cysteine residues.

While practical applications for alliinase and myrosinase may soon become a reality, the next chapter will consider yet another enzyme that has been turned into a practical application and a true commercial success story. Bromelain represents a mixture of different cysteine endopeptidases, which is used against minor injuries. It is readily available from the pineapple plant and has been converted into a commercial product which is sold in many countries worldwide over the counter to promote the healing of injuries. Recent research has surfaced a number of other potential applications for this natural enzyme product, which will be considered in the next chapter, together with the question of whether it is possible to apply such an enzyme systemically. Here, the latest evidence is rather exciting. It sheds doubt on the traditional point of view that enzymes cannot be administered orally for systemic therapy because of their degradation in the stomach and difficulties to enter the bloodstream. If confirmed, these findings may open up a whole new field of redox active product research. Here, readily available plant enzymes are likely to be at the forefront of a new wave of therapeutically active natural agents.

Chapter 18
Systemic Enzyme Therapy: Fact or Fiction? A Review with Focus on Bromelains, Proteolytic Enzymes from the Pineapple Plant

Peter Meiser, Zhanjie Xu, Gilbert Kirsch and Claus Jacob

Keywords Anti-inflammatory • Bromelain • Cancer • Therapeutic enzymes • Myrosinase

1 Introduction: Enzymes as Therapeutic Agents

Speaking of enzymes and therapy, one might think of (small) molecules that aim at the inhibition of catalytic sites of enzymes in the first instance. Indeed, this approach represents the majority of pharmacological activities used for treatment of various diseases, be it by the inhibition of the transpeptidase enzyme in bacteria (e.g., by penicillin), the inhibition of cyclooxygenase as important factor in inflammation (e.g., by non-steroidal anti-inflammatory drugs such as ibuprofen), of proton pumps in the stomach (e.g., by omeprazole), or the inhibition of tyrosine kinases involved in cancer progression (e.g., by imatinib).

In many cases disease pathogenesis indeed is induced or enhanced through an overshooting enzyme activity, and thus therapy profits from the inhibition of one particular enzyme. In some cases however, the selected inhibition of 'just one

P. Meiser (✉) · Z. Xu
URSAPHARM Arzneimittel GmbH, Industriestraße 35,
66129 Saarbruecken, Germany
e-mail: peter.meiser@ursapharm.de

Z. Xu · G. Kirsch
Laboratoire d'Ingénierie Moléculaire et Biochimie Pharmacologique,
Université de Lorraine 1, Boulevard Arago, 57070 Metz, France

C. Jacob
Bioorganische Chemie, Universitaet des Saarlandes,
66041 Saarbrücken, Germany

C. Jacob et al. (eds.), *Recent Advances in Redox Active Plant and Microbial Products*,
DOI: 10.1007/978-94-017-8953-0_18,

enzyme' might not be sufficient to address an overall complex disease condition.[1] In fact, the drawbacks associated with this traditionally propagated 'one-target-approach' are more and more being recognized, and natural products in particular return to the spotlight of the search for new lead structures that address several targets at once (see also Chaps. 2 and 5).

But is it possible to use enzymes themselves as therapeutic agents?

What might look a little peculiar at a first glance, indeed has a long tradition (see also Explanatory Box 1). The Bible reports the cure of the Jewish King Hiskia after treatment of his ulcer with figs puree (Second Book of Kings, 20th Chapter, 7th Verse)—the 'active ingredient' of this therapeutic measure most presumably was the enzyme ficin, a proteolytic enzyme. Similarly, Native Americans used cut fruits like pineapples for the treatment of open wounds—again, proteolytic enzymes like bromelain were certainly responsible for the therapeutic benefit.

Explanatory Box 1: Myrosinase

Several chapters of this book consider practical uses of enzymes. Indeed, such biocatalysts are powerful, highly effective, and selective catalysts which have turned out to be quite useful for many diverse practical applications. In the field of pharmaceutical research, this combination of efficiency (at low doses) and selectivity is of particular interest and has given rise to a number of innovative approaches. As far as the development of selective cytotoxic or antimicrobial agents is concerned, some enzymes seem to provide particularly promising solutions as they are part of natural-antimicrobial-defense systems which occur in edible plants, i.e., are non-toxic to humans. Such plant-based (redox) defense systems resemble the human defense based on myeloperoxidase, which catalytically generates hypochlorous acid as a potent weapon against many microorganisms.

When considering such plant defense systems, two substrate/enzyme combinations resulting in highly active agents stand out. The first one is the alliin/alliinase system from garlic which generates highly aggressive allicin, and has been discussed already. The second one is the glucosionolate/

[1] To put it right: the inhibition of only one enzyme through the action of a highly selective active ingredient is not necessarily and always equivalent to a highly restricted pharmacodynamic action. For example, inhibitors of the mentioned tyrosine kinase influence several intracellular processes by inhibiting the signaling pathway at an early stage. Still these inhibitors act much more selectively as compared to classical chemotherapeutics. Generally spoken, the more downstream the interference within a particular signaling cascade and the more selective the inhibitor, the more selective the pharmacodynamic activity will be.

myrosinase system, which is found in many plants of the *Brassicaceae* family and generates a range of highly reactive and biologically active species, such as isothiocyanates (RNCS), thiocyanates (RSCN), and nitrils (RCN). Rather well-known plants such as mustard, garden cress, rape seed, broccoli, cabbage, daikon, and wasabi contain this potent defense system against predators. Indeed, the rather distinct smell and taste of these plants and products derived from them is mostly due to the presence of such highly active species. In the case of mustard, the chemical name 'mustard oil' actually refers to allyl isothiocyanate (C_3H_5NCS). From a biochemical point of view, compounds such as mustard oil react fast and effectively with thiol and amine groups in biomolecules, are able to regulate or inhibit many proteins and enzymes and hence may cause cell death in some cell types or organisms not able to counteract such an insult. As some cancer cells and many microorganisms seem to be particularly sensitive to isothiocyanates and thiocyanates, these natural, redox active metabolites are of particular interest in anticancer and antimicrobial drug development.

As many of the compounds discussed here (including allyl isothiocyanate) are smelly, chemically unstable, and often overly reactive (and hence unselective) to be applied in practice, it is possible to generate such aggressive species using a simple glucosinolate/myrosinase combination. Indeed, myrosinase is fairly non-specific concerning the precise chemical structure of its various substrates, and hence it is possible to select and/or design various natural and artificial substrates which will then be converted to a palette of distinct isothiocyanates, thiocyanates, and nitriles. In the future, the use of such binary, two-component systems based on natural defense systems of edible plants is likely to increase, as it enables the generation of the reactive species at the time and place of action. Such considerations are not limited to alliinase and myrosinase, of course, yet those two enzymes stand out as they are derived from two widely used culinary plants, are fairly stable and easy to handle, and also provide some flexibility with regard to their specific substrates and hence products generated.

Of course, these and further anecdotal reports cannot be considered as a proof-of-concept for such therapeutic measures, and using cut or mashed fruits for the treatment of open wounds raises some microbiological issues as well. Nonetheless, more recent research substantiates the usefulness of topical application of proteolytic enzymes. Bromelain applied as an ointment or gel proved to be useful for the removal of necrotic tissues from wounds or burned skin ('debridement'), without affecting the healthy tissue (Pavan et al. 2012). At closer inspection, making use of enzymes for wound cleansing and wound healing is actually also behind the safe, simple, and effective treatment of chronic wounds with maggots (Gilead et al. 2012).

As a matter of fact, enzymes are not only used for the treatment of topical disorders, but also for the therapy of systemic diseases. One important example is the supplementation of pancreatic enzymes in patients suffering from pancreatic insufficiency, which occurs for instance as a consequence of cystic fibrosis. These enzymes are swallowed, and exhibit their beneficial—in this case even vital—effects in the small intestine. Thus the pancreatic enzymes have to pass the stomach without being degraded, which is usually accomplished by using an acid-resistant coating, but do not have to enter the systemic circulation for exhibiting their activity.

Furthermore, enzymes are used for thrombolytic therapy, for treatment of myocardial infarction, thromboembolic strokes, or deep vein thrombosis. Here, thrombolytic enzyme drugs such as alteplase, streptokinase, urokinase, or reteplase are used to clear blocked arteries in order to restore perfusion of the affected tissue (Menon et al. 2004). These drugs are administered by injection or infusion, thus the therapeutic agent reaches the target location directly via the bloodstream.

Going one step further, enzymes are even used for systemic therapy after oral administration. Obviously, the prerequisite for the manifestation of a pharmacological effect appears to be the intact gastrointestinal passage and systemic absorption of such enzymes. To postulate the uptake of large intact macromolecules such as proteolytic enzymes challenges a key existing dogma, since the human intestinal epithelium has traditionally been described as nonpermeable to proteins. Nonetheless, there is reason to believe that this is indeed possible. The following overview focuses on bromelains, proteolytic enzymes from the pineapple plant, since this phytotherapeutic is very well characterized and most of preclinical and clinical trials investigating the effects of systemic enzyme therapy have been conducted with bromelain or mixtures of enzymes that also contained bromelain.

2 Biochemical Properties of Bromelain

Bromelain is gained from aqueous extracts from the stem and the fruit of pineapple by centrifugation, ultrafiltration, and subsequent lyophilization. The resulting mixture contains a complex cocktail of cysteine endopeptidases, other enzymes such as phosphatases, glucosidases, cellulases, as well as low molecular weight compounds and several protease inhibitors (Maurer 2001). Among the proteases described are stem bromelain (mainly in the stem), fruit bromelain (mainly in fruits), and ananain. Stem and fruit bromelain both occur in the stem and the fruit of the plant, but in differing amounts. Thus bromelain gained from the fruit is not identical to bromelain gained from the stem. This is important to mention, since commercial production of bromelain only started after the discovery that the stem of the pineapple plant contains more bromelain than the expensive fruit

Fig. 1 Bromelain is gained from stems of the pineapple plant

(Heinicke and Gortner 1957), and therefore only bromelain from the stem is used in therapy, as it is more cost-effective (Fig. 1).

The enzymatic activity of bromelain covers a wide pH range of 5.5–8.0 (Maurer 2001), which might be attributable to distinct pH optima of the single proteases. Enzymatic activity of bromelain is determined using different substrates such as casein (Fédération Internationale Pharmaceutique, F.I.P. unit) or gelatin (gelatin digestion units, GDU).

The main proteases of bromelain have been (bio-)chemically characterized, and molecular weights greater than 24 kDa have been determined (Harrach et al. 1995). This brings us back to the question, "How can such a large molecule—being for instance 133-fold heavier than acetyl-salicylic acid (Aspirin®)—be absorbed after oral intake and have a systemic effect?"

3 Absorption and Bioavailability

Well, the short answer is: We do not know yet exactly how. As a matter of fact, we do not even know whether absorption of bromelain is indeed necessary for a systemic effect.

To start from the beginning: Several independent investigations confirm the absorption of bromelain. After intraduodenal application of ^{125}I-labeled bromelain to adult rats, 40 % of bromelain in a high molecular form could be detected in blood and lymph during a 6 h observation period by antibromelain antibodies using the agar-double diffusion technique (Seifert et al. 1979). Similarly, the presence of undegraded bromelain in plasma of healthy human volunteers was shown after oral intake by precipitation with antibromelain antibodies followed by gel electrophoresis and immunodetection (Castell et al. 1997).

The presence of high molecular weight components of bromelain, however, does not necessarily prove the presence of the *active* enzyme. Furthermore, proteases present in blood are rapidly captured by antiproteases such as α_2-macroglobulin and α_1-antitrypsin. It appears, however, that bromelain may not have to occur as the free enzyme in blood, since bromelain retains its proteolytic capacity even in presence of α_2-macroglobulin and α_1-antitrypsin.

Still, what had to be proven in these studies was that bromelain retains its proteolytic activity in blood after oral intake and passage through the intestinal cell wall. This proof was accomplished in an elaborative set of experiments by capturing the active enzyme from blood of healthy volunteers by use of antibromelain antibodies and subsequent fluorimetric testing for proteolytic activity against a specific fluorogenic substrate (Castell et al. 1997).

Using a different approach—one might call it 'effect kinetics', since the effect itself is 'measured', rather than the effect-causing agent—and as early as in 1964, Miller and Opher tested the proteolytic activity of the blood serum before and after a 3-day therapy with bromelain in 11 healthy volunteers. All of the probands showed an increased proteolytic activity to digest the substrate casein after bromelain-therapy (Miller and Opher 1964).

A more recent study also aimed to measure the effect of bromelain after oral intake, and in contrast to the above-mentioned investigation, clinically relevant parameters were assessed. Since bromelain is known to exhibit anti-inflammatory as well as immunomodulating activities, the release of various cytokines was monitored in healthy humans after a single intake of bromelain or placebo. Subsequently, a significant shift of the circadian profiles of interferon gamma and trends for interleukin-5 and interleukin-10 were observed even after a single intake of bromelain (Muller et al. 2013).

Taken together, these studies unequivocally demonstrate that orally administered bromelain causes a systemic effect, either by reaching the blood circulation [suggested by (Castell et al. 1997)] or by triggering an indirect cascade that finally ends in a systemic (proteolytical, immunological etc.,) effect. Possible targets for such a trigger could be so-called protease-activated receptors (Borrelli et al. 2011; Reddy and Lerner 2010) or as would seem logical, cells from the mucosa associated lymphatic tissue (Muller et al. 2013).

Recently, a bromelain-specific peptide (DYGAVNEVK) was identified by LC-MS/MS analysis in plasma of mice treated intraperitoneally with bromelain. This unique peptide was proposed to serve as a suitable biomarker for a future standardization of the therapeutic dose (Secor et al. 2012).

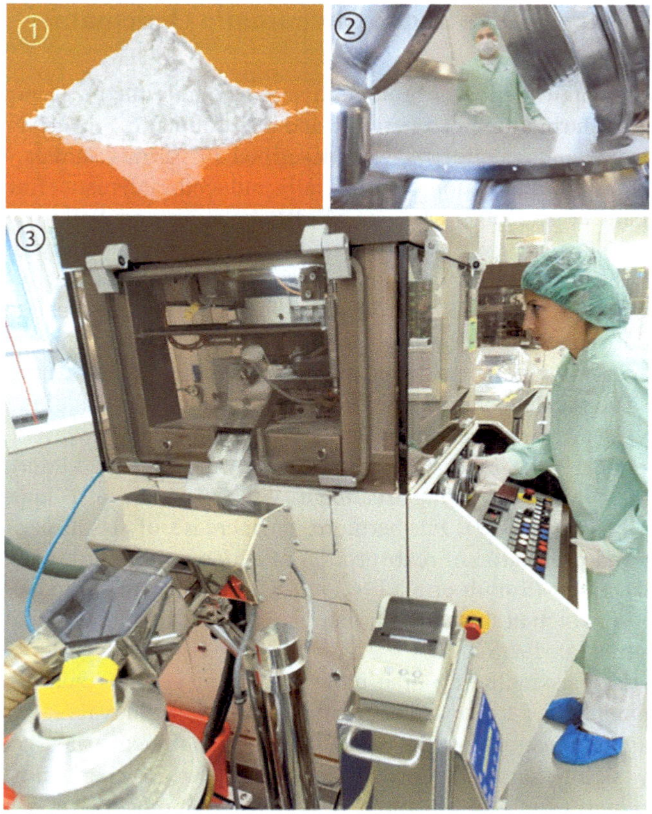

Fig. 2 Bromelain base powder is pressed into tablets. For protection against denaturation bromelain tablets should carry a gastro-resistant coating (pictures from manufacturing facilities kindly provided by B. Roeder, URSAPHARM Arzneimittel GmbH)

Hence, there remains the question of how bromelain could be absorbed across the gut wall (if absorption is indeed necessary for its medicinal function according to the above-mentioned observations). *In vitro* studies using Caco-2 cell monolayers have been performed by measuring the changes of epithelial barrier function. The studies indicate that epithelial electrical resistance decreases time- and concentration-dependent after application of bromelain. Furthermore, transport of macromolecular markers over the cell barrier was significantly enhanced. Thus uptake of orally administered enzymes might indeed occur by *self-enhanced paracellular transport* (Kolac et al. 1996).

While such issues may be quite enthralling from the scientific point of view, even more important are the effects and benefits of bromelain as far as its medicinal uses are concerned. These more applied aspects will be addressed in the following sections (Fig. 2).

4 Acute Postoperative and Posttraumatic Oedemata

The anti-oedematous action of bromelain is presumably the most important therapeutic benefit during the postoperative and posttraumatic convalescence period. This pharmacodynamic effect is well described in various studies, which generally focus on the alleviation of artificially induced oedemata (Enomoto et al. 1968; Gaspani et al. 2002; Moss et al. 1963; Uhlig and Seifert 1981). Here, bromelain showed an even stronger oedema-protective efficacy as compared to drugs such as acetyl-salicylic acid or indometacin (Netti et al. 1966). An oedema-protective and oedema-reducing effect is clinically desirable, particularly in case of acute posttraumatic and postoperative oedemata, and is therefore an important parameter in many of the clinical studies performed with bromelain. The mechanism on which this anti-oedematous effect is based is not yet entirely clear. The proteolytic activity of bromelain, however, is essential (Enomoto et al. 1968; Shigei et al. 1967) and lowering of the kininogen concentration in the plasma following treatment with bromelain also appears to play a role (Ohishi et al. 1979). Furthermore, an increase of tissue permeability by fibrinolysis and promotion of reabsorption of oedema fluid into blood circulation has been discussed (Smyth et al. 1962) as well as the induction of an anti-protease response which at the site of inflammation may counteract local proteolytic activity (Netti et al. 1966).

A meta-analysis including nine controlled clinical trials addressed the efficacy of bromelain for treatment of traumata. An overall advantage of bromelain therapy was demonstrated in all studies, and five clinical trials could prove a statistically significant reduction of oedemata. Two further studies observed statistically significant effects on other clinical parameters such as reduction of hematoma. Thus, taken together, efficacy of bromelain on the protection and reduction of oedemata today can be considered as scientifically proven (van Eimeren 1994).

To evaluate the extent of an oedema or the anti-oedematous effect of a drug, respectively, is indeed quite challenging. Some studies use pragmatic approaches, e.g., by measuring the distance of two (or more) fixed points. Using the latter method, by measuring the trago-pogonion distance in patients before as well as at day 1, day 3, day 5, and day 7 after tooth exodontia clearly demonstrated the effectiveness of bromelain in this particular indication (Inchingolo et al. 2010). More accurate methods, however, are needed in order to ascertain such effects and to avoid ‘false-negative’ results. For example, measuring a distance between two fixed points will indeed overlook swellings that are not in close proximity to the measured line and in this case, a heretical effect will be underestimated (or not seen at all). Sophisticated approaches thus will aim at the three-dimensional capture of the swollen tissue and its surroundings, e.g., by using ultrasound or optical scanning techniques.

5 Joint Disorders

Joints are built up from complex structures of connective tissue (borne, cartilage, synovia, capsules, and ligaments) that all together form a functional unit. A metabolic imbalance of one structure always has an impact on neighboring structures that finally leads to impairment of the overall function of the joint.

Joint diseases have differing pathogeneses and prognoses: by way of example, while arthritis is characterized by a chronical inflammation of the joint tissue, arthrosis (osteoarthritis) is hallmarked by mechanical degeneration caused by overweight, job- or sports-related overstressing, and structure defects of the musculoskeletal system. On the other hand, symptoms like swellings or inflammatory processes—be it causative to the disease or established secondary during the course of disease progression—occur in both disease pathologies. Bromelain has both anti-oedematous and anti-inflammatory properties, and therefore a therapy with bromelain was investigated in the context of both diseases. Indeed, an improvement of symptoms was reported both for the treatment of rheumatoid arthritis (Cohen and Goldman 1964) and arthrosis (Brien et al. 2004).

Meanwhile an early and 'harsh' intervention in patients suffering from rheumatoid arthritis has been established in therapy (e.g., by the use of methotrexate, azathioprine, cyclosporine, or TNF-α-inhibitors). With significant improvements in patient prognosis, conventional therapies are limited for the treatment of degenerative arthrosis. Therefore bromelain appears to be a more reasonable therapeutic option for the supportive treatment of arthrosis, in particular with regard to the potential for reduction in the use of non-steroidal anti-inflammatory drugs.

Bromelain's anti-inflammatory action is well described and has been confirmed by new scientific evidence (Yuan et al. 2006). Various studies indicate a direct influence on immunologically important cell types (CD4+ and CD8+ T lymphocytes, CD4+/CD8+ T-cell ratio) and changes in the cytokine pattern (Hale et al. 2005; Huang et al. 2008; Muller et al. 2013; Ogino et al. 1996; Secor et al. 2005, 2008). Influencing the pattern of cell surface markers of peripheral blood lymphocytes, monocytes, and granulocytes relates to changes in cellular adhesion and activation. Anti-inflammatory effects of bromelain therefore appear to be based at least in part on changes in leucocyte migration and activation (Fitzhugh et al. 2008; Hale et al. 2002).

Furthermore, bromelain appears to impact on the plasma kallikrein system and thus seems to be able to inhibit the production of the pro-inflammatory bradykinin (Kumakura et al. 1988) and to interfere with the arachidonic acid cascade by reducing the levels of prostaglandin E_2 and thromboxane A_2 (Vellini et al. 1986). So far, the mechanisms of action of the effects described are not fully understood. Cleavage of cell surface proteins could trigger intracellular cascades that lead to the various documented effects, and interference with the anti-protease system might play a role as well. Notably, bromelain acts at several levels against the inflammatory process, and rather by modulating than by suppressing the immune function (Berg et al. 2005; Muller et al. 2013).

6 Rhinosinusitis

Inflammations of the nose and the paranasal sinuses are usually associated with viral infections. Only about 0.5–10 % of acute upper respiratory tract infections are caused or complicated by bacterial infections, thus antibiotic therapy will not be beneficial in most patients suffering from these infections (Fahey et al. 1998; Leggett 2004). While acute rhinosinusitis can be considered as a more severe form of a common cold infection (more severe and prolonged symptoms), a chronic rhinosinusitis is characterized through immunological and morphological changes such as mucosal basement membrane thickening, goblet cell hyperplasia, subendothelial oedema, and increase in neutrophils, eosinophils, mast cells, and basophils (Fokkens et al. 2007).

Herbal medicines are frequently used for treatment of nasal and paranasal infections, and 32 % of patients with chronic rhinosinusitis have used herbal therapies alone or as adjunctive treatment for their condition. A recent assessment of several phytotherapeutics with regard to their efficacy in treatment of rhinosinusitis has come to the conclusion that the overall evidence for an effectiveness of herbal medicines is limited. Concerning bromelain, by contrast, this particular review states that adjunctive use of this drug significantly improves some symptoms of acute rhinosinusitis (Guo et al. 2006). Furthermore, taking a look at the pathophysiological changes occurring in case of chronic forms of rhinosinusitis, bromelain appears to be an ideal candidate for the treatment of this complex condition. In particular, decongesting effects on subendothelial edema, the anti-inflammatory mode of action and modulation of the immunological imbalance might contribute to an improvement of symptoms and a slow-down of disease progression. Indeed, a recent pilot test with bromelain in patients with chronic rhinosinusitis, reporting on a treatment period of three months, showed improvement in total symptom scores and sinonasal outcome test (patient's quality of life) as well as in an objective evaluation of the nasal mucosa by a qualified physician (Buttner et al. 2013). While results available currently are promising, further clinical trials will obviously be necessary to substantiate the efficacy of bromelain in the treatment of rhinosinusitis.

7 Cancer

Bromelain acts on multiple cellular and molecular targets, as pointed out above. Several pathophysiological pathways important for carcinogenesis and cancer malignancy are or may be influenced by the treatment with bromelain. Even though the overall knowledge and evidence for use of bromelain in cancer therapy is limited, a number of physicians recommend bromelain as a treatment measure, and of course the Internet serves as a broad platform for the presentation of potential health benefits. It should be stated that bromelain should not be used as a substitute for conventional therapy but rather may be considered as an *adjuvant* measure in certain types of cancer. The current evidence and rationale for the use of bromelain in this indication is presented and discussed in the following.

An inflammatory microenvironment promotes the development of tumors, aids in proliferation and survival of malignant cells, and promotes angiogenesis and metastasis, among others (Mantovani et al. 2008). Pivotal players in cancer-related inflammation are NF-κB and *cox*-2 as one of its target genes as well as TGF-β. Downregulation of both NF-κB and COX-2 by treatment with bromelain could be demonstrated in skin tumors and human monocytic leukemia cell lines (Bhui et al. 2009, 2012; Huang et al. 2008; Kalra et al. 2008). TGF-β expression was also reduced in patients with elevated TGF-β blood levels after oral administration of bromelain, meanwhile basal TGF-β levels were not affected in healthy volunteers (Desser et al. 2001). Reducing elevated TGF-β levels might decrease cancer-induced immune suppression and inflammation and reactivate the natural immune response. Immunomodulating effects might further activate the healthy immune system to ensure an adequate response to pathogens and cellular stresses (Chobotova et al. 2010). Oral application of bromelain to breast cancer patients stimulated the deficient monocytic cytotoxicity, which may indirectly contribute to the proposed antitumor activity (Eckert et al. 1999).

As described before, bromelain's impact on the immune system is presumably *inter alia* mediated by cleavage of cell surface markers. CD44 plays an important role among these markers in cancer progression, since this adhesion molecule is directly involved in cancer growth and metastasis. Bromelain preferentially cleaves off CD44 (Eckert et al. 1999; Harrach et al. 1994) and thus might contribute to the inhibition of one of the first steps of the metastatic process. Influencing the tumor's microenvironment might moreover facilitate attack by cells of the immune system and increase the efficacy of chemotherapeutic drugs (Pillai et al. 2013). Stimulation of neutrophils to produce Reactive Oxygen Species (ROS) might further increase tumor cell killing properties of the immune system (Brakebusch et al. 2001). Additionally, bromelain-induced depletion of intracellular glutathione and generation of ROS in skin tumor cell lines has been described (Bhui et al. 2012). The therapeutic concept of controlling and modulating intracellular levels of ROS is elaborated in detail in various chapters of this book and represents one of its central themes.

In several studies applying different cancer cell lines, a significant reduction of cell growth and activation of apoptosis could be observed. Among the effects observed were activation of the caspase system, increase in expression of p53 and Bax as well as a decrease in activity of cell survival regulators such as Akt, Erk, and Bcl-2 (Amini et al. 2013; Chobotova et al. 2010; Dhandayuthapani et al. 2012; Pavan et al. 2012).

Finally, inhibition of an overshooting activity of endogenous proteases, such as cathepsins, via the induction of antiproteinase production might contribute to antitumoral activity. Alternatively, an induction of anti-bromelain-antibodies which may also inhibit the activity of cathepsins has been proposed with regard to the protective effects observed for bromelain (Baez et al. 2007).

Thus, taken together, bromelain appears to exhibit very promising properties for the adjuvant treatment of cancer. More research and in particular large randomized controlled clinical trials are required, however, in order to substantiate the scientific basis for its application. Direct effects on cancer cells, such as the induction of apoptosis, will have to be scrutinized in a clinical setting, and susceptibility of different cancer cell types must be evaluated in more detail. On the other hand, interference with cell surface markers (CD44) and immunomodulating effects were reported even after oral use and thus raise hopes for the further development of new approaches for cancer treatment involving bromelain (Eckert et al. 1999; Muller et al. 2013).

And last but not least, well-documented biological activities of bromelain provide plausible reasoning for its use as part of an adjuvant therapy. In this context, anti-oedematous effects and the anti-inflammatory mode of action might decrease side effects of cancer therapy, such as lymphedema and mucositis, and thus increase adherence to and efficiency of cancer therapy.

8 Safety

Bromelain is considered to be non-toxic and in general as a safe drug. The U.S. Food and Drug Administration graded bromelain as food additive that is 'generally recognized as safe' (FDA 2001). Indeed, besides being used as active pharmaceutical ingredient, bromelain enjoys great popularity as a meat tenderizer. No LD_{50} could be determined in several animal species. Oral doses of up to 10 g/kg, and doses up to 750 mg/kg per day over a six month period did not result in any observable toxic effects in dogs (Taussig et al. 1975). Based on literature data, orally taken bromelain is well tolerated even in high doses of up to 3 g per day over a long period (Brien et al. 2004; Castell et al. 1997; Taussig and Batkin 1988). Possible side effects include gastrointestinal disorders such as stomach trouble and diarrhea and allergic reactions which might manifest themselves as eczema and asthma-like symptoms. Patients with a known intolerance against pineapple should therefore not take bromelain-containing supplements or medications.

Since effects on blood coagulation parameters have been described for bromelain (Maurer 2001) potential interactions with anticoagulants and thrombocyte aggregation inhibitors should be taken into consideration. Therapeutically used doses of up to 3.000 F.I.P. units per day over a ten day period, however, did not significantly affect blood coagulation parameters (Eckert et al. 1999) and recent investigations on the effects of peri- or postoperative combination therapy with low molecular weight heparins and bromelain in patients that underwent cruciate ligament- or coxarthrosis surgery did not indicate an increased risk of hemorrhage either (Johann et al. 2011). Thus, the clinical relevance of this potential interaction demands further exploration.

9 Conclusions and Outlook

In this chapter, bromelain, a crude multi-component extract isolated from the stems of the pineapple plant, took us on a tour through rather unusual therapeutic terrain and pharmacological mechanisms. Bromelain acts on several targets that address

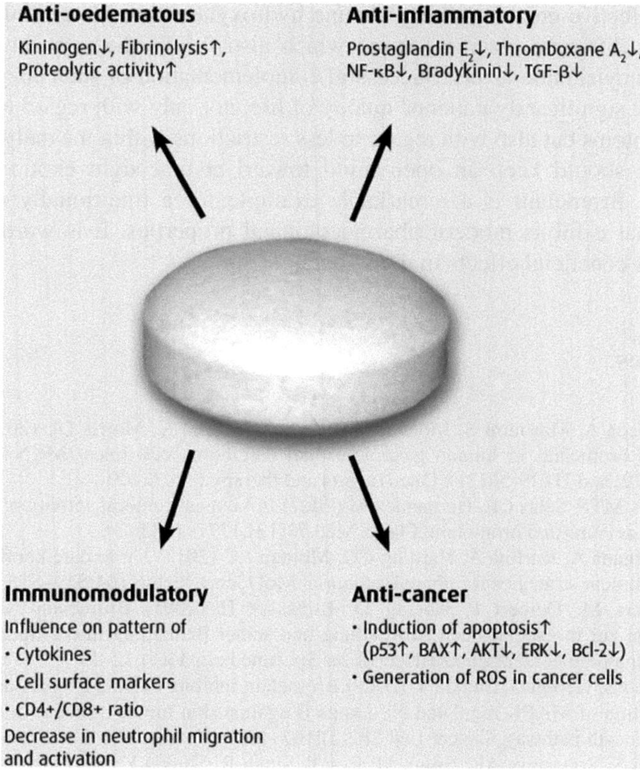

Fig. 3 Pharmacological effects described for bromelain

oedematous, inflammatory and immunological processes. Therefore many indications have been treated with or are conceivable to be treated by use of this phytomedicine. Indeed, more questions may have been raised than answered during this journey, and many of these questions are unlikely to be answered in the near future. On the other hand, the same holds true for complex disease conditions, in which neither the pathogenesis nor important factors for disease progression are comprehensively understood. Therefore it appears to be essential to continue (a) to use the potential of natural products that are well established in medical therapy for decades or even centuries and (b) to evaluate why and how these phytotherapeutics exert their beneficial effects. The latter is important not only for herbal medicine research but also for the design of novel drugs in general. Furthermore, there seems to be a considerable potential for the use of enzymes in therapy. For example, enzyme defects that result in maldigestion of nutrients and accumulation of toxic intermediate products may be complemented by supplementation of these enzymes. One example for such an approach is the treatment of patients with phenylketonuria with (PEGylated) phenylalanine ammonia lyase (Belanger-Quintana et al. 2011). In this case, the defective enzyme (phenylalanine hydroxylase) is not itself supplemented but rather a different enzyme is used which also fulfills the relevant function to decrease phenylalanine levels. Successful complementation of such enzyme defects will enhance significantly patients' quality of life, not only with regard to the reduction of symptoms but also with regard to less restrictions within the daily diet.

Thus we should keep an open mind toward at first sight exotic therapeutic approaches. Bromelain is a remarkable example for a traditionally used herbal medicine that exhibits modern pharmacological properties. It is worth to further elaborate its beneficial effects in daily therapy (Fig. 3).

References

Amini A, Ehteda A, Masoumi S, Moghaddam, Akhter J, Pillai K, Morris DL (2013) Cytotoxic effects of bromelain in human gastrointestinal carcinoma cell lines (MKN45, KATO-III, HT29-5F12, and HT29-5M21). OncoTargets and therapy 6:403–409

Baez R, Lopes MTP, Salas CE, Hernandez M (2007) In vivo antitumoral activity of stem pineapple (*Ananas comosus*) bromelain. Planta Med 73(13):1377–1383

Belanger-Quintana A, Burlina A, Harding CO, Muntau AC (2011) Up to date knowledge on different treatment strategies for phenylketonuria. Mol Genet Metab 104:S19–S25

Berg A, Peters M, Deibert P, Koenig D, Birnesser H (2005) Bromelain- Übersicht und Diskussion zur therapeutischen Anwendung und seiner Bedeutung in der Sportmedizin und Sporttraumatologie. Deutsche Zeitschrift für Sportmedizin 56(1):12–19

Bhui K, Prasad S, George J, Shukla Y (2009) Bromelain inhibits COX-2 expression by blocking the activation of MAPK regulated NF-kappa B against skin tumor-initiation triggering mitochondrial death pathway. Cancer Lett 282(2):167–176

Bhui K, Tyagi S, Srivastava AK, Singh M, Roy P, Singh R, Shukla Y (2012) Bromelain inhibits nuclear factor kappa-B translocation, driving human epidermoid carcinoma A431 and melanoma A375 cells through G2/M arrest to apoptosis. Mol Carcinogen 51(3):231–243

Borrelli F, Capasso R, Severino B, Fiorino F, Aviello G, De Rosa G, Mazzella M, Romano B, Capasso F, Fasolino I, Izzo AA (2011) Inhibitory effects of bromelain, a cysteine protease derived from pineapple stem (*Ananas comosus*), on intestinal motility in mice. Neurogastroent Motil 23(8):745–E331

Brakebusch M, Wintergerst U, Petropoulou T, Notheis G, Husfeld L, Belohradsky BH, Adam D (2001) Bromelain is an accelerator of phagocytosis, respiratory burst and killing of candida albicans by human granulocytes and monocytes. Eur J Med Res 6(5):193–200

Brien S, Lewith G, Walker A, Hicks SM, Middleton D (2004) Bromelain as a treatment for osteoarthritis: a review of clinical studies. Evid-Based Compl Alt 1(3):251–257

Buttner L, Achilles N, Bohm M, Shah-Hosseini K, Mosges R (2013) Efficacy and tolerability of bromelain in patients with chronic rhinosinusitis—a pilot study. B-Ent 9(3):217–225

Castell JV, Friedrich G, Kuhn CS, Poppe GE (1997) Intestinal absorption of undegraded proteins in men: presence of bromelain in plasma after oral intake. Am J Physiol-Gastr L 273(1):G139–G146

Chobotova K, Vernallis AB, Majid FAA (2010) Bromelain's activity and potential as an anti-cancer agent: current evidence and perspectives. Cancer Lett 290(2):148–156

Cohen A, Goldman J (1964) Bromelains therapy in rheumatoid arthritis. Pennsylvania Med J 67:27–30

Desser L, Holomanova D, Zavadova E, Pavelka K, Mohr T, Herbacek I (2001) Oral therapy with proteolytic enzymes decreases excessive TGF-beta levels in human blood. Cancer Chemoth Pharm 47:S10–S15

Dhandayuthapani S, Perez HD, Paroulek A, Chinnakkannu P, Kandalam U, Jaffe M, Rathinavelu A (2012) Bromelain-induced apoptosis in GI-101A breast cancer cells. J Med Food 15(4):344–349

Eckert K, Grabowska E, Stange R, Schneider U, Eschmann K, Maurer HR (1999) Effects of oral bromelain administration on the impaired immunocytotoxicity of mononuclear cells from mammary tumor patients. Oncol Rep 6(6):1191–1199

Enomoto T, Mineshita S, Shigei T (1968) Protective effect of stem bromelain against adrenaline pulmonary edema, and its dependence on the proteolytic activity. Jpn J Pharmacol 18(2):260–265

Fahey T, Stocks N, Thomas T (1998) Systematic review of the treatment of upper respiratory tract infection. Arch Dis Child 79(3):225–230

FDA (2001) Partial list of enzyme preparations that are used in foods (GRAS). Center for Food Safety and Applied Nutrition

Fitzhugh DJ, Shan SQ, Dewhirst MW, Hate LP (2008) Bromelain treatment decreases neutrophil migration to sites of inflammation. Clin Immunol 128(1):66–74

Fokkens W, Lund V, Mullol J (2007) European position paper on rhinosinusitis and nasal polyps 2007. Rhinology 20 1–136

Gaspani L, Limiroli E, Ferrario P, Bianchi M (2002) In vivo and in vitro effects of bromelain on PGE(2) and SP concentrations in the inflammatory exudate in rats. Pharmacology 65(2):83–86

Gilead L, Mumcuoglu KY, Ingber A (2012) The use of maggot debridement therapy in the treatment of chronic wounds in hospitalised and ambulatory patients. J Wound Care 21(2):78

Guo R, Canter PH, Ernst E (2006) Herbal medicines for the treatment of rhinosinusitis: a systematic review. Otolaryng Head Neck 135(4):496–506

Hale LP, Greer PK, Sempowski GD (2002) Bromelain treatment alters leukocyte expression of cell surface molecules involved in cellular adhesion and activation. Clin Immunol 104(2):183–190

Hale LP, Greer PK, Trinh CT, Gottfried AR (2005) Treatment with oral bromelain decreases colonic inflammation in the IL-10-deficient murine model of inflammatory bowel disease. Clin Immunol 116(2):135–142

Harrach T, Eckert K, Schulzeforster K, Nuck R, Grunow D, Maurer HR (1995) Isolation and partial characterization of basic proteinases from stem bromelain. J Protein Chem 14(1):41–52

Harrach T, Gebauer F, Eckert K, Kunze R, Maurer HR (1994) Bromelain Proteinases Modulate the Cd44 Expression on Human Molt-4/8 Leukemia and Sk-Mel-28 Melanoma-Cells in Vitro. Int J Oncol 5(3):485–488

Heinicke R, Gortner W (1957) Stem bromelain, a new protease preparation from pineapple plants. Econ Bot 11:225–234

Huang JR, Wu CC, Hou RCW, Jeng KC (2008) Bromelain inhibits lipopolysaccharide-induced cytokine production in human THP-1 monocytes via the removal of CD14. Immunol Invest 37(4):263–277

Inchingolo F, Tatullo M, Marrelli M, Inchingolo AM, Picciariello V, Inchingolo AD, Dipalma G, Vermesan D, Cagiano R (2010) Clinical trial with bromelain in third molar exodontia. Eur Rev Med Pharmaco 14(9):771–774

Johann K, Eschmann K, Meiser P (2011) Keine klinische Evidenz für ein erhöhtes Blutungsrisiko durch Bromelain bei perioperativem Einsatz. Sportverl Sportschad 25(2):108–113

Kalra N, Bhui K, Roy P, Srivastava S, George J, Prasad S, Shukla Y (2008) Regulation of p53, nuclear factor KB and cyclooxygenase-2 expression by bromelain through targeting mitogen-activated protein kinase pathway in mouse skin. Toxicol Appl Pharm 226(1):30–37

Kolac C, Streichhan P, Lehr CM (1996) Oral bioavailability of proteolytic enzymes. Eur J Pharm Biopharm 42(4):222–232

Kumakura S, Yamashita M, Tsurufuji S (1988) Effect of bromelain on kaolin-induced inflammation in rats. Eur J Pharmacol 150(3):295–301

Leggett JE (2004) Acute sinusitis: when-and when not-to prescribe antibiotics. Postgrad Med 115(1):13–19

Mantovani A, Allavena P, Sica A, Balkwill F (2008) Cancer-related inflammation. Nature 454(7203):436–444

Maurer HR (2001) Bromelain: biochemistry, pharmacology and medical use. Cell Mol Life Sci 58(9):1234–1245

Menon V, Harrington RA, Hochman JS, Cannon CP, Goodman SD, Wilcox RG, Schunemann HJ, Ohman EM (2004) Thrombolysis and adjunctive therapy in acute myocardial infarction. Chest 126(3):549s–575s

Miller JM, Opher AW (1964) Increased proteolytic activity of human blood serum after oral administration of bromelain. Exp Med Surg 22(4):277

Moss JN, Frazier CV, Martin GJ (1963) Bromelains. The pharmacology of the enzymes. Arch Int Pharmacodyn Ther 145:166–189

Muller S, Marz R, Schmolz M, Drewelow B, Eschmann K, Meiser P (2013) Placebo-controlled randomized clinical trial on the immunomodulating activities of low- and high-dose bromelain after oral administration new evidence on the antiinflammatory mode of action of bromelain. Phytother Res 27(2):199–204

Netti C, Bandi G, Pecile A (1966) Anti-inflammatory action of proteolytic enzymes of animal, vegetable or bacterial origin, administered orally compared with that of known antiphlogistic compounds. Il Farmaco Ed Pr 27(8):453–466

Ogino M, Majima M, Kawamura M, Hatanaka K, Saito M (1996) Increased migration of neutrophils to granulocyte-colony stimulating factor in rat carrageenin-induced pleurisy: roles of complement, bradykinin, and inducible cyclooxygenase-2. Inflamm Res 45(7):335–346

Ohishi S, Uchida Y, Ueno A, Katori M (1979) Bromelain, a thiolprotease from pineapple stem, depletes high molecular-weight kininogen by activation of hageman-factor (factor-XII). Thromb Res 14(4–5):665–672

Pavan R, Jain S, Shraddha, Kumar A (2012) Properties and therapeutic application of bromelain: a review. Biotechnol Res Int 2012:976203

Pillai K, Akhter J, Chua TC, Morris DL (2013) Anticancer property of bromelain with therapeutic potential in malignant peritoneal mesothelioma. Cancer Invest 31(4):241–250

Reddy VB, Lerner EA (2010) Plant cysteine proteases that evoke itch activate protease-activated receptors. Brit J Dermatol 163(3):532–535

Secor ER, Carson WF, Cloutier MM, Guernsey LA, Schramm CM, Wu CA, Thrall RS (2005) Bromelain exerts anti-inflammatory effects in an ovalbumin-induced murine model of allergic airway disease. Cell Immunol 237(1):68–75

Secor ER, Carson WF, Singh A, Pensa M, Guernsey LA, Schramm CM, Thrall RS (2008) Oral bromelain attenuates inflammation in an ovalbumin-induced murine model of asthma. Evid-Based Compl Alt 5(1):61–69

Secor ER, Szczepanek SM, Singh A, Guernsey L, Natarajan P, Rezaul K, Han DK, Thrall RS, Silbart LK (2012) LC-MS/MS Identification of a bromelain peptide biomarker from *Ananas comosus*. pp 1–10 Merr. Evid-Based Compl Alt.\

Seifert J, Ganser R, Brendel W (1979) Absorption of a proteolytic-enzyme originating from plants out of the gastrointestinal-tract into blood and lymph of rats. Z Gastroenterol 17(1):1–8

Shigei T, Sakuma A, Nishiwaki T (1967) A study on the protective effect of bromelain, crude pineapple proteases, against adrenaline-induced pulmonary edema in rats. Jpn Heart J 8(6):718–720

Smyth RD, Brennan R, Martin GJ (1962) Systemic biochemical changes following the oral administration of a proteolytic enzyme, bromelain. Arch Int Pharmacodyn Ther 136:230–236

Taussig SJ, Batkin S (1988) Bromelain, the enzyme complex of pineapple (*Ananas comosus*) and its clinical application. An update. J Ethnopharmacol 22(2):191–203

Taussig SJ, Yokoyama MM, Chinen A, Onari K, Yamakido M (1975) Bromelain: a proteolytic enzyme and its clinical application. A review. Hiroshima J Med Sci 24(2–3):185–193

Uhlig G, Seifert J (1981) The effect of proteolytic enzymes (traumanase) on posttraumatic edema. Fortschr Med 99(15):554–556

van Eimeren W (1994) Therapie traumatisch verursachter schwellungen. Thieme Verlag, Stuttgart, Deutschland

Vellini M, Desideri D, Milanese A, Omini C, Daffonchio L, Hernandez A, Brunelli G (1986) Possible involvement of eicosanoids in the pharmacological action of bromelain. Arzneimittel-Forsch 36(1):110–112

Yuan G, Wahlqvist ML, He G, Yang M, Li D (2006) Natural products and antiinflammatory activity. Asia Pacific J Clinical Nutr 15(2):143–152

Authors Biography

Peter Meiser (born 1977) has been trained as a pharmaceutical scientist at the University of Saarland, Saarbruecken, Germany, spending one year abroad at the University of Mary Washington, in 2000. In 2003 he obtained his degree in Pharmacy and in 2008 his PhD in Pharmaceutical Biotechnology (under the supervision of Prof. Rolf Mueller). Peter also holds two certificates of advanced training as pharmaceutical specialist for pharmaceutical analytics and drug information.

Peter is Head of the Medical-Scientific Department in General Medicine at URSAPHARM Arzneimittel GmbH. He has recently participated as industrial supervisor and as a member of the supervisory board in the "RedCat" Network and currently coordinates a research project on Bromelain publicly sponsored by the German Federal Ministry of Education and Research. Peter has published several papers on the biochemistry and pharmacological activity of natural products. Besides his commitment to this field of more basic research, he also regularly supervises diverse clinical trial projects.

Zhanjie Xu (born 1984) received his Bachelor degree in Chemical Engineering and Technology from Qingdao University of Science and Technology, China, in 2006. He then obtained a Master degree in Organic Chemistry from the University Paul Verlaine, Metz, France. He subsequently joined URSAPHARM Arzneimittel GmbH as an Early Stage Researcher of the EU Marie Curie Initial Training Network "RedCat". Zhanjie is just about to complete his PhD at the University of Lorraine under the supervision of Prof. Gilbert Kirsch.

Gilbert Kirsch (born 1947) has been trained as an organic chemist at the Universities of Strasbourg and Metz. He started his academic career in 1973 at the University of Metz (now University of Lorraine) where he currently holds the position of Professor of Organic Chemistry. He has been a postdoc at Oak Ridge National Laboratory (TN) in the Nuclear Medicine Group and was also an invited scientist at Kodak (Rochester, NY) as well as invited professor at the University of Minho (Portugal) and Emory University (Atlanta, GA).

Gilbert's interests reside in heterocyclic chemistry, especially in the field of five-membered aromatic systems (thiophenes, selenophenes, tellurophenes, thiazoles, selenazoles) and their benzo-condensed derivatives. Lately, he developed synthetic work in the field of coumarins, looking at biological activities (CDC25 phosphatase inhibition). From his research, he published over 200 papers, wrote different chapters in books, like Patai's Functional Group Series, in Houben-Weyl, in Chemistry of Heterocyclic Compounds (J. Wiley Interscience) and in Springer's Selenium and Tellurium Chemistry. He holds also few patents in the field of heart imaging and sulfur-containing tire additives. Gilbert has coordinated an Interreg program (acronym "Corena") on natural compounds for Medicine and Agriculture, and has participated to the EU ITN Marie Curie program "RedCat". He is participating in the regional programs Bioprolor and Biocaptech and in a national French ANR program on para-hydrogen.

Claus Jacob (born 1969) has been trained as a synthetic (in)organic and biological chemist at the Universities of Kaiserslautern, Leicester, Oxford, and Harvard. He graduated with a first class B.Sc. (Hons.) degree from the University of Leicester in 1993, and with a D.Phil. from the University of Oxford in 1997 ("Genetic engineering of redox active enzymes", supervisor Prof. Allen Hill FRS). He subsequently joined the institute of Prof. Bert Vallee at Harvard Medical School as a Feodor Lynen Fellow (Alexander von Humboldt-Foundation) to study processes controlling intracellular zinc homeostasis. During this time, he also obtained a Magister Artium degree in Philosophy, History, and Psychology from the University of Hagen in Germany (M.A. dissertation on Protochemistry as constructivist foundation of chemistry). He left the US in 1999 to spend some time with Prof. Helmut Sies at the Heinrich-Heine-University in Duesseldorf, Germany, as part of a BASF Research Fellowship from the German Merit Foundation.

Claus started his independent scientific career as lecturer at the University of Exeter in the UK in 1999 and in 2005 moved to the University of Saarland where he currently holds the position of Professor of Bioorganic Chemistry. Claus is an expert in redox active compounds and their impact on biological systems and to date has published over 100 publications in this field. Over the years, his research has focussed on Reactive Sulfur Species (RSS) and the cellular thiolstat, terms his team has introduced in 2001 and 2010, respectively. Besides his strong interest in redox active sulfur, Claus has also developed an active research program on synthetic 'sensor/effector' redox modulators based on selenium and tellurium, on redox active plant metabolites, and on nanoscopic redox particles. His research includes synthetic and analytical chemistry, biological activity studies, and 'intracellular' diagnostics to decipher and map out intracellular events and mode(s) of actions. Claus has coordinated the EU Marie Curie Initial Training Network

“RedCat” (2008–2012), has been a partner in the technology transfer project “Corena” (2009–2012), and is currently in charge of the natural products project “NutriOx”.

Throughout the years, Claus has undertaken many projects to become a highly skilled undertaker, but never a true philosopher, yet his more philosophical and cunning linguistic outpourings are famous and he still maintains a keen interest in various aspects related to the philosophy of chemistry.

Part XVIII
Connecting Section Between Chapters 18 and 19

We have now reached the final chapter of this book, which will provide a brief summary and some perspectives on redox active secondary metabolites, in research as well as product development, as part of existing as well as likely future applications. Here, we will take account of a rather diverse field of redox active compounds and also enzymes, and of applications, which range from food supplements to agricultural and industrial uses. As before, we will place our focus on the unusual and exciting without necessarily ignoring the more trivial and obvious. It is the declared aim of this final chapter to raise interest, to stimulate debate, and above all to encourage young researchers to join the ongoing and future efforts to harvest the 'green gold' provided by plants, fungi, and bacteria in a responsible and scientifically sound manner. Once more, we emphasize our belief that redox active secondary metabolites are a lot more—and more powerful—than just antioxidants and food supplements. Besides the many unconventional applications described in the previous chapters, we also call as witness the various small- and medium-sized enterprises, which have successfully explored the facet-rich spectrum of such metabolites, their processing, application, and commercialization. Here, the close relationship between academic research on the one hand, and commercial product development on the other has often been the bedrock of success. In return, the success of these commercial products is slowly, but certainly moving the field of natural product research back to the center of rigorous scientific investigation, and relieves the researchers involved in this field from traditional suspicions of hocus-pocus, which have been raised against them quite frequently during the last decade or two.

In the end, we will close the final chapter and the book with a few personal views of the editors. Here, each of them will provide his own prediction of the most exciting developments, which in his opinion are likely to occur in the field of redox active secondary metabolite research during the next ten years or so. Only time can tell who ultimately will have been right and who will have been wrong.

Chapter 19
Some Personal Conclusions

Claus Jacob, Torsten Burkholz, Gilbert Kirsch, Alan Slusarenko and Paul G. Winyard

After a firework display of molecules, biological activities, research, and product development it is now time to conclude, to take stock, and look ahead at the developments we are likely to face in this field in the next decade or two. Here, it is worthwhile to recall the beginnings of this book, and to remember our intention to reflect the various research interests of the EU "RedCat" project consortium. Indeed, when confronted with the idea of a book on this topic, our first thought was "oh no, not yet another book on antioxidants". Luckily, however, one of the editors at the time was—and apparently still is—not just a researcher in this field, but also a passionate practitioner. Rumor has it that this particular editor is actually addicted to such substances, and probably has already tried most of them on himself. Thanks to him, we were therefore literally able to smell the vast and diverse range of redox active secondary metabolites from numerous sources, and with distinct effects on the body. This encouraged us to move ahead with the book project and focus on some of the more exotic products and applications, and to sniff out any emerging compounds, activities, or emerging fields of research (such as nanocrystals) that may ultimately bear considerable promise in research as well as in the form of concrete product development.

C. Jacob (✉) · T. Burkholz
Division of Bioorganic Chemistry, School of Pharmacy, Universitaet des Saarlandes, 66123 Saarbruecken, Germany
e-mail: c.jacob@mx.uni-saarland.de

G. Kirsch
SRSMC UMR 7565 (former Laboratoire d'Ingénierie Moléculaire et Biochimie Pharmacologique), Université de Lorraine, 57070 Metz, France

A. Slusarenko
Department of Plant Physiology, Molecular Phytopathology, Institut fuer Biologie III, RWTH Aachen, 52056 Aachen, Germany

P. G. Winyard
University of Exeter Medical School, St. Luke's Campus, Exeter EX1 2LU, UK
e-mail: p.g.winyard@exeter.ac.uk

C. Jacob et al. (eds.), *Recent Advances in Redox Active Plant and Microbial Products*, DOI: 10.1007/978-94-017-8953-0_19, © Springer Science+Business Media Dordrecht 2014

Hence, one of the primary aims of this book has been to illustrate the enormous diversity and potential which is associated with many aspects of redox active secondary metabolites. Indeed, 'diversity' is the correct category when approaching such substances. As the various chapters have shown, a vast diversity exists when it comes to the natural sources of such metabolites, which can be found in a multiplicity of organisms, ranging from simple bacteria rich in bacterial thiols to plants armed with defense systems relying on sophisticated two-component substrate/enzyme warheads. Diversity also exists, of course, with respect to the chemical structure and physicochemical properties of such metabolites. Here, we have indulged in exciting and often slightly exotic chemical molecules, including lesser known thiosulfinates, polysulfanes, oligomeric flavonoids, and synthetic hybrid conjugates. We have even learned that cows not only give milk but also give rise to some processed metabolites, such as phytols. Diversity also applies to the vast spectrum of biological activities which are firmly linked to many of the compounds considered, including cardiovascular protection, induction of apoptosis in cancer cells, and antimicrobial activities. In fact, one of the primary aims of this book has been to demonstrate that redox active secondary metabolites are not just 'antioxidants', but a lot more facet-rich in their activity and their potential applications. Indeed, diversity is also the key to commercial success, and with the help of several examples we have been able to illustrate that such a success in the highly competitive market of (processed) natural products indeed becomes possible once academic research and industrial development join forces to explore and subsequently exploit such products in a scientific and responsible manner.

Based on the recent progress in this field, especially in the context of technological advances and commercial applications, it is now possible to consider some of the developments likely to occur in the next couple of years. Here, several factors come together and need to be considered in earnest. First of all, the recent scientific momentum in this field, from bioanalytics and 'intracellular diagnostics' to nanotechnology, bodes well for future discoveries and ever more widespread and sophisticated applications of such natural products. This innovative climate will also ensure that rigorous research into redox active metabolites will become more widely accepted and lose some of its more unfortunate associations with Folk Medicine and make-believe. In fact, the demand for such natural products is likely to increase significantly during the next decade or two, as (a) these products will soon be in a position to substitute for synthetic and often dangerous chemicals, especially in the field of agriculture, and (b) redox active compounds, in a nutshell, are poised to address physiological redox imbalances which are a hallmark of an aging society. We will consider these two issues in more detail later.

Obviously, there are many other good reasons to consider the field of redox active metabolites in the (near) future. Applications in the field of cosmetics, weight control, and more balanced diets come to mind. We will therefore conclude this book with individual, and hence rather personal predictions by the editors of this book. In doing so, we will provide space for more speculative ideas, and also allow a closer look at this topic from different angles and corners of the spectrum of scientific disciplines that have contributed to this book. According to taste, we will be faced on our journey into the future with the personal predictions of Claus Jacob (who represents 'A' as in '*Allium*'), Torsten Burkholz (*engl.* Burkwood), Gilbert Kirsch (whose surname translates to 'cherry'), Alan Slusarenko (who appears to be into 'raspberries' these days) and, *nomen est omen*, Paul G. Winyard.

During the years, there have been several real and fictitious characters, who have stimulated our interest in redox active metabolite research. Together, these characters reflect various aspects of such metabolites and, probably unwillingly, have assisted us in pursuing our own research in this. Needless to say, these persons, real or fictitious, will continue to inspire and stimulate our research in this field in the future. The first one is "Bad Bob", a character played by Stacy Keach in the 1972 Western "The Life and Times of Judge Roy Bean", who almost exclusively feeds on whole raw onions. To us, "Bad Bob" reflects all the nutritional—and often highly beneficial—aspects associated with redox active secondary metabolites. Indeed, mild natural remedies, such as onion teas, are currently experiencing a certain renaissance as an alternative to serious medication, e.g., against common colds or slight gastrointestinal disagreements. Similar developments are likely in other fields of nutrition, for instance in the area of proanthocyanidine-based inhibitors of digestive enzymes in the gut and the potential uses of such products in 'medication-free' forms of weight control. And, of course, the demand for antioxidant preparations is likely to increase dramatically in the future, first because of the significant loss of the body's own antioxidant defenses in older age (an aspect important in chemoprevention in an aging society), and second because nanotechnology now provides the necessary means to render these antioxidants bio-available. In parallel, other aspects of chemoprevention, such as cardiovascular protection by resveratrol (in the form of red (nonalcoholic) wine or as food supplement) will gain prominence. An increased future demand is also likely regarding cosmetic applications of such products. Other nutritional aspects associated with this class of compounds will slowly but certainly emerge. Here, some pioneering research recently conducted on xanthohumol and its effects on epigenetic processes provides a short glimpse of the kind of research and development we may expect in this particular field in the near future. It should also be remembered—and this is yet another potential 'growth area' for natural products—that many compounds isolated from plants and bacteria may *act* on redox processes whilst not actually being redox active *per se*. For example, there is scope for the discovery of novel plant-derived inhibitors of ROS-generating enzymes such as NADPH oxidase, myeloperoxidase, and nitric oxide synthase.

Twenty years after the appearance of "Bad Bob" on the silver screen, his real-life counterpart, Eric Block from the University at Albany, published a landmark paper on Angewandte Chemie entitled "The Organosulfur Chemistry of the Genus *Allium* and Its Importance to the Organic Chemistry of Sulfur", which in many ways forms the bedrock of current research in the field of natural organosulfur compounds. Like "Bad Bob", Eric Block is well known to his friends and colleagues to relish *Allium*-rich food, from the classical leek-and-potato soup as a starter to garlic-flavored mussels as a main dish (with a shallot salad on the side), and the yet to be invented garlic-flavored ice crèmes for dessert. Importantly, his particular taste pioneers aspects of a balanced diet which is becoming ever more important, especially in Germany, and here among the older generation. In contrast to salt and other unhealthy seasonings currently used to spice up daily food in many European countries and the US, natural spices rich in 'beneficial' substances will provide a real alternative. Indeed, garlic, onions, shallots, and mustard are prime candidates to constitute and flavor food, and are currently being joined by curcuma (rich in curcumin) and grape seed flour (rich in proanthocyanidins). This trend in the higher (and lower cuisine) is likely to gather considerable steam within the next few years.

Coincidentally, 1992 was also the year when the film "Medicine Man" with Lorraine Bracco and Sir Sean Connery in the title role was released. Besides confirming the rather eccentric dress code typical of many natural product researchers, the story told in this film exemplifies many aspects of natural product research. Undeniably, the search for new therapeutic agents often leads to exotic places and equally exotic plants, whose biologically active ingredients bear considerable potential in drug development. Here, redox active metabolites, such as thiosulfinates, polysulfanes, and oligomeric flavonoids armed with numerous redox centers in one molecule, are likely to attract considerable attention in the future. Those compounds seem to exhibit a pronounced biological activity, often combined with an amazing selectively, and could provide new leads in the field of cancer research, against inflammatory diseases and microbial infections. The results obtained so far, for instance against cancer cells or in animal models of scleroderma, are extremely promising. Once followed up with the necessary vigor, these studies are likely to provide considerable new insights into redox modulation, intracellular redox responses, and potential applications. Importantly, most target organisms cannot evade the action of such compounds by simple resistance mechanisms. It therefore appears that certain organosulfur compounds may eventually be able to outpace the use of classical mitosis inhibitors (such as taxols, vinca alkaloids) in cancer research and the use of classical antimicrobial agents as novel anti-infectious treatments. At the same time, more sophisticated approaches to deliver or generate Reactive Sulfur Species *in situ* may be developed, for instance based on the two-component alliin/alliinase or glucosinolate/myrosinase substrate/enzyme systems.

The final defining moment, and its associated story, is related to a visit to an Armenian vintner a couple of years ago. There, massive amounts of processed, squeezed-out grapes could be found as apparent 'waste' on a gigantic heap in the backyard. Funnily, this mountain of waste, if processed further, was probably worth considerably more than all the wine in the tank. Here, scientific knowledge of the active ingredients contained within the squeezed-out grapes (such as resveratrol and diverse flavonoids in the skin, grape seed oil, and flour in the seeds), of their respective biological activities and of methods to *process* and *refine* such materials for potential applications, could easily have turned this heap of waste into a pile of gold. In the future, it is likely that considerably more attention will be paid to reaping the benefits of such agricultural 'waste products', especially in developing countries. Here, agricultural uses of such (processed, refined) materials, e.g., in the form of plant protective agents and 'green pesticides' are likely to complement the more medical, nutritional, or cosmetic applications mentioned before.

In any case, the future of redox active secondary metabolite research and product development is bright. We will witness new scientific insights, more widespread practical applications, and an ever expanding palette of suitable products based on these natural compounds. Importantly, we will probably also change our various attitudes to such products, from their daily consumption to a more responsible use of the natural sources and resources and our existing concept of plant 'waste'.

Editors Biography

Claus Jacob (born 1969) has been trained as a synthetic (in)organic and biological chemist at the Universities of Kaiserslautern, Leicester, Oxford, and Harvard. He graduated with a 1st class B.Sc. (Hons.) degree from the University of Leicester in 1993, and with a D.Phil. from the University of Oxford in 1997 ("Genetic engineering of redox active enzymes", supervisor Prof. Allen Hill FRS). He subsequently joined the institute of Prof. Bert Vallee at Harvard Medical School as a Feodor Lynen Fellow (Alexander von Humboldt-Foundation) to study processes controlling intracellular zinc homeostasis. During this time, he also obtained a Magister Artium degree in Philosophy, History, and Psychology from the University of Hagen in Germany (M.A. dissertation on Protochemistry as constructivist foundation of chemistry). He left the US in 1999 to spend some time with Prof. Helmut Sies at the Heinrich-Heine-University in Duesseldorf, Germany, as part of a BASF Research Fellowship from the German Merit Foundation.

Claus started his independent scientific career as lecturer at the University of Exeter in the UK in 1999 and in 2005 moved to the University of Saarland where he currently holds the position of Professor of Bioorganic Chemistry. Claus is an expert in redox active compounds and their impact on biological systems and to date has published over 100 publications in this field. Over the years, his research has focused on Reactive Sulfur Species (RSS) and the cellular thiolstat, terms his team has introduced in 2001 and 2010, respectively. Besides his strong interest in redox active sulfur, Claus has also developed an active research program on synthetic 'sensor/effector' redox modulators based on selenium and tellurium, on redox active plant metabolites and on nanoscopic redox particles. His research includes synthetic and analytical chemistry, biological activity studies and 'intracellular diagnostics' to decipher and map out intracellular events and mode(s) of actions. Claus has coordinated the EU Marie Curie Initial Training Network

C. Jacob et al. (eds.), *Recent Advances in Redox Active Plant and Microbial Products*,
DOI: 10.1007/978-94-017-8953-0, © Springer Science+Business Media Dordrecht 2014

"RedCat" (2008–2012), has been a partner in the technology transfer project "Corena" (2009–2012) and is currently partially in charge of the natural products project "NutriOx".

Throughout the years, Claus has undertaken many projects to become a highly skilled undertaker, but never a true philosopher, yet his more philosophical and cunning linguistic outpourings are famous and he still maintains a keen interest in various aspects related to the philosophy of chemistry.

Gilbert Kirsch (born 1947) has been trained as an organic chemist at the Universities of Strasbourg and Metz. He started his academic carrier in 1973 at the University of Metz (now University of Lorraine) where he currently holds the position of Professor of Organic Chemistry. He has been a postdoc at Oak Ridge National Laboratory (TN) in the Nuclear Medicine Group and was also an invited scientist at Kodak (Rochester, NY) as well as invited professor at the University of Minho (Portugal) and Emory University (Atlanta, GA).

Gilbert's interests include heterocyclic chemistry, especially in the field of five-membered aromatic systems (thiophenes, selenophenes, tellurophenes, thiazoles, selenazoles) and their benzo-condensed derivatives. Lately, he developed synthetic work in the field of coumarins, looking at biological activities (CDC25 phosphatase inhibition). From his research, he published over 200 papers, wrote different chapters in books, like Patai's Functional Group Series, in Houben-Weyl, in Chemistry of Heterocyclic Compounds (J. Wiley Interscience), and in Springer's Selenium and Tellurium Chemistry. He holds also a few patents in the field of heart imaging and sulfur-containing tire additives. Gilbert has coordinated an Interreg program (acronym "Corena") on natural compounds for Medicine and Agriculture, and has participated to the EU ITN Marie Curie programme "RedCat". He is participating in the regional programs Bioprolor and Biocaptech and in a national French ANR program on para-hydrogen.

Alan Slusarenko is Head of the Plant Physiology Department at RWTH Aachen University. His research has centered on resistance mechanisms of *Arabidopsis* to infection and more recently on Natural Products in Plant Protection. Alan obtained a PhD in Plant Pathology from Imperial College in 1981 and was a lecturer in the Department of Plant Biology at Hull University in the UK from 1983 until moving in 1988 to an Assistant Professorship in Molecular Plant Pathology at the University of Zuerich in Switzerland and subsequently in 1995 to the Chair of Plant Physiology at RWTH Aachen in Germany.

Paul Winyard is Professor of Experimental Medicine at the University of Exeter Medical School (formerly Peninsula College of Medicine and Dentistry), Exeter, UK, where he has been based since 2002. Previously, he held the same title at St Bartholomew's and the Royal London School of Medicine and Dentistry, London, and was a Visiting Professor at the University of California, San Francisco (2000–2001). Paul originally trained as a biochemist, and his current research interests center on the role of oxidative/nitrative stress and redox signaling in chronic inflammatory diseases such as rheumatoid arthritis. Paul leads an internationally recognized research group, having published over 200 research papers in the field of Oxidative Stress in inflammation. In particular, his research has focused on the development of novel free radical-related assays and therapeutic strategies, and the translation of these developments into clinical diagnostic assays and pre-clinical and early phase clinical studies.

Paul is a co-inventor in relation to seven patents, and is a Senior Editor of the *Journal of Inflammation*. He is also a member of the editorial boards of *Redox Biology*, *Frontiers in Oxidant Physiology*, *Frontiers in Inflammation*, *Current Pharmaceutical Design* and the *Open Inflammation Journal*. Paul is a committee member of the *British Inflammation Research Association* (*BIRAs*), and serves on a number of UK and European research grant awarding committees.

Torsten Burkholz (born 1979) has been trained as an inorganic and medicinal chemist at the University of Saarland in Saarbruecken, Germany, graduating with a German "Diplom" in chemistry. After completing his PhD studies at the University of Saarland in the fields of Chemistry and Pharmacy under the supervision of Prof. Claus Jacob in 2010, he joined the European Marie Curie Initial Training Network "RedCat" as postdoctoral Experienced Researcher, conducting research in the field of Cell Biology in the group of Prof. Dr. Paul G. Winyard at the Peninsula College of Medicine and Dentistry, Exeter, UK. In 2012 Torsten moved back to the University of Saarland where he currently holds the position of an "Akademischer Rat" in Bioorganic Chemistry.

As part of this senior position, Torsten is managing the research laboratory of Bioorganic Chemistry, as well as the relevant teaching and the examinations of undergraduate students. Together with Prof. Dr. Claus Jacob, he has to date published over 30 publications in the field of Oxidative Stress, chalcogen containing natural compounds and their biological activity and, more recently, on the cellular thiolstat. Torsten's ongoing research includes synthetic and analytical chemistry, biological activity studies and 'intracellular' diagnostics.

In 2013, Torsten was appointed as Visiting Professor at the University of Applied Sciences Kaiserslautern, where he is lecturing Pharmacology. In the same year, he established his own small company named "Dr. Burkholz Life Science Consulting UG" which provides scientific consultations for small and medium sized companies in the field of nutrition, food supplements and natural compounds. His company also offers training and consulting in Inorganic and Analytical Chemistry, as well as in Physics and Pharmacology.

Index

C. Jacob et al. (eds.), *Recent Advances in Redox Active Plant and Microbial Products*, DOI: 10.1007/978-94-017-8953-0, © Springer Science+Business Media Dordrecht 2014

MIX
Papier aus verantwortungsvollen Quellen
Paper from responsible sources
FSC® C105338

If you have any concerns about our products,
you can contact us on
ProductSafety@springernature.com

In case Publisher is established outside the EU,
the EU authorized representative is:
Springer Nature Customer Service Center GmbH
Europaplatz 3, 69115 Heidelberg, Germany

Printed by Libri Plureos GmbH
in Hamburg, Germany